Introduction to Mathematics

INTRODUCTION TO MATHEMATICS

R. A. Good

UNIVERSITY OF MARYLAND

Harcourt, Brace & World, Inc.

NEW YORK / CHICAGO / SAN FRANCISCO / ATLANTA

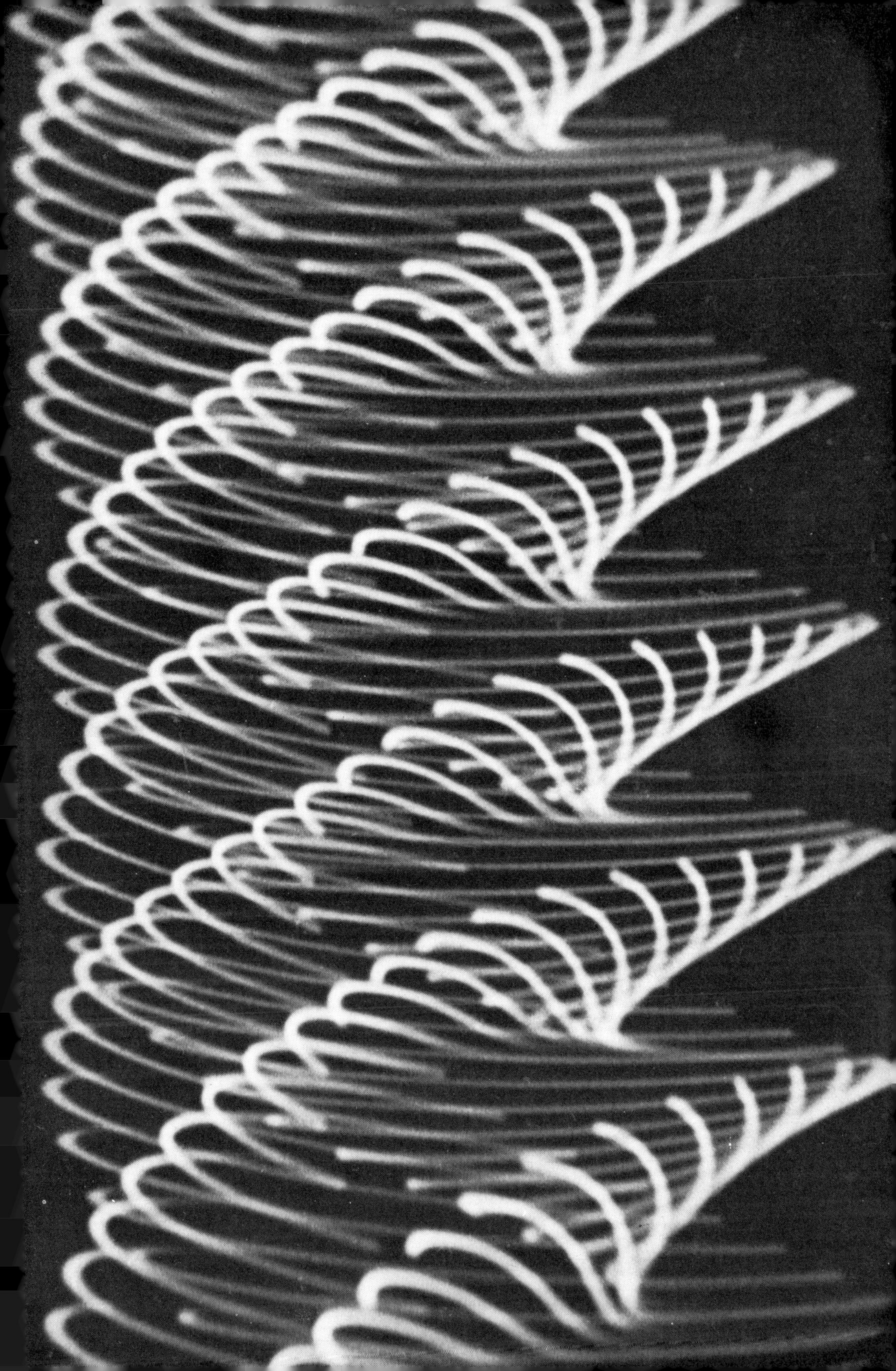

The photograms on the cover, title page, and pages 3, 149, and 329 are photographic images based on mathematical formulas and created by either oscillating a light pendulum in front of an open camera or moving a camera in front of a lighted object. Courtesy of Dr. Herbert W. Franke, from *Kunst und Konstruktion: Physik und Mathematik als fotografisches Experiment* (Munich: F. Bruckmann, KG, 1957).

Introduction to Mathematics
R. A. Good

© 1966 by Harcourt, Brace & World, Inc.

All rights reserved. No part of this publication may be reproduced or transmitted in any form or by any means, electronic or mechanical, including photocopy, recording, or any information storage and retrieval system, without permission in writing from the publisher.

Library of Congress Catalog Card Number: 66-16744

Printed in the United States of America

To Our Twins

Preface

Until recently, the application of mathematics was associated with only certain disciplines, such as the physical sciences and engineering. Today, however, the increased use of high-speed computing machinery and statistical techniques has made it necessary for students of the biological sciences, the social sciences, business, education, and the liberal arts to understand the techniques of solving linear problems and the fundamentals of probability theory. Further, the frequent occurrence of the function concept requires that any student who will one day use mathematics in his vocation be familiar with the properties of elementary functions.

To meet the needs of these first-year college students, I have written this book for a two-semester introductory mathematics course. Since many of these students will not continue in mathematics, the book is appropriate for a terminal course. However, the treatment is "open ended" so that the student may proceed with further mathematical study. The only prerequisite is two years of high-school mathematics.

The topics in this book fall naturally into three parts: linear mathematics, probability, and elementary functions—the three subjects that will be most useful to these students. The book begins with an extension of the student's high-school study of linear equations. This section emphasizes systems of linear relations, both equalities and inequalities, which are solved by a systematic algorithm, such as that used by computing-machine programmers. In addition, for the benefit of students of economics, business, and other social sciences, the language of vectors and matrices is exploited.

The second section provides a detailed study of probability that lays a foundation for future courses in statistics. The approach is axiomatic, and the treatment is limited to finite probability spaces. Extensive preparation for the probability exercises is provided by combinatorial topics.

The treatment of elementary functions in the third section stresses a few basic properties for functions and for the relationships among functions. Throughout this section, the property of a function and the geometrical interpretation of the property are persistently related.

Except for Chapter 1, which builds on the student's high-school background, concepts are introduced only as they are motivated by previous work or immediate need. To focus the student's attention on basic principles, nonessential aspects of topics have been omitted. For example, if one approach to solving a problem is fully adequate, the many alternative approaches are not presented. Yet if the alternative methods illustrate different fundamental notions, they are, of course, included. Another feature that has been suggested by the student audience is the style of presentation, which is less formal than that of many mathematics textbooks. Also with the student in mind, I have included numerous examples and have often introduced words and ideas in these examples before they are formally defined.

I appreciate very much the help I have received in preparing this textbook: the skillful services, at different stages during the manuscript preparation, of three typists, Mrs. Frances DiCarlo, Mrs. Ruth Ginter, and Mrs. Ouida Taylor; the friendly criticism of numerous interested readers of the preliminary drafts; and financial assistance from the National Science Foundation during a part of the work.

R. A. Good

College Park, Maryland

Contents

PART I

Linear Mathematics 2

1

Linear Functions 4

1	Introduction	4
2	Linear function	5
3	Graph of a linear function	7
4	Slope	9
5	Root of a linear equation	13
6	Graph of a linear equation	16
7	Lines with the same slope	18
8	Lines with different slopes	19
9	Proportionality	23
10	Distance between a point and the origin	26
11	Perpendicular lines	27
12	Linear equation	30

2

Systems of Linear Equations 35

1	Introduction	35
2	Systems with one root	35
3	Three-dimensional coordinate system	42
4	Coordinates of a given point	46
5	Vectors	46
6	Operations involving vectors	47
7	Matrices	53

8	Matrix multiplication	57
9	Gauss elimination method in matrix terminology	63
10	Graph of a linear equation in three dimensions	68
11	Plane determined by three points	73
12	Geometry of a system of linear equations in x,y,z	75
13	Systems for which the number of roots is not one	81
14	The Gauss elimination method	88

3

Linear Transformations 94

1	Vector space	94
2	Linear transformation	94
3	A vector whose image is given	100
4	Composite of linear transformations	102
5	Inverse of a linear transformation	107
6	Inverse of a matrix	113

4

Linear Inequalities 120

1	Introduction	120
2	Inequalities	120
3	Inequalities and the arithmetical operations	122
4	Summary of properties	125
5	Halfplane	126
6	System of linear inequalities	129
7	Convex polyhedral set	134
8	Extreme point	138
9	Linear programming	144

PART II

Probability 148

5

Sets and Logic 150

1	Introduction	150
2	Set, member, belongs to	150

3	Subset	152
4	Intersection, union, complement	154
5	Number of elements	161
6	And, not, or	163
7	Compound statements	167
8	If . . . , then . . . ; if and only if	172
9	Sentences pertaining to the outcome of an activity	175
10	Truth set	179
11	Boolean algebra	181
12	Proofs in Boolean algebra	186
13	Distributivity in a Boolean algebra	191
14	De Morgan's laws	194

6

Counting 198

1	Counting	198
2	Partition	198
3	Number of elements in a union	200
4	Application	203
5	Tree diagrams	207
6	Number of permutations	208
7	Summary of counting principles	211
8	Number of permutations of a finite set	214
9	Ordered partition with two cells	217
10	Number of subsets with a specified number of elements	222
11	Properties of the binomial coefficients	225
12	Summary	228

7

Powers and Sequences 230

1	Introduction	230
2	The binomial theorem	230
3	Binomial coefficients and the binomial theorem	235
4	Powers	238
5	Rational exponents	241
6	Sequence	244
7	Arithmetical sequence	245
8	Geometrical sequence	247
9	Formulas for sequences	248
10	The summation notation	251
11	Sum of numbers in a sequence	253

Contents   xi

8
Probability 259

1	Introduction	259
2	Probability in the finite case	260
3	More examples	263
4	Basic principles in probability	265
5	The connective NOT	266
6	Equiprobable measure	270
7	Logically inconsistent pair of statements	270
8	The connective OR	274
9	Illustrations	278
10	Conditional probability	283
11	Independent pair of statements	287
12	A posteriori probability	291
13	Repeated trials	293
14	Bernoulli experiments	293
15	Assignment of measures	296
16	Summary of probability	299
17	Random variable	302
18	Expected value of a random variable	303
19	Properties of expected value	307
20	Mean, variance, and standard deviation	309
21	Direct calculation of variance	311
22	Alternate calculation of variance	312
23	Properties of variance	314
24	Summary of mean and variance	319
25	Binomial distribution	320
26	Samples	322
27	Median and mode	325

PART III
Elementary Functions 328

9
Functions and Their Graphs 330

1	Introduction	330
2	Congruent parabolas	331

3	Other examples	335
4	Function	337
5	Domain, range, image	342
6	Graph	346
7	The absolute-value function	348
8	Increasing function	351
9	Summary	355

10

Distance 356

1	Introduction	356
2	Distance and absolute value	356
3	Distance between points	360
4	Distance in three-dimensional space	365
5	Definition of parabola	367
6	Equation for parabola	369
7	Translation	372
8	Quadratic function	376
9	Graph of a quadratic function	378
10	Midpoint	382
11	Perpendicular-bisector	383
12	Circle	387
13	Equation for a circle	389
14	Sphere	395
15	Summary	398

11

Inverse and Rational Functions 401

1	Introduction	401
2	Inverse functions	402
3	Geometric interpretation	407
4	Inverses for certain functions	410
5	Reflections	414
6	Even rational functions	422
7	Other rational functions	427

12

Transcendental Functions 431

1	Introduction	431
2	Functions of exponential type	431
3	The exponential function	434

4	Applications	438
5	Inverse of an increasing function	444
6	The logarithmic function	445
7	Applications	448
8	Normal probability distribution	456
9	Standard normal variable	462
10	Binomial and normal distributions	464
11	Winding function	467
12	Sine and cosine functions	471
13	Properties of sine and cosine	474
14	Addition formulas	479
15	Double	485
16	Applications of the addition formulas	487
17	Tangent function	492
18	Summary	496
19	Angle	496
20	Rotation matrix	498
21	Triangles and trigonometry	502

Appendix 507
  A Table of Functional Values 508
  Answers to Selected Exercises 513

Index 541

Introduction to Mathematics

PART I
Linear Mathematics

Linear Functions

Systems of Linear Equations

Linear Transformations

Linear Inequalities

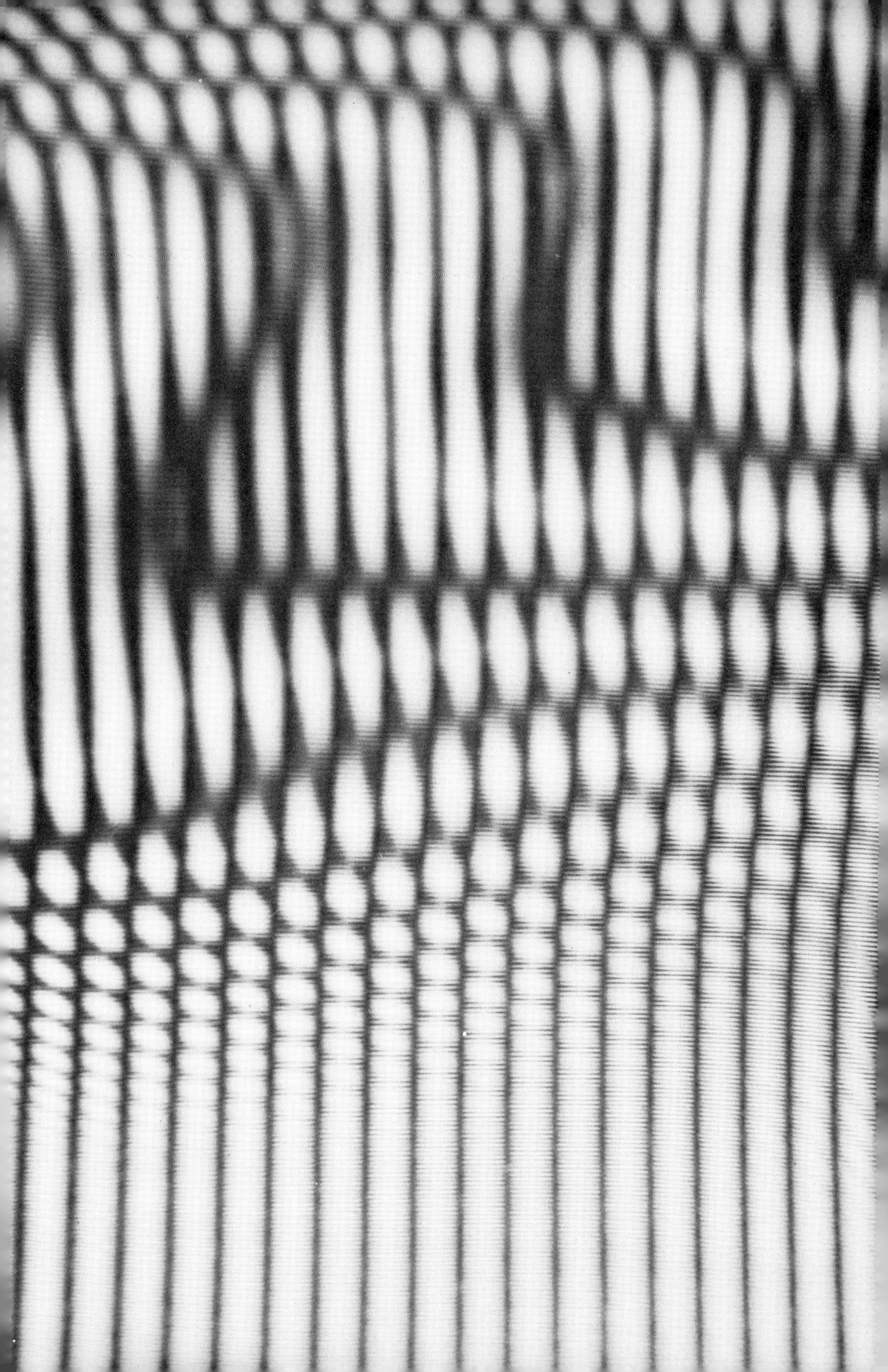

Linear Functions

1 Introduction

The concept of a function has paramount importance in contemporary mathematics. Broadly speaking, functions express relationships between classes, that is, correspondences associating to each object of one class an object of the other. Typical examples include: (i) the relation between time and space given by the position of the center of gravity of a certain satellite in orbit, (ii) the correspondence associating to each land parcel in a certain community its local real estate tax for the current fiscal year, (iii) the assignment of numbers to triangles prescribed by the area function. Mathematics, both as a study itself and in its applications to other fields of knowledge, relies heavily upon the extremely broad notion of a function. Consequently, this book is largely devoted, directly or indirectly, to studying functions of various types and their properties.

Although the terminology may not be familiar to the reader, he has certainly studied many functions during his previous mathematical training. Indeed, considerable emphasis in his earlier work has been placed on a simple but extremely useful type of function, namely, the linear function. We shall begin the present chapter by discussing linear functions in preparation for subsequent treatment of linear equations and linear inequalities.

A helpful aid to understanding a function is provided by a picture called the graph of the function. In particular, the graph of a linear function is a line. Indeed, the latter statement explains the choice of the word "linear." We will examine in detail the relationship between a linear function and the line that is its graph.

We suppose that the reader is familiar with the basic notion of a coordinate system in a plane. Each point in the plane represented in Fig. 1a corresponds to an ordered pair of real numbers. Thus, in Fig. 1b, the point P has the ordered pair of coordinates (a, b), where a represents the distance measured horizontally from the vertical coordinate axis to P and where b represents the distance vertically from the horizontal coordinate axis to P. In particular, the point A, which from the origin is three units to the left and two units up, pictures the ordered pair of numbers, $(-3, 2)$.

Figure 1   (a)   (b)

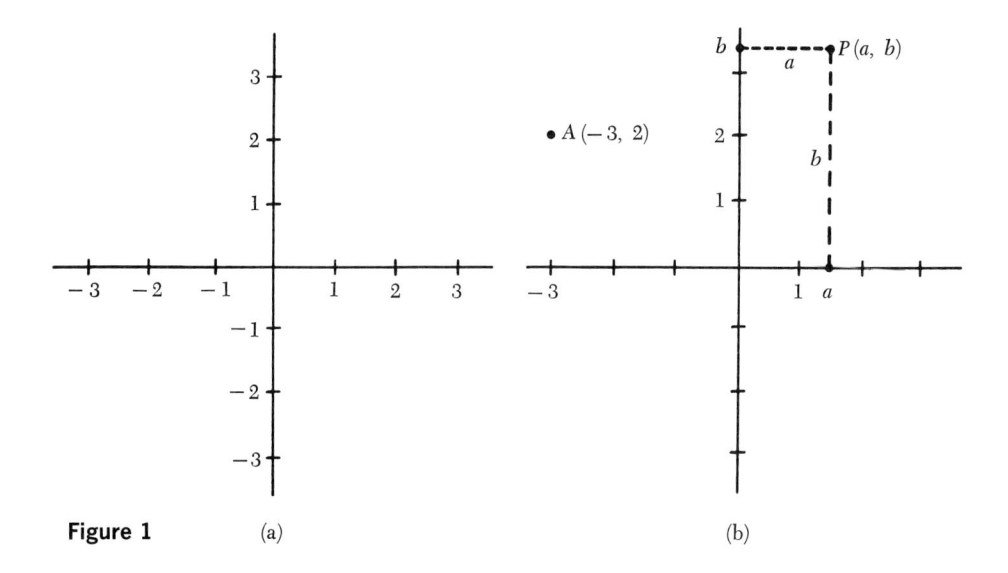

We shall see that a linear function is a certain collection of ordered pairs of numbers. Thus, its graph is the corresponding set of points on the coordinate system. The geometry will enable us to visualize the analytical properties of our functions, and the analysis will enable us to use algebraic techniques for handling geometric problems.

2 Linear function

We begin our study with an example.

> **Example 1** Consider the formula $2x + 1$. If we pick a number as a value of x, the formula gives us instructions for calculating. Specifically, we multiply the given number by 2, and to the resulting product we add 1.
> Picking 4 as a value of x, we follow the instructions to obtain the number $2 \cdot 4 + 1 = 9$. We may speak of 9 as the *image* of 4 according to our formula. If we choose 0 as a value for x, we find that its image, $2 \cdot 0 + 1$, is 1. Likewise, the image of $-\frac{2}{3}$ is $-\frac{1}{3}$, because $2(-\frac{2}{3}) + 1 = -\frac{1}{3}$.
> Our formula enables us to assign to any number an image. The image assigned to a number p is the number $2p + 1$.

> **Example 2** Consider another formula, namely, $-3x + 2$. This formula also provides a set of instructions for computing the image of a number. To find the image, we multiply the given number by

2 / Linear function   5

−3, and then to the product we add 2. These instructions are different from those in the preceding example, and we may expect that the image we obtain for a given number is different from the image we found above for the same given number. According to the formula $-3x + 2$, the image of 4 is -10, because $-3 \cdot 4 + 2 = -10$. We may think of the formula as matching -10 with the number 4. This matching may be indicated by the pair $(4, -10)$. If we choose 1 as a value of x, the corresponding number that we compute according to the formula $-3x + 2$ is $-3 \cdot 1 + 2 = -1$. We may think of -1 as being paired with 1 according to the formula $-3x + 2$. This pairing is suggested by the *ordered pair* $(1, -1)$.

In the ordered pair $(1, -1)$, the selected number 1 appears first, at the left; the image, namely, -1, appears second, at the right. Another ordered pair that is found from the formula $-3x + 2$ is the pair $(-\frac{1}{3}, 3)$, because 3 is the image of $-\frac{1}{3}$; indeed $-3(-\frac{1}{3}) + 2 = 3$.

The relationships in the two examples above illustrate the notion of a linear function. A linear function may be given by a formula of a certain type. The formula enables us to calculate, for any given number, the image of that number. The given number and its image form an ordered pair. All these ordered pairs collectively constitute the linear function.

A frequent convenience in discussing a linear function is an appropriate symbolism. Suppose we use a letter such as L to name a particular linear function, and suppose we select a number p. Then the symbol $L(p)$ is used for the image of p. The above discussion has paved the way for the following.

Definition Let m be any real number different from zero, and let b be any real number. Then $mx + b$ is a *formula* for a linear function L; thus, $L(x) = mx + b$. The *linear function* is the set of all ordered pairs of the form $(p, mp + b)$ for real numbers p. For each real number p, the number $mp + b$ is the *image* of p under the linear function L. The ordered pair $(p, mp + b)$ belongs to the linear function.

We note that the image of p, namely, $mp + b$, is obtainable by replacing x in the formula $mx + b$ by p and performing the computation.

For each linear function there is a rule for determining an image for every real number. The rule is conveniently prescribed by a formula; to find an image we "substitute in" the formula. For a linear function the formula is of a particularly simple type, namely, $mx + b$: the instructions are to multiply the given number [x] by the nonzero constant m and then to add the constant b.

Example 1 (reviewed)   The formula $2x + 1$ is the special case of the general formula $mx + b$ in which $m = 2$ and $b = 1$. If $L(x) = 2x + 1$, then $L(4) = 2 \cdot 4 + 1 = 9$ and $L(-1) = -1$. The ordered pair $(-\frac{2}{3}, -\frac{1}{3})$ belongs to L.

Example 2 (reviewed) The formula $-3x + 2$ is the special case in which $m = -3$ and $b = 2$. If we denote this linear function by K (to distinguish it from the linear function in Example 1), then $K(x) = -3x + 2$. Thus $K(4) = -10$. The ordered pair $(-\frac{5}{3}, 7)$ belongs to K, because $K(-\frac{5}{3}) = -3(-\frac{5}{3}) + 2 = 7$.

Example 3 A linear function is given by the formula $-\sqrt{3}x$. This is the special case in which $m = -\sqrt{3}$ and $b = 0$. Notice that, although we require m, the coefficient of x, to be different from zero, we do not impose such a restriction on b.

Example 4 Consider the linear function N given by the formula $N(x) = 5x - 3$. The image of the number p is $5p - 3$. What number has image 7? This question asks what number p satisfies the condition $5p - 3 = 7$. Since $5p = 10$, the answer is 2. We check by noting that $N(2) = 5 \cdot 2 - 3 = 7$.

EXERCISES A

1. Suppose that $L(x) = 5x - 8$. Find each of the following numbers.
   (a) $L(2)$. 2 (b) $L(0)$. -8 (c) $L(-32)$. -168

2. Suppose that L is the linear function given by the formula $L(x) = \frac{1}{2}x + 3$.
   (a) What is the image of 10 under L? $L(10)$
   (b) Find $L(-\frac{4}{5})$.
   (c) What is the image of -8 under L?

3. Let $K(x) = -1.6x + 0.48$. Find the image under K of each of the following numbers.
   (a) 3.1.   (b) 0.3.   (c) -2.5.

4. Suppose that, for every number p, the image of p under a linear function K is $3p - 2$. Which of the following ordered pairs belong to K? $K = 3p-2$
   (a) $(3, 7)$.   (b) $(0, -2)$.   (c) $(\frac{2}{3}, 0)$.   (d) $(-1, 5)$.   (e) $(1, 1)$.

5. Let $N(y) = 1 - 3y$.
   (a) What is the image of 0 under N?
   (b) What number has 0 as its image?
   (c) What number has -11 as its image?

3 Graph of a linear function

The graph of a linear function on a two-dimensional coordinate system is a (straight) line.

Example 1 (continued) Consider again the linear function L given by $L(x) = 2x + 1$. We have already noted some of the ordered pairs that belong to L; they are $(4, 9)$, $(0, 1)$, $(-\frac{2}{3}, -\frac{1}{3})$, $(-1, -1)$.

3 / Graph of a linear function

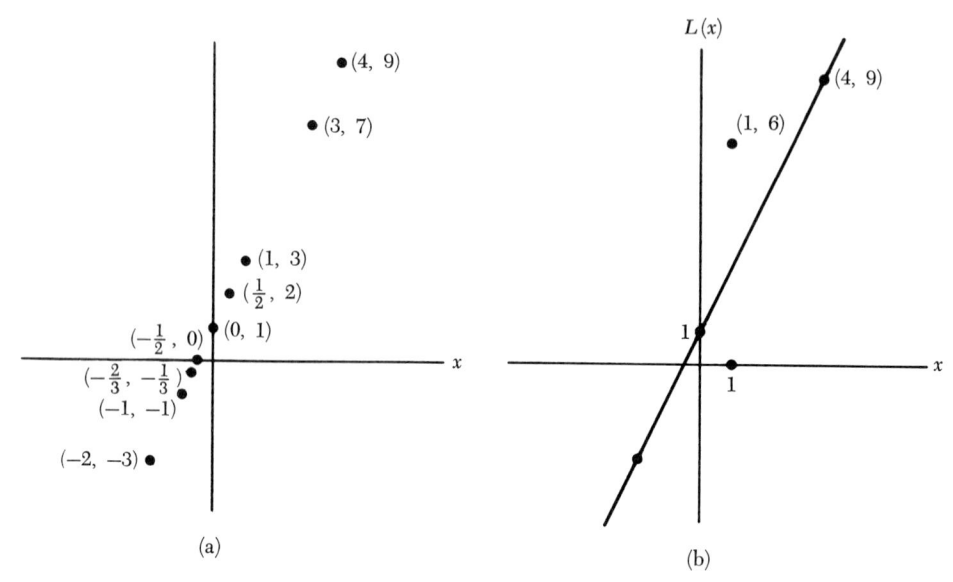

Figure 2

Others include $(3, 7)$, $(1, 3)$, $(-2, -3)$, $(\frac{1}{2}, 2)$, $(-\frac{1}{2}, 0)$. In Fig. 2a we have plotted these ordered pairs as points on a (two-dimensional) coordinate system.

If we imagine plotting each point whose coordinates form an ordered pair belonging to L, the collection of all such points is a line.

The line shown in Fig. 2b is the graph of the linear function L. The point $(4, 9)$ belongs to the graph because the ordered pair $(4, 9)$ belongs to L. The point $(1, 6)$ is not a point of the graph, because $L(1) = 2 \cdot 1 + 1$ is not 6.

On a two-dimensional coordinate system, the graph of a linear function consists of all points (c, d) whose coordinates belong to the function. Thus the statement that a point (c, d) lies on the graph of the linear function L means that $L(c) = d$.

Example 2 (continued) The graph of the linear function K given by the formula $K(x) = -3x + 2$ is a line (see Fig. 3). Some of its points are $(4, -10)$, $(-\frac{5}{3}, 7)$, and $(1, -1)$. The point $(3, 4)$ shown in Fig. 3 does not lie on the graph of K, because $K(3) = -3 \cdot 3 + 2 = -7$; hence, $K(3)$ is not 4.

A convenient safety measure to avoid mistakes when sketching (or drawing) the graph of a linear function is to plot at least three widely spaced points on the graph. The collinearity of the plotted points then provides a check on the accuracy of the work.

8   Linear Functions

Figure 3

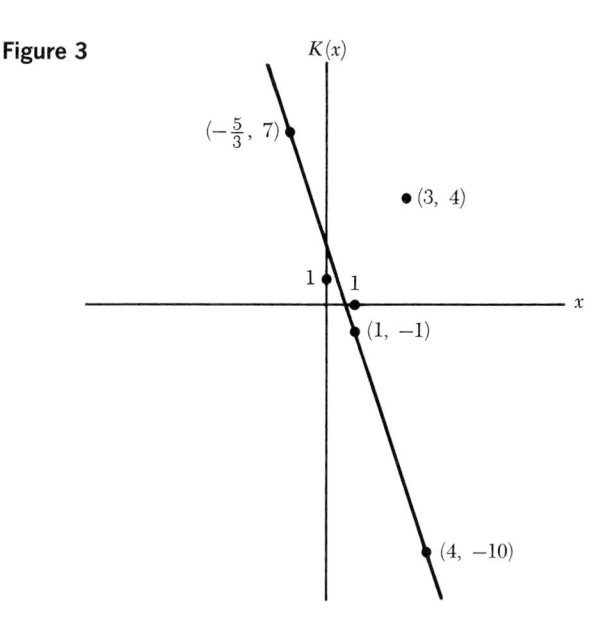

4 Slope

Definition If a line is the graph of a linear function L given by the formula $L(x) = mx + b$, then the number m is called the *slope* of the line.

A line with a positive slope slants from lower left to upper right on the coordinate system, while a line with negative slope slants from upper left to lower right. (See Fig. 4.) Suppose that $m > 0$. Then as x increases, mx also increases, and hence $mx + b$ increases; thus the graph of $mx + b$ suggests an upward path (Fig. 4a). On the other hand, suppose that $m < 0$. Then as x increases, mx decreases, hence $mx + b$ decreases, and the graph of $mx + b$ appears like a downward path (from left to right, see Fig. 4b).

Figure 4 **(a)** A line with positive slope; **(b)** a line with negative slope.

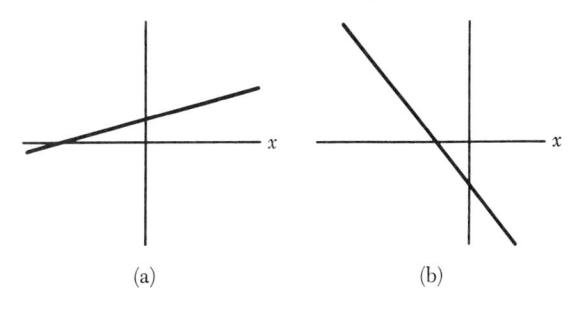

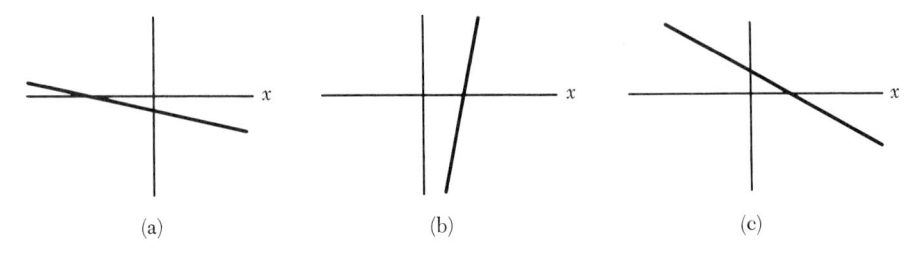

Figure 5
(a) A line with slope of small absolute value; (b) a line with slope of large absolute value; (c) a line with slope of moderate absolute value.

The absolute value (or numerical size) of the slope is a measure of the steepness of the line on the coordinate system. A line whose slope is near zero is nearly flat, or level, while a steep line has a numerically large slope. See Fig. 5.

Example 1 (continued) The graph of the linear function L, where $L(x) = 2x + 1$, is a line with slope 2. The line (see Fig. 2b) slants from lower left to upper right, and its steepness is moderate.

Two points on this line are $(0, 1)$ and $(1, 3)$. We notice (see Fig. 6) that these two points are one unit apart horizontally: their respective x-coordinates are 0 and 1. The respective images of these two numbers, namely, 1 and 3, are two units apart. This number, 2, is the slope of the line. Figure 6 shows how the vertical distance 2 matches the one unit of horizontal distance between the points.

More generally, let p be any real number. Its image under the function L is

$$2p + 1.$$

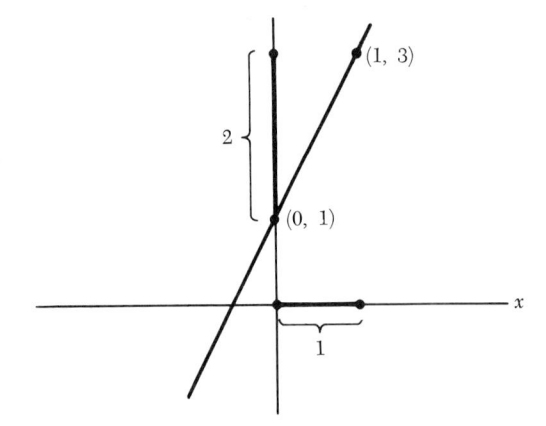

Figure 6

10   Linear Functions

Figure 7

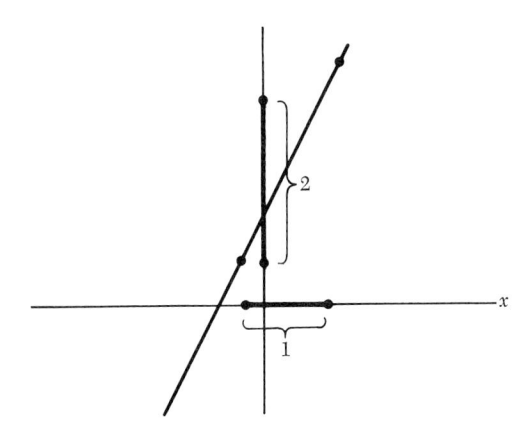

Now consider the number that is one more than p, namely, $p + 1$. Its image under L is

$$2 \cdot (p + 1) + 1 = 2p + 3.$$

Note that $(2p + 3) - (2p + 1) = 2$; thus, the difference between the images is 2, the same number as the slope of the line. Figure 7 shows that, for any two points on the line that are one unit apart horizontally, the corresponding vertical distance between the points is 2.

The preceding example suggests the following interpretation of the slope of a line. Consider any line that is the graph of a linear function. Suppose that the linear function is L and that its formula is $mx + b$. Let p be any number. Then, the image of p is

$$mp + b.$$

The number $p + 1$, which is one more than p, has image

$$m \cdot (p + 1) + b = mp + m + b = (mp + b) + m.$$

A change of one unit in the number,

$$\text{from} \quad p \quad \text{to} \quad p + 1,$$

produces a change of m units in the image,

$$\text{from} \quad (mp + b) \quad \text{to} \quad (mp + b) + m.$$

Thus, we have the following result.

Theorem 1 *If a line is the graph of a linear function, then the slope m of the line is the distance—vertically—from any point on the line to the point on the line one unit toward the right.*

Figure 8

Figure 9

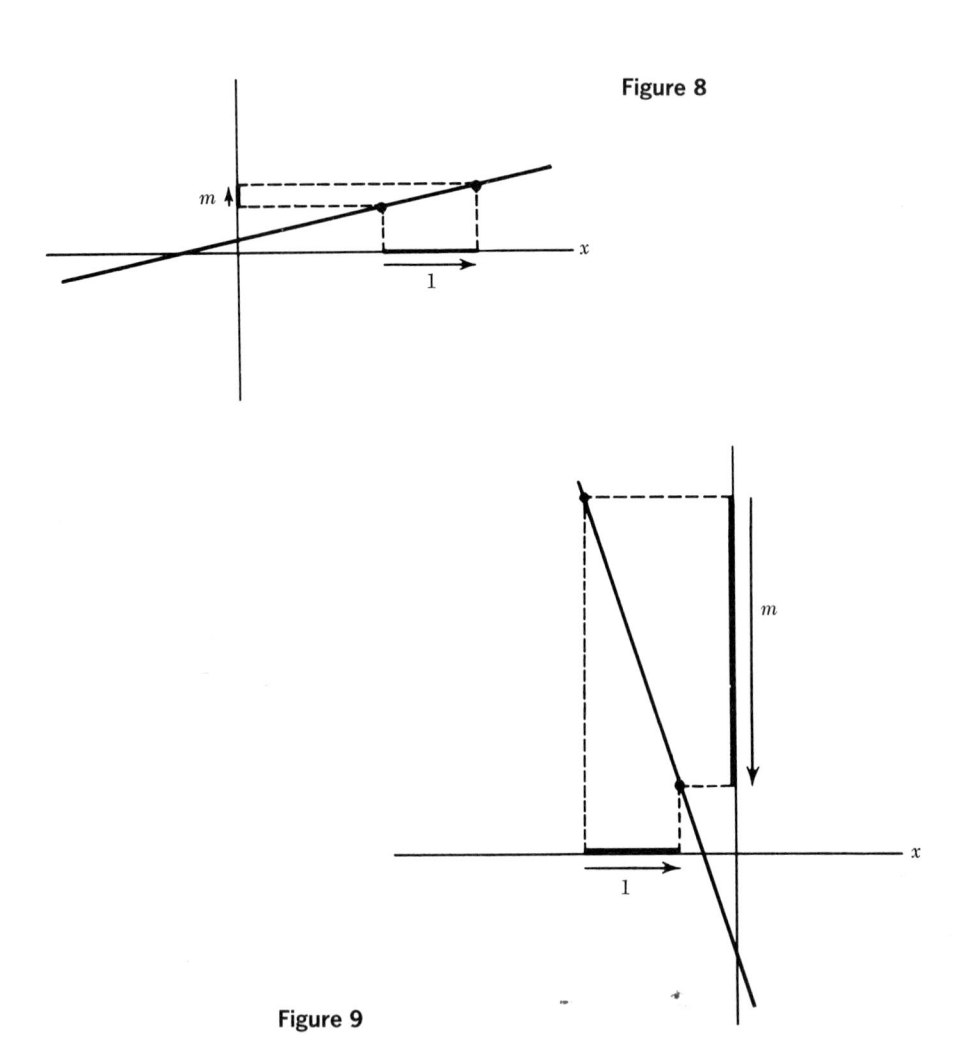

Figure 8 illustrates the case where m is a small positive number. Figure 9 portrays the situation where m is negative. Note that Theorem 1 confirms our earlier claims concerning the lines in Figs. 4 and 5.

EXERCISES B

1. Sketch the graph of the linear function L, where $L(x) = 2x - 1$.
2. Sketch the graph of the linear function given by the formula $-2x - 1$.
3. Sketch the graph of the linear function L, where $L(x) = \frac{1}{4}x$.
4. Sketch the graph of the linear function given by the formula $\frac{2}{3}x - \frac{4}{3}$.
5. Explain why the point $(1, 1)$ lies on the line that is the graph of the function K given by $K(x) = 6x - 5$.
6. Find the image of the number 1 under the linear function L, where $L(x) = mx + b$.

$$L(1) = m + b$$

12   Linear Functions

7. ~~class~~ If the point $(0, y)$ [annotated $(x, A(x))$] belongs to the line that is the graph of the function A given by $A(x) = 2 - 3x$, find the number y. $= 2$

8. Find the number z, if $(0, z)$ is an ordered pair belonging to the function given by the formula $(x/4) - \frac{2}{7}$.

9. Let the linear function L be given by the formula $L(x) = mx + b$, and let p be a number.
   (a) If the point (p, y) belongs to the graph of L, find y.
   (b) If the point $(p + 1, z)$ belongs to the graph of L, find z.
   (c) Find $z - y$.

10. For each of the following, the graph of the linear function given by the formula is a line. What is the slope of the line in each case?
    (a) $5x + 2$. (b) $-\frac{1}{4}x + \frac{2}{3}$. (c) $2.83x + 0.17$.
    (d) $2 - 3x$. (e) $\frac{5}{3}x$. (f) $-8 - x$.

11. ~~work~~ The point $(2, 3)$ lies on a certain line. Another point on the line is $(2 + 1, 3 + 7)$. What is the slope of the line? [HINT: The answer can be given extremely easily.] 7

12. ~~class~~ Two points on a certain line are $(p, 4)$ and $(p + 1, 1)$. What is the slope of the line? $\frac{-3}{1}$

13. ~~work~~ The point $(6.82, 12)$ belongs to a line with slope -12. If the point $(7.82, y)$ also belongs to the line, find the number y.

14. For each of the following numbers, describe (in words) the appearance of a line whose slope is the given number.
    (a) 8.5. (b) -0.13. (c) $\frac{6}{5}$. (d) Ten billion.

15. Find a linear function L such that $L(0) = 0$.

16. Find a linear function K such that its graph is a line with slope -8.

Class: $(3, 6)$ & $(7, 3)$ $m = ?$
$m = \frac{6-3}{3-7} = -\frac{3}{4}$

5 Root of a linear equation

Solve for x

Two elementary computational problems arise in the study of linear functions. Suppose that a linear function is given. The first problem is, "Given a number, find its image." We solve this problem by computing the image according to the formula for the linear function.

The second problem is, "Given the image of a number, find the number." This problem may be solved by the techniques of handling linear equations.

Example 5 Let A be the linear function given by the formula $3x - 5$. If we are given a number -2, then we compute its image as follows:

$$A(-2) = 3 \cdot (-2) - 5 = -6 - 5 = -11.$$

On the other hand, if we are told that the image of a certain number s under this linear function A is 7 and are asked to find s, we proceed as follows. The given information states that

$$3s - 5 = 7.$$

Hence,
$$3s = 12$$
and
$$s = 4.$$

In a straightforward manner we verify that the image of 4, namely, $3 \cdot 4 - 5$, is 7.

In the former situation, having been given the first number -2 in the ordered pair, we find that $(-2, -11)$ belongs to A. In the latter situation, having been given the second number 7 in the ordered pair, we find that $(4, 7)$ belongs to A.

Suppose that L is any linear function and that L is given by the formula $mx + b$, where m and b are real numbers such that m is not zero. If c is a given number, we evaluate $L(c)$ by substituting c for x in the formula and calculating. The result is
$$L(c) = m \cdot c + b.$$
The ordered pair $(c, mc + b)$ belongs to L.

If d is a given number, we may inquire whether d is the image of any number under L, and, if it is, what number (or numbers) has d as image. Rephrased, the question becomes, "Find x such that $L(x) = d$." To answer the question, we solve the equation
$$mx + b = d$$
for x. After the intermediate step
$$mx = d - b,$$
we find that
$$x = \frac{d-b}{m}.$$
Thus, d can be the image of only one number under L, namely, $(d - b)/m$. The calculation
$$L\left(\frac{d-b}{m}\right) = m \cdot \frac{d-b}{m} + b = (d - b) + b = d$$
shows that d actually is the image of $(d - b)/m$ under L. The ordered pair $\left(\frac{d-b}{m}, d\right)$ belongs to L.

We may note that we obtained exactly one solution in the preceding paragraph because m is not zero. This restriction on m guarantees that the expression $(d - b)/m$ has a well-defined meaning. The following theorem summarizes our results.

Theorem 2 Let m and b be real numbers such that *m is not zero*, and let the linear function L be given by the formula $L(x) = mx + b$. For every real number d, there belongs to L exactly one ordered pair of the type $(*, d)$; it is the pair $\left(\dfrac{d-b}{m}, d\right)$.

The problem of identifying a number whose image is given has an important special case. This is the case in which the image is zero.

Let L be a linear function, say the function given by the formula $mx + b$. The task of solving the linear equation

$$mx + b = 0$$

is the same as the problem of identifying the number whose image under L is zero. Applying Theorem 2 to the case where $d = 0$, we find that the *root* of the equation is $-b/m$. This result is easily verified as follows:

$$L\left(-\frac{b}{m}\right) = m \cdot \left(-\frac{b}{m}\right) + b = (-b) + b = 0.$$

Since $\left(-\dfrac{b}{m}, 0\right)$ belongs to the linear function L, the graph of L contains the point with coordinates $\left(-\dfrac{b}{m}, 0\right)$. Any point with second coordinate 0 lies on the x-coordinate axis. Therefore $\left(-\dfrac{b}{m}, 0\right)$ is the point of intersection of the graph of L and the horizontal coordinate axis. Figure 10 shows the intersection.

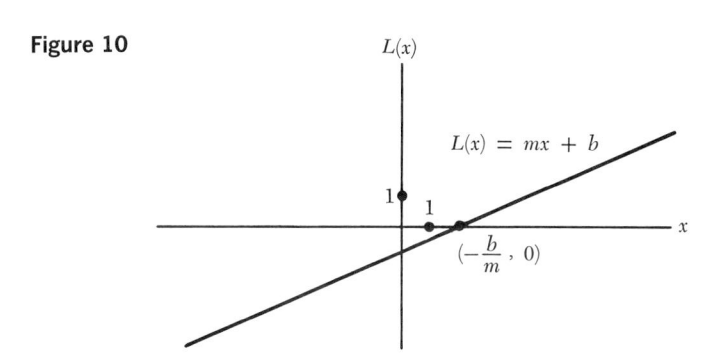

Figure 10

Example 6 Consider the linear function given by the formula $-4x + 20$. The root of the equation obtained by setting the formula equal to zero, namely,

$$-4x + 20 = 0,$$

is the number 5, since $-4 \cdot 5 + 20$ is zero.

EXERCISES C

1. If the function L is given by $L(t) = 3t - 6$, find the numbers c and d, where $c = L(0)$ and $0 = L(d)$.   [handwritten: work; $L(0) = -6$; $L(d) = 0$, $t = 2$]

2. If each of the points $(0, c)$ and $(d, 0)$ lies on the line that is the graph of the linear function given by the formula $(4t/3) - \frac{2}{3}$, find c and d.

3. If the ordered pair $(2, 5)$ belongs to the linear function L, where $L(x) = (n/2)x - 1$, find n.   [handwritten: class; $L(2) = 5 = \frac{n}{2} \cdot 2 - 1 = n - 1$, $n = 6$]

4. Find the root of each of the following linear equations.
   (a) $-3x + 2 = 0$.   [handwritten: $x = \frac{2}{3}$]   (b) $-1.8 - 5.4x = 0$.   (c) $43x = 0$.

5. The graph of a linear function contains the points $(0, 4)$ and $(6, 0)$. What is the root of the linear equation obtained by setting equal to zero the formula for the function?   [handwritten: 6]

6. If the function K is given by $K(x) = (1/a)x - (1/b)$, and if g is a number such that $K(g) = (1/a) + (1/b)$, find the number g.

[handwritten: Lect]

6 Graph of a linear equation

We often use the letter y to denote $mx + b$. The line that is the graph of the linear function given by the formula $mx + b$ is also called the graph of the equation $y = mx + b$. Any equation that is equivalent to $y = mx + b$ has the same line as its graph.

Example 7 The equation $2x + 3y = 6$ is equivalent to the equation $3y = -2x + 6$ and is equivalent to the equation $y = -\frac{2}{3}x + 2$. The graph of the equation $2x + 3y = 6$ is the line that is the graph of the linear function L, where

$$L(x) = -\tfrac{2}{3}x + 2.$$

[handwritten in margin: $7x - 2y = 5$]

The line has slope $-\frac{2}{3}$. The line contains each of the points $(3, 0)$ and $(0, 2)$. The point $(3, 0)$ we obtained by replacing y by 0 in the equation $2x + 3y = 6$ and solving for x. The point $(0, 2)$ we obtained by replacing x by 0 and solving for y. Note that 2 is the image of 0 under L and that 0 is the image of 3 under L. (See Fig. 11.)

Example 8 The graph of the equation $4x - 5y + 10 = 0$ is a line. If we replace y by 0 and solve for x in the resulting equation $4x + 10 = 0$, we find that the line contains the point $(-\frac{5}{2}, 0)$. If we replace x by 0 and solve for y in the resulting equation $-5y + 10 = 0$, we identify the point $(0, 2)$ as belonging to the line.

The given equation $4x - 5y + 10 = 0$ is equivalent to $5y = 4x + 10$ and also equivalent to $y = \frac{4}{5}x + 2$. The slope of the line is $\frac{4}{5}$. The line is the graph of the linear function given by the formula $\frac{4}{5}x + 2$. (See Fig. 12.)

16   Linear Functions

Figure 11

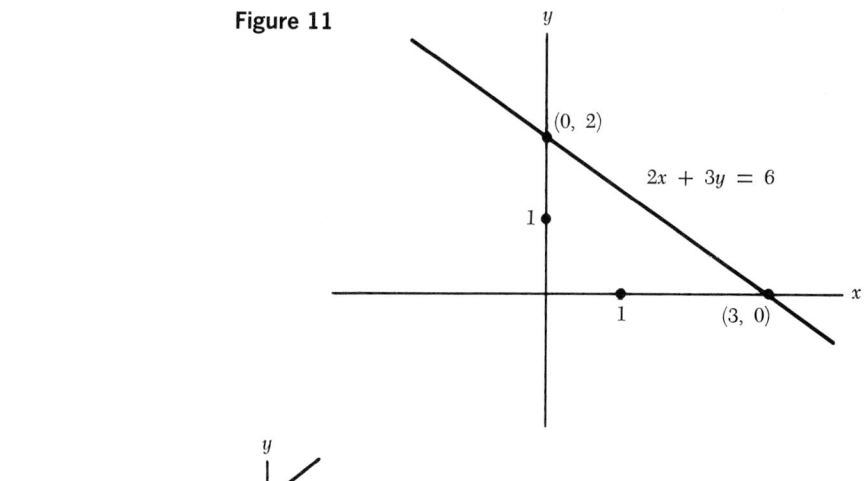

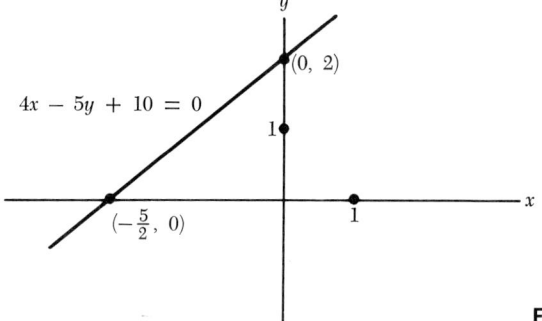

(0, 2)

$2x + 3y = 6$

(3, 0)

$4x - 5y + 10 = 0$

(0, 2)

$(-\frac{5}{2}, 0)$

Figure 12

EXERCISES D

1. For each of the following, sketch the line given by the equation.
   (a) $y = -x + 3$.
   (b) $y = \frac{x-1}{2}$.
   (c) $3y = 8 - 5x$.
   (d) $x + y = 0$.

2. If each of the points $(2, g)$ and $(h, 2)$ lies on the line given by the equation $y = -6x + 8$, find the number $g - h$. [HINT: Find g first, h next, and then $g - h$.]

3. Solve the equation $\frac{1}{2}x + 3 = \frac{11}{2}$.

4. Find the root of the equation $\frac{5}{6}w - \frac{3}{4} = \frac{2}{3}w + \frac{7}{8}$.

5. Find the root of the equation $2.1x - 1.4 = 8.1 - 1.7x$.

6. For each of the following, find the linear function whose graph is the line given by the equation.
   (a) $2x + 3y = 4$.
   (b) $x - y + 1 = 0$.
   (c) $32x + 21y = 0$.

7. For each of the following, find the slope of the line given by the equation.
   (a) $2y = 1 - 5x$.
   (b) $3x - 4y + 5 = 0$.
   (c) $-5x = 2 - 8y$.
   (d) $6x - 5y = 1$.

8. Let each of the numbers a and b be different from zero. Find the slope of the line given by the equation $bx + ay + ab = 0$.

9. If the point $(1, 3)$ lies on the line given by $y = mx + 4$, find the number m.

7 Lines with the same slope

Although two different equations can be equivalent and therefore have the same graph, any two different linear functions have different graphs. In the present and next sections we wish to clarify this statement and to study possible relationships between the graphs of different linear functions.

Let K and L be two different linear functions. Suppose that L is given by the formula $L(x) = mx + b$ and that K is given by the formula $K(x) = nx + g$.

First consider the case in which $m = n$. If it were true that $b = g$, then K and L would be the same, contrary to hypothesis. Thus, $b \neq g$. For any number c,

$$L(c) = mc + b$$

and

$$K(c) = nc + g = mc + g.$$

Since b and g are different, $L(c) \neq K(c)$. Every number has an image under L that is different from its image under K. This means that there is no ordered pair that belongs to both L and K.

Let us interpret the preceding paragraph geometrically. The hypothesis $m = n$ means that the graph of the equation $y = mx + b$ and the graph of the equation $y = nx + g$ are lines with the same slope. The conclusion that L and K have no ordered pair in common means that the lines have no point in common. In a plane, two lines with no common point are parallel. As a summary, we have the following result.

Theorem 3 If two lines, which are the graphs of different linear functions, have the same slope, then the lines are parallel.

Figure 13 shows two illustrations of a pair of lines with the same slope. We note that Theorem 3 fits consistently with our early interpretation for the slope of a line: the slope is a measure of steepness, and its sign identifies the upward or downward slant. Thus, it seems reasonable that two different lines with the same slope should have the same direction and be parallel to one another.

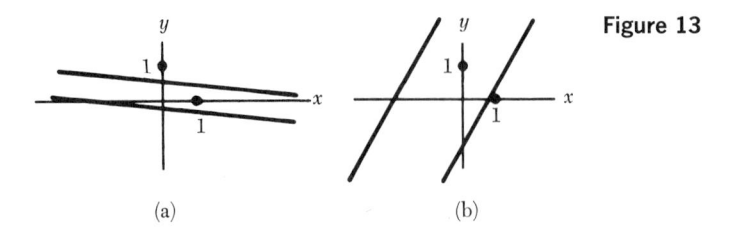

Figure 13

(a)   (b)

18   Linear Functions

Example 9 Each of the lines $y = \frac{2}{3}x + 1$ and $y = \frac{2}{3}x - 2$ has slope $\frac{2}{3}$. An algebraic attempt to find a point that belongs to both lines is doomed to failure. In order to understand how such a venture miscarries, let us try to find a point, say (c, d), that lies on each of the lines. Since (c, d) is on the line $y = \frac{2}{3}x + 1$, we obtain the statement $d = \frac{2}{3}c + 1$; hence,

$$d - \tfrac{2}{3}c = 1.$$

Since (c, d) is on the line $y = \frac{2}{3}x - 2$, we obtain the statement $d = \frac{2}{3}c - 2$; hence,

$$d - \tfrac{2}{3}c = -2.$$

The number $d - \frac{2}{3}c$ cannot be equal to both 1 and -2. We have reached a contradiction in our attempt to find a point of intersection for the two lines. The lines are parallel. (See Fig. 14.)

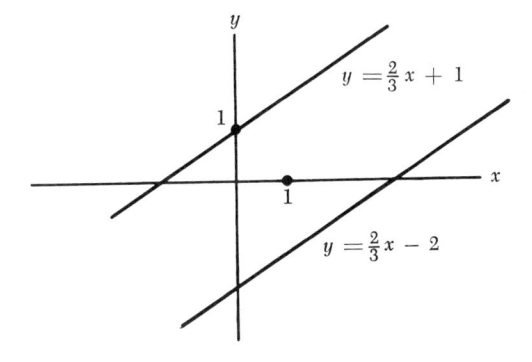

Figure 14

8 Lines with different slopes

Again let K and L be different linear functions given by the respective formulas $K(x) = nx + g$ and $L(x) = mx + b$. Now consider the case in which $m \neq n$. The graphs of K and L are lines that do not have the same slope. We hope to show that these lines have exactly one point of intersection and therefore are not parallel to each other. When we achieve this goal, then we will have a criterion for deciding whether two lines which are the graphs of linear functions are parallel. The lines are parallel to each other if and only if they have the same slope.

To illustrate what we hope to show, let us first consider a specific example.

Example 10 Let A and B be the linear functions given by $A(x) = 2x - 1$ and $B(x) = -x + 5$. Is there an ordered pair (p, q) belonging to both A and B? If so, what is it?

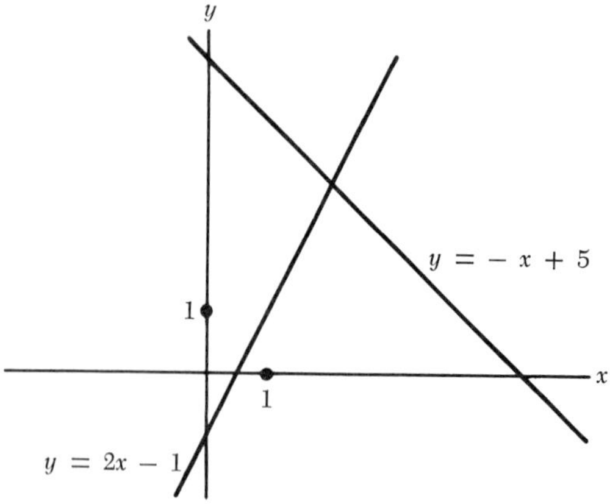

Figure 15

[Figure 15: Graph showing lines $y = 2x - 1$ and $y = -x + 5$ intersecting]

Figure 15 suggests that there is a point of intersection (p, q) for the two lines that are the respective graphs of A and B. We find (p, q) by an algebraic method. Since (p, q) belongs to A,

$$q = 2p - 1.$$

Also, since (p, q) belongs to B,

$$q = -p + 5.$$

Since q is equal to both $2p - 1$ and $-p + 5$, we obtain

$$2p - 1 = -p + 5.$$

Hence, $3p = 6$, and

$$p = 2.$$

Finally,

$$q = -p + 5 = -2 + 5 = 3.$$

The ordered pair $(2, 3)$ belongs to both A and B. The point of intersection for the two lines in Fig. 15 is $(2, 3)$.

The method illustrated in Example 10 can be applied to the general situation. An ordered pair (p, q) that belongs to the linear function K given by $K(x) = nx + g$ satisfies the condition

$$q = np + g.$$

If (p, q) also belongs to L given by $L(x) = mx + b$, then

$$q = mp + b.$$

Since each of $np + g$ and $mp + b$ is equal to q, they are equal to each other:

$$np + g = mp + b;$$

hence,
$$(n - m)p = b - g.$$

We now exploit the hypothesis that m and n are different. The number $n - m$ therefore is not zero, and we find that
$$p = \frac{b - g}{n - m}.$$

Thus,
$$q = np + g = n\frac{b - g}{n - m} + g = \frac{nb - ng + ng - mg}{n - m} = \frac{nb - mg}{n - m}.$$

By straightforward algebraic manipulation (see Exercise 20 below) the image of
$$\frac{b - g}{n - m}$$
under each of the linear functions K and L is
$$\frac{nb - mg}{n - m}.$$

Thus, the two linear functions K and L have exactly one ordered pair in common, namely,
$$\left(\frac{b - g}{n - m}, \frac{nb - mg}{n - m}\right).$$

The following summary is a companion to Theorem 3.

> **Theorem 4** If two lines, which are the graphs of linear functions, have different slopes, then the lines have exactly one point in common.

A line on a coordinate system is often determined by two bits of information. Consider the following two examples.

> **Example 11** Find a linear function such that its graph has slope $-\frac{3}{10}$ and such that the ordered pair $(5, -\frac{1}{2})$ belongs to it. A formula for the linear function is of the following type:
> $$mx + b.$$

The number m is the slope of the graph and hence is $-\frac{3}{10}$. Thus, the formula is of the type
$$-\tfrac{3}{10}x + b.$$

The remaining problem is to determine b. Since $(5, -\frac{1}{2})$ belongs to the linear function, we obtain the statement
$$-\tfrac{3}{10} \cdot 5 + b = -\tfrac{1}{2}.$$

8 / Lines with different slopes

Hence, $b = 1$. The desired linear function is given by the formula
$$-\tfrac{3}{10}x + 1.$$

Example 12 Find a linear function L such that $L(2) = 1$ and $L(-1) = 10$. A formula for L is of the type $L(x) = mx + b$, where m and b are to be determined. From the given data, using 2 for x gives us
$$2m + b = 1,$$
and using -1 for x gives us
$$-m + b = 10.$$
From these two statements, we find (by subtraction) that
$$3m = -9$$
and (by division) that
$$m = -3;$$
then
$$b = 1 - 2m = 1 - 2(-3) = 7.$$
Thus, $L(x) = -3x + 7$.

To verify that this function does possess the desired properties, we note that the images of 2 and -1 are $-3 \cdot 2 + 7 = 1$ and $-3(-1) + 7 = 10$, respectively.

EXERCISES E

1. Two linear functions are given by the respective formulas $584x - 1782$ and $584x - 1872$. Their graphs are a pair of _____ lines. Fill the blank with the most appropriate word.

2. If the line whose equation is $3ax + 2y = 1$ is parallel to the line $y = 4 - x$, find the number a.

3. If the line given by $y = mx + b$ has slope 2 and contains the point $(-1, 5)$, find the numbers m and b.

4. If each of the points $(3, 1)$ and $(2, 4)$ lies on the line whose equation is $y = mx + b$, find m and b.

5. If the linear function L given by $L(x) = nx + c$ contains each of the ordered pairs $(9, 5)$ and $(5, 9)$, find n and c.

6. Find a linear function L such that $L(-1) = 1$ and such that the graph of L has slope 3.

7. Find a linear function K such that $K(\tfrac{1}{2}) = \tfrac{1}{2}$ and $K(-\tfrac{1}{2}) = \tfrac{3}{2}$.

8. Find an equation of the line containing the point $(0, 0)$ and having slope $-\tfrac{9}{8}$.

9. Let p, q be given numbers neither of which is zero. Find an equation of the line containing the points $(0, 0)$ and (p, q).

22 Linear Functions

10. If the graph of a linear function L is parallel to the line $x + 2y = 5$ and if $L(1) = 0$, find $L(4)$.

11. If the graph of a linear function L is parallel to the line $y = 15x + 27$, and if $L(1) = 80$, and if c is a number different from zero, find $L(c + 1) - L(c)$. [HINT: This problem, done thoughtfully, requires very little computation.]

12. On the same coordinate system, sketch the three lines given by the respective equations $y = 2x + 1$, $y = 2x - 1$, $y = 2x - 3$.

13. By the graphical method, identify the point that lies on the line $x - 2y = 1$ and lies on the line $2x + y = 7$.

14. Find graphically a point that belongs to each of the lines $2x - y = 3$ and $x + y = 0$.

15. Suppose that the linear functions A and B are given by $A(t) = t + 3$ and $B(t) = 5t + 7$, respectively. If p and q are two numbers such that $q = A(p)$ and $q = B(p)$, find p.

16. Analyze and discuss the following information. Two linear functions A and B are given by $A(t) = 150t - 149$ and $B(x) = 150x + 149$, respectively; two numbers p and q satisfy the conditions that $q = A(p)$ and $q = B(p)$.

17. Analyze and discuss the following information. The linear function D has all the following properties: each of the ordered pairs $(1, 2)$ and $(2, 1)$ belongs to D, and $D(0) = 0$.

18. If the ordered pair (p, q) belongs to each of the linear functions A and D given by $A(x) = mx + b$ and $D(x) = kx + c$, respectively, find p and then find q.

19. In essay form, compare and/or contrast the two lines that are the graphs of the respective equations $3y = 2 + 17x$ and $3y = 2 - 17x$.

20. Let K and L be linear functions given by the respective formulas $K(x) = nx + g$ and $L(x) = mx + b$, where the numbers m and n are different from zero and from each other.

   (a) Verify that the image of $\dfrac{b - g}{n - m}$ under K is $\dfrac{nb - mg}{n - m}$.

   (b) Show that the ordered pair $\left(\dfrac{b - g}{n - m}, \dfrac{nb - mg}{n - m}\right)$ does belong to L.

9 Proportionality

Special terminology is occasionally applied to the case of a linear function to which the ordered pair $(0, 0)$ belongs. Let m be a real number not zero. If the product mx is denoted by y, we may say that y *varies as* x. Alternatively, we sometimes say that y is *proportional to* x, and we call the number m the *constant of proportionality*.

Example 13 The statement that y is proportional to x with 6 as the constant of proportionality means that $y = 6x$.

Example 14 Suppose that u is proportional to v; suppose that $u = 9$ if $v = 6$. Find u if $v = 18$.

One method of solution is the following. By hypothesis, there is a number m such that $u = mv$. From the given data, $9 = m \cdot 6$. Hence, $m = \frac{3}{2}$, and $u = \frac{3}{2}v$. If $v = 18$, then $u = \frac{3}{2} \cdot 18 = 27$.

Another method of solution is the following. Since 18 is triple 6, the desired value of u is triple 9, namely, 27.

The second method of solution in Example 14 may be applied to the general case. Consider any linear function L to which $(0, 0)$ belongs. A linear function has a formula of the type $L(x) = mx + b$. Since $L(0) = 0$, we find that

$$0 = m \cdot 0 + b = b.$$

Thus, L is more simply given by

$$L(x) = mx.$$

Suppose now that we are given the information that the ordered pair (c, d) belongs to L, where $c \neq 0$. In other words, we know that $L(c) = d$. Suppose we are asked to find $L(p)$.

The statement $L(c) = d$ means that $d = mc$. Thus,

$$m = \frac{d}{c}.$$

Then,

$$L(p) = mp = \frac{d}{c}p.$$

Now we notice that

$$\frac{d}{c}p = d \cdot \frac{p}{c}.$$

The number p/c is the ratio of the two known values of x. In Example 14, $p = 18$ and $c = 6$, giving $p/c = 3$; this explains why we obtain the desired answer by "tripling."

In general, if y is proportional to x, then the ratio of any two nonzero values of x is the same as the ratio of the corresponding values of y.

In many practical applications of proportionality, only positive values of x and of y have a sensible meaning.

Example 15 A distance between two points may be expressed as x inches or as y centimeters. In this situation, y is proportional to x. The constant of proportionality, established by public law, is 2.54. How many centimeters long is a yardstick?

Since $y = (2.54)x$, if $x = 36$, then $y = (2.54)(36)$. A yardstick is 91.44 centimeters long.

24   Linear Functions

EXERCISES F

1. If y is proportional to x, and if $y = 10$ in case $x = 3$, find:
   (a) y, in case $x = 30$.
   (b) y, in case $x = 0$.
   (c) x, in case $y = 2$.

2. Suppose that y is proportional to x with 3 as the constant of proportionality. Explain why x is proportional to y, and tell what the constant of proportionality is.

3. If u varies as v, show that v varies as u.

4. Suppose that one peso in Mexican money is equivalent to eight cents in United States money. If the value of an article is either y pesos or x dollars, show that y is proportional to x, and show that the constant of proportionality is $\frac{25}{2}$.

5. Suppose that one mark in West German money is equivalent to one quarter in United States money. If the value of an article is either z marks or x dollars, show that x is proportional to z, and find the constant of proportionality.

6. Using the notation of Exercises 4 and 5, explain why y is proportional to z, and identify the proportionality constant.

7. If u varies as v, and if v varies as w, show that u varies as w.

8. Let the measure of an angle be either d degrees or f right angles. Show that f is proportional to d, and find the constant of proportionality.

9. The area of a rectangular region may be expressed either as t square feet or as s square meters. Hence t is proportional to s.
   (a) What is the constant of proportionality? [HINT: See Example 15.]
   (b) Find t, in case $s = 40$.
   (c) What is the area, in square meters, of a lot which measures 18,000 square feet?

10. The distance traveled by a body moving in a fixed direction during a given period of time is proportional to the speed of the body. Suppose that a body moving at a speed of 20 centimeters per second travels one meter.
    (a) How far does the body travel if the speed is increased by 10%?
    (b) How fast does the body move if the distance traveled is only half as much as in the original situation?

11. A machine manufactures 300 kilograms of a substance in one hour. Let n be the number of kilograms of the substance produced in t hours of machine operating time.
    (a) Explain why n is proportional to t.
    (b) What is the constant of proportionality?
    (c) How much machine time is required to produce 1000 kilograms of the substance?

12. A store regularly sells articles at a retail price 40% above the wholesale price paid by the store. During its August sale the store reduces the retail price on all its merchandise by one-quarter.
    (a) Show that the August retail price of an article is proportional to the wholesale price of the article.
    (b) Find the constant of proportionality.

13. The temperature of a body is either t degrees Centigrade or u degrees Fahrenheit. If L is the linear function such that $t = L(u)$ for all meaningful temperatures, then $L(32) = 0$ and $L(212) = 100$.
    (a) Find the formula for the linear function L.
    (b) Decide whether t is proportional to u, and (if so) what the constant of proportionality is; justify your decisions.

14. In a simple type of electrical circuit the voltage varies as the resistance.
    (a) If the number of ohms of resistance is doubled, what is the effect on the number of volts?
    (b) If the number of ohms of resistance is increased to ten times its former value, how is the number of volts affected?
    (c) What change in the number of ohms of resistance would cause a 25% decrease in the voltage?

10 Distance between a point and the origin

An important result in plane geometry is the theorem of Pythagoras and its converse. Let the lengths of the sides of a triangle be the numbers a, b, h. The two sides of lengths a and b are perpendicular if and only if $a^2 + b^2 = h^2$. We use this characterization of a right triangle in our next topics.

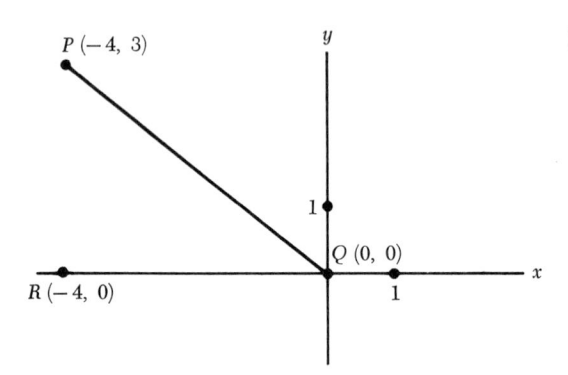

Figure 16

Example 16 Find the distance between the point $(0, 0)$ and the point $(-4, 3)$. Let the points $(-4, 3)$, $(0, 0)$, $(-4, 0)$ be called P, Q, R, respectively. (See Fig. 16.) These three points are the vertices of a triangle. The distance between Q and R is 4. Since P and R lie on a line that is parallel to the y-coordinate axis, the distance between them may be determined by examining their respective y-coordinates, 3 and 0. The distance is 3. Let h be the desired distance between P and Q. Since the triangle with vertices P, Q, R is a right triangle, $4^2 + 3^2 = h^2$. Thus $h^2 = 16 + 9 = 25$, and $h = 5$. The point $(-4, 3)$ is five units from the origin.

In a coordinate plane, let (c, d) be any point not lying on a coordinate axis. We wish to show that the distance between the point (c, d) and the point $(0, 0)$ is the positive number

$$(c^2 + d^2)^{1/2}$$

26 Linear Functions

(which is sometimes expressed as $\sqrt{c^2 + d^2}$). Following the pattern of Example 16, we consider the triangle whose vertices are $Q = (0, 0)$, $P = (c, d)$, and $R = (c, 0)$. The side of the triangle joining Q and R has length either c or $-c$ (depending on whether c is positive or negative). The side of the triangle joining R and P has length either d or $-d$ (according as d is positive or negative). If h is the desired distance between P and Q, then the theorem of Pythagoras assures us that $h^2 = (\pm c)^2 + (\pm d)^2$. More simply,

$$h^2 = c^2 + d^2.$$

Then h is the positive number whose square is $c^2 + d^2$. The positive "square root" is $(c^2 + d^2)^{1/2}$.

EXERCISES G

1. Find the distance between the point $(0, 0)$ and each of the following points:
   (a) $(3, -4)$.   (b) $(-5, 12)$.   (c) $(-6, -8)$.
   (d) $(1, -1)$.   (e) $(-\sqrt{2}, \sqrt{2})$.   (f) $(1, 2)$.

2. Find the distance between the point $(0, 0)$ and the point of intersection of the two lines whose respective equations are $6x + 5y + 9 = 0$ and $y = 2x + 11$. (-4, 3)

3. Find the distance between each of the following pairs of points. [HINT: A graph may be a revealing aid.]
   (a) $(1, 0)$, $(7, 0)$.   (b) $(1, 5)$, $(7, 5)$.
   (c) $(0, -1)$, $(0, 8)$.   (d) $(1, -1)$, $(1, 8)$.
   (e) $(0, a)$, $(0, b)$, in case a is greater than b.
   (f) $(1, c)$, $(1, d)$, in case d is greater than c.

4. If d is a negative number, find the distance between $(0, 0)$ and the point $(-6d, 8d)$.

5. The three sides of a triangle are respectively contained in the lines which are the graphs of the linear functions $x + 3$, $5x - 1$, $-7x - 13$. Draw a sketch of the triangle, labeling each vertex with its pair of coordinates.

11 Perpendicular lines

Let L and K be two different linear functions. Suppose that L and K are given by the respective formulas $L(x) = mx + b$ and $K(x) = nx + g$. The graphs of L and K are, respectively, lines. We recall that we have a simple technique for deciding whether these lines are parallel: the lines are parallel if and only if $m = n$.

We now seek a criterion for deciding whether the graph of L and the graph of K are perpendicular lines. Since perpendicularity is a comparison of directions, and since the directions of the two lines are given respectively by their slopes m and n, it seems reasonable that the criterion we are seeking will involve m and n in some manner.

Figure 17

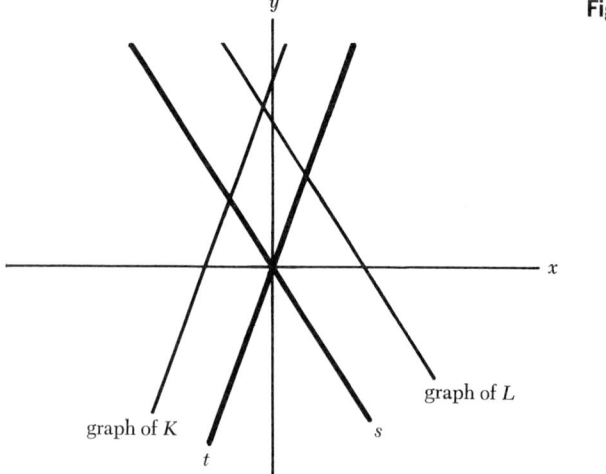

There is a line that contains the origin (0, 0) and that has the same slope as the graph of L has; call this line s; the line s and the graph of L are parallel (or perhaps coincide). Likewise, there is a line that contains the origin (0, 0) and that has the same slope as the graph of K has; call this line t; then t and the graph of K are parallel (perhaps, coincident). (See Fig. 17.)

The question whether the graphs of L and K are perpendicular has the same answer as the question whether s and t are perpendicular lines.

Now s is the graph of the equation $y = mx$, since the slope of s is m. Hence, s contains the point (1, m). Likewise, t is the graph of the equation $y = nx$ and contains the point (1, n). (See Fig. 18.)

The square of the distance between (0, 0) and (1, m) is

$$1 + m^2.$$

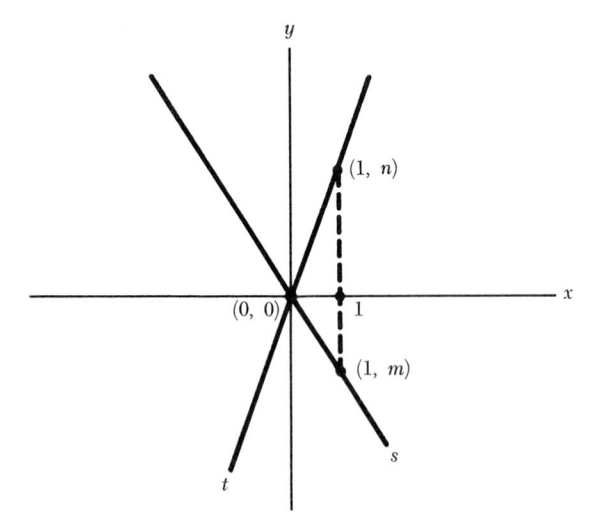

Figure 18

28   Linear Functions

The square of the distance between $(0, 0)$ and $(1, n)$ is
$$1 + n^2.$$
Since $(1, m)$ and $(1, n)$ lie on a line parallel to the y-coordinate axis, the distance between them is either $n - m$ or $m - n$ (depending on whether n is greater than m or not). (In Fig. 18, n is greater than m, and the distance is $n - m$.) In either case, the square of the distance between $(1, m)$ and $(1, n)$ is
$$(n - m)^2.$$
[Note that $(n - m)^2$ and $(m - n)^2$ are the same.]

According to the theorem of Pythagoras and its converse, a criterion for deciding whether s and t are perpendicular lines is given by whether the following statement is true concerning the lengths of the sides of the triangle with vertices $(0, 0)$, $(1, m)$, $(1, n)$.
$$(1 + m^2) + (1 + n^2) = (n - m)^2.$$
This equation is equivalent to
$$2 + m^2 + n^2 = n^2 + m^2 - 2mn,$$
hence to
$$2 = -2mn,$$
and therefore to
$$mn = -1.$$

We have now found the criterion we have been seeking. We state the result as follows.

> **Theorem 5** On a coordinate system two lines that are the respective graphs of linear functions are perpendicular to each other if and only if the product of their slopes is the number -1.

> **Example 17** Two lines whose respective slopes are 3 and $-\frac{1}{3}$ are perpendicular to one another, because the product $3 \cdot (-\frac{1}{3})$ is the number -1.

> **Example 18** A line that is perpendicular to the graph of the linear function given by the formula $\frac{2}{5}x + \frac{1}{3}$ has slope $-\frac{5}{2}$. We obtained the value $-\frac{5}{2}$ by noting that the slope of the given line is $\frac{2}{5}$ and therefore the desired slope m satisfies the condition $m(\frac{2}{5}) = -1$.

> **Example 19** An equation of a certain line is $8x - 3y = 41$. Find an equation of the line r that is perpendicular to the given line and contains the point $(1, 6)$.
> The given equation is equivalent to $3y = 8x - 41$ or to
> $$y = \tfrac{8}{3}x - \tfrac{41}{3}.$$

The slope of the given line is $\frac{8}{3}$. The line r has slope $-\frac{3}{8}$. Thus, an equation for the line r has the form

$$y = -\tfrac{3}{8}x + b,$$

where the number b remains to be determined. Since the line r contains $(1, 6)$, we obtain the statement

$$6 = -\tfrac{3}{8} \cdot 1 + b.$$

Thus, $b = \frac{51}{8}$. An equation for the line r is

$$y = -\tfrac{3}{8}x + \tfrac{51}{8}.$$

Alternative equations that are also acceptable answers are $8y = -3x + 51$ or $3x + 8y = 51$.

If two lines that are the graphs of linear functions are perpendicular to one another, then the product of their slopes is the number -1. Thus, one of the slopes is positive and the other is negative. Notice that this fits with our interpretation of slope. Of two perpendicular lines (neither of which is parallel to a coordinate axis), one should slant upward and the other slant downward.

12 Linear equation

Definition A linear equation in x and y is an equation of the form

$$cx + dy + e = 0,$$

in which c, d, e are real numbers subject to the restriction that one or both of the numbers c, d are different from zero.

The graph of any linear equation in x and y is a line. Before proving this assertion, let us examine a few illustrations. Example 7 (in Section 6) discusses the equation $2x + 3y = 6$. This equation is equivalent to

$$2x + 3y - 6 = 0,$$

and the latter fits the pattern described above, with the numbers 2, 3, -6 playing the respective roles of c, d, e. We observed in Example 7 that the graph of the equation is the line shown in Fig. 11.

Example 20 The definition of a linear equation in x and y permits the number d to be zero. In this case, c must be different from zero. Thus, if $c = 2$, $d = 0$, $e = -5$, we have the linear equation $2x + 0y - 5 = 0$, or more simply,

$$2x - 5 = 0.$$

This equation is equivalent to $2x = 5$ and hence to
$$x = \tfrac{5}{2}.$$
The graph of the equation consists of all the points that have $\tfrac{5}{2}$ as their x-coordinate. These points form the vertical line that is $\tfrac{5}{2}$ units to the right of the y-coordinate axis. See Fig. 19.

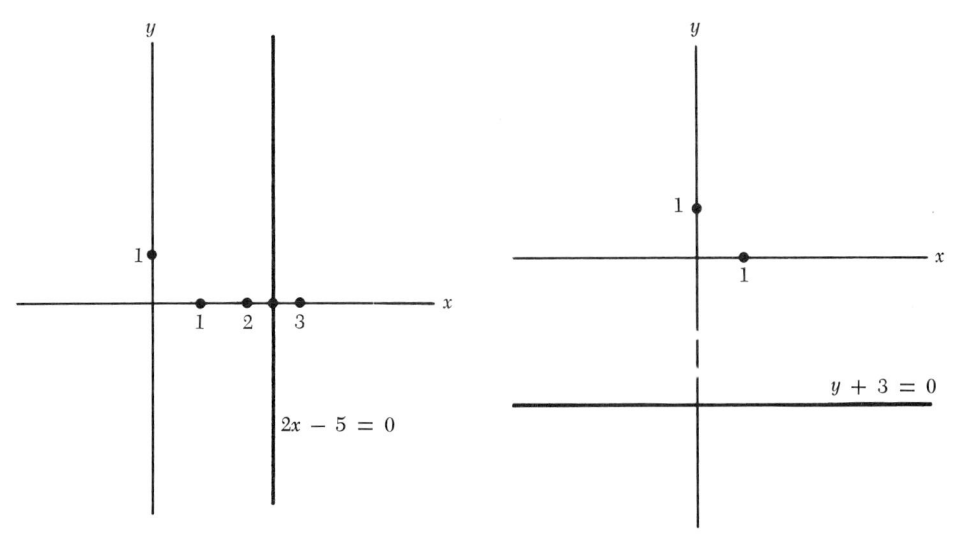

Figure 19

Figure 20

Example 21 The description of a linear equation in x and y also permits c to be zero, provided d is not zero. In case $c = 0$, $d = 1$, $e = 3$, we have $0x + 1y + 3 = 0$. The graph of this equation is the set of all points satisfying the condition that
$$y = -3,$$
namely, the line that is horizontal and that is 3 units below the x-coordinate axis. See Fig. 20.

We are now ready to justify the use of the adjective "linear" in the phrase "linear equation."

Theorem 6 On a two-dimensional coordinate system, the graph of every linear equation is a line.

PROOF Let any linear equation in the coordinates x and y be given. It has the form
$$cx + dy + e = 0,$$
where at least one of the numbers c, d is different from zero. We consider three cases.

(i) Suppose neither c nor d is zero. Then the above equation is equivalent to the equation

$$y = -\frac{c}{d}x - \frac{e}{d},$$

and the graph is the same as the graph of the linear function

$$-\frac{c}{d}x - \frac{e}{d}.$$

(ii) Suppose $c = 0$. Then d is not zero, and the equation $cx + dy + e = 0$ is equivalent to

$$y = -\frac{e}{d}.$$

The graph is a line parallel to the x-coordinate axis (or perhaps is the axis itself).

(iii) Suppose $d = 0$. Then c is not zero, and the equation $cx + dy + e = 0$ is equivalent to

$$x = -\frac{e}{c}.$$

The graph is a line parallel to the y-coordinate axis (or coincides with the axis).

In each case, the graph is a line.

The next theorem is the converse of Theorem 6.

Theorem 7 Every line on a two-dimensional coordinate system is the graph of a linear equation.

PROOF Let any line in the coordinate plane be given. From its many points, choose any two, say A and B. Let their coordinates be (p, q) and (r, s), respectively. Note that the letters p, q, r, s, representing the coordinates of specific points, denote constants. Since the points (p, q) and (r, s) are different, we conclude that one or both of the following statements is true: p and r are different numbers; q and s are different numbers. Hence, we find that one or both of the differences $s - q$, $p - r$ are not zero. Thus,

$$(s - q)x + (p - r)y = ps - qr$$

fulfills the requirements for being a linear equation in x and y. By Theorem 6, the graph of this linear equation is a line. By direct substitution, the coordinates (p, q) of the point A satisfy the equation, because

$$(s - q)p + (p - r)q = ps - pq + pq - qr = ps - qr.$$

Likewise, the coordinates (r, s) of the point B satisfy the equation. Therefore, the graph of the linear equation is the line that contains both points A and B. That line, however, is the same as the line originally given. Thus the given line is the graph of the constructed linear equation in x and y.

Example 22 Find an equation for the line that contains the points $(2, 6)$ and $(2, -3)$.

The question reminds us of Examples 11 and 12 (in Section 8). We wish to observe that the method of Section 8 is not applicable here. Suppose that there is an equation of the type $y = mx + b$ for the line. Then

$$6 = 2m + b$$

because $(2, 6)$ is a point of the line, and

$$-3 = 2m + b$$

because the line contains the point $(2, -3)$. By subtraction, these two equations yield the false conclusion that $9 = 0$. Thus, the previous method is not pertinent in this example.

In fact, any algebraic approach to this problem is too ponderous. A brief inspection of Fig. 21, which shows the two given points, reveals that the line joining them is the vertical line whose equation is

$$x = 2.$$

The experienced problem-solver judiciously mixes algebraic and geometric methods by using at each stage the most efficient technique available to him.

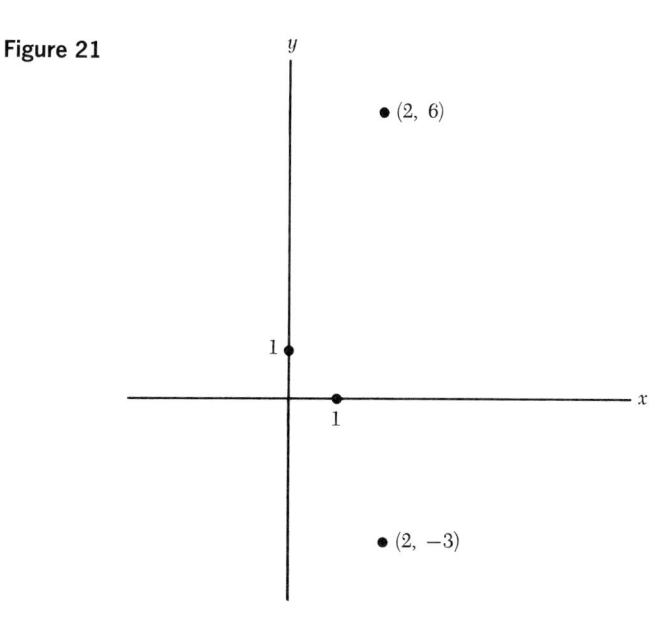

Figure 21

EXERCISES H

1. The respective slopes of two lines are the following pairs of numbers. In each case, decide whether the lines are perpendicular, or parallel, or neither.
   (a) $2, -2$.   (b) $\frac{1}{2}, \frac{1}{2}$.   (c) $\frac{1}{3}, -3$.   (d) $1, 1$.   (e) $1, -1$.   (f) $-\frac{1}{3}, -3$.

2. What is the slope of a line that is perpendicular to the graph of the equation $3x + 2y = 0$?

3. The point P with coordinates $(1, -3)$ belongs to the line $2x + y + 1 = 0$. A line containing P and perpendicular to the given line is the graph of a linear function A. Find $A(4)$.

4. The point Q has coordinates $(2, \frac{1}{3})$. A line contains Q and is perpendicular to the line $3x - y + 2 = 0$. Find an equation for the line described.

5. Suppose that a and b are positive numbers. The two lines whose respective equations are $ax + by = a + b$ and $bx + uy = 1$ are perpendicular to each other. Find the number u.

6. The points P, Q, and R have coordinates $(1, 3)$, $(-4, 1)$, and $(0, -7)$, respectively. Find an equation for the line that contains R and that is perpendicular to the line containing P and Q.

7. Let a, b, c, d, e, f be six different numbers. Find an equation for the line that contains the point (a, b) and that is perpendicular to the line containing the points (c, d) and (e, f).

8. Use the theorem of Pythagoras to verify that the three points $(0, 0)$, $(1, \frac{5}{4})$, $(1, -\frac{4}{5})$ are vertices of a right triangle.

9. Use the theorem of Pythagoras to decide whether the three points $(0, 0)$, $(0.9, 1)$, $(-1.1, 1)$ are the vertices of a right triangle. Justify your decision.

10. For each of the following pairs of points, find an equation of the line joining the two points. In the latter parts, let a, b be positive numbers.
    (a) $(3, 7), (9, 1)$.   (b) $(4, 12), (4, 5)$.   (c) $(3, 8), (-3, 8)$.   (d) $(2, 9), (0, 0)$.
    (e) $(\frac{3}{8}, 1), (\frac{2}{5}, 1)$.   (f) $(-6, \frac{1}{4}), (-6, \frac{2}{3})$.   (g) $(a, b), (-a, b)$.   (h) $(a, b), (a, -b)$.
    (i) $(a, b), (-a, -b)$.   (j) $(-a, b), (-a, -b)$.

11. The management of a factory finds that the cost of producing and selling its product is \$30 for each ton produced in a week. There are fixed expenses of \$7500 for operating the factory during the week. Let t be the number of tons of the product manufactured in a week, let c be the number of dollars in the cost of production and sales during the week, and let g be the total number of dollars of weekly expense for the factory. Suppose that the entire output of the factory is sold at a price of \$50 per ton. Let s be the number of dollars of gross sales revenue.
    (a) Show that c is proportional to t.
    (b) Give the constant of proportionality in part (a).
    (c) Show that s is proportional to t.
    (d) What is the constant of proportionality in part (c)?
    (e) Express g as a linear function of t.
    (f) Interpret the difference $s - g$ from the economic viewpoint.
    (g) What value of t is a root of the equation $s = g$?
    (h) Interpret part (g) economically.
    (i) How many tons of the product must the factory produce in order to make a gross profit of \$6000 for the week?

12. Repeat Exercise 11 with the following modification. The selling price of the product is \$45 per ton.

34   Linear Functions

Systems of Linear Equations

1 Introduction

There are a number of valuable methods for solving systems of linear equations. Some of them are particularly advantageous in special situations. However, rather than study several methods, we prefer to emphasize just one. Of course, if we restrict our attention to just one procedure, we want to choose a method that will handle all problems, and we would prefer a method that is not only universally applicable but also efficient. We have chosen the (complete) Gauss elimination method.

The *Gauss elimination method* may be applied to any given system of linear equations—regardless of whether the number of equations is the same as the number of unknowns or not. This method reveals whether or not the system has a root. If there is exactly one root, the method yields it. If there are many roots, the method finds all of them in a convenient form. Furthermore, the method involves a step-by-step pattern that makes it not only relatively easy to learn but also readily adaptable to programming for high-speed computing machinery.

To aid our understanding of the set of roots for a system of linear equations, we shall invoke geometric ideas. Various situations arise according as the number of roots of the system is zero, or one, or more than one. Since some characteristic features do not occur in two-dimensional systems, we shall introduce three-dimensional geometry. Vector and matrix terminology will be helpful in discussing and analyzing systems in general.

2 Systems with one root

In each of the next few examples, the Gauss elimination method is applied to a system of linear equations that has (exactly) one root. In later sections of this chapter we shall encounter other systems.

For illustrative purposes while we become familiar with the method, we shall use simple examples. It is possible that in some simple situations the

Gauss elimination method is not the easiest or quickest procedure. The student should bear in mind the long-range goals announced in Section 1 and should exploit these simple examples to become acquainted with the basic steps in the method.

Example 1 Find numbers x and y such that

(a)
(b)
$$\begin{cases} 2x - y = -2, \\ -3x + 2y = 5. \end{cases}$$

We multiply each member of (a) by $\frac{1}{2}$, and replace (a) by the resulting equivalent equation,

(c)
$$x - \tfrac{1}{2}y = -1.$$

To each member of (b) we add 3 times the corresponding member of (c); in the resulting left member the term involving x is $(-3x) + 3(x)$ and may be omitted, the term involving y is $(2y) + 3(-\tfrac{1}{2}y) = \tfrac{1}{2}y$, while in the right member the constant is $(5) + 3(-1) = 2$; thus, the simplified equation is

(d)
$$\tfrac{1}{2}y = 2.$$

We multiply each member of (d) by 2, and replace (d) by the resulting equivalent equation,

(e)
$$y = 4.$$

To each member of (c) we add $\frac{1}{2}$ times the corresponding member of (e), obtaining

$$(x - \tfrac{1}{2}y) + \tfrac{1}{2} \cdot y = -1 + \tfrac{1}{2} \cdot 4,$$

or

(f)
$$x = 1.$$

The desired numbers x and y are 1 and 4, respectively. Indeed, if we replace x by 1 and y by 4 in the left member of (a), we obtain $2 \cdot 1 - 4$, which is the same as the number -2 on the right; likewise, if we substitute 1 and 4 for x and y, respectively, the left member of (b) becomes $-3 \cdot 1 + 2 \cdot 4$, which agrees with the number 5 in the right member of (b).

In actual practice, we perhaps write only steps (a, b, c, d, e, f). However, in order to help acquire an understanding of our method, we rewrite the solution in a fashion that suggests the pattern for the method.

(a)
(b)
$$\begin{cases} 2x - y = -2, \\ -3x + 2y = 5. \end{cases}$$

(c)
(b)
$$\begin{cases} x - \tfrac{1}{2}y = -1, \\ -3x + 2y = 5. \end{cases}$$

$$\begin{matrix}(c)\\(d)\end{matrix} \qquad \begin{cases} x - \tfrac{1}{2}y = -1, \\ \phantom{x -{}}\tfrac{1}{2}y = 2. \end{cases}$$

$$\begin{matrix}(c)\\(e)\end{matrix} \qquad \begin{cases} x - \tfrac{1}{2}y = -1, \\ \phantom{x - \tfrac{1}{2}}y = 4. \end{cases}$$

$$\begin{matrix}(f)\\(e)\end{matrix} \qquad \begin{cases} x \phantom{- \tfrac{1}{2}y}= 1, \\ \phantom{x - \tfrac{1}{2}}y = 4. \end{cases}$$

The root of the system is $(x, y) = (1, 4)$.

Our steps were as follows. In one equation make the coefficient of x equal to 1, in every other equation make the coefficient of x equal to 0 (that is, "eliminate" x). Repeat the preceding two stages for the coefficients of y, without affecting the coefficients of x that have already been selected nicely. At the end, read the root of the system.

Let us examine more closely the step to "read the root" at the end. Consider the final system displayed above, namely,

$$\begin{matrix}(f)\\(e)\end{matrix} \qquad \begin{cases} x = 1, \\ y = 4. \end{cases}$$

For any root of this system, we "read" that x must be 1 and that y must be 4. Conversely, the ordered pair $(x, y) = (1, 4)$ does satisfy this system. In fact, the extreme simplicity of the system of equations makes almost too apparent the fact that the one and only root is $(1, 4)$. Now a vital feature of our method is that successive steps of the procedure make no change whatever in the root or roots (if any). Consequently, the sole root of the final system,

$$\begin{matrix}(f)\\(e)\end{matrix} \qquad \begin{cases} x = 1, \\ y = 4, \end{cases}$$

is also the one and only root of the original system,

$$\begin{matrix}(a)\\(b)\end{matrix} \qquad \begin{cases} 2x - y = -2, \\ -3x + 2y = 5. \end{cases}$$

It is in this sense that whatever root we "read" at the end agrees precisely with the desired root (and that there are no other desired roots).

Example 2 Find a linear function L such that $L(3) = 5$ and $L(-2) = 6$.

There are numbers m and b such that $L(x) = mx + b$. Using the prescribed conditions, we obtain the statements $3m + b = 5$ and $-2m + b = 6$. We use the elimination method to find m and b.

$$\begin{cases} b + 3m = 5, \\ b - 2m = 6. \end{cases}$$

$$\begin{cases} b + 3m = 5, \\ \phantom{b +{}} -5m = 1. \end{cases}$$

2 / Systems with one root

$$\begin{cases} b + 3m = 5, \\ m = -\tfrac{1}{5}. \end{cases}$$

$$\begin{cases} b = \tfrac{28}{5}, \\ m = -\tfrac{1}{5}. \end{cases}$$

Hence, L is given by the formula $(-x + 28)/5$. We verify that the formula $(-x + 28)/5$ assigns to 3 the image $(-3 + 28)/5 = 5$ and associates with -2 the image $[-(-2) + 28]/5 = 6$. Thus, the required properties are possessed by the linear function given by our formula.

Notice that the pattern of steps outlined in Example 1 has been followed in determining b and m.

Example 3 Find three numbers x, y, z such that

$$\begin{cases} 3x + 2y - z = 3, \\ -2x + y + 3z = 5, \\ 4x - y - 2z = -1. \end{cases}$$

Our plan of solution is the following. In one equation make the coefficient of x equal to 1; in every other equation make the coefficient of x equal to 0 (thereby eliminating x). Repeat for the coefficients of y, without changing the coefficients of x. Repeat for the coefficients of z, without affecting the coefficients of x or of y. At the end, read the root of the system of equations. Observe how this program is accomplished in a stepwise manner. Explanations for some of the steps are given below.

(a)
(b)
(c)
$$\begin{cases} x + \tfrac{2}{3}y - \tfrac{1}{3}z = 1, \\ -2x + y + 3z = 5, \\ 4x - y - 2z = -1. \end{cases}$$

(d)
(e)
$$\begin{cases} x + \tfrac{2}{3}y - \tfrac{1}{3}z = 1, \\ \tfrac{7}{3}y + \tfrac{7}{3}z = 7, \\ -\tfrac{11}{3}y - \tfrac{2}{3}z = -5. \end{cases}$$

(f)
$$\begin{cases} x + \tfrac{2}{3}y - \tfrac{1}{3}z = 1, \\ y + z = 3, \\ -\tfrac{11}{3}y - \tfrac{2}{3}z = -5. \end{cases}$$

(g)
$$\begin{cases} x - z = -1, \\ y + z = 3, \\ 3z = 6. \end{cases}$$

$$\begin{cases} x - z = -1, \\ y + z = 3, \\ z = 2. \end{cases}$$

38 Systems of Linear Equations

(h)
(i)
(j)
$$\begin{cases} x & = 1, \\ y & = 1, \\ z = 2. \end{cases}$$

The root is $(x, y, z) = (1, 1, 2)$.

In (a) the coefficient of x is 1. We obtain (a) by multiplying each member of the first given equation by $\frac{1}{3}$. We could have achieved the first step in our program in any of various other ways; for example, we might have added the corresponding members of the first and second given equations.

By adding to the members of (b) twice the corresponding members of (a), we obtain (d). Likewise, we obtain (e) by adding to the members of (c) the corresponding members of (a) each multiplied by -4.

We obtain (f) by multiplying each member of (d) by $\frac{3}{7}$.

To each member of (a) we add the product of $-\frac{2}{3}$ and the corresponding member of (f); this gives us (g).

Stepwise, we continue our program. Finally, from steps (h, i, j) it is easy to recognize the desired values of x, y, z.

We again emphasize an important aspect of our method. The desired numbers x, y, z that satisfy the original system of equations

$$\begin{cases} 3x + 2y - z = 3, \\ -2x + y + 3z = 5, \\ 4x - y - 2z = -1, \end{cases}$$

are precisely the same as the numbers that satisfy the final system

(h)
(i)
(j)
$$\begin{cases} x & = 1, \\ y & = 1, \\ z = 2. \end{cases}$$

Since the latter system has exactly one root, namely, $(x, y, z) = (1, 1, 2)$, this is the one and only root for the original system. After the system of equations has been converted into the extremely simple form of (h), (i), (j), we are able to "read" the root of the original system.

The proof that the Gauss elimination method enables us to solve a system of equations as claimed will be given in the final section of this chapter. Meanwhile, we shall freely use the method in order to gain better familiarity with its assorted capabilities.

Example 4 Find four numbers x, y, z, t such that

$$\begin{cases} x + 3y + 2z + 7t = -5, \\ 3x + 7y + 2z + 11t = 1, \\ 2x + 5y + 3z + 12t = -7, \\ -x + z + t = -6. \end{cases}$$

Our procedure follows the same pattern we used in the preceding examples. In one equation we make the coefficient of x equal to 1; from every other equation we eliminate x by making its coefficient equal to 0. We repeat for the coefficients of y, but without altering the already attained "nice" coefficients for x. We repeat for the coefficients of z, without destroying the previously achieved goals. We again repeat for the coefficients of t, without changing the coefficients of x or y or z. Finally, we read the root of the system. Observe how the following steps fit this pattern. Comments for some of the steps are given below.

(a)
(b)
(c)
(d)
$$\begin{cases} x + 3y + 2z + 7t = -5, \\ 3x + 7y + 2z + 11t = 1, \\ 2x + 5y + 3z + 12t = -7, \\ -x + z + t = -6. \end{cases}$$

(e)
(f)
(g)
$$\begin{cases} x + 3y + 2z + 7t = -5, \\ -2y - 4z - 10t = 16, \\ -y - z - 2t = 3, \\ 3y + 3z + 8t = -11. \end{cases}$$

(h)
$$\begin{cases} x + 3y + 2z + 7t = -5, \\ y + 2z + 5t = -8, \\ -y - z - 2t = 3, \\ 3y + 3z + 8t = -11. \end{cases}$$

(i)

(j)
$$\begin{cases} x - 4z - 8t = 19, \\ y + 2z + 5t = -8, \\ z + 3t = -5, \\ -3z - 7t = 13. \end{cases}$$

(k)
(l)

(m)
$$\begin{cases} x + 4t = -1, \\ y - t = 2, \\ z + 3t = -5, \\ 2t = -2. \end{cases}$$

$$\begin{cases} x + 4t = -1, \\ y - t = 2, \\ z + 3t = -5, \\ t = -1. \end{cases}$$

(n)

(o)
(p)
(q)
(r)
$$\begin{cases} x = 3, \\ y = 1, \\ z = -2, \\ t = -1. \end{cases}$$

The root is $(x, y, z, t) = (3, 1, -2, -1)$.

Systems of Linear Equations

The coefficient of x in the first equation of the given system is 1. Consequently, no adjustment is needed to obtain (a).

By adding to each member of (b) the corresponding member of (a) multiplied by -3, we obtain (e). Likewise, by adding to each member of (c) the respective member of (a) multiplied by -2, we obtain (f). We obtain (g) by adding the corresponding members of (d) and (a).

We obtain (h) from (e) by multiplying each member by $-\frac{1}{2}$.

We find (i) by adding to each member of (a) the product of -3 and the corresponding member of (h). We find (j) by adding respective members of (f) and (h).

Since the coefficient of z in (j) is 1, no adjustment is necessary before steps (k, l, m), in which z is eliminated from the other equations.

We obtain (n) from (m) by halving each member.

Stepwise, we continue. Finally, from steps (o, p, q, r) we easily identify the desired values of x, y, z, t, respectively.

Example 5 On a coordinate system, the graph of the equation $6x - 11y = 50$ meets the line $7x + 13y = 17$ at the point E. Find the coordinates of E.

Figure 1

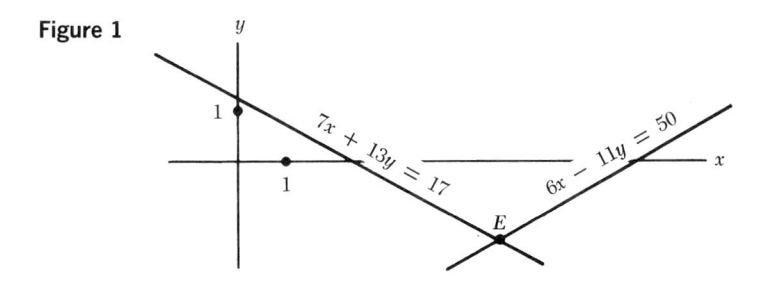

Figure 1 shows the two lines and the point E. We solve algebraically for the coordinates of E.

(a)
(b)
$$\begin{cases} 6x - 11y = 50, \\ 7x + 13y = 17. \end{cases}$$

(c)
$$\begin{cases} x + 24y = -33, \\ -155y = 248. \end{cases}$$

$$\begin{cases} x = \frac{27}{5}, \\ y = -\frac{8}{5}. \end{cases}$$

The point E is $(\frac{27}{5}, -\frac{8}{5})$.

We avoided fractions at an early stage by subtracting the members of (a) from the respective members of (b) in order to obtain (c).

EXERCISES A

1. Use the Gauss elimination method in order to solve each of the following systems.
   (a) $x + 3y = -1,$
       $2x + 3y = 4.$
   (b) $3u - 2v = 4,$
       $2u + 3v = 5.$
   (c) $4x + y - 2z = -6,$
       $-3x + 2z = 1,$
       $x + y + 4z = 3.$
   (d) $2a + 4b - c = -7,$
       $-5a - 3b + 2c = 6,$
       $3a - b + 4c = 11.$
   (e) $w - x + 2y - z = 0,$
       $w + x - y + z = 2,$
       $2w - x + 3y - z = 1,$
       $w + 2x - y + z = 1.$

2. On a coordinate system, the point Q belongs to each of the lines $5x + 12y = 13$ and $3x + 7y = 10$. Use the Gauss elimination method to find the coordinates of Q.

3. Find a linear function K such that $K(1.3) = 1.8$ and $K(2.8) = 3.6$. Use the Gauss elimination method.

4. Suppose that a and b are positive numbers. Use the Gauss elimination method to find x and y such that $ax - by = a^2 + b^2$, $bx + ay = a^2 + b^2$.

5. Let a, b, c, d, e, f be numbers such that $ae \neq bd$. Use the Gauss elimination method to solve the following system: $ax + by = c$, $dx + ey = f$.
   (a) First, consider the case in which $a \neq 0$.
   (b) Second, consider the case in which $a = 0$, and show that the same result obtained in part (a) is applicable here.

6. Use the Gauss elimination method to find the ordered pair of numbers (x, y) such that $x + 2y = 5044$, $y + 2x = 4760$.

7. Use the Gauss elimination method in an attempt to find a root of the system $4x - 6y - 9 = 0$, $9y - 6x + 8 = 0$. Then explain, in geometrical language, the results you obtained.

3 Three-dimensional coordinate system

A root of a system of linear equations in three unknowns is an *ordered triple* of numbers. We desire a geometric interpretation for an ordered triple.

The graph of an ordered pair of numbers is a point in the coordinate plane. We plot the point by referring to two coordinate axes, which are perpendicular to each other. Whereas a two-dimensional coordinate system suffices to represent ordered pairs of numbers, we use a three-dimensional coordinate system to picture ordered triples of numbers. Note the analogy between the familiar two-dimensional situation and the three-dimensional case which we now introduce.

Consider three lines in space that have one point in common and that do not lie in a single plane. Customarily we select lines each of which is perpendicular to each of the others. We may think of one line as having a right-

42 Systems of Linear Equations

Figure 2

Figure 3

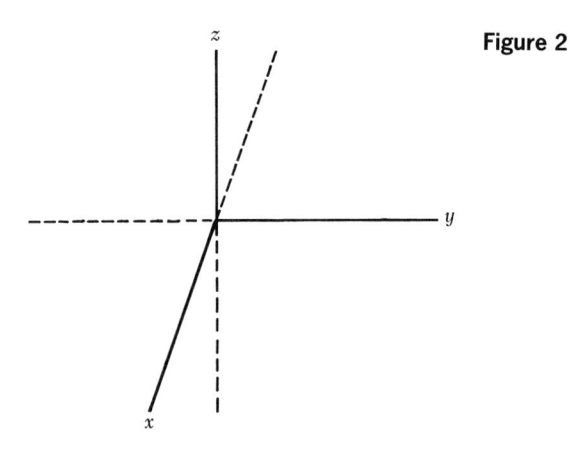

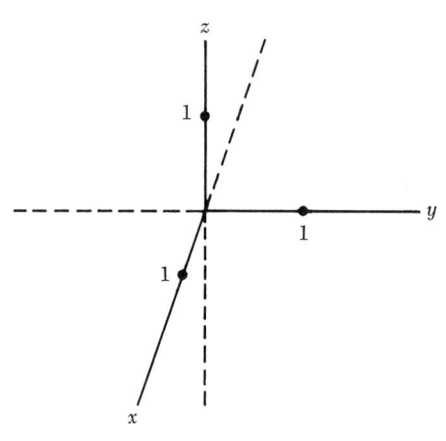

left direction, another as having a front-back direction, and a third as having an up-down direction. Facility in drawing diagrams is gained only by patience and experience.

Suppose we label the three lines as shown in Fig. 2. We choose a unit of measuring distance. We convert each of the lines into a coordinate axis by selecting the common intersection point as the origin for each axis and by picking the unit point on the respective axes as one unit "in front of," "to the right of," or "above" the common origin. See Fig. 3.

Having established a three-dimensional coordinate system in space, we may represent an ordered triple of real numbers by a point in space referred to the coordinate system. Given an ordered triple (a, b, c), we measure from the origin a units parallel to the x-coordinate axis (forward or backward according as a is positive or negative), then measure b units parallel to the y-coordinate axis (right or left according as b is positive or negative), and finally measure c units parallel to the z-coordinate axis (up or down according as c is positive or negative). The point we reach in this manner is said to have coordinates (a, b, c), relative to the chosen coordinate system.

Figure 4

(a)   (b)

Figure 5

(a)   (b)

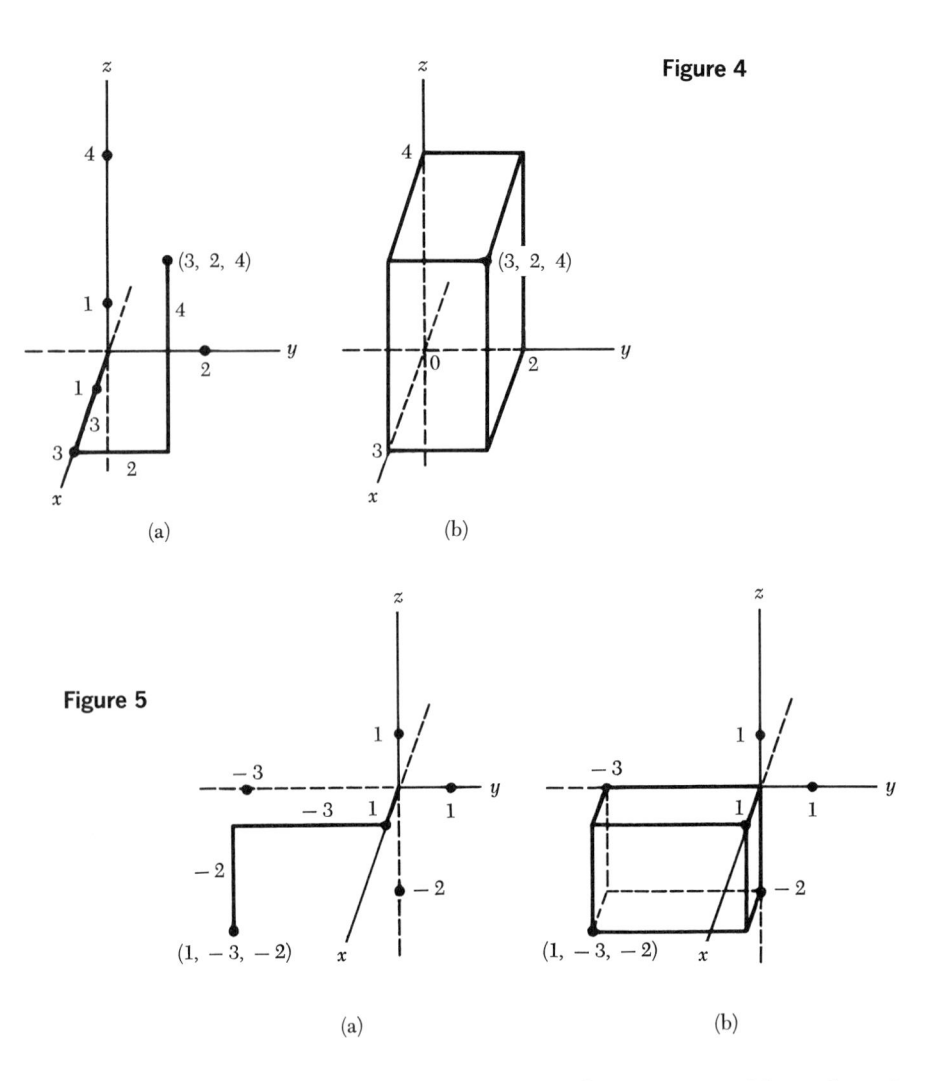

Example 6 The point (3, 2, 4) may be represented by either of the styles shown in Fig. 4.

Example 7 Graph the ordered triple $(1, -3, -2)$.
Each diagram in Fig. 5 shows the point whose coordinates are $(1, -3, -2)$.

Each two of the coordinate axes lie in a so-called *coordinate plane*. For example, the *x*-coordinate axis and the *y*-coordinate axis determine the *xy*-coordinate plane.

This plane is characterized by the statement that it consists of all points whose third coordinate (that is, *z*-coordinate) is 0. Thus, the *xy*-coordinate plane is described by the equation $z = 0$. We may call the plane the graph of the equation $z = 0$.

Example 8 The graph of the equation $z = 3$ consists of all points whose third coordinate is 3. The graph is a plane parallel to the xy-coordinate plane and three units above it. See Fig. 6.

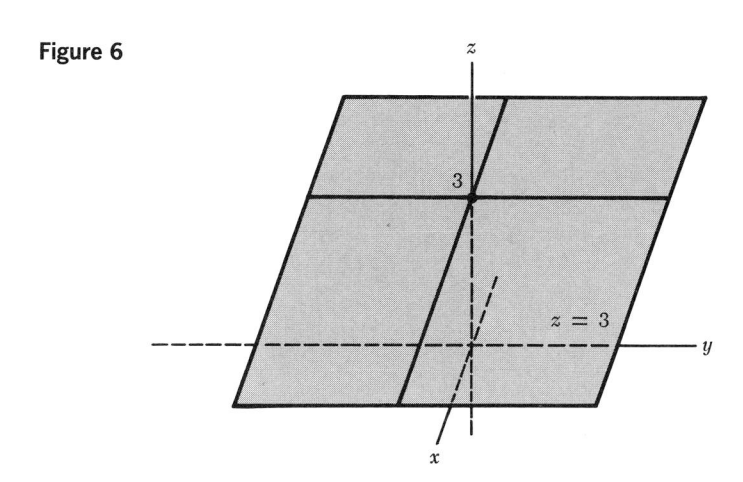

Figure 6

Each point on the y-coordinate axis has both its x-coordinate (forward or backward displacement) and its z-coordinate (upward or downward displacement) equal to 0. Thus the y-coordinate axis is characterized by the two equations $x = 0$ and $z = 0$.

EXERCISES B

1. Plot each of the following points on a three-dimensional coordinate system. (Make a separate diagram for each part.)
   (a) $(1, 6, 3)$. (b) $(2, -4, 1)$. (c) $(0, 3, -2)$. (d) $(0, 2, 0)$.
   (e) $(-1, 4, 0)$. (f) $(5, -7, -3)$. (g) $(1, 0, -2)$. (h) $(0, 0, -\frac{5}{3})$.

2. For each of the following, describe in words the location, on a three-dimensional coordinate system, of the points whose coordinates satisfy the condition.
   (a) Coordinates are of the form $(0, 0, z)$ for some number z.
   (b) Coordinates are of the type $(0, 1, z)$ for some number z.
   (c) Coordinates are of the form $(x, 0, 0)$ for some number x.
   (d) Coordinates are of the type $(x, 3, 2)$ for some number x.
   (e) Coordinates have the form $(0, y, z)$ where y and z are numbers.

3. On a three-dimensional coordinate system, where is a point (x, y, z) located in case $y = 2$?

4. Where is a point (x, y, z) located on a coordinate system in case $x = -1$?

5. Describe the collection of all points (x, y, z) on a coordinate system that satisfy the equation $z = 10$.

6. Describe the set of all points (x, y, z) that satisfy the system $y = -2$, $z = 4$.

7. What equation is satisfied by the points that lie four units to the right of the xz-coordinate plane?

8. Find an equation that is satisfied by all points lying $\frac{1}{2}$ unit to the left of the xz-coordinate plane.

9. Give an equation for the plane which is parallel to, and $\frac{5}{3}$ units behind, the yz-coordinate plane.

4 Coordinates of a given point

To each ordered triple of numbers there is associated a point in space. Conversely, to each point in space there is assigned an ordered triple of real numbers. If a point P is given, its coordinates may be determined by reversing the procedure in Section 3 for graphing a given triple. The line through P parallel to the z-coordinate axis pierces the xy-coordinate plane at some point, say Q (see Fig. 7). Relative to the two-dimensional coordinate system formed by the x-coordinate axis and the y-coordinate axis, the point Q has a pair of coordinates (a, b). If c is the number of units parallel to the z-coordinate axis that we need to measure from Q to P (where c is negative in case P is below Q), then the coordinates of P are (a, b, c).

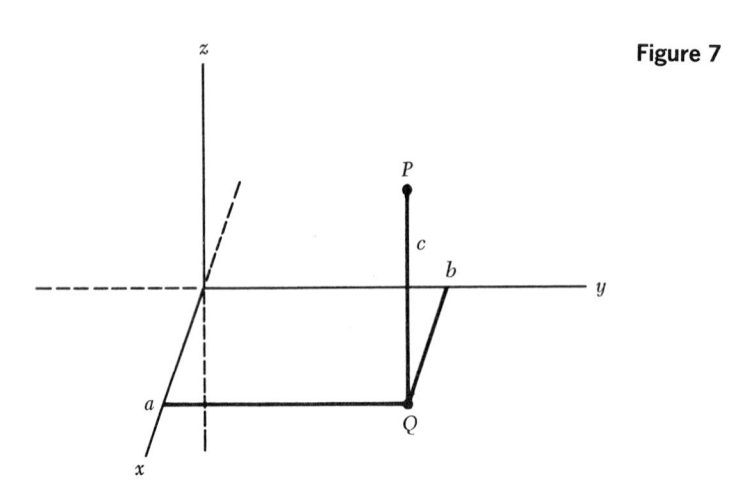

Figure 7

5 Vectors

We have often considered an ordered pair or an ordered triple of numbers. Frequently, an ordered array of numbers is called a *vector*. Each of the individual numbers is a *component* of the vector. Thus, an ordered triple of numbers is a vector with three components. An example of a vector with five components is $(2, -\frac{1}{2}, 0, -3, \sqrt{2})$.

46   Systems of Linear Equations

It is not essential that the components of a vector be written in a row fashion, although this familiar arrangement is perhaps most convenient for writing or printing. We shall soon see situations in which other arrangements are particularly useful.

Definition Let n be a natural number, and let $a_1, a_2, \ldots, a_n$ be numbers (not necessarily distinct). Then the ordered n-tuple $(a_1, a_2, \ldots, a_n)$ is a *vector* with n components. For each subscript i, the number a_i is the ith *component* of the vector.

We read a_i as "a, sub i," or simply as "a, i" in case the experienced listener understands from the context that i is a subscript. The subscript notation is a convenient device for naming many numbers without exhausting so many different letters of the alphabet. We often use a single letter such as V to denote a vector. A vector with n components is sometimes called *n-dimensional*.

To say that vectors U and V are the same, or are equal to each other, means that, first, U has the same number of components as V, and second, each component of U is the same as the corresponding component of V.

Thus, $(a_1, a_2, a_3) = (c, d, e)$ means that $a_1 = c$ and $a_2 = d$ and $a_3 = e$. Each three-dimensional vector is different from every two-dimensional vector.

An important elementary application of vectors is their use in recording data.

Example 9 Each applicant for a license gives his age in years, his height in inches, and his weight in pounds. The data for a particular applicant are conveniently summarized by the three-dimensional vector (27, 73, 168).

Example 10 For each player in a baseball game the scorer tabulates the number of times at bat, number of runs, number of hits, number of putouts, number of assists, number of errors. The six-dimensional vector (4, 1, 2, 3, 1, 0) gives the data for a certain second baseman in a game.

6 Operations involving vectors

Although the tabulation of data in a concise form is a worthwhile use of vectors, of much greater significance is the use of vectors in performing algebraic manipulations. The study of vectors, which now has applications to many varied situations, has been strongly influenced in its history by its importance in the physical sciences. Several important concepts in physics are vectorial in nature; two such examples are force and displacement. This latter application may serve to motivate our definitions for the algebraic operations involving vectors.

The first operation we wish to discuss is addition. If two vectors have the same number of components, then we may add the vectors.

Example 11 One application for a two-dimensional vector is to describe a displacement in a plane. Consider the vector (2, 1). We have customarily thought of the ordered pair (2, 1) as identifying a point in the plane (see Fig. 8). Another interpretation is to imagine that (2, 1) describes a movement, or displacement, of an object two

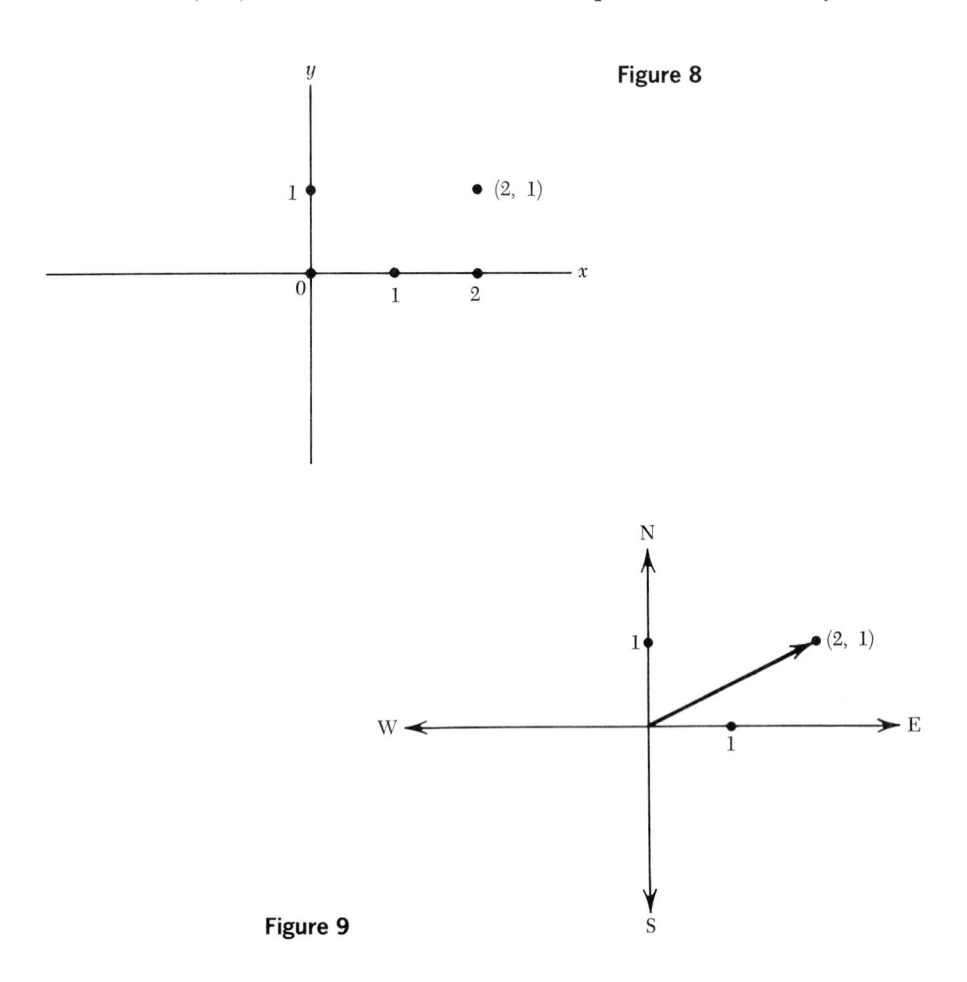

Figure 8

Figure 9

units to the right and one unit up. Geographically, we can think of a movement two units east and one unit north. Such a displacement is sometimes shown on a diagram by an arrow (see Fig. 9).

Now consider another displacement, say three units west and two units north. Since 3 units west may be rephrased as -3 units east, our displacement can be specified by the vector $(-3, 2)$. What is the net effect if we first move an object according to the vector

48   Systems of Linear Equations

Figure 10

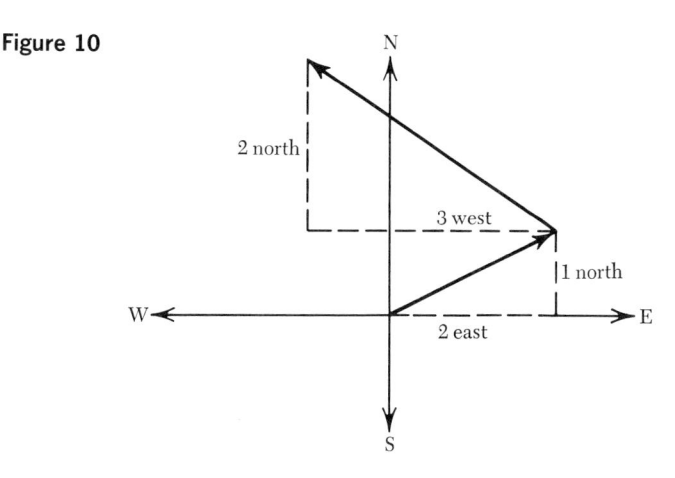

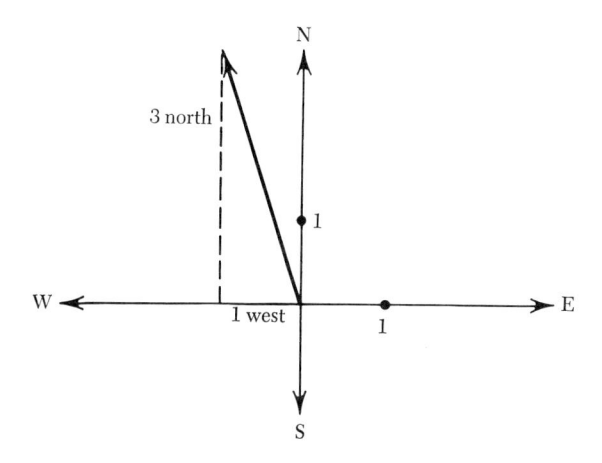

Figure 11

(2, 1) and afterward displace it according to $(-3, 2)$ (see Fig. 10)? The total effect is a movement of -1 unit east and 3 units north, a displacement given by the vector $(-1, 3)$ (see Fig. 11). We note that the first component -1 of this vector is the sum of the two easterly movements and that the second component 3 is the sum of the second components of the two given displacement vectors.

The preceding example suggests the operation of addition for vectors.

Definition Let U and V be two vectors with the same number n of components. The *sum* of U and V, denoted by $U + V$, is also a vector with n components. Each component of $U + V$ is the sum of the corresponding components from the vectors U and V.

Example 12 The sum of $(3, 5, 8, 14)$ and $(1, 12, 2, -4)$ is
$$(3 + 1, 5 + 12, 8 + 2, 14 - 4) = (4, 17, 10, 10).$$

6 / Operations involving vectors   **49**

Example 13 The vectors $(1, 0)$ and $(0, 1, 2)$ do not have a sum.

In symbols, if the n-dimensional vectors U and V are given by $U = (a_1, a_2, \ldots, a_n)$ and $V = (b_1, b_2, \ldots, b_n)$, then their sum $U + V$ is

$$(a_1, \ldots, a_n) + (b_1, \ldots, b_n) = (a_1 + b_1, \ldots, a_n + b_n).$$

We may shorten our description of the addition process by saying that we add "componentwise."

The zero vector with n components is the vector $(0, 0, \ldots, 0)$, each component of which is 0. It is straightforward computation to show that the sum of the zero vector and any vector V is V.

Now, we introduce another algebraic operation. In this case we do not, as in addition, combine two given vectors; instead, we combine a given number and a given vector, and we obtain as their "product" a vector.

Example 14 If the vector $(2, 1)$ is interpreted as a displacement, as in Example 11, then doubling this displacement (that is, applying the movement twice) yields the total displacement $(2, 1) + (2, 1) = (4, 2)$, which we may take as two times $(2, 1)$. In the same manner, tripling the displacement $(2, 1)$ gives $3 \cdot (2, 1) = (6, 3)$.

This notion of multiplication may be extended to the product of any number and any vector. In accordance with the following definition, we multiply the number by the vector "componentwise."

Definition The *product* of a number and a vector is a vector having the same number of components as the given vector. Each component in the product is obtained by multiplying the given number and the corresponding component in the given vector.

In symbols, the product cU of the number c and the vector

$$U = (a_1, a_2, \ldots, a_n)$$

is the vector

$$(ca_1, ca_2, \ldots, ca_n).$$

Example 15 The product of 3 and $(4, -2)$ is $(12, -6)$.

Example 16
$$-\tfrac{1}{2}(4, -1, 0, 7) = (-2, \tfrac{1}{2}, 0, -\tfrac{7}{2}).$$

Let V be a two-dimensional nonzero vector, say $V = (c, d)$. Let b be a number. The product bV is (bc, bd). On a coordinate system, the point (c, d) and the origin determine a line. (See Figs. 12a and 12b.) This line has equation

$$dx = cy.$$

50 Systems of Linear Equations

Figure 12

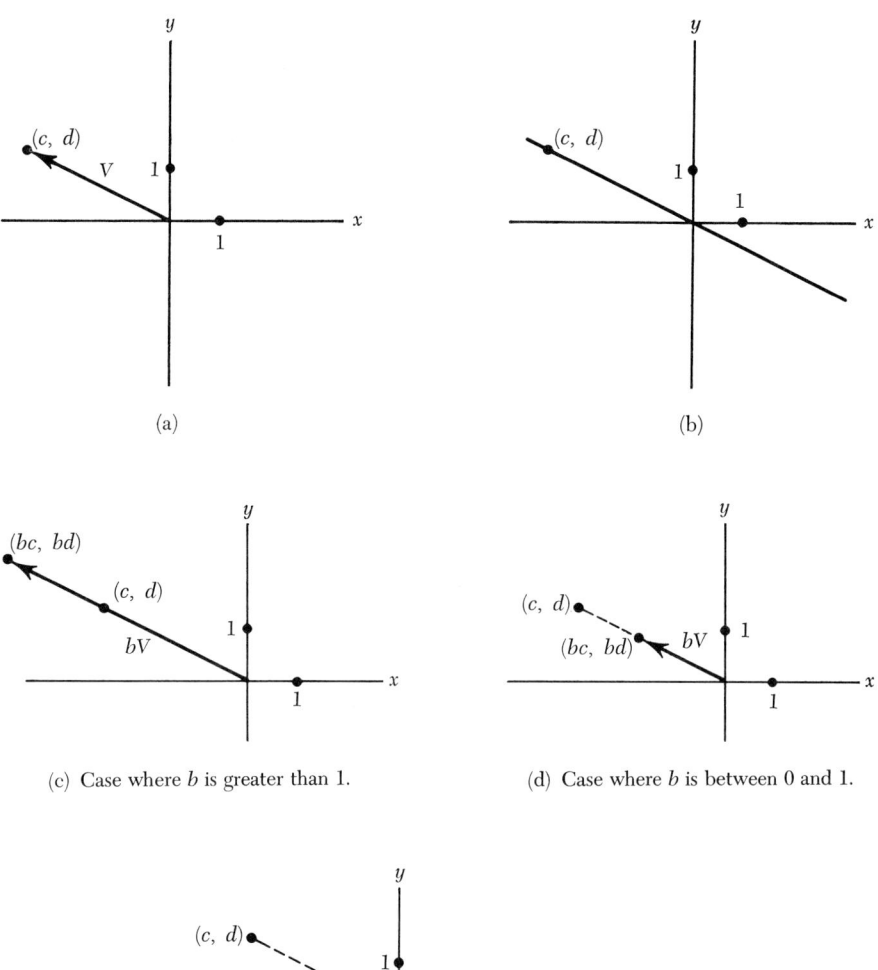

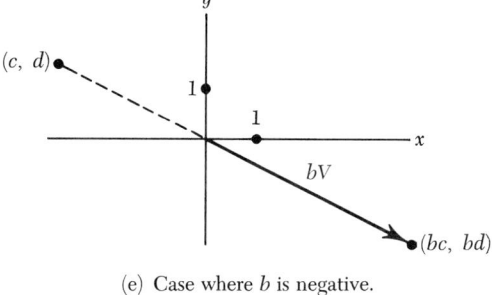

(a)

(b)

(c) Case where b is greater than 1.

(d) Case where b is between 0 and 1.

(e) Case where b is negative.

Consequently, the point (bc, bd) lies on the line, since its coordinates satisfy the equation of the line; indeed,

$$d \cdot bc = c \cdot bd.$$

In Figs. 12a, 12c, 12d, and 12e, the arrows that depict the vector V and the vector bV are collinear.

Example 17 Sealed bids by three competitors for an international project are quoted in London by the vector (30000, 28000, 31000), in which the amounts are in pounds sterling. If the exchange rate is $2.80 = £1$, express the bid vector in United States money. We obtain the desired vector by multiplying the number 2.8 and the given vector; the result is quoted in New York as (84000, 78400, 86800).

EXERCISES C

1. In each of the following, find the sum of the two given vectors.
   (a) $(5, 8), (-1, 4)$.　　(b) $(0.82, -1.23), (-0.16, 0.74)$.
   (c) $(1, 0, 9), (2, -1, 6)$.　(d) $(\frac{2}{3}, -\frac{4}{5}, \frac{3}{2}, 0), (-\frac{1}{6}, \frac{3}{10}, -\frac{1}{4}, 2)$.

2. Find each of the following products of a number and a vector.
   (a) $7(8, -2)$.　　(b) $-\frac{1}{2}(5, -3, 1)$.　　(c) $12(\frac{1}{3}, \frac{1}{4}, -\frac{1}{2}, \frac{5}{6})$.

3. Sketch each of the following vectors.
   (a) $(-3.7, 2.1)$.　　(b) $(1, -2, -2)$.　　(c) $(5, 4, -3)$.

4. The respective components of a certain five-dimensional vector V are the heights, in feet, of the players on a basketball team. Give an interpretation for the vector $12V$.

5. A businessman handles four kinds of merchandise. His current inventory is given by the vector (280, 30, 52, 164), where the ith component is the number of boxes of the ith commodity on hand. A truck delivers additional supplies given by the vector (20, 75, 15, 0). Find the inventory vector after the delivery.

6. During the following week, the businessman in Exercise 5 sold merchandise as given by the vector (120, 25, 35, 80). Find the inventory vector at the end of the week.

7. At the end of the week, the businessman in Exercise 6 received shipments that increased his inventory by 50% in each commodity. Find the inventory vector after the arrival of the shipments.

8. Let U, V, W be the vectors given by $U = (9, 7)$, $V = (-1, 4)$, $W = (0, 0, 1)$. For each of the following, decide whether the expression represents a vector and, if so, find the vector in simplified form.
   (a) $2U + (-5)V$.　(b) $6W$.　　　　(c) $5U + 7W$.
   (d) $U + (-1)U$.　(e) $\frac{1}{2}(U + V)$.　(f) UV.

9. Find a vector W such that $(11, 3) + W = (8, 7)$.

10. Find a vector V such that $(6, -2, 1) + 2V = (8, 1, -7)$.

11. Find the sum of the vector $\begin{bmatrix} 6 \\ -1 \end{bmatrix}$ and the vector $\begin{bmatrix} -3 \\ 2 \end{bmatrix}$.

12. Find a vector W such that $\begin{bmatrix} a \\ b \\ c \end{bmatrix} + W = \begin{bmatrix} d \\ e \\ f \end{bmatrix}$.

13. Let U and V be n-dimensional vectors, say $U = (a_1, \ldots, a_n)$ and $V = (b_1, \ldots, b_n)$.
    (a) Show that the sum $U + V$ is the same vector as the sum $V + U$.

(b) Show that $(0, 0, \ldots, 0) + V = V$.
(c) Find the vector $(-1)V$.
(d) Find the sum of the vectors V and $(-1)V$.

7 Matrices

Start

We next introduce the notion of a matrix. Of major importance in our work is the possibility of multiplying two matrices. Before discussing multiplication in the next section, we wish to observe certain resemblances between matrices and vectors. Like a vector, a matrix is an ordered array of numbers. However, the components of a matrix are always arranged in a rectangular pattern.

Example 18 An example of a matrix is

$$\begin{bmatrix} 2 & 0 & 4 & -1 \\ 1 & \frac{2}{3} & -5 & \sqrt{2} \\ -\frac{1}{4} & 60 & -4 & 0 \end{bmatrix}.$$

Each of the 12 numbers in this array is often called an *element* of the matrix (rather than a "component"). The 12 elements are arranged in a rectangular fashion with three *rows* (horizontally through the array) and four *columns* (vertically through the array). We call the matrix a three-by-four (written 3×4) matrix.

The matrix

$$\begin{bmatrix} a_1 & a_2 \\ b_1 & b_2 \\ c_1 & c_2 \end{bmatrix}$$

is a 3×2 matrix. Its three rows are numbered from the top of the array toward the bottom: first row is $[a_1 \ a_2]$, second row is $[b_1 \ b_2]$, etc. The columns are numbered from left to right: first column is

$$\begin{bmatrix} a_1 \\ b_1 \\ c_1 \end{bmatrix}.$$

In matrix notation it is customary to identify an element of a matrix A by locating its position in the array, that is, by naming the row and the column to which the element belongs. The element in the third row and second column of A may be denoted by using a double subscript on the symbol a; thus, a_{32} (which we may read as "a, sub three, two" or perhaps more simply as "a, three, two"). In this symbolism, a_{ij} is the element in the ith row and jth column of the matrix A.

Definition Let m and n be natural numbers, and let $a_{11}, \ldots, a_{1n}, a_{21}, \ldots, a_{mn}$ be numbers (not necessarily distinct). The m-by-n *matrix* A is the ordered array

$$A = \begin{bmatrix} a_{11} & a_{12} & \cdots & a_{1n} \\ a_{21} & a_{22} & \cdots & a_{2n} \\ \cdot & \cdot & \cdots & \cdot \\ \cdot & \cdot & a_{ij} & \cdot \\ \cdot & \cdot & \cdots & \cdot \\ a_{m1} & a_{m2} & \cdots & a_{mn} \end{bmatrix}.$$

The use of dots in the array indicates that there may be rows or columns not explicitly written; on the other hand, there is no inference that m, the number of rows, must be as large as 3. The notation is intended to suggest m rows and n columns, no matter what the natural numbers m and n may be.

A particularly important special case is the case $n = 1$. A matrix with only one column, such as

$$\begin{bmatrix} b_1 \\ b_2 \\ \vdots \\ b_m \end{bmatrix},$$

may be called a *column matrix*. Another useful special case occurs if $m = 1$. A matrix, such as $[c_1 \ c_2 \ \cdots \ c_n]$, that has a single row may be described as a *row matrix*.

Two matrices are considered different unless they are essentially the same rectangular array. More formally, to say that a pair of matrices are equal to one another means that, first, they have the same number of rows, second, they have the same number of columns, and third, in every position the element in one array is equal to the element in the corresponding position of the other array. In particular, a matrix each of whose elements is the number 0 may be called a *zero matrix*.

Just as for vectors, the significance of matrices in our work lies in the algebraic operations that we may perform. The first two operations involving matrices that we introduce resemble corresponding operations involving vectors because in each case the calculation is "componentwise."

Definition A pair of matrices that have the same number of rows and the same number of columns may be added. Their *sum* is the matrix each element of which is the sum of the elements in the respective positions of the given matrices.

Definition A number and a matrix may be multiplied. The *product* is the matrix each of whose elements is the product of the given number and the corresponding element in the given matrix.

Example 19 The sum of the matrices

$$\begin{bmatrix} 5 & 0 & 1 \\ -2 & 1 & 9 \\ 3 & 0 & 6 \end{bmatrix} \text{ and } \begin{bmatrix} -2 & 8 & 2 \\ 12 & 16 & 20 \\ -100 & 200 & 300 \end{bmatrix}$$

is

$$\begin{bmatrix} 3 & 8 & 3 \\ 10 & 17 & 29 \\ -97 & 200 & 306 \end{bmatrix}.$$

The matrices

$$\begin{bmatrix} 1 & 2 & 3 \end{bmatrix} \text{ and } \begin{bmatrix} 4 \\ 6 \\ 5 \end{bmatrix}$$

do not have a sum.

The product of the number 5 and the matrix $\begin{bmatrix} 3 & 4 \\ -1 & 2 \end{bmatrix}$ is

$$\begin{bmatrix} 15 & 20 \\ -5 & 10 \end{bmatrix}.$$

Since addition and multiplication by a number are componentwise operations both for vectors and for matrices, the familiar rules for computations with numbers are applicable to these processes. By contrast, the corresponding remark does not apply to the multiplicative operation that we shall discuss in the next section.

EXERCISES D

class
1. Let B be the matrix

$$\begin{bmatrix} -3 & -1 & 0 & 1 \\ 2 & 4 & -2 & 5 \\ 3 & -6 & 4 & 7 \end{bmatrix}.$$

(a) What is the element b_{31} in the third row and first column of B?
(b) Name the element b_{24}.
(c) What matrix is the first row of B?
(d) Find the sum of the elements in the third column of B.
(e) Find the sum of the elements b_{21} and b_{32}.

2. Explain the distinction between a 3×4 matrix and a 4×3 matrix.

3. Suppose that A, B, C are the matrices given by

$$A = \begin{bmatrix} -1 & 3 & 2 \\ 0 & 4 & -5 \end{bmatrix}, \quad B = \begin{bmatrix} 6 & -2 & 7 \\ 1 & 9 & 8 \end{bmatrix}, \quad C = \begin{bmatrix} 0 & 0 \\ 1 & 0 \\ 0 & 1 \end{bmatrix}.$$

For each of the following, decide whether the expression is a matrix, and if it is find the matrix in simplified form.
(a) $5A$.
(b) $(-3)C$.
(c) $A + C$.
(d) $B + A$.
(e) $3A + 2B$.
(f) $\frac{1}{3}A + (-\frac{1}{2})B$.
(g) $C + C$.

4. Find a 2×2 matrix M such that

$$\begin{bmatrix} 3 & -2 \\ 5 & 1 \end{bmatrix} + M = \begin{bmatrix} 4 & 0 \\ 3 & -2 \end{bmatrix}.$$

5. Find a 3×2 matrix N such that

$$\begin{bmatrix} 5 & 1 \\ 3 & -2 \\ -1 & 4 \end{bmatrix} + 3N = \begin{bmatrix} 2 & 7 \\ 0 & 1 \\ 2 & 5 \end{bmatrix}.$$

6. Suppose that A, B, C are $m \times n$ matrices and that a_{ij}, b_{ij}, c_{ij} are, respectively, the elements in row i and column j, for each i and each j.
(a) What is the element in the ith row and jth column of $A + B$?
(b) What is the element in the ith row and jth column of $B + A$?
(c) Explain why $A + B$ is the same matrix as $B + A$.
(d) What element is in the ith row and jth column of the sum $(A + B) + C$?
(e) Show that, in the sum $A + (B + C)$, the element in row i and column j is $a_{ij} + (b_{ij} + c_{ij})$.
(f) Explain why $(A + B) + C$ is the same as $A + (B + C)$.
(g) What is the element in row i and column j of the matrix M such that $A + M = B$?

7. According to Newtonian mechanics, a force vector W acting on a body with mass m causes the body to move with an acceleration given by the vector $(1/m)W$.
(a) Find the acceleration vector in case a body of mass 10 is subjected to a force $(28, 35, 7)$.
(b) Find the total acceleration in case the additional force $(7, 15, 18)$ is applied to the same body.

56 Systems of Linear Equations

8 Matrix multiplication

The basic feature in the operation of multiplying two matrices is the following. A sum of products of numbers may be expressed as the product of a row matrix and a column matrix. A sum of two products, like $3(-5) + (-2)4$, is the product of the row matrix $[3 \quad -2]$ and the column matrix $\begin{bmatrix} -5 \\ 4 \end{bmatrix}$. A sum of three products, such as $ax + by + cz$, is the product of $[a \quad b \quad c]$ by $\begin{bmatrix} x \\ y \\ z \end{bmatrix}$.

These examples illustrate the following definition.

Definition The *product* of the row matrix
$$A = [a_1 \quad a_2 \quad \ldots \quad a_n]$$
with n columns and the column matrix
$$B = \begin{bmatrix} b_1 \\ b_2 \\ \vdots \\ b_n \end{bmatrix}$$
with n rows is the number[1]
$$AB = a_1 b_1 + a_2 b_2 + \cdots + a_n b_n.$$

Note that the number is the sum of the products obtained by multiplying each element of the row matrix by the "corresponding" element of the column matrix; here the word "corresponding" means that we pair the top element of B with the left element of A, the second element of B with the second element of A, and so on. In order that the correspondence fit, it is of course mandatory that the number of elements in B be the same as the number of elements in A.

Example 20 The product of $[5 \quad 2 \quad -1]$ by $\begin{bmatrix} 4 \\ 3 \\ 8 \end{bmatrix}$ is 18, since
$$5 \cdot 4 + 2 \cdot 3 + (-1) \cdot 8 = 18.$$

---
[1] Strictly speaking, the product is the 1×1 matrix whose sole element is this number. But we often do not distinguish between a number f and the one-element matrix $[f]$.

We have not yet described what we mean by the product of

the column matrix $\begin{bmatrix} 4 \\ 3 \\ 8 \end{bmatrix}$ and the row matrix $[5 \quad 2 \quad -1]$.

This product (to be explained later) is not 18. This remark serves as a warning that, so far, we have discussed the product of a row matrix by a column matrix; the row matrix is the left-hand factor in the product.

Example 21

$$2x + 9y - 17 = [2 \quad 9 \quad -17] \begin{bmatrix} x \\ y \\ 1 \end{bmatrix}.$$

Example 22 The equation $3w - 2x + 4y + 11z = 6$ is equivalent to the equation

$$[3 \quad -2 \quad 4 \quad 11] \begin{bmatrix} w \\ x \\ y \\ z \end{bmatrix} = 6.$$

Example 23 Consider the system of equations

$$\begin{cases} 2x - 3y + 4z = 4, \\ x \phantom{{}-3y} - 8z = -5, \\ 5x + 9y - 7z = 26. \end{cases}$$

These three equations are equivalent, respectively, to the equations

$$[2 \quad -3 \quad 4] \begin{bmatrix} x \\ y \\ z \end{bmatrix} = 4,$$

$$[1 \quad 0 \quad -8] \begin{bmatrix} x \\ y \\ z \end{bmatrix} = -5,$$

$$[5 \quad 9 \quad -7] \begin{bmatrix} x \\ y \\ z \end{bmatrix} = 26.$$

58 Systems of Linear Equations

We may condense the preceding as follows: [3×3] × [3×1] = [3×1]

$$\begin{bmatrix} 2 & -3 & 4 \\ 1 & 0 & -8 \\ 5 & 9 & -7 \end{bmatrix} \begin{bmatrix} x \\ y \\ z \end{bmatrix} = \begin{bmatrix} 4 \\ -5 \\ 26 \end{bmatrix}.$$

In order to interpret this latter statement, we need to introduce more fully the idea of a product of two matrices.

Definition Let A and B be two matrices. Suppose that A is an $m \times n$ matrix and that B is an $n \times p$ matrix. (Note that the number of columns in A is the same as the number of rows in B.) The *product* of A and B, denoted by AB, is the $m \times p$ matrix whose elements are determined as follows: The element in the ith row and jth column of AB is the product of the ith row of A and the jth column of B.

Example 24 Let A and B be the matrices [m×n] × [n×p] = [m×p]

{2×3}
$$A = \begin{bmatrix} 1 & 4 & 5 \\ -2 & 3 & 1 \end{bmatrix}, \quad B = \begin{bmatrix} 8 & 5 \\ -7 & 1 \\ 6 & 9 \end{bmatrix}.$$
[3×2]

Since A is a 2×3 matrix and B is a 3×2 matrix, the product is a 2×2 matrix (in the notation of the above definition $m = 2$, $n = 3$, $p = 2$). The element in the first row and first column of AB is the product

$$[1 \quad 4 \quad 5]\begin{bmatrix} 8 \\ -7 \\ 6 \end{bmatrix},$$

namely, $8 - 28 + 30 = 10$. Thus,

$$AB = \begin{bmatrix} 10 & * \\ * & * \end{bmatrix},$$

where the three asterisks indicate entries that we have not yet calculated. The element in the first row and second column (the upper right corner in the array) is the product of the first row in A and the second column in B, namely,

$$[1 \quad 4 \quad 5]\begin{bmatrix} 5 \\ 1 \\ 9 \end{bmatrix} = 5 + 4 + 45 = 54.$$

8 / Matrix multiplication

The element in the lower left corner (in second row and first column) is the product

$$[-2 \quad 3 \quad 1]\begin{bmatrix} 8 \\ -7 \\ 6 \end{bmatrix} = -16 - 21 + 6 = -31.$$

Finally,

$$[-2 \quad 3 \quad 1]\begin{bmatrix} 5 \\ 1 \\ 9 \end{bmatrix} = -10 + 3 + 9 = 2$$

is the element in the second row and second column of AB. These calculations show that

$$AB = \begin{bmatrix} 10 & 54 \\ -31 & 2 \end{bmatrix}.$$

Example 25 If

$$C = \begin{bmatrix} -2 & 9 \\ 6 & 5 \end{bmatrix} \text{ and } D = \begin{bmatrix} 4 & 3 \\ 0 & 1 \end{bmatrix},$$

verify that

$$CD = \begin{bmatrix} -8 & 3 \\ 24 & 23 \end{bmatrix}.$$

The element 3 in the first row and second column of CD is the product of the first row $[-2 \quad 9]$ in C and the second column $\begin{bmatrix} 3 \\ 1 \end{bmatrix}$ in D; indeed,

$$[-2 \quad 9]\begin{bmatrix} 3 \\ 1 \end{bmatrix} = (-2) \cdot 3 + 9 \cdot 1 = 3.$$

The other three entries in the product CD may be checked similarly.

For the same two matrices C and D, we may multiply D and C. The product DC is

$$\begin{bmatrix} 4 & 3 \\ 0 & 1 \end{bmatrix}\begin{bmatrix} -2 & 9 \\ 6 & 5 \end{bmatrix} = \begin{bmatrix} 10 & 51 \\ 6 & 5 \end{bmatrix}.$$

Note that the product DC and the product CD are different!

In general, a product of two matrices depends upon which of the matrices is the left factor and which is the right factor in the product. In formal language, matrix multiplication is not a commutative operation. A little reflection about the method of multiplying two matrices makes our observation highly plausible.

Consider the element in the first row and first column of a product $\begin{bmatrix} \square \\ \end{bmatrix}$. In the product AB, this element comes from the top row of A and the left column of B; while in the product BA, the element in this upper left corner position comes from the top row of B and the left column of A; there is little reason to expect that the element in AB should agree with the corresponding element in BA.

Furthermore, it is possible to have two matrices F and G such that the product FG is a matrix but such that the product of G and F is not defined. For example, if F is a 1×3 matrix and G is a 3×3 matrix, then FG is a 1×3 matrix but GF has no meaning.

EXERCISES E

1. Calculate each of the following products of matrices.

   (a) $[3 \quad 5]\begin{bmatrix} 2 \\ 4 \end{bmatrix}$.

   (b) $[1 \quad -3 \quad 2]\begin{bmatrix} 5 \\ 7 \\ 6 \end{bmatrix}$.

   (c) $[a \quad b \quad c]\begin{bmatrix} 2 \\ -1 \\ 0 \end{bmatrix}$.

   (d) $[\tfrac{1}{2} \quad \tfrac{2}{3} \quad -\tfrac{1}{4} \quad \tfrac{7}{6}]\begin{bmatrix} x \\ w \\ v \\ u \end{bmatrix}$.

2. Write each of the following equations in an equivalent form involving a product of matrices.
   (a) $6a - 5b = 11$.   (b) $3x - 2y + 4z = 12$.

3. On a two-dimensional coordinate system a line contains the points $(1, 5)$ and $(-2, -7)$. Write an equation for the line in a form that exhibits a product of matrices.

4. Let (a, b) be a vector with nonzero components. What is the slope of the line whose equation is
$$[a \quad b]\begin{bmatrix} x \\ y \end{bmatrix} = c?$$

5. Each of the following is an equation of a line. Which pair, if any, of the lines are perpendicular to one another?

   $[4 \quad -2]\begin{bmatrix} x \\ y \end{bmatrix} = 7.$ $[5 \quad 10]\begin{bmatrix} x \\ y \end{bmatrix} = 9.$

   $[-6 \quad 3]\begin{bmatrix} x \\ y \end{bmatrix} = 8.$ $[7 \quad 14]\begin{bmatrix} x \\ y \end{bmatrix} = 6.$

6. Let a, b, c, d, e, f be nonzero numbers. Consider the lines whose respective equations are

$$[a \ b]\begin{bmatrix}x\\y\end{bmatrix} = e \quad \text{and} \quad [c \ d]\begin{bmatrix}x\\y\end{bmatrix} = f.$$

   (a) Formulate, in terms of the given numbers, a criterion for deciding whether the given lines are perpendicular to one another.
   (b) Formulate, in terms of the given numbers, a criterion for determining whether the given lines are parallel to each other.
   (c) Formulate, in terms of the vectors (a, b, e) and (c, d, f), a criterion for deciding whether the given lines are the same.

7. Calculate each of the following products of matrices.

   (a) $\begin{bmatrix}5 & 3\\2 & 6\end{bmatrix}\begin{bmatrix}4 & 8\\7 & 1\end{bmatrix}.$

   (b) $\begin{bmatrix}-2 & 0\\3 & 1\end{bmatrix}\begin{bmatrix}3 & 2 & 1\\4 & -5 & -3\end{bmatrix}.$

   (c) $\begin{bmatrix}1 & -2 & 3\\-4 & 5 & -6\\7 & -8 & 9\end{bmatrix}\begin{bmatrix}3 & 2 & 1\\2 & -1 & 3\\1 & -3 & -2\end{bmatrix}.$

   (d) $\begin{bmatrix}2 & 0 & 0 & -3\\1 & 4 & -5 & 2\\0 & -6 & 1 & 7\\3 & -1 & 0 & -4\end{bmatrix}\begin{bmatrix}2\\1\\6\\-1\end{bmatrix}.$

   (e) $[1 \ 6]\begin{bmatrix}4\\-1\end{bmatrix}.$

   (f) $\begin{bmatrix}4\\-1\end{bmatrix}[1 \ 6].$

8. Let A and B be the matrices given by

   $$A = \begin{bmatrix}1 & 1\\1 & 1\end{bmatrix} \quad \text{and} \quad B = \begin{bmatrix}1 & 1\\-1 & -1\end{bmatrix}.$$

   (a) Show that $BA = 2B$.
   (b) Compute the product AB.

9. Calculate the product

$$\begin{bmatrix}3 & -1 & 2\\2 & 4 & -3\\5 & 0 & 2\end{bmatrix}\begin{bmatrix}x\\y\\z\end{bmatrix}.$$

10. Express each of the following as a product of a matrix whose elements are numbers and the matrix $\begin{bmatrix}x\\y\end{bmatrix}$.

   (a) $\begin{bmatrix}x - 2y\\3x + y\\-4x - 3y\\-x + 5y\end{bmatrix}.$

   (b) $\begin{bmatrix}2x + y\\-x + y\\3x\\4y\end{bmatrix}.$

62 Systems of Linear Equations

9 Gauss elimination method in matrix terminology

Example 26 Let

$$X = \begin{bmatrix} x \\ y \\ z \end{bmatrix}.$$

Give a detailed interpretation of the equation

$$\begin{bmatrix} 2 & 1 & 1 \\ -2 & 3 & 0 \\ 3 & 5 & 2 \end{bmatrix} X = \begin{bmatrix} 0 \\ 1 \\ -1 \end{bmatrix}.$$

The product in the left member of the equation may be found; it is

$$\begin{bmatrix} 2x + y + z \\ -2x + 3y \\ 3x + 5y + 2z \end{bmatrix}.$$

Since an equality of matrices means that their respective elements are equal, this single equation involving matrices is equivalent to the following system of three equations:

$$\begin{cases} 2x + y + z = 0, \\ -2x + 3y = 1, \\ 3x + 5y + 2z = -1. \end{cases}$$

Example 26 (continued) Let

$$A = \begin{bmatrix} 2 & 1 & 1 \\ -2 & 3 & 0 \\ 3 & 5 & 2 \end{bmatrix}, \quad B = \begin{bmatrix} 0 \\ 1 \\ -1 \end{bmatrix}, \quad \text{and} \quad X = \begin{bmatrix} x \\ y \\ z \end{bmatrix}.$$

Solve the equation $AX = B$.

We have just noted that the equation $AX = B$ is equivalent to a system of simultaneous equations in x, y, z. Consequently, we may apply the Gauss elimination method to solve the given equation. A review of the method reveals that the successive decisions about what to do at each step are based on the coefficients of x, y, z. This observation suggests that we can adapt the method to the terminology of matrices.

First, let us write the solution of the system of equations, using our method.

$$\begin{cases} 2x + y + z = 0, \\ -2x + 3y = 1, \\ 3x + 5y + 2z = -1. \end{cases}$$

$$\begin{cases} x + \tfrac{1}{2}y + \tfrac{1}{2}z = 0, \\ 4y + z = 1, \\ \tfrac{7}{2}y + \tfrac{1}{2}z = -1. \end{cases}$$

$$\begin{cases} x + \tfrac{3}{8}z = -\tfrac{1}{8}, \\ y + \tfrac{1}{4}z = \tfrac{1}{4}, \\ -\tfrac{3}{8}z = -\tfrac{15}{8}. \end{cases}$$

$$\begin{cases} x = -2, \\ y = -1, \\ z = 5. \end{cases}$$

The solution is $(x, y, z) = (-2, -1, 5)$.

Next, let us rewrite the above solution, using matrices.

$$\begin{bmatrix} 2 & 1 & 1 \\ -2 & 3 & 0 \\ 3 & 5 & 2 \end{bmatrix} X = \begin{bmatrix} 0 \\ 1 \\ -1 \end{bmatrix}.$$

$$\begin{bmatrix} 1 & \tfrac{1}{2} & \tfrac{1}{2} \\ 0 & 4 & 1 \\ 0 & \tfrac{7}{2} & \tfrac{1}{2} \end{bmatrix} X = \begin{bmatrix} 0 \\ 1 \\ -1 \end{bmatrix}.$$

$$\begin{bmatrix} 1 & 0 & \tfrac{3}{8} \\ 0 & 1 & \tfrac{1}{4} \\ 0 & 0 & -\tfrac{3}{8} \end{bmatrix} X = \begin{bmatrix} -\tfrac{1}{8} \\ \tfrac{1}{4} \\ -\tfrac{15}{8} \end{bmatrix}.$$

$$\begin{bmatrix} 1 & 0 & 0 \\ 0 & 1 & 0 \\ 0 & 0 & 1 \end{bmatrix} X = \begin{bmatrix} -2 \\ -1 \\ 5 \end{bmatrix}.$$

The solution is

$$X = \begin{bmatrix} -2 \\ -1 \\ 5 \end{bmatrix}.$$

Refer again to the plan of attack in the Gauss elimination method, as outlined in Examples 1, 3, and 4 (in Section 2). Expressed in matrix notation, we manipulate the various rows of the two known matrices, the 3 × 3 matrix, which is the coefficient of X, and the 3 × 1 matrix in the right-hand member of the equation. The rows of the two matrices are handled together. It seems reasonable to shorten the writing further by using the schematic device of a single matrix, sometimes referred to as the *augmented matrix*. To keep in mind the distinctive character of one of the columns in the augmented matrix, we may indicate a separation. In this form, the solution of the above equation $AX = B$ becomes

$$\begin{bmatrix} 2 & 1 & 1 & | & 0 \\ -2 & 3 & 0 & | & 1 \\ 3 & 5 & 2 & | & -1 \end{bmatrix},$$

$$\begin{bmatrix} 1 & \frac{1}{2} & \frac{1}{2} & | & 0 \\ 0 & 4 & 1 & | & 1 \\ 0 & \frac{7}{2} & \frac{1}{2} & | & -1 \end{bmatrix},$$

$$\begin{bmatrix} 1 & 0 & \frac{3}{8} & | & -\frac{1}{8} \\ 0 & 1 & \frac{1}{4} & | & \frac{1}{4} \\ 0 & 0 & -\frac{3}{8} & | & -\frac{15}{8} \end{bmatrix},$$

$$\begin{bmatrix} 1 & 0 & 0 & | & -2 \\ 0 & 1 & 0 & | & -1 \\ 0 & 0 & 1 & | & 5 \end{bmatrix}.$$

From the final augmented matrix, we read the root,

$$X = \begin{bmatrix} -2 \\ -1 \\ 5 \end{bmatrix}.$$

Formulated in matrix language, the stepwise procedure in the Gauss elimination method is as follows:

In the augmented matrix, we attempt to obtain in the first column an element 1 and all the remaining elements 0. Then, in the next column we try to obtain an element 1 and all other elements 0; furthermore, we do this without changing the first column. Next we do likewise in the third column, seeking to obtain an entry 1 and all other entries 0, but without affecting the previous columns. We continue this process for as many columns as appear to the left of the separation mark. The maneuvers which we may use to reach our goal are the following manipulations on the rows in the augmented matrix.

(a) We may add to any row of a matrix the product of any number and any other row.

(b) We may replace any row of a matrix by the product of any nonzero number and that same row.

Each of these steps applied to a matrix usually changes it into a different matrix. Our aim is to change the given augmented matrix into an augmented matrix that will permit us to recognize the root (or roots) of the given equation.

Example 27 Let A be a 4×4 matrix whose elements are numbers, let B be a 4×1 matrix whose elements are numbers, and let

$$X = \begin{bmatrix} w \\ x \\ y \\ z \end{bmatrix}.$$

Suppose that the Gauss elimination method is applied to the equation $AX = B$, and suppose that the initial augmented matrix $[A \mid B]$ is gradually changed over to the augmented matrix

$$\left[\begin{array}{cccc|c} 1 & 0 & 0 & 0 & 5 \\ 0 & 1 & 0 & 0 & 7 \\ 0 & 0 & 1 & 0 & -2 \\ 0 & 0 & 0 & 1 & 1 \end{array}\right].$$

At this stage, we may read the root of the equation $AX = B$; it is

$$X = \begin{bmatrix} 5 \\ 7 \\ -2 \\ 1 \end{bmatrix}.$$

Indeed, the final augmented matrix that is displayed above represents the system of equations

$$\begin{cases} 1w + 0x + 0y + 0z = 5, \\ 0w + 1x + 0y + 0z = 7, \\ 0w + 0x + 1y + 0z = -2, \\ 0w + 0x + 0y + 1z = 1, \end{cases}$$

and the root of this system is $(w, x, y, z) = (5, 7, -2, 1)$.

The preceding example illustrates how the solution of a matrix equation of the type $AX = B$ can be easily read in case the augmented matrix can be transformed into a special form. In not every case is it possible, however, to achieve the special form that we have been so fortunate to obtain in our examples. We will discuss other possibilities later.

EXERCISES F

1. Express the matrix equation

$$\begin{bmatrix} 1 & 2 & -1 \\ 4 & -1 & 2 \end{bmatrix} \begin{bmatrix} x \\ y \\ z \end{bmatrix} = \begin{bmatrix} 9 \\ 21 \end{bmatrix}$$

in the form of a system of linear equations.

2. Express, in the form of a single equation involving matrices, the following system of equations.

$$3w + 2x - y + 4z = 8,$$
$$2w + 3y - z = 6,$$
$$5x + z - 6y = 1.$$

3. Use the Gauss elimination method to find the root X of the equation

$$\begin{bmatrix} 3 & 2 \\ -5 & 4 \end{bmatrix} X = \begin{bmatrix} 4 \\ -58 \end{bmatrix}.$$

4. Use the Gauss elimination method and the idea of augmented matrices in order to find the root X of each of the following equations.

(a) $\begin{bmatrix} 7 & 4 \\ 8 & 5 \end{bmatrix} X = \begin{bmatrix} 5 \\ 2 \end{bmatrix}.$

(b) $\begin{bmatrix} 1 & 3 & 2 \\ 2 & 1 & 6 \\ 1 & 2 & 1 \end{bmatrix} X = \begin{bmatrix} -2 \\ 4 \\ 1 \end{bmatrix}.$

(c) $\begin{bmatrix} 1 & 3 & 2 \\ 1 & 3 & 4 \\ 2 & 1 & 1 \end{bmatrix} X = \begin{bmatrix} 5 \\ 3 \\ -7 \end{bmatrix}.$

(d) $\begin{bmatrix} 2 & 1 & 0 & 0 \\ 1 & 2 & 1 & 0 \\ 0 & 1 & 2 & 1 \\ 0 & 0 & 1 & 2 \end{bmatrix} X = \begin{bmatrix} 4 \\ 2 \\ -5 \\ -5 \end{bmatrix}.$

5. Let A be the matrix $\begin{bmatrix} 5 & 4 \\ 13 & 11 \end{bmatrix}$. Find the root of each of the following equations, using the Gauss elimination method and writing the successive steps in the notation of augmented matrices.

(a) $AX = \begin{bmatrix} 1 \\ 0 \end{bmatrix}.$ (b) $AW = \begin{bmatrix} 0 \\ 1 \end{bmatrix}.$ (c) $AV = \begin{bmatrix} 1 \\ 1 \end{bmatrix}.$

9 / Gauss elimination method   67

(d) $AU = \begin{bmatrix} a \\ b \end{bmatrix}$, where a and b are given numbers.

6. In the notation of Exercise 5, verify each of the following results.
   (a) $V = X + W$.
   (b) $U = aX + bW$.

7. Let A be an $m \times n$ matrix whose elements are numbers, let B be an $m \times 1$ matrix whose elements are numbers, and let k be a number which is not zero. Suppose that M is an $n \times 1$ matrix such that $AM = B$. Find a root of the equation $A \cdot X = kB$.

8. Suppose that A, D, F are the matrices given by

$$A = \begin{bmatrix} a & b \\ c & d \end{bmatrix}, \quad D = \begin{bmatrix} h \\ k \end{bmatrix}, \quad F = \begin{bmatrix} p \\ q \end{bmatrix}.$$

   (a) Find the product AD.
   (b) Find the product AF.
   (c) Find the sum $AD + AF$.
   (d) Find the sum $D + F$.
   (e) Find the product of A and the matrix $D + F$.
   (f) Compare the sum $AD + AF$ with the product $A(D + F)$.

9. Let k be a number, and let R, U, G be the matrices given by

$$R = [r_1 \ r_2 \ r_3 \ r_4], \quad U = [u_1 \ u_2 \ u_3 \ u_4], \quad G = \begin{bmatrix} g_1 \\ g_2 \\ g_3 \\ g_4 \end{bmatrix}.$$

   (a) Find the product RG.
   (b) Find the product UG.
   (c) Find $k(RG)$.
   (d) Find the sum $UG + kRG$.
   (e) Find the sum $U + kR$.
   (f) Find the product of $U + kR$ and the matrix G.
   (g) Compare the sum $UG + kRG$ with the product $(U + kR) \cdot G$.

10 Graph of a linear equation in three dimensions

We have previously observed the advantage in visualizing geometrically an algebraic equation. We now pursue this tie-in with geometry by considering graphs in three-dimensional space.

On a two-dimensional coordinate system, the graph of a linear equation is a line. As we recall from Section 12 of Chapter 1, a linear equation in x and y is an equation having the form

$$ax + by = d,$$

where a, b, d are numbers such that one or both of the numbers a, b are different from zero. We may exploit the language of matrices and rephrase this description. A linear equation in x and y is an equation of the type

$$[a \quad b]\begin{bmatrix} x \\ y \end{bmatrix} = d,$$

where a, b, d are numbers such that $[a \quad b]$ is not the zero matrix. A linear equation imposes one restriction on the numbers x and y. A point on the graph of the equation corresponds to a vector that is restricted by (that is, must satisfy) the equation. We sometimes say that such a vector has one fewer *degrees of freedom* than an unrestricted vector. Note that the line that is the graph of the equation has dimension one, which is one less than the dimension of the plane itself.

We now investigate the analogous situation in three dimensions. A linear equation in x, y, z is an equation having the form

$$ax + by + cz = d,$$

where a, b, c, d are numbers such that $[a \quad b \quad c]$ is not the zero matrix. Although it is permissible for one or two of the three numbers a, b, c to be zero, we exclude the possibility that all of them are zero. Now an equation such as

$$[a \quad b \quad c]\begin{bmatrix} x \\ y \\ z \end{bmatrix} = d$$

imposes one restriction on the vector $\begin{bmatrix} x \\ y \\ z \end{bmatrix}$. The vectors that satisfy the equation have one fewer degrees of freedom than do unrestricted vectors. It seems reasonable that the graph of the equation should have dimension one less than the dimension of the entire space. Since we are now considering three-dimensional space, the graph of a linear equation should be a two-dimensional set. Furthermore, the graph of a linear equation on a three-dimensional coordinate system should play the role a line plays on a two-dimensional coordinate system. This role is played by a *plane*.

On a three-dimensional coordinate system the graph of a linear equation is a plane.

Example 28 The graph of the equation

$$4x + 2y + 3z = 12$$

is a plane. If a point belonging to the graph has its second coordinate 0 and its third coordinate 0, then its first coordinate x satisfies the statement $4x + 2 \cdot 0 + 3 \cdot 0 = 12$; hence, the point $(3, 0, 0)$ belongs to the graph. We also find that the points $(0, 6, 0)$ and $(0, 0, 4)$ belong

Figure 13

Figure 14

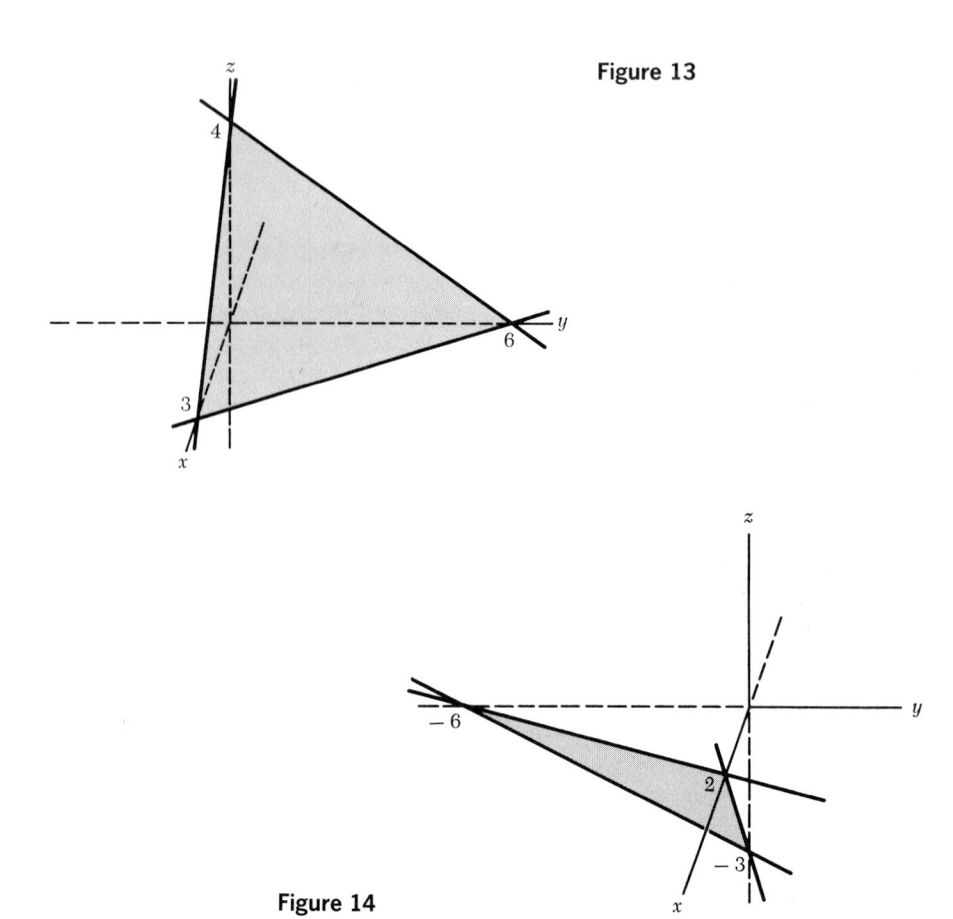

to the graph. Since these three points are noncollinear, they determine a plane. The plane that contains them is the graph we seek. (See Fig. 13, which suggests this plane.)

The plane portrayed by Fig. 13 is the plane containing the triangle whose vertices are the points identified on the respective coordinate axes.

Example 29 The graph of the equation $3x - y - 2z = 6$ is a plane. Among the points belonging to the plane are $(2, 0, 0)$, $(0, -6, 0)$, $(0, 0, -3)$; each of these three points is identified by asking what point on the plane has two coordinates equal to zero. The graph is shown in Fig. 14.

Example 30 A plane contains the z-coordinate axis and the point $(1, 4, 0)$. What is an equation for the plane?

For any point in the plane (see Fig. 15), the line that contains the point and is parallel to the z-coordinate axis is also in the plane. Points

70   Systems of Linear Equations

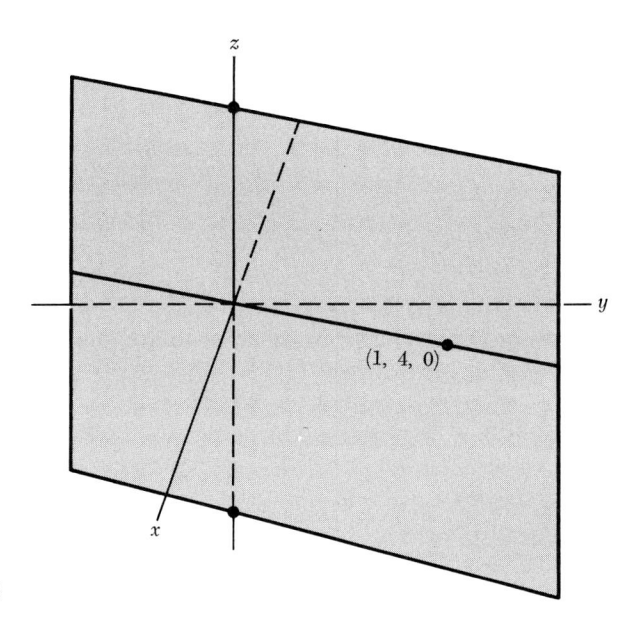

Figure 15

on the line are distinguished from the given point only by a different third coordinate. Thus, the third coordinate of a point is immaterial in determining whether the point belongs to the graph. In other words, the equation we want does not involve the coordinate z. Thus, the desired equation has the form

$$ax + by = d.$$

Since the point $(0, 0, 0)$ belongs to the plane,

$$a \cdot 0 + b \cdot 0 = d,$$

and $d = 0$. Since the point $(1, 4, 0)$ belongs to the plane,

$$a \cdot 1 + b \cdot 4 = 0,$$

or $a = -4b$. Thus the equation has the form $-4bx + by = 0$. Any choice of a nonzero number for b is acceptable; for simplicity, we choose $b = 1$. An equation for the plane is

$$y - 4x = 0.$$

Notice that this equation is a linear equation in x, y, z (even though z does not explicitly appear). The equation can be expressed as

$$\begin{bmatrix} -4 & 1 & 0 \end{bmatrix} \begin{bmatrix} x \\ y \\ z \end{bmatrix} = 0.$$

10 / Graph of a linear equation

The preceding example emphasizes the importance of context. Consider an equation such as $y - 4x = 0$. On a three-dimensional coordinate system, this equation is a simplified version of

$$[-4 \quad 1 \quad 0] \begin{bmatrix} x \\ y \\ z \end{bmatrix} = 0,$$

and the graph of the equation is a plane, as pictured in Fig. 15. By contrast, on a two-dimensional coordinate system, the same equation has a graph which is a line.

EXERCISES G

1. On a single three-dimensional coordinate system, do all of the following.
   (a) Plot the point (5, 2, 1).
   (b) Plot the point (1, 6, 4).
   (c) Draw the segment whose endpoints are (5, 2, 1) and (1, 6, 4).

2. On a single three-dimensional coordinate system, do all of the following.
   (a) Plot the point $(2, -4, 5)$.
   (b) Plot the point (3, 4, 1).
   (c) Draw the line containing each of the plotted points.

3. On a two-dimensional coordinate system, show all the points (x, y) such that y is positive.

4. On a three-dimensional coordinate system, show all the points (x, y, z) such that y is positive.

5. For each of the following, sketch the graph of the equation on a three-dimensional coordinate system.
   (a) $2x + 3y + 4z = 12$.   (b) $2x + 3y - 4z = 12$.   (c) $2x - 3y - 4z = 12$.

6. For each of the following, sketch the graph of the equation on a three-dimensional coordinate system.
   (a) $2x + y = 6$.   (b) $2y + z = 6$.   (c) $x + z = 4$.
   (d) $y = -3$.   (e) $z = 4$.   (f) $x = -1$.

7. On a single three-dimensional coordinate system, do all three parts of this exercise.
   (a) Sketch the graph of the equation $2x + y + 5z = 10$.
   (b) Sketch the graph of the equation $2x - y + 5z = 10$.
   (c) Draw the line of intersection of the two planes.

8. Sketch the graph of each of the following. (Use a different coordinate system in each part.)

   (a) $[4 \quad -2 \quad 3] \begin{bmatrix} x \\ y \\ z \end{bmatrix} = 10.$ (b) $[-3 \quad 2 \quad 4 \quad 9] \begin{bmatrix} x \\ y \\ z \\ 1 \end{bmatrix} = 0.$

72   Systems of Linear Equations

9. Sketch the graph of each of the following.

(a) $[2\ 0]\begin{bmatrix} x \\ y \end{bmatrix} = 5.$ (b) $[2\ 0\ 0]\begin{bmatrix} x \\ y \\ z \end{bmatrix} = 5.$ (c) $[2][x] = 5.$

11 Plane determined by three points

Example 31 Find an equation for the plane that contains the three points $(3, 1, 7)$, $(1, 7, 5)$, and $(-6, -2, 4)$.

The problem, in three dimensions, of finding a plane containing three given points is analogous to the problem, in two dimensions, of finding a line containing two given points. Consequently, we use the same method of attack.

One method that is often (but not always[2]) successful is the two-dimensional case is the following. Suppose that the line is represented by an equation of the form $y = b + mx$; using the given information, determine b and m. We adopt an analogous plan in our three-dimensional situation.

Suppose the plane has an equation of the form

$$z = d + ex + fy,$$

where d, e, f are numbers to be determined. The coordinates of each of the three given points satisfy the equation. Hence,

$$\begin{cases} 7 = d + e \cdot 3 + f \cdot 1, \\ 5 = d + e \cdot 1 + f \cdot 7, \\ 4 = d + e(-6) + f(-2). \end{cases}$$

We proceed to solve this system of equations for (d, e, f), using schematic matrix notation.

$$\begin{bmatrix} 1 & 3 & 1 & | & 7 \\ 1 & 1 & 7 & | & 5 \\ 1 & -6 & -2 & | & 4 \end{bmatrix},$$

$$\begin{bmatrix} 1 & 3 & 1 & | & 7 \\ 0 & -2 & 6 & | & -2 \\ 0 & -9 & -3 & | & -3 \end{bmatrix},$$

---

[2] Compare Example 22 in Section 12 of Chapter 1.

$$\begin{bmatrix} 1 & 0 & 10 & | & 4 \\ 0 & 1 & -3 & | & 1 \\ 0 & 0 & -30 & | & 6 \end{bmatrix},$$

$$\begin{bmatrix} 1 & 0 & 0 & | & 6 \\ 0 & 1 & 0 & | & \frac{2}{5} \\ 0 & 0 & 1 & | & -\frac{1}{5} \end{bmatrix}.$$

Since $(d, e, f) = (6, \frac{2}{5}, -\frac{1}{5})$, an equation for the plane is

$$z = 6 + \tfrac{2}{5}x - \tfrac{1}{5}y.$$

If we wish, we may prefer an equation that does not involve fractional coefficients, such as

$$-2x + y + 5z = 30.$$

It is important to check that each of the three given ordered triples actually does satisfy the equation we have found. For example,

$$-2 \cdot 1 + 7 + 5 \cdot 5 = -2 + 7 + 25 = 30;$$

the details for the other two triples are left to the reader to verify. The three given points and the plane are shown in Fig. 16.

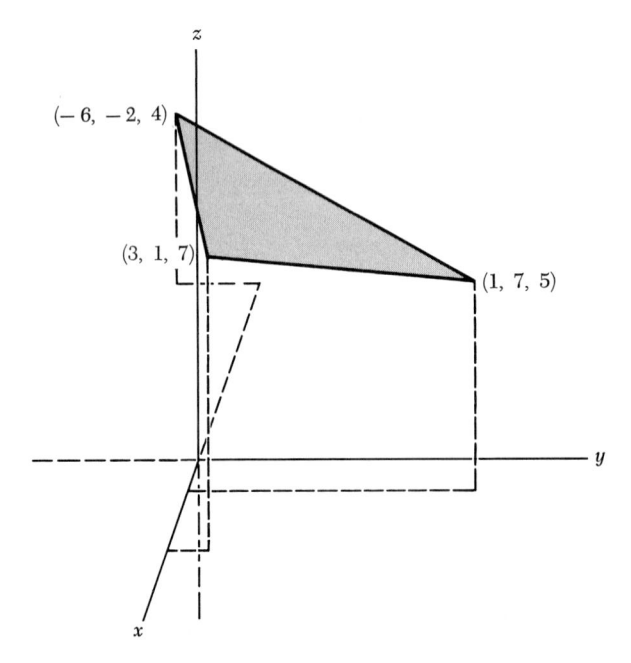

Figure 16

74   Systems of Linear Equations

EXERCISES H

1. In each of the following, find an equation for the plane that contains the three specified points.
   (a) (2, 0, 1), (1, 5, 2), (0, −1, 3).
   (b) (0, 0, 0), (5, 3, 9), (2, 7, 1).
   (c) (−1, 3, 1), (2, −1, 1), (3, 4, −2).
   (d) (0, 0, 2), (0, 4, 1), (3, 0, −2).

2. Let a, b, c be nonzero numbers. Find an equation for the plane that contains each of the points $(a, 0, 0)$, $(0, b, 0)$, $(0, 0, c)$.

3. Let a, b, c be nonzero numbers. Find an equation for the plane that contains each of the points $(a, b, 0)$, $(a, 0, c)$, $(0, b, c)$.

4. Suppose that we wish to find an equation for the plane that contains each of the points (8, 12, 5), (4, 18, 7), (18, −3, 1).
   (a) Our usual attack on this type of problem has been to assume that the plane has an equation of the form $z = d + ex + fy$ and to proceed to find the numbers d, e, f. Show that this attack is not a successful method in the present problem.
   (b) Modify the approach suitably, and find the equation requested.

5. Repeat Exercise 4, modified to apply to the plane containing the three points (2, 3, 1), (10, 13, 7), (−2, −2, 2).

12 Geometry of a system of linear equations in x, y, z

On a three-dimensional coordinate system, the graph of a linear equation is a plane. Suppose that

$$ax + by + cz = d,$$

in which the vector (a, b, c) is not the zero vector, is a linear equation under consideration. Suppose that (p, q, r) is an ordered triple of numbers. The criterion for deciding whether or not the point (p, q, r) belongs to the plane $ax + by + cz = d$ is whether or not the triple (p, q, r) is a root of the equation $ax + by + cz = d$, that is, whether or not the following statement is true:

$$ap + bq + cr = d.$$

A point belongs to each of two planes if its coordinates satisfy equations for the respective planes. The same remark applies to any number of planes. The converse is also true: If a point belongs to several planes, then the ordered triple of its coordinates is a root of the equation for each of the individual planes.

We are now able to formulate the geometric interpretation for a familiar algebraic problem. Suppose we are given a system of simultaneous linear equations in x, y, z. What does it mean, geometrically, to find a root of the system? Since the graph of each equation is a plane, the graph of a root of the system is a point belonging to all the planes. If there is one root, the planes have one point of intersection. If there is no root of the system, the planes have no point in common. If there are many roots for the system, the planes have many points of intersection.

Example 32 Find each point of intersection of the three planes whose respective equations are
$$2x - 3y + 5z = 7,$$
$$3x + y + z = 13,$$
$$4x + 2y - 3z = 7.$$

We solve the system of equations simultaneously.

$$\begin{bmatrix} 2 & -3 & 5 & | & 7 \\ 3 & 1 & 1 & | & 13 \\ 4 & 2 & -3 & | & 7 \end{bmatrix},$$

$$\begin{bmatrix} 1 & -\frac{3}{2} & \frac{5}{2} & | & \frac{7}{2} \\ 0 & \frac{11}{2} & -\frac{13}{2} & | & \frac{5}{2} \\ 0 & 8 & -13 & | & -7 \end{bmatrix},$$

$$\begin{bmatrix} 1 & 0 & \frac{8}{11} & | & \frac{46}{11} \\ 0 & 1 & -\frac{13}{11} & | & \frac{5}{11} \\ 0 & 0 & -\frac{39}{11} & | & -\frac{117}{11} \end{bmatrix},$$

$$\begin{bmatrix} 1 & 0 & 0 & | & 2 \\ 0 & 1 & 0 & | & 4 \\ 0 & 0 & 1 & | & 3 \end{bmatrix}.$$

The three planes have exactly one point in common, namely, (2, 4, 3). As a check, we calculate the product

$$\begin{bmatrix} 2 & -3 & 5 \\ 3 & 1 & 1 \\ 4 & 2 & -3 \end{bmatrix} \begin{bmatrix} 2 \\ 4 \\ 3 \end{bmatrix}$$

and verify that the product agrees with $\begin{bmatrix} 7 \\ 13 \\ 7 \end{bmatrix}$.

The three planes and their single point of intersection are shown in Fig. 17. Each pair of the planes have a line in common. The three lines of intersection contain the common point (2, 4, 3).

Figure 17

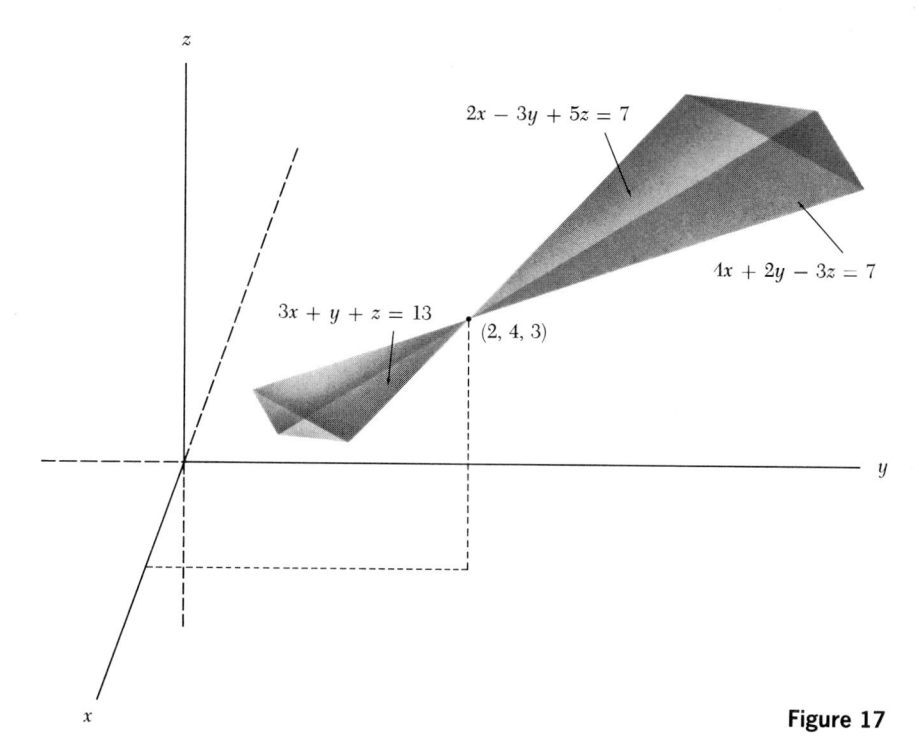

$$2x - 3y + 5z = 7$$
$$4x + 2y - 3z = 7$$
$$3x + y + z = 13$$
$$(2, 4, 3)$$

It may happen that several given planes do not have any point in common. This is a statement in geometric language corresponding to an algebraic statement that a system of linear equations may have no root. An advantage of the Gauss elimination method for solving a system of equations is that the method reveals (rather quickly) the lack of a root in case the system has no solution.

Example 33 Solve the system

$$\begin{cases} x - 3y - 5z = 1, \\ -3x + 7y + 9z = 1, \\ x + 4z = 16. \end{cases}$$

Our method yields the following augmented matrices in succession.

$$\begin{bmatrix} 1 & -3 & -5 & | & 1 \\ -3 & 7 & 9 & | & 1 \\ 1 & 0 & 4 & | & 16 \end{bmatrix},$$

$$\begin{bmatrix} 1 & -3 & -5 & | & 1 \\ 0 & -2 & -6 & | & 4 \\ 0 & 3 & 9 & | & 15 \end{bmatrix},$$

$$\begin{bmatrix} 1 & -3 & -5 & 1 \\ 0 & 1 & 3 & -2 \\ 0 & 0 & 0 & 21 \end{bmatrix}.$$

We interpret the final row in the preceding augmented matrix. It states that

$$0x + 0y + 0z = 21.$$

No choice of numbers x, y, z satisfies this condition. Recalling our claim (made in Section 2 and to be justified in Section 14) that successive steps in the Gauss elimination method make no change whatever in the root or roots, if any, for a system, we conclude in the present example that no ordered triple of numbers is a root of all three of the given equations.

The graph of each of the three linear equations is a plane. (See Fig. 18.) Each pair of the three planes intersect in a line; the three lines thus determined are parallel to one another. No point belongs to all three planes.

Figure 18

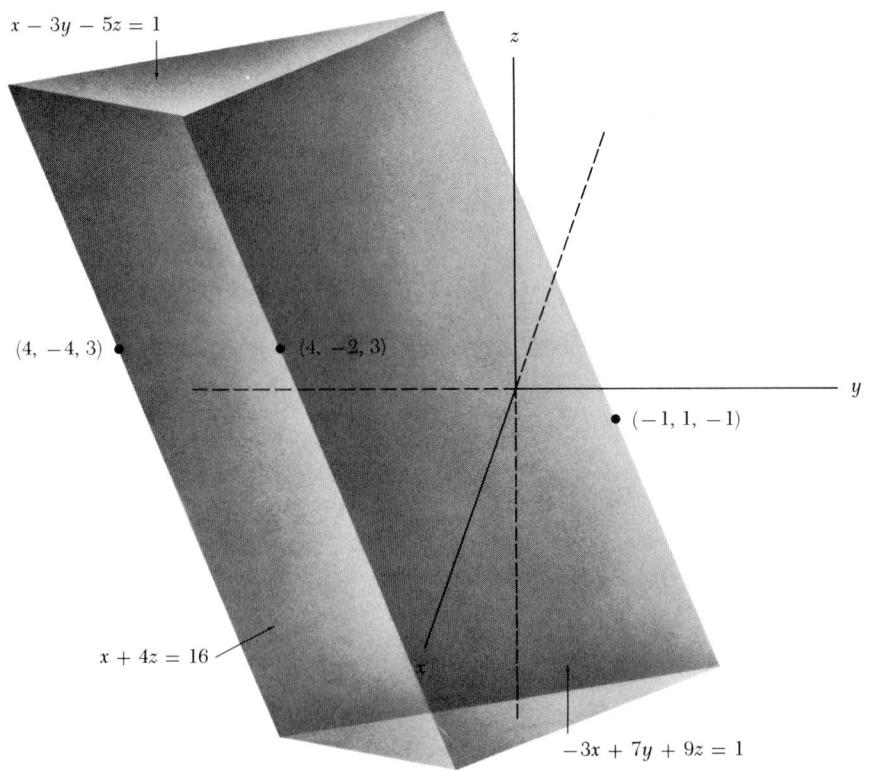

78   Systems of Linear Equations

Example 34 Solve the system

$$\begin{cases} x - 3y - 5z = 1, \\ -3x + 7y + 9z = 1, \\ x + 4z = -5. \end{cases}$$

Note that, although this system strongly resembles the system in the previous example, the results we obtain are quite different in the two cases.

$$\begin{bmatrix} 1 & -3 & -5 & | & 1 \\ -3 & 7 & 9 & | & 1 \\ 1 & 0 & 4 & | & -5 \end{bmatrix},$$

$$\begin{bmatrix} 1 & -3 & -5 & | & 1 \\ 0 & -2 & -6 & | & 4 \\ 0 & 3 & 9 & | & -6 \end{bmatrix},$$

$$\begin{bmatrix} 1 & 0 & 4 & | & -5 \\ 0 & 1 & 3 & | & -2 \\ 0 & 0 & 0 & | & 0 \end{bmatrix}.$$

The final row in the preceding augmented matrix may be interpreted as stating that

$$0x + 0y + 0z = 0.$$

Since this information imposes no restriction on x, y, z, we may, in seeking the roots of the system, ignore the row and focus our attention on the remaining rows of the augmented matrix.

The system of equations has many roots. The final augmented matrix states that

$$\begin{cases} x + 4z = -5, \\ y + 3z = -2. \end{cases}$$

We restate this result in another form:

$$\begin{cases} x = -4z - 5, \\ y = -3z - 2. \end{cases}$$

If z is assigned any value whatsoever, say p, then $x = -4p - 5$ and $y = -3p - 2$. That is,

$$\begin{bmatrix} x \\ y \\ z \end{bmatrix} = \begin{bmatrix} -4p - 5 \\ -3p - 2 \\ p \end{bmatrix} = \begin{bmatrix} -4p \\ -3p \\ p \end{bmatrix} + \begin{bmatrix} -5 \\ -2 \\ 0 \end{bmatrix} = p\begin{bmatrix} -4 \\ -3 \\ 1 \end{bmatrix} + \begin{bmatrix} -5 \\ -2 \\ 0 \end{bmatrix}.$$

12 / Geometry of a system of equations

A convenient form in which to express the answer to the problem of solving the given system

$$\begin{bmatrix} 1 & -3 & -5 \\ -3 & 7 & 9 \\ 1 & 0 & 4 \end{bmatrix} \cdot X = \begin{bmatrix} 1 \\ 1 \\ -5 \end{bmatrix}$$

is

$$X = p \begin{bmatrix} -4 \\ -3 \\ 1 \end{bmatrix} + \begin{bmatrix} -5 \\ -2 \\ 0 \end{bmatrix} \quad \text{for any number } p.$$

Since each choice of p gives a root X for the equation, there are infinitely many different roots. Geometrically, these roots form a line that is contained in all three of the given planes. (See Fig. 19.)

Figure 19

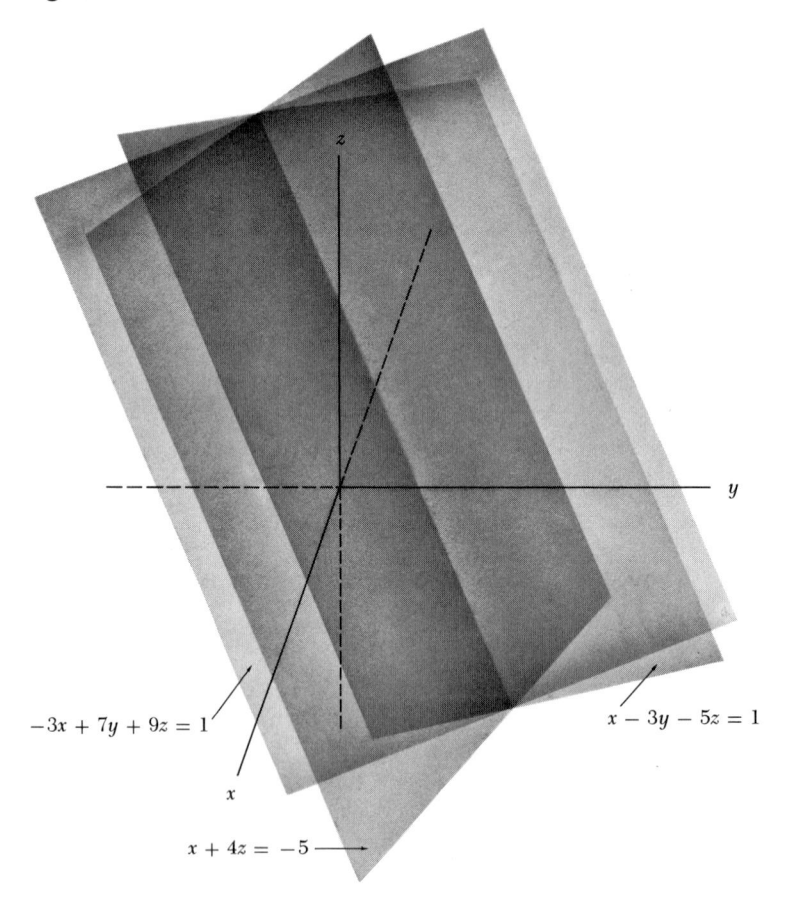

$-3x + 7y + 9z = 1$

$x - 3y - 5z = 1$

$x + 4z = -5$

80   Systems of Linear Equations

13. Systems for which the number of roots is not one

Example 35 Solve the system $AX = B$, where — *no solution*

$$A = \begin{bmatrix} 1 & -1 & 3 & 0 & 1 & -2 \\ 2 & -2 & 6 & 1 & 0 & -1 \\ -1 & 1 & -3 & 3 & -7 & 11 \\ 3 & -3 & 9 & -1 & 5 & -9 \end{bmatrix} \quad \text{and} \quad B = \begin{bmatrix} 4 \\ 7 \\ -5 \\ 13 \end{bmatrix}.$$

We apply the Gauss elimination method, using the augmented matrix notation.

$$\begin{bmatrix} 1 & -1 & 3 & 0 & 1 & -2 & \vdots & 4 \\ 2 & -2 & 6 & 1 & 0 & -1 & \vdots & 7 \\ -1 & 1 & -3 & 3 & -7 & 11 & \vdots & -5 \\ 3 & -3 & 9 & -1 & 5 & -9 & \vdots & 13 \end{bmatrix},$$

$$\begin{bmatrix} 1 & -1 & 3 & 0 & 1 & -2 & \vdots & 4 \\ 0 & 0 & 0 & 1 & -2 & 3 & \vdots & -1 \\ 0 & 0 & 0 & 3 & -6 & 9 & \vdots & -1 \\ 0 & 0 & 0 & -1 & 2 & -3 & \vdots & 1 \end{bmatrix},$$

$$\begin{bmatrix} 1 & -1 & 3 & 0 & 1 & -2 & \vdots & 4 \\ 0 & 0 & 0 & 1 & -2 & 3 & \vdots & -1 \\ 0 & 0 & 0 & 0 & 0 & 0 & \vdots & \boxed{2} \\ 0 & 0 & 0 & 0 & 0 & 0 & \vdots & 0 \end{bmatrix}.$$

The third row in the final matrix reveals that the given system has no root. Indeed, it states that the six components of X, which we may denote by u, v, w, x, y, z, satisfy the condition

$$0u + 0v + 0w + 0x + 0y + 0z = 2.$$

However, no sextuple of numbers (u, v, w, x, y, z) meets this requirement.

Example 36 Solve the equation $AX = B$, where

$$A = \begin{bmatrix} 1 & 2 & 2 & -1 \\ 2 & 4 & -1 & 8 \\ 1 & 2 & -5 & 13 \\ 3 & 6 & 4 & 1 \end{bmatrix} \quad \text{and} \quad B = \begin{bmatrix} 19 \\ 8 \\ -23 \\ 45 \end{bmatrix}.$$

As usual, we form the augmented matrix $[A \mid B]$ and apply the Gauss elimination method.

$$\begin{bmatrix} 1 & 2 & 2 & -1 & 19 \\ 2 & 4 & -1 & 8 & 8 \\ 1 & 2 & -5 & 13 & -23 \\ 3 & 6 & 4 & 1 & 45 \end{bmatrix},$$

$$\begin{bmatrix} 1 & 2 & 2 & -1 & 19 \\ 0 & 0 & -5 & 10 & -30 \\ 0 & 0 & -7 & 14 & -42 \\ 0 & 0 & -2 & 4 & -12 \end{bmatrix},$$

$$\begin{bmatrix} 1 & 2 & 0 & 3 & 7 \\ 0 & 0 & 1 & -2 & 6 \\ 0 & 0 & 0 & 0 & 0 \\ 0 & 0 & 0 & 0 & 0 \end{bmatrix}.$$

If $X = \begin{bmatrix} w \\ x \\ y \\ z \end{bmatrix}$, then we have found that

$$\begin{cases} w + 2x \phantom{{}+y} + 3z = 7, \\ \phantom{w + 2x +{}} y - 2z = 6. \end{cases}$$

These results may be expressed

$$\begin{cases} w = -2x - 3z + 7, \\ y = \phantom{-2x -{}} 2z + 6. \end{cases}$$

We may assign to the unknown x any value we please, say a number p, and assign to the unknown z any number whatsoever, say q; if we do so, the corresponding values of w and y are determined: $w = -2p - 3q + 7$ and $y = 2q + 6$. Thus,

$$X = \begin{bmatrix} -2p - 3q + 7 \\ p \\ 2q + 6 \\ q \end{bmatrix} = p\begin{bmatrix} -2 \\ 1 \\ 0 \\ 0 \end{bmatrix} + q\begin{bmatrix} -3 \\ 0 \\ 2 \\ 1 \end{bmatrix} + \begin{bmatrix} 7 \\ 0 \\ 6 \\ 0 \end{bmatrix}.$$

Systems of Linear Equations

The latter form for the answer, namely,

$$X = p \begin{bmatrix} -2 \\ 1 \\ 0 \\ 0 \end{bmatrix} + q \begin{bmatrix} -3 \\ 0 \\ 2 \\ 1 \end{bmatrix} + \begin{bmatrix} 7 \\ 0 \\ 6 \\ 0 \end{bmatrix},$$

expresses X in terms of two *parameters*, p and q. Each of these parameters may assume any value, and the column matrix X that results from any such assignment is a root of the equation $AX = B$.

In particular, the root

$$\begin{bmatrix} 7 \\ 0 \\ 6 \\ 0 \end{bmatrix}$$

is obtained by setting $p = 0 = q$; the root

$$\begin{bmatrix} 4 \\ 0 \\ 8 \\ 1 \end{bmatrix}$$

is obtained by choosing $p = 0$ and $q = 1$; the root

$$\begin{bmatrix} 1 \\ 3 \\ 6 \\ 0 \end{bmatrix}$$

arises if $p = 3$ and $q = 0$; each of the many other roots can be found by appropriately selecting p and q. The reader should verify that each of the three particular column matrices cited above actually is a root.

To check that all of the claimed roots really do satisfy the equation, we may proceed as follows. First, by direct computation, we find that

$$A \cdot \begin{bmatrix} -2 \\ 1 \\ 0 \\ 0 \end{bmatrix}$$

13 / Systems for which the number of roots is not one

is the zero matrix; second, also by straightforward calculation, we find that

$$A \cdot \begin{bmatrix} -3 \\ 0 \\ 2 \\ 1 \end{bmatrix}$$

is the zero matrix; we have already checked that

$$A \cdot \begin{bmatrix} 7 \\ 0 \\ 6 \\ 0 \end{bmatrix}$$

is B. Hence,

$$A \cdot \left\{ p \begin{bmatrix} -2 \\ 1 \\ 0 \\ 0 \end{bmatrix} + q \begin{bmatrix} -3 \\ 0 \\ 2 \\ 1 \end{bmatrix} + \begin{bmatrix} 7 \\ 0 \\ 6 \\ 0 \end{bmatrix} \right\} = pA \begin{bmatrix} -2 \\ 1 \\ 0 \\ 0 \end{bmatrix} + qA \begin{bmatrix} -3 \\ 0 \\ 2 \\ 1 \end{bmatrix} + A \begin{bmatrix} 7 \\ 0 \\ 6 \\ 0 \end{bmatrix}$$

$$= p \begin{bmatrix} 0 \\ 0 \\ 0 \\ 0 \end{bmatrix} + q \begin{bmatrix} 0 \\ 0 \\ 0 \\ 0 \end{bmatrix} + B$$

$$= B.$$

Notice how the product simplifies to B, regardless of what the number p is, and regardless of what number is chosen for q.

In case an equation of the type $AX = B$ has many roots, there are many acceptable ways for expressing the set of roots. In some situations one acceptable form may be more convenient than others. We may pass from one form to another by exploiting the properties of numbers and matrices.

Example 37 Suppose that the set of roots of a certain equation is the set of matrices

$$\begin{bmatrix} 2 \\ -1 \end{bmatrix} + p \begin{bmatrix} \frac{1}{2} \\ \frac{2}{3} \end{bmatrix} \quad \text{for all numbers } p.$$

84 Systems of Linear Equations

Since
$$\begin{bmatrix} \frac{1}{2} \\ \frac{2}{3} \end{bmatrix} = \tfrac{1}{6} \begin{bmatrix} 3 \\ 4 \end{bmatrix},$$
we may rewrite
$$\begin{bmatrix} 2 \\ -1 \end{bmatrix} + p \begin{bmatrix} \frac{1}{2} \\ \frac{2}{3} \end{bmatrix} \quad \text{as} \quad \begin{bmatrix} 2 \\ -1 \end{bmatrix} + \frac{p}{6} \begin{bmatrix} 3 \\ 4 \end{bmatrix}.$$

Let us introduce $q = p/6$. As we allow p to assume any real number as a value, q does also. Thus, the set of matrices
$$\begin{bmatrix} 2 \\ -1 \end{bmatrix} + p \begin{bmatrix} \frac{1}{2} \\ \frac{2}{3} \end{bmatrix} \quad \text{for all numbers } p$$
is the same as the set of matrices
$$\begin{bmatrix} 2 \\ -1 \end{bmatrix} + q \begin{bmatrix} 3 \\ 4 \end{bmatrix} \quad \text{for all numbers } q.$$

Example 38 Show that the set of vectors
$$(1, 3, 7) + p(2, 1, 0) \quad \text{for all numbers } p$$
is the same as the set of vectors
$$(5, 5, 7) + q(2, 1, 0) \quad \text{for all numbers } q.$$

As we allow p to take on any real number as value, $p - 2$ does also. Abbreviate $p - 2$ by q. Then $p = 2 + q$. Thus, the vector
$$\begin{aligned}(1, 3, 7) + p(2, 1, 0) &= (1, 3, 7) + (2 + q) \cdot (2, 1, 0) \\ &= (1, 3, 7) + 2(2, 1, 0) + q(2, 1, 0) \\ &= (1 + 2 \cdot 2, 3 + 2 \cdot 1, 7 + 2 \cdot 0) + q(2, 1, 0) \\ &= (5, 5, 7) + q(2, 1, 0).\end{aligned}$$

The set of all vectors of the type $(1, 3, 7) + p(2, 1, 0)$ is the same as the set of all vectors of the style $(5, 5, 7) + q(2, 1, 0)$.

Example 39 Consider the set of vectors
$$(1, 3, 1) + p(3, 2, -1) \quad \text{for all numbers } p.$$
Express this set in a fashion that explicitly shows the vector in the set with second component equal to zero.

In the given form the second component is $3 + 2p$. The choice of p which makes the second component zero is $-\frac{3}{2}$. We introduce k by the relation $p = k - \frac{3}{2}$. Then
$$\begin{aligned}(1, 3, 1) + p(3, 2, -1) &= (1, 3, 1) + (k - \tfrac{3}{2})(3, 2, -1) \\ &= (1, 3, 1) - (\tfrac{9}{2}, 3, -\tfrac{3}{2}) + k(3, 2, -1) \\ &= (-\tfrac{7}{2}, 0, \tfrac{5}{2}) + k(3, 2, -1).\end{aligned}$$

13 / Systems for which the number of roots is not one   85

The given set of vectors is the same as the set of vectors

$$(-\tfrac{7}{2}, 0, \tfrac{5}{2}) + k(3, 2, -1) \quad \text{for all numbers } k.$$

EXERCISES I

1. On a three-dimensional coordinate system, let $x - 2y = 6$ be the equation of a plane P, let $3y + z = 3$ be the equation of a plane Q, and let $3x + 2z = 12$ be the equation of a plane R.
   (a) Solve simultaneously the two equations for planes P and Q, and show that the intersection of the two planes is the line consisting of the points $(6, 0, 3) + k(2, 1, -3)$ for all real numbers k.
   (b) Draw a diagram, showing the planes P and Q and their line of intersection.
   (c) Solve simultaneously the two equations for planes P and R, and show that the intersection of the two planes is the line consisting of the points $(0, -3, 6) + h(2, 1, -3)$ for all real numbers h.
   (d) Draw a picture, showing the planes P and R and their line of intersection.
   (e) Solve simultaneously the two equations for planes Q and R.
   (f) Draw a diagram, showing the planes Q and R and their line of intersection.
   (g) On a single-coordinate system, draw all three lines of intersection, as found in parts (a), (c), (e).
   (h) The three lines of intersection are parallel to one another. Explain why the three given equations have no root in common.

2. On a three-dimensional coordinate system, let $x - 2y = 6$ be the equation of a plane P, let $3y + z = 3$ be the equation of a plane Q, and let $3x + 2z = 24$ be the equation of a plane S. Let M be the line of intersection of planes P and Q; note that M was found in Exercise 1a.
   (a) Solve simultaneously the two equations for planes P and S, and show that the intersection of the two planes is the line M.
   (b) Draw a diagram, showing the planes P and S and their line of intersection.
   (c) Solve simultaneously the two equations for planes Q and S, and show that the intersection of the two planes is the line M.
   (d) Draw a diagram, showing all three planes P, Q, S and the line M, which lies in each of the three planes.

3. On a three-dimensional coordinate system, let P, Q, R, S be the planes whose respective equations are $2x - y = 4$, $x + z = 3$, $y + 2z = 0$, $y + 2z = 2$.
   (a) Show algebraically that the equations for the three planes P, Q, R have no root in common.
   (b) Show geometrically that the three planes P, Q, R have no point in common.
   (c) Find all roots for the system of equations for the three planes P, Q, S.
   (d) Draw a diagram, showing the three planes P, Q, S and their intersection.

4. Each of the following is the final augmented matrix to appear in the process of solving an equation of the type $AX = B$. In each case, answer whichever of the two following questions is pertinent: (i) if the equation has no root, explain how this information is revealed; (ii) if the equation has one or more roots, express all of the roots in a convenient form.

(a) $\begin{bmatrix} 1 & 0 & 0 & 0 & | & -2 \\ 0 & 1 & 0 & 0 & | & 0 \\ 0 & 0 & 1 & 0 & | & 0 \\ 0 & 0 & 0 & 1 & | & 81 \end{bmatrix}.$

(b) $\begin{bmatrix} 1 & 0 & 0 & 1 & | & 0 \\ 0 & 1 & 3 & 0 & | & 0 \\ 0 & 0 & 0 & 0 & | & 1 \end{bmatrix}.$

(c) $\begin{bmatrix} 1 & 0 & 0 & | & 0 \\ 0 & 1 & 0 & | & 0 \\ 0 & 0 & 1 & | & 0 \end{bmatrix}$.

(d) $\begin{bmatrix} 1 & 0 & 0 & 2 & | & 3 \\ 0 & 1 & 0 & 1 & | & -4 \\ 0 & 0 & 0 & 0 & | & 0 \end{bmatrix}$.

(e) $\begin{bmatrix} 1 & -3 & 0 & | & 5 \\ 0 & 0 & 1 & | & -2 \\ 0 & 0 & 0 & | & 0 \end{bmatrix}$.

(f) $\begin{bmatrix} 1 & 0 & -\frac{1}{2} & 3 & | & \frac{7}{2} \\ 0 & 1 & -3 & -\frac{2}{3} & | & \frac{5}{3} \\ 0 & 0 & 0 & 0 & | & 0 \\ 0 & 0 & 0 & 0 & | & -\frac{4}{5} \\ 0 & 0 & 0 & 0 & | & 0 \end{bmatrix}$.

(g) $\begin{bmatrix} 1 & 0 & -1 & 0 & 3 & | & 5 \\ 0 & 1 & 2 & 0 & -1 & | & -2 \\ 0 & 0 & 0 & 1 & 2 & | & 6 \end{bmatrix}$.

5. Let k be a real number, and let $(x, y) = (-3, -1) + k(1, 2)$.
   (a) Choose several (at least five) different values for k; for each of the chosen numbers k, calculate the ordered pair (x, y).
   (b) On a two-dimensional coordinate system, plot all the points obtained in part (a).
   (c) What geometrical figure is formed by the points (x, y) for all possible choices of k?
   (d) Show that the same figure is formed by the points $(0, 5) + p(1, 2)$ for all numbers p.

6. Let k be a real number, and let $(x, y) = (2, 5) + k(-1, 0)$.
   (a) Choose several different values for k; for each choice, calculate the corresponding vector (x, y).
   (b) On a two-dimensional coordinate system, plot all the points obtained in part (a).
   (c) What type of a geometrical figure is the set of points (x, y) for all possible choices of k?

7. On a two-dimensional coordinate system the set of points
$$(5, -2) + p(3, 4) \quad \text{for all possible numbers } p$$
is a line. What is the slope of the line?

8. Let a, b, c, d be nonzero numbers. What is the slope of the line consisting of the points $(a, b) + p(c, d)$ for all real numbers p?

9. Let q be a real number, and let $(x, y, z) = (1, 0, 0) + q(1, 2, 1)$.
   (a) Choose five different values for q; on a three-dimensional coordinate system plot the corresponding five points.
   (b) What geometrical figure is the set of points (x, y, z) for all values of q?
   (c) Show that the same figure is the set of points $(0, -2, -1) + t(1, 2, 1)$ for all numbers t.

10. The solution of a certain matrix equation is the set of all column matrices
$$\begin{bmatrix} 3 \\ 7 \\ -4 \end{bmatrix} + p \begin{bmatrix} 2 \\ -1 \\ 3 \end{bmatrix}, \quad \text{where } p \text{ is a real number.}$$

What, on a three-dimensional coordinate system, is the graph of the equation?

13 / Systems for which the number of roots is not one

14 The Gauss elimination method

We are now ready to summarize the Gauss elimination method for solving an equation of the type $AX = B$, where A is a given $m \times n$ matrix, B is a given $m \times 1$ matrix, and where the $n \times 1$ matrix X is to be found. We form the (augmented) matrix $[A \mid B]$ with m rows and $n + 1$ columns. We transform $[A \mid B]$ by a sequence of steps, each step being applied to the matrix that results from the preceding step. Each step is a manipulation on the rows of the matrix of one of the two following types:

(a) Multiplication of a row by a nonzero number.
(b) Addition, to a row, of the product of a number and any other row.

By such steps we seek to obtain

$$\begin{bmatrix} 1 \\ 0 \\ 0 \\ \vdots \\ 0 \end{bmatrix}$$

as a column of the matrix (that is, a column with first element 1 and every other element 0); furthermore, we want this column as far to the left as possible (in fact, preferably the first column). Without disturbing the results already achieved, we next attempt to obtain

$$\begin{bmatrix} 0 \\ 1 \\ 0 \\ \vdots \\ 0 \end{bmatrix}$$

as a column of the matrix (that is, a column with second element 1 and every other element 0); furthermore, we want this column as far to the left as possible. Without disturbing results already achieved, we attempt to obtain

$$\begin{bmatrix} 0 \\ 0 \\ 1 \\ 0 \\ \vdots \\ 0 \end{bmatrix}$$

as a column (a column with third element 1 and every other element 0); again we want this as far to the left as possible. We continue in this pattern as long as we can. When finished, we interpret the final augmented matrix.

Definition If n is a natural number, the $n \times n$ matrix

$$\begin{bmatrix} 1 & 0 & 0 & \cdots & 0 \\ 0 & 1 & 0 & \cdots & 0 \\ 0 & 0 & 1 & \cdots & 0 \\ & & \cdots & & \\ 0 & 0 & 0 & \cdots & 1 \end{bmatrix}$$

is called the *identity matrix* and is often denoted by I. Each column in I has a single element that is not zero: in the jth column the jth element is 1.

In case A is an $n \times n$ matrix and B is an $n \times 1$ matrix, the elimination method applied to the equation $AX = B$ often yields at the final stage the augmented matrix

$$\begin{bmatrix} I & \vdots & c_1 \\ & \vdots & \vdots \\ & \vdots & c_n \end{bmatrix} = \begin{bmatrix} 1 & 0 & \cdots & 0 & \vdots & c_1 \\ 0 & 1 & \cdots & 0 & \vdots & c_2 \\ & & \cdots & & \vdots & \vdots \\ 0 & 0 & \cdots & 1 & \vdots & c_n \end{bmatrix}.$$

In this case, the equation $AX = B$ has exactly one root, namely,

$$C = \begin{bmatrix} c_1 \\ c_2 \\ \vdots \\ c_n \end{bmatrix}.$$

Indeed, if we interpret the final stage, we have $IX = C$; since $IX = X$ (which may be verified in a straightforward manner), the sole root is C. Notice that this is the type of situation that has occurred frequently in our examples.

If at any stage in the elimination method we obtain a row in the augmented matrix of the type $[0 \ 0 \ \cdots \ 0 \mid g]$, where $g \neq 0$, then we may stop the process and immediately conclude that the equation has no root at all. Illustrations of this situation have been met in Examples 33 and 35 (in Sections 12 and 13, respectively). In discussing those examples, we explained the reason for the assertion that there is no root.

Another outcome is possible: we may obtain one or more zero rows in the augmented matrix. A zero row is essentially noninformative. Consequently, insofar as solving the equation is concerned, such a row may be subsequently ignored. Typical illustrations have been provided by Examples 34 and 36 (in Sections 12 and 13, respectively).

In brief, by the Gauss elimination method we proceed to transform the given equation into a form that permits us, by inspection, to determine whether there is a root and, if so, to read all the roots. The transformation is a stepwise procedure. The justification for the method is that each step in the process preserves the set of roots for the equation. At each successive stage the root or roots (if any) are the same as at the preceding stage; none has been lost and none has been introduced in the maneuver from the previous stage. This claim is substantiated by the following theorem.

Theorem 1 Let the augmented matrix $[A^* \mid B^*]$ appear at some stage in an application of the Gauss elimination method. Let it be transformed into the augmented matrix $[A^\dagger \mid B^\dagger]$ by a step of one of the two following types:
(a) Multiplication of a row by a nonzero number.
(b) Addition, to a row, of the product of a number and any other row.

Then the equation $A^*X = B^*$ has precisely the same roots as the equation $A^\dagger X = B^\dagger$.

In particular, if $A^*X = B^*$ has no root, then $A^\dagger X = B^\dagger$ has no root; if $A^*X = B^*$ has exactly one root, then $A^\dagger X = B^\dagger$ has the same root and no others; if $A^*X = B^*$ has many roots, then $A^\dagger X = B^\dagger$ also has these same roots and no others.

PROOF First, consider a step of type (a). Suppose that in the augmented matrix $[A^* \mid B^*]$ the row

$$[r_1 \quad r_2 \quad \cdots \quad r_n \mid s] = [R \mid s]$$

is multiplied by the nonzero number k in order to obtain the new row

$$[kr_1 \quad kr_2 \quad \cdots \quad kr_n \mid ks] = [kR \mid ks]$$

of the augmented matrix $[A^\dagger \mid B^\dagger]$. If the equation $A^*X = B^*$ has a root H, then,

$$r_1 h_1 + r_2 h_2 + \cdots + r_n h_n = s,$$

or $RH = s$. Multiplying both members by k, we obtain

$$kr_1 h_1 + kr_2 h_2 + \cdots + kr_n h_n = ks,$$

or $(kR)H = (ks)$. Since every other row of the augmented matrix $[A^\dagger \mid B^\dagger]$ is a copy of the corresponding row in $[A^* \mid B^*]$, H satisfies the new equation $A^\dagger X = B^\dagger$.

On the other hand, if G is a root of $A^\dagger X = B^\dagger$, then $(kR)G = ks$. Since $k \neq 0$, we may divide each member of this equation by k; we obtain $RG = s$. Inasmuch as all the other rows of $[A^* \mid B^*]$ are copies of the respective rows in $[A^\dagger \mid B^\dagger]$, we find that G is a root of $A^*X = B^*$. In summary, the roots of $A^*X = B^*$ are the same as the roots of $A^{\dagger}X = B^{\dagger}$.

Second, consider a step of type (b). Suppose that the row $[U \mid v]$ in the augmented matrix

$$[A^* \mid B^*] = \begin{bmatrix} & \cdots & & \vdots & \cdot \\ r_1 & \cdots & r_n & \vdots & s \\ & \cdots & & \vdots & \cdot \\ u_1 & \cdots & u_n & \vdots & v \\ & \cdots & & \vdots & \cdot \end{bmatrix}$$

is replaced by $[U + kR \mid v + ks]$ to form the new row in the augmented matrix

$$[A^{\dagger} \mid B^{\dagger}] = \begin{bmatrix} & \cdots & & \vdots & \cdot \\ r_1 & \cdots & r_n & \vdots & s \\ & \cdots & & \vdots & \cdot \\ u_1 + kr_1 & \cdots & u_n + kr_n & \vdots & v + ks \\ & \cdots & & \vdots & \cdot \end{bmatrix}.$$

If the equation $A^*X = B^*$ has a root H, then

$$RH = s$$

and

$$UH = v.$$

Hence, $(kR)H = ks$ and

$$(U + kR)H = UH + kRH = v + ks.$$

Since every other row in the augmented matrix $[A^{\dagger} \mid B^{\dagger}]$ is a copy of the corresponding row in $[A^* \mid B^*]$, we find that H satisfies the equation $A^{\dagger}X = B^{\dagger}$.

On the other hand, if G is a root of $A^{\dagger}X = B^{\dagger}$, then

$$RG = s$$

and

$$(U + kR)G = v + ks.$$

Multiplying both members of the former equation by k, we obtain

$$kRG = ks.$$

Now subtraction yields $(U + kR)G - kRG = (v + ks) - ks$, or, more simply,

$$UG = v.$$

14 / The Gauss elimination method

Since all the other rows of $[A^* \mid B^*]$ are copies of the respective rows in $[A^\dagger \mid B^\dagger]$, we conclude that G is also a root of $A^*X = B^*$. In summary, the set of roots for $A^*X = B^*$ is the same as the set of roots for $A^\dagger X = B^\dagger$.

On the basis of Theorem 1, each step in the Gauss elimination method preserves the set of roots for the equation. Consequently, the roots at the final stage are the same as the roots at the original stage.

EXERCISES J

1. Solve each of the following equations. [HINT: By the word "solve," we mean the following: If the equation has exactly one root, find it; if the equation has no root, discover this information and announce it; if the equation has more than one root, find all of them, expressing the various possibilities in terms of one or more parameters.]

(a) $\begin{bmatrix} 1 & 3 & -2 \\ 2 & -1 & 1 \\ 1 & 10 & -7 \end{bmatrix} X = \begin{bmatrix} 1 \\ 0 \\ 0 \end{bmatrix}$.

(b) $\begin{bmatrix} 1 & 3 & -2 \\ 2 & -1 & 1 \\ 1 & 10 & -7 \end{bmatrix} X = \begin{bmatrix} 1 \\ 1 \\ 2 \end{bmatrix}$.

(c) $\begin{bmatrix} 1 & 2 & -3 & 1 \\ 2 & 5 & -4 & 0 \\ -1 & 1 & 2 & -3 \end{bmatrix} X = \begin{bmatrix} 0 \\ 1 \\ 0 \end{bmatrix}$.

(d) $\begin{bmatrix} 6 & -4 \\ -3 & 3 \\ 5 & 2 \end{bmatrix} X = \begin{bmatrix} 2 \\ 6 \\ 39 \end{bmatrix}$.

(e) $\begin{bmatrix} 1 & 0 & 2 & 2 & -1 \\ -1 & 1 & -3 & -7 & 4 \\ 2 & 1 & 5 & 5 & -3 \\ -3 & 3 & -4 & -6 & 2 \\ -2 & -1 & -2 & 4 & -3 \end{bmatrix} X = \begin{bmatrix} 2 \\ -3 \\ 8 \\ -5 \\ -3 \end{bmatrix}$.

(f) $\begin{bmatrix} 3 & 1 & 0 & 0 \\ 1 & 2 & 1 & 0 \\ 0 & 1 & 2 & 1 \\ 0 & 0 & 1 & 3 \end{bmatrix} X = \begin{bmatrix} -3 \\ 0 \\ 0 \\ -5 \end{bmatrix}$.

92   Systems of Linear Equations

(g) $\begin{bmatrix} 1 & 0 & 2 & -1 & 3 \\ 2 & 1 & 4 & -1 & 6 \\ 1 & -3 & 2 & -4 & 3 \\ 4 & 3 & 8 & -1 & 12 \\ 3 & -2 & 6 & -5 & 9 \end{bmatrix} X = \begin{bmatrix} 8 \\ 21 \\ -7 \\ 47 \\ 14 \end{bmatrix}$.

(h) $\begin{bmatrix} 1 & 1 & -2 & 0 & -2 \\ 1 & -1 & -2 & -2 & 2 \\ 2 & -1 & -4 & -3 & 2 \\ -1 & 3 & 2 & 4 & -6 \\ 3 & -2 & -6 & -5 & 4 \end{bmatrix} X = \begin{bmatrix} 3 \\ 3 \\ 6 \\ -3 \\ -6 \end{bmatrix}$.

(i) $\begin{bmatrix} 1 & 2 & 3 & -3 & 0 \\ 1 & -1 & -3 & -3 & 3 \\ 3 & -1 & -5 & -9 & 7 \\ -2 & 3 & 8 & 6 & -7 \end{bmatrix} X = \begin{bmatrix} 2 \\ -1 \\ -1 \\ 3 \end{bmatrix}$.

2. Find an equation of a plane (if there is any) that contains each of the points $(1, -2, 1)$, $(2, 3, 1)$, $(-2, 1, 7)$, $(1, 4, 3)$.

3. Find an equation of a plane (if there is any) that contains each of the points $(1, 4, 1)$, $(2, -1, 0)$, $(1, 2, 2)$, $(-1, 4, -2)$.

4. Find an equation of a plane (if there is any) that contains each of the points $(7, 5, -3)$, $(8, 3, -1)$, $(5, 9, -7)$, $(10, -1, 3)$.

5. Find an equation of a plane (if there is any) that contains each of the points $(1, 5, -2)$ and $(-2, 1, 1)$.

Linear Transformations 3

1 Vector space

Let n be a natural number. Consider the collection of all n-dimensional vectors. For convenience in this discussion, we write the components of a vector in a column; we may alternatively think of the vector as an $n \times 1$ column matrix. We have already seen how we may add two vectors in the collection and how we may multiply a number by a vector. The result of each of these operations is also a vector in the set. We have a mathematical system which we wish to study further.

> **Definition** The set of all n-dimensional vectors (expressed as column matrices), together with the operation of addition and the operation of multiplication by a number, constitutes a mathematical system known as the (column) *vector space* of dimension n.

Our use of the word "space" is borrowed from geometry. Indeed, we have already noted that the ordinary space in which we live may be described by the system of three-dimensional vectors. The familiar two-dimensional coordinate system is a means for converting a plane into a vector space of dimension two.

2 Linear transformation

The n-dimensional vector space consists of ordered n-tuples of numbers, whereas the familiar mathematical system of ordinary numbers consists of single numbers. We wish to extend to vectors some of the notions about linear functions we studied in Chapter 1. A linear function is a particular way of assigning to every number a certain number; the assignment is provided by a formula of the type $mx + b$. We consider the special case given by mx, and extend the idea to a vector space.

Let A be a fixed $n \times n$ matrix. If U is a vector in the n-dimensional vector space, then we may multiply A by the column matrix U to obtain the product AU. The product is a vector in the vector space. We have described a way of assigning to each vector a vector (usually a different vector, but not necessarily). This assignment gives a *linear transformation* on the vector space. Suppose we denote the linear transformation by $\mathcal{T}$. Then the image of a vector U, denoted by $\mathcal{T}(U)$, is given by the formula $\mathcal{T}(U) = AU$.

[Observe the analogy with the notion of a linear function on the set of real numbers: the image of a number x under the linear function L may be given by a formula $L(x) = mx$.]

Example 1 Let $n = 2$, and let

$$A = \begin{bmatrix} -1 & -3 \\ 1 & 2 \end{bmatrix}.$$

The image of a vector $\begin{bmatrix} a \\ b \end{bmatrix}$ under the linear transformation $\mathcal{T}$ defined by $\mathcal{T}(U) = AU$ is

$$A \cdot \begin{bmatrix} a \\ b \end{bmatrix} = \begin{bmatrix} -1 & -3 \\ 1 & 2 \end{bmatrix} \begin{bmatrix} a \\ b \end{bmatrix} = \begin{bmatrix} -a - 3b \\ a + 2b \end{bmatrix}.$$

In particular, the image of $\begin{bmatrix} 3 \\ -4 \end{bmatrix}$ is

$$A \cdot \begin{bmatrix} 3 \\ -4 \end{bmatrix} = \begin{bmatrix} -1 & -3 \\ 1 & 2 \end{bmatrix} \begin{bmatrix} 3 \\ -4 \end{bmatrix} = \begin{bmatrix} 9 \\ -5 \end{bmatrix}.$$

Each fixed $n \times n$ matrix A determines a linear transformation $\mathcal{T}$ on the n-dimensional vector space. Different matrices define different linear transformations, for two different matrices A and B will yield different products AU and BU for appropriate vectors U.

There are two characteristic properties of a linear transformation. If U and V are vectors in the space on which $\mathcal{T}$ is a linear transformation, then $\mathcal{T}(U + V) = \mathcal{T}(U) + \mathcal{T}(V)$. If k is a number and if U is a vector in the space on which $\mathcal{T}$ is a linear transformation, then $\mathcal{T}(kU) = k\mathcal{T}(U)$.

The first of these properties states the relationship between a linear transformation and the operation of addition of vectors. The image of a sum of vectors is the same as the sum of their images. The second of the properties describes the relationship between a linear transformation and the other operation in a vector space, namely, multiplication of a number by a vector. The image of the product of a number and a vector is the same as the product of the number by the image of the vector.

In order to establish these properties, we utilize the properties of matrices. Let the linear transformation $\mathfrak{T}$ be determined by the matrix A. Then $\mathfrak{T}(U) = AU$ and $\mathfrak{T}(V) = AV$; hence,

$$\mathfrak{T}(U) + \mathfrak{T}(V) = AU + AV.$$

Now,

$$AU + AV = A(U + V).$$

On the other hand,

$$\mathfrak{T}(U + V) = A \cdot (U + V).$$

Thus, $\mathfrak{T}(U + V)$ is the same as $\mathfrak{T}(U) + \mathfrak{T}(V)$.

To verify the second property, we note that

$$k \cdot \mathfrak{T}(U) = k \cdot AU.$$

Since k is a number,

$$k \cdot AU = A \cdot (kU).$$

On the other hand,

$$\mathfrak{T}(kU) = A \cdot kU.$$

Hence, $\mathfrak{T}(kU) = k\mathfrak{T}(U)$.

Example 2 Let $\mathfrak{T}$ be the linear transformation that assigns to each vector

$$\begin{bmatrix} x \\ y \end{bmatrix} \quad \text{the image} \quad \begin{bmatrix} 5x + 2y \\ 2x + y \end{bmatrix}.$$

We identify the matrix A associated with $\mathfrak{T}$ by recognizing that

$$\begin{bmatrix} 5x + 2y \\ 2x + y \end{bmatrix} = \begin{bmatrix} 5 & 2 \\ 2 & 1 \end{bmatrix} \begin{bmatrix} x \\ y \end{bmatrix}.$$

Thus, $A = \begin{bmatrix} 5 & 2 \\ 2 & 1 \end{bmatrix}$.

The image of $U = \begin{bmatrix} 2 \\ -1 \end{bmatrix}$ under this $\mathfrak{T}$ is

$$AU = \begin{bmatrix} 5 & 2 \\ 2 & 1 \end{bmatrix} \begin{bmatrix} 2 \\ -1 \end{bmatrix} = \begin{bmatrix} 8 \\ 3 \end{bmatrix}.$$

The image of $V = \begin{bmatrix} -1 \\ 4 \end{bmatrix}$ under $\mathfrak{T}$ is

$$\mathcal{T}(V) = \begin{bmatrix} 5 & 2 \\ 2 & 1 \end{bmatrix} \begin{bmatrix} -1 \\ 4 \end{bmatrix} = \begin{bmatrix} 3 \\ 2 \end{bmatrix}.$$

The sum $U + V$ is $\begin{bmatrix} 2 \\ -1 \end{bmatrix} + \begin{bmatrix} -1 \\ 4 \end{bmatrix} = \begin{bmatrix} 1 \\ 3 \end{bmatrix}$, and the image of $U + V$ is

$$\begin{bmatrix} 5 & 2 \\ 2 & 1 \end{bmatrix} \begin{bmatrix} 1 \\ 3 \end{bmatrix} = \begin{bmatrix} 11 \\ 5 \end{bmatrix}.$$

On the other hand, the sum of the images of U and of V is

$$\begin{bmatrix} 8 \\ 3 \end{bmatrix} + \begin{bmatrix} 3 \\ 2 \end{bmatrix} = \begin{bmatrix} 11 \\ 5 \end{bmatrix}.$$

We have verified that, for the chosen U, V, $\mathcal{T}$, the statement $\mathcal{T}(U + V) = \mathcal{T}(U) + \mathcal{T}(V)$ is true.
If $k = 3$ and if U and $\mathcal{T}$ are the same as before, then

$$k \cdot \mathcal{T}(U) = 3 \begin{bmatrix} 8 \\ 3 \end{bmatrix} = \begin{bmatrix} 24 \\ 9 \end{bmatrix}.$$

Also,

$$kU = 3 \begin{bmatrix} 2 \\ -1 \end{bmatrix} = \begin{bmatrix} 6 \\ -3 \end{bmatrix},$$

and the image of kU under $\mathcal{T}$ is

$$A \cdot kU = \begin{bmatrix} 5 & 2 \\ 2 & 1 \end{bmatrix} \begin{bmatrix} 6 \\ -3 \end{bmatrix} = \begin{bmatrix} 24 \\ 9 \end{bmatrix}.$$

Thus, we have checked, in this instance, that $\mathcal{T}(kU) = k\mathcal{T}(U)$.

We summarize the highlights of this section in the following definition and theorem.

Definition A linear transformation $\mathcal{T}$ on an n-dimensional vector space is a rule that assigns to each vector U in the space the image $A \cdot U$, where A is a fixed $n \times n$ matrix.

Theorem 1 Every linear transformation on a vector space satisfies the following two conditions.
  (a) For every vector U and every vector V in the space,
      $\mathcal{T}(U + V) = \mathcal{T}(U) + \mathcal{T}(V)$.
  (b) For every vector U in the space and every number k,
      $\mathcal{T}(kU) = k \cdot \mathcal{T}(U)$.

EXERCISES A

1. Let $\mathcal{T}$ be the linear transformation given by the formula

$$\mathcal{T}(U) = \begin{bmatrix} 2 & -1 \\ -5 & 3 \end{bmatrix} U.$$

Let V, W, P be the vectors given by

$$V = \begin{bmatrix} 1 \\ 4 \end{bmatrix}, \quad W = \begin{bmatrix} 5 \\ 2 \end{bmatrix}, \quad P = \begin{bmatrix} -\frac{1}{4} \\ \frac{3}{4} \end{bmatrix}.$$

(a) Find $\mathcal{T}(V)$.   (b) Find $\mathcal{T}(W)$.   (c) Find the sum $\mathcal{T}(V) + \mathcal{T}(W)$.
(d) Find the sum $V + W$, and then find $\mathcal{T}(V + W)$.
(e) Compare $\mathcal{T}(V) + \mathcal{T}(W)$ and $\mathcal{T}(V + W)$.
(f) Find $\mathcal{T}(P)$.

2. Let the linear transformation $\mathcal{T}$ be given by

$$\mathcal{T}(V) = \begin{bmatrix} 3 & 5 & 3 \\ 5 & 3 & 2 \\ -1 & 2 & 1 \end{bmatrix} V.$$

Let

$$U = \begin{bmatrix} -1 \\ 2 \\ -1 \end{bmatrix} \quad \text{and} \quad W = \begin{bmatrix} 0 \\ 1 \\ 2 \end{bmatrix}.$$

(a) Find $\mathcal{T}(U)$.   (b) Find $\mathcal{T}(W)$.   (c) Find the sum $\mathcal{T}(U) + \mathcal{T}(W)$.
(d) Find the sum $U + W$, and then find $\mathcal{T}(U + W)$.
(e) Compare $\mathcal{T}(U) + \mathcal{T}(W)$ and $\mathcal{T}(U + W)$.
(f) Find $4U$, and then find $\mathcal{T}(4U)$.
(g) Compare $\mathcal{T}(4U)$ and $4 \cdot \mathcal{T}(U)$.

3. Let A be the matrix $\begin{bmatrix} a & b \\ c & d \end{bmatrix}$, let U be the column matrix $\begin{bmatrix} p \\ q \end{bmatrix}$, and let k be a number.

(a) Find kU, and then find $A \cdot (kU)$.
(b) Find AU, and then find $k \cdot (AU)$.
(c) Show that $A(kU)$ is the same as $k(AU)$.

[REMARK: This result proves that, if $\mathcal{T}$ is a linear transformation on a two-dimensional vector space, then $\mathcal{T}(kU) = k\mathcal{T}(U)$ for every vector U and every number k. The same method establishes the result for any other dimensionality and thereby proves Theorem 1b.]

4. Let A be the matrix $\begin{bmatrix} a & b \\ c & d \end{bmatrix}$, let U be the column matrix $\begin{bmatrix} p \\ q \end{bmatrix}$, and let W be the column matrix $\begin{bmatrix} s \\ t \end{bmatrix}$.

(a) Find $U + W$, and then find $A \cdot (U + W)$.
(b) Find AU and AW, and then find $AU + AW$.

(c) Show that $A \cdot (U + W)$ is the same as $AU + AW$.

[REMARK: This result proves in the two-dimensional case that if $\mathcal{T}$ is a linear transformation on a vector space, then $\mathcal{T}(U + W) = \mathcal{T}(U) + \mathcal{T}(W)$ for every vector U and every vector W. The same method establishes the result in the case of any other dimension and thus proves Theorem 1a.]

5. Let $\mathcal{T}$ be the linear transformation on a three-dimensional column vector space given by the matrix

$$M = \begin{bmatrix} 1 & 2 & -1 \\ 3 & -1 & -4 \\ 1 & 16 & 1 \end{bmatrix}.$$

Let

$$X = \begin{bmatrix} 1 \\ 0 \\ 0 \end{bmatrix}, \quad Y = \begin{bmatrix} 0 \\ 1 \\ 0 \end{bmatrix}, \quad Z = \begin{bmatrix} 0 \\ 0 \\ 1 \end{bmatrix}.$$

(a) Find $\mathcal{T}(X)$. Which column of the matrix M is the same as $\mathcal{T}(X)$?
(b) Find $\mathcal{T}(Y)$. Which column of the matrix M is $\mathcal{T}(Y)$?
(c) Find $\mathcal{T}(Z)$. Which column of M is $\mathcal{T}(Z)$?
(d) Find $X + Y$, and then find $\mathcal{T}(X + Y)$.
(e) Is $\mathcal{T}(X + Y)$ the sum of two column matrices chosen from the columns of M? Explain fully.

(f) Explain why the image of the column vector $\begin{bmatrix} 0 \\ 1 \\ -2 \end{bmatrix}$ under the linear transformation $\mathcal{T}$ is the difference between the second column of the matrix M and twice the third column of M.

(g) Explain why the image of the column vector $\begin{bmatrix} 1 \\ 1 \\ 1 \end{bmatrix}$ is the sum of the three columns of the matrix M.

6. Let $a, b, c, d, e, f, g, h, i$ be numbers, and let $\mathcal{T}$ be the linear transformation on a three-dimensional column vector space given by the matrix

$$N = \begin{bmatrix} a & b & c \\ d & e & f \\ g & h & i \end{bmatrix}.$$

Let

$$X = \begin{bmatrix} 1 \\ 0 \\ 0 \end{bmatrix}, \quad Y = \begin{bmatrix} 0 \\ 1 \\ 0 \end{bmatrix}, \quad Z = \begin{bmatrix} 0 \\ 0 \\ 1 \end{bmatrix}.$$

(a) Find $\mathcal{T}(X)$. Which column of the matrix N is the same as $\mathcal{T}(X)$?
(b) Find $\mathcal{T}(Y)$. Which column of the matrix N is $\mathcal{T}(Y)$?
(c) Find $\mathcal{T}(Z)$. Which column of N is $\mathcal{T}(Z)$?

3 A vector whose image is given

Example 3 Using the linear transformation given in Example 2, we may ask whether there is a vector

$$\begin{bmatrix} x \\ y \end{bmatrix} \text{ whose image is } \begin{bmatrix} 4 \\ 7 \end{bmatrix}.$$

If so, we may ask what the vector is (or, in case there are several such vectors, what they are). Since the image of

$$\begin{bmatrix} x \\ y \end{bmatrix} \text{ is } \begin{bmatrix} 5x + 2y \\ 2x + y \end{bmatrix},$$

the questions are essentially the problem of solving the equation

$$A \cdot X = \begin{bmatrix} 4 \\ 7 \end{bmatrix}.$$

We proceed to solve in the usual manner.

$$\begin{bmatrix} 5 & 2 & | & 4 \\ 2 & 1 & | & 7 \end{bmatrix},$$

$$\begin{bmatrix} 1 & \frac{2}{5} & | & \frac{4}{5} \\ 0 & \frac{1}{5} & | & \frac{27}{5} \end{bmatrix},$$

$$\begin{bmatrix} 1 & 0 & | & -10 \\ 0 & 1 & | & 27 \end{bmatrix}.$$

There is only one vector whose image is $\begin{bmatrix} 4 \\ 7 \end{bmatrix}$; it is $\begin{bmatrix} -10 \\ 27 \end{bmatrix}$.

In general, suppose we have a linear transformation $\mathfrak{T}$ on a vector space and a vector W in the space. There may, or may not, be a vector whose image is W. In some cases there may be exactly one such vector, while in other cases there may be infinitely many. The problem of finding every vector whose image is W is the same as the problem of solving the equation $A \cdot U = W$, where A is the matrix that determines the linear transformation and U is the unknown vector we are seeking. All of our work in Chapter 2 concerning the solution of a system of linear equations or concerning the solution of a matrix equation may be rephrased in the language of linear transformations.

EXERCISES B

1. Let $\mathcal{T}$ be the linear transformation given by the formula
$$\mathcal{T}(U) = \begin{bmatrix} 2 & -1 \\ -5 & 3 \end{bmatrix} U.$$

   (a) Find the vector V such that $\mathcal{T}(V) = \begin{bmatrix} 2 \\ 3 \end{bmatrix}$.

   (b) Find X, where $\mathcal{T}(X) = \begin{bmatrix} 1 \\ 0 \end{bmatrix}$.

   (c) Solve the following equation for Y: $\mathcal{T}(Y) = \begin{bmatrix} 0 \\ 1 \end{bmatrix}$.

   (d) Suppose that a and b are known numbers. Find the root Z of the equation $\mathcal{T}(Z) = \begin{bmatrix} a \\ b \end{bmatrix}$.

   (e) Verify that your answer in part (d) reduces to your answer in part (a) in case $a = 2$ and $b = 3$.

   (f) Verify that your root Z in part (d) becomes your root X in part (b) in case $a = 1$ and $b = 0$.

   (g) Express your root Z in part (d) as a product $M \begin{bmatrix} a \\ b \end{bmatrix}$, where M is a matrix whose elements are numbers.

2. Let $\mathcal{T}$ be the linear transformation given by the matrix
$$\begin{bmatrix} 3 & 5 & 3 \\ 5 & 3 & 2 \\ -1 & 2 & 1 \end{bmatrix}.$$

   (a) Find the vector X such that
$$\mathcal{T}(X) = \begin{bmatrix} 1 \\ 0 \\ 0 \end{bmatrix}.$$

   (b) Find the vector Y whose image under $\mathcal{T}$ is $\begin{bmatrix} 0 \\ 1 \\ 0 \end{bmatrix}$.

   (c) Find the root Z of the equation
$$\mathcal{T}(Z) = \begin{bmatrix} 0 \\ 0 \\ 1 \end{bmatrix}.$$

   (d) Suppose that a, b, c are known numbers. Find the vector W such that
$$\mathcal{T}(W) = \begin{bmatrix} a \\ b \\ c \end{bmatrix}.$$

3 / A vector whose image is given

(e) Verify that your answer in part (d) reduces to your answer in part (b) in case $a = 0$, $b = 1$, $c = 0$.

(f) Express your vector W as a product of the type $M \cdot \begin{bmatrix} a \\ b \\ c \end{bmatrix}$, where M is a matrix to be recognized from your answer in part (d).

(g) Compare the first column of the matrix M obtained in part (f) and the column matrix X obtained in part (a).

(h) Compare the root Z obtained in part (c) and the third column of the matrix M obtained in part (f).

3. Repeat Exercise 2 with the following modification: the linear transformation $\mathfrak{T}$ is given by the matrix

$$\begin{bmatrix} 1 & -2 & 4 \\ -1 & 4 & -3 \\ -2 & 4 & -7 \end{bmatrix}.$$

4. Let A be a 4×4 matrix; let $\mathfrak{S}$ be the linear transformation given by $\mathfrak{S}(U) = AU$; let d, e, f, g be numbers; let W, X, Y, Z be vectors such that

$$\mathfrak{S}(W) = \begin{bmatrix} 1 \\ 0 \\ 0 \\ 0 \end{bmatrix}, \quad \mathfrak{S}(X) = \begin{bmatrix} 0 \\ 1 \\ 0 \\ 0 \end{bmatrix}, \quad \mathfrak{S}(Y) = \begin{bmatrix} 0 \\ 0 \\ 1 \\ 0 \end{bmatrix}, \quad \mathfrak{S}(Z) = \begin{bmatrix} 0 \\ 0 \\ 0 \\ 1 \end{bmatrix}.$$

Explain why

$$\mathfrak{S}(dW + eX + fY + gZ) = \begin{bmatrix} d \\ e \\ f \\ g \end{bmatrix}.$$

5. Each of parts (a), (b), (c) in Exercise 2 (or in Exercise 3) may be solved by the Gauss elimination method. The solution of each part involves a sequence of augmented matrices. Discuss the marked resemblance that the successive augmented matrices in each part of the exercise have with the corresponding augmented matrices in every other part. Note in particular the entries on the left of the separation marks in the respective augmented matrices, and explain the observed phenomenon.

4 Composite of linear transformations

Suppose that $\mathfrak{T}$ and $\mathfrak{S}$ are two linear transformations on the same vector space. Let $\mathfrak{T}$ be given by the matrix A, and let E be the matrix determining the linear transformation $\mathfrak{S}$. Thus, for each vector U in the space, $\mathfrak{T}(U) = A \cdot U$; likewise, for every vector V, $\mathfrak{S}(V)$ is the vector $E \cdot V$.

If we are given a vector U, we may apply the linear transformation $\mathfrak{T}$ to U and obtain the vector $\mathfrak{T}(U)$. This vector $\mathfrak{T}(U)$ has an image under the linear transformation $\mathfrak{S}$, namely, $\mathfrak{S}(\mathfrak{T}(U))$. If we assign the vector $\mathfrak{S}(\mathfrak{T}(U))$ to the original vector U, we have described a new linear transformation on the vector space. We denote this new linear transformation by $\mathfrak{S}\mathfrak{T}$. Note that, in order to find the image of a vector U under $\mathfrak{S}\mathfrak{T}$, we first apply $\mathfrak{T}$ to U and afterward apply $\mathfrak{S}$ to $\mathfrak{T}(U)$. Thus, $(\mathfrak{S}\mathfrak{T})(U) = \mathfrak{S}(\mathfrak{T}(U))$.

Definition Let $\mathfrak{T}$ and $\mathfrak{S}$ be linear transformations on the same vector space. The linear transformation $\mathfrak{S}\mathfrak{T}$ that assigns to each vector U in the space the image $\mathfrak{S}(\mathfrak{T}(U))$ may be called the *composite* (or sometimes the *product*) of the given linear transformations.

The composite linear transformation $\mathfrak{S}\mathfrak{T}$ has, of course, a matrix associated with it. What is this matrix? How is it related to the matrices E and A, which, respectively, determine the given linear transformations $\mathfrak{S}$ and $\mathfrak{T}$? The answer is that the matrix associated with the composite $\mathfrak{S}\mathfrak{T}$ is the product EA of the matrices E and A. We proceed to show this for the two-dimensional case. An analogous discussion establishes the result for other dimensions (in which case it is highly desirable to select a suitable notation).

Theorem 2 Let $\mathfrak{T}$ and $\mathfrak{S}$ be linear transformations on a two-dimensional vector space. Then the matrix associated with the composite $\mathfrak{S}\mathfrak{T}$ is the product of the matrix associated with $\mathfrak{S}$ and the matrix associated with $\mathfrak{T}$.

PROOF Suppose that the linear transformations $\mathfrak{T}$ and $\mathfrak{S}$ are given by the respective matrices

$$A = \begin{bmatrix} a & b \\ c & d \end{bmatrix} \quad \text{and} \quad E = \begin{bmatrix} p & q \\ r & s \end{bmatrix}.$$

We wish to show that the image of any vector

$$U = \begin{bmatrix} x \\ y \end{bmatrix}$$

under the linear transformation $\mathfrak{S}\mathfrak{T}$ is the product of the matrix EA by the column matrix U. To find the image of U under $\mathfrak{S}\mathfrak{T}$, we first find

$$\mathfrak{T}(U) = \begin{bmatrix} a & b \\ c & d \end{bmatrix} \begin{bmatrix} x \\ y \end{bmatrix} = \begin{bmatrix} ax + by \\ cx + dy \end{bmatrix}$$

and then we find the image of $\mathfrak{T}(U)$ under $\mathfrak{S}$, namely,

$$S(\mathcal{T}(U)) = E \cdot \begin{bmatrix} ax + by \\ cx + dy \end{bmatrix} = \begin{bmatrix} p & q \\ r & s \end{bmatrix} \begin{bmatrix} ax + by \\ cx + dy \end{bmatrix}$$

$$= \begin{bmatrix} pax + pby + qcx + qdy \\ rax + rby + scx + sdy \end{bmatrix}.$$

The latter matrix is $S\mathcal{T}(U)$. On the other hand, EA is the matrix

$$\begin{bmatrix} p & q \\ r & s \end{bmatrix} \begin{bmatrix} a & b \\ c & d \end{bmatrix} = \begin{bmatrix} pa + qc & pb + qd \\ ra + sc & rb + sd \end{bmatrix};$$

the product of EA by $\begin{bmatrix} x \\ y \end{bmatrix}$ is

$$EA \cdot U = \begin{bmatrix} pa + qc & pb + qd \\ ra + sc & rb + sd \end{bmatrix} \begin{bmatrix} x \\ y \end{bmatrix} = \begin{bmatrix} pax + qcx + pby + qdy \\ rax + scx + rby + sdy \end{bmatrix}.$$

A comparison of the two vectors $S\mathcal{T}(U)$ and $EA \cdot U$ shows that they are the same—the expanded forms are identical except for arrangement of terms in the sums. Thus, the matrix product EA gives the composite linear transformation $S\mathcal{T}$.

The preceding discussion provides a clue explaining one of the principal reasons why the operation of multiplying matrices is defined as it is (see Section 8 of Chapter 2). The operation is chosen in order that a composite of linear transformations should be associated with the product of their respective associated matrices. The only way in which we may achieve this tie-in connecting the two operations (the combination of linear transformations and the combination of matrices) is to define matrix multiplication as we have done.

Example 4 Let $\mathcal{T}$ be the linear transformation determined by the matrix

$$A = \begin{bmatrix} 4 & -1 \\ -3 & 2 \end{bmatrix},$$

let S be the linear transformation given by the matrix

$$E = \begin{bmatrix} 3 & 5 \\ 0 & -1 \end{bmatrix},$$

and let U be the vector $\begin{bmatrix} -1 \\ 3 \end{bmatrix}$. Then,

$$\mathcal{T}(U) = A \cdot U = \begin{bmatrix} 4 & -1 \\ -3 & 2 \end{bmatrix} \begin{bmatrix} -1 \\ 3 \end{bmatrix} = \begin{bmatrix} -7 \\ 9 \end{bmatrix}.$$

Linear Transformations

and
$$S(T(U)) = E \cdot \begin{bmatrix} -7 \\ 9 \end{bmatrix} = \begin{bmatrix} 3 & 5 \\ 0 & -1 \end{bmatrix} \begin{bmatrix} -7 \\ 9 \end{bmatrix} = \begin{bmatrix} 24 \\ -9 \end{bmatrix}.$$

Also,
$$EA = \begin{bmatrix} 3 & 5 \\ 0 & -1 \end{bmatrix} \begin{bmatrix} 4 & -1 \\ -3 & 2 \end{bmatrix} = \begin{bmatrix} -3 & 7 \\ 3 & -2 \end{bmatrix}$$

and
$$EA \cdot U = \begin{bmatrix} -3 & 7 \\ 3 & -2 \end{bmatrix} \begin{bmatrix} -1 \\ 3 \end{bmatrix} = \begin{bmatrix} 24 \\ -9 \end{bmatrix}.$$

We observe that the same vector $\begin{bmatrix} 24 \\ -9 \end{bmatrix}$ is obtained by either computational route; it is $E(AU)$ and it is also $(EA)U$.

The observation that the product of E and AU is the same as the product of EA and U is an illustration of a general principle. Multiplication of matrices has the *associative* property. If B, C, D are matrices such that the products $(BC)D$ and $B(CD)$ are defined, then the two products are the same—and we commonly write simply BCD.

By contrast, we have already noted (in Section 8 of Chapter 2) that, if B and C are matrices such that the products BC and CB are defined, then BC and CB may be different. This suggests that, if T and S are linear transformations on a vector space, then the composite TS may be a different linear transformation from the composite ST.

Example 5 If S, T, U have the same meanings as in Example 4, then

$$S(U) = E \cdot U = \begin{bmatrix} 3 & 5 \\ 0 & -1 \end{bmatrix} \begin{bmatrix} -1 \\ 3 \end{bmatrix} = \begin{bmatrix} 12 \\ -3 \end{bmatrix}$$

and

$$T(S(U)) = A \cdot \begin{bmatrix} 12 \\ -3 \end{bmatrix} = \begin{bmatrix} 4 & -1 \\ -3 & 2 \end{bmatrix} \begin{bmatrix} 12 \\ -3 \end{bmatrix} = \begin{bmatrix} 51 \\ -42 \end{bmatrix}.$$

Note that $TS(U)$ is not the same as $ST(U)$. Also,

$$AE = \begin{bmatrix} 4 & -1 \\ -3 & 2 \end{bmatrix} \begin{bmatrix} 3 & 5 \\ 0 & -1 \end{bmatrix} = \begin{bmatrix} 12 & 21 \\ -9 & -17 \end{bmatrix},$$

and AE is not the same as EA. However,

$$(AE) \cdot U = \begin{bmatrix} 12 & 21 \\ -9 & -17 \end{bmatrix} \begin{bmatrix} -1 \\ 3 \end{bmatrix} = \begin{bmatrix} 51 \\ -42 \end{bmatrix},$$

4 / Composite of linear transformations

which is the same as $\mathcal{TS}(U)$, giving us another illustration verifying Theorem 2.

EXERCISES C

1. Let the linear transformations $\mathcal{T}$ and $\mathcal{S}$ on the two-dimensional vector space be given, respectively, by

$$\mathcal{T}(U) = \begin{bmatrix} 3 & -1 \\ -5 & 2 \end{bmatrix} U \quad \text{and} \quad \mathcal{S}(V) = \begin{bmatrix} 2 & 1 \\ 5 & 3 \end{bmatrix} V.$$

   (a) Find the image of the vector $\begin{bmatrix} 8 \\ -7 \end{bmatrix}$ under the transformation $\mathcal{S}$.
   (b) Find the image under $\mathcal{T}$ of the vector obtained in part (a).
   (c) Find the image under $\mathcal{ST}$ of the vector $\begin{bmatrix} -7 \\ 8 \end{bmatrix}$.

2. Let $\mathcal{T}$ and $\mathcal{S}$ be linear transformations given as follows: for every vector $\begin{bmatrix} a \\ b \end{bmatrix}$, its image under $\mathcal{T}$ is $\begin{bmatrix} 2a \\ a \end{bmatrix}$ and its image under $\mathcal{S}$ is $\begin{bmatrix} a \\ b \end{bmatrix}$.
   (a) What matrix is associated with the linear transformation $\mathcal{T}$?
   (b) What matrix is associated with $\mathcal{S}$?
   (c) What matrix is associated with the composite $\mathcal{ST}$?
   (d) What matrix is associated with the composite $\mathcal{TT}$?

3. Let $\mathcal{T}$ and $\mathcal{S}$ be linear transformations on a three-dimensional vector space given by the respective matrices

$$\begin{bmatrix} 2 & 0 & -1 \\ 1 & 3 & -2 \\ -3 & 1 & 2 \end{bmatrix} \quad \text{and} \quad \begin{bmatrix} 3 & -1 & -2 \\ 0 & 1 & 0 \\ -2 & 0 & 1 \end{bmatrix}.$$

   Let

$$U = \begin{bmatrix} 4 \\ 1 \\ 7 \end{bmatrix} \quad \text{and} \quad W = \begin{bmatrix} 3 \\ -1 \\ 5 \end{bmatrix}.$$

   (a) Find $\mathcal{T}(U)$, and then find $\mathcal{S}(\mathcal{T}(U))$.
   (b) Find $\mathcal{S}(U)$, and then find $\mathcal{T}(\mathcal{S}(U))$.
   (c) Find $\mathcal{S}(\mathcal{T}(W)) - \mathcal{T}(\mathcal{S}(W))$.
   (d) Find $\mathcal{S}(\mathcal{S}(U)) - \mathcal{T}(\mathcal{T}(W))$.
   (e) Find the matrix associated with the composite linear transformation $\mathcal{TT}$.
   (f) Calculate the matrix associated with the composite $\mathcal{SS}$.

4. Find a matrix M (if there is any) such that

$$\begin{bmatrix} 3 & -1 \\ 2 & 1 \end{bmatrix} M = \begin{bmatrix} 5 & -6 \\ 5 & 1 \end{bmatrix}.$$

5. Find a matrix N (if there is any) such that

$$N \begin{bmatrix} 3 & -1 \\ 2 & 1 \end{bmatrix} = \begin{bmatrix} 5 & -6 \\ 5 & 1 \end{bmatrix}.$$

6. Consider the 2×2 identity matrix

$$I = \begin{bmatrix} 1 & 0 \\ 0 & 1 \end{bmatrix}$$

and a two-dimensional vector

$$W = \begin{bmatrix} a \\ b \end{bmatrix}.$$

Show that W is the image of itself under the linear transformation associated with the matrix I.

7. Let I be the 3×3 identity matrix

$$\begin{bmatrix} 1 & 0 & 0 \\ 0 & 1 & 0 \\ 0 & 0 & 1 \end{bmatrix}$$

and let W be a three-dimensional vector $\begin{bmatrix} a \\ b \\ c \end{bmatrix}$. Find the image of W under the linear transformation associated with I.

5 Inverse of a linear transformation

Definition The linear transformation on a vector space that assigns to each vector U in the space the vector U itself is called the *identity transformation*.

Theorem 3 Let n be a natural number, and let I be the $n \times n$ identity matrix. On an n-dimensional vector space, the linear transformation associated with the matrix I is the identity transformation.

PROOF By straightforward calculation, the product of I and an n-dimensional column vector

$$U = \begin{bmatrix} a_1 \\ a_2 \\ \vdots \\ a_n \end{bmatrix}$$

is

$$I \cdot U = \begin{bmatrix} 1 & 0 & \cdots & 0 \\ 0 & 1 & \cdots & 0 \\ & & \cdots & \\ 0 & 0 & \cdots & 1 \end{bmatrix} \begin{bmatrix} a_1 \\ a_2 \\ \vdots \\ a_n \end{bmatrix} = \begin{bmatrix} a_1 \\ a_2 \\ \vdots \\ a_n \end{bmatrix},$$

namely, the vector U itself.

We are now ready to link together the solution of equations (Section 3) and the composite of linear transformations (Section 4).

Example 6 Consider the linear transformation $\mathfrak{T}$ that assigns to each vector

$$\begin{bmatrix} x \\ y \end{bmatrix} \quad \text{the image} \quad \begin{bmatrix} -x - y \\ 3x + 2y \end{bmatrix}.$$

The matrix A associated with $\mathfrak{T}$ is

$$\begin{bmatrix} -1 & -1 \\ 3 & 2 \end{bmatrix}.$$

Suppose that a and b are known numbers and that C is the vector $\begin{bmatrix} a \\ b \end{bmatrix}$.

Does the equation $\mathfrak{T}(V) = C$ have a root? If so, find each root. This problem is a restatement of the question, "What vector, or vectors, if any, have C as image under the linear transformation $\mathfrak{T}$?"

To solve the equation

$$A \cdot \begin{bmatrix} x \\ y \end{bmatrix} = \begin{bmatrix} a \\ b \end{bmatrix},$$

we use the following familiar method.

$$\left[\begin{array}{cc|c} -1 & -1 & a \\ 3 & 2 & b \end{array}\right],$$

$$\left[\begin{array}{cc|c} 1 & 1 & -a \\ 0 & -1 & 3a + b \end{array}\right],$$

$$\left[\begin{array}{cc|c} 1 & 0 & 2a + b \\ 0 & 1 & -3a - b \end{array}\right].$$

There is only one vector whose image is C; it is

$$\begin{bmatrix} 2a + b \\ -3a - b \end{bmatrix}.$$

We express this vector as a matrix product:
$$\begin{bmatrix} 2a+b \\ -3a-b \end{bmatrix} = \begin{bmatrix} 2 & 1 \\ -3 & -1 \end{bmatrix} \begin{bmatrix} a \\ b \end{bmatrix}.$$

If we denote the matrix $\begin{bmatrix} 2 & 1 \\ -3 & -1 \end{bmatrix}$ by B, then the root of the equation $AV = C$ is BC.

By direct computation we observe that
$$AB = \begin{bmatrix} -1 & -1 \\ 3 & 2 \end{bmatrix} \begin{bmatrix} 2 & 1 \\ -3 & -1 \end{bmatrix} = \begin{bmatrix} 1 & 0 \\ 0 & 1 \end{bmatrix}$$
and that
$$BA = \begin{bmatrix} 2 & 1 \\ -3 & -1 \end{bmatrix} \begin{bmatrix} -1 & -1 \\ 3 & 2 \end{bmatrix} = \begin{bmatrix} 1 & 0 \\ 0 & 1 \end{bmatrix}.$$

Thus, $AB = I = BA$, where I is the identity matrix. If $\mathcal{R}$ is the linear transformation determined by the matrix B, then each of the composites $\mathcal{TR}$ and $\mathcal{RT}$ is the identity transformation that maps every vector in the space onto itself.

We have already observed that if $\mathcal{T}(V) = C$, then $V = BC = \mathcal{R}(C)$. We now show the converse. Suppose that $V = \mathcal{R}(C)$. The image of $\begin{bmatrix} a \\ b \end{bmatrix}$ under $\mathcal{R}$ is

$$V = B \cdot \begin{bmatrix} a \\ b \end{bmatrix} = \begin{bmatrix} 2 & 1 \\ -3 & -1 \end{bmatrix} \begin{bmatrix} a \\ b \end{bmatrix} = \begin{bmatrix} 2a+b \\ -3a-b \end{bmatrix},$$

and hence
$$\mathcal{T}(V) = A \cdot V = \begin{bmatrix} -1 & -1 \\ 3 & 2 \end{bmatrix} \begin{bmatrix} 2a+b \\ -3a-b \end{bmatrix} = \begin{bmatrix} a \\ b \end{bmatrix} = C.$$

Thus, the conditions that $\mathcal{T}(V) = C$ and that $\mathcal{R}(C) = V$ are equivalent.

Example 7 Let A be the matrix
$$\begin{bmatrix} 1 & 1 & 3 \\ 2 & 3 & 8 \\ 3 & 1 & 5 \end{bmatrix},$$

and let $\mathcal{T}$ be the linear transformation determined by A. Suppose that a, b, c are known numbers and that D is the vector
$$\begin{bmatrix} a \\ b \\ c \end{bmatrix}.$$

Does the equation $\mathcal{T}(V) = D$ have a root? If so, find each root. We use the elimination method:

$$\begin{bmatrix} 1 & 1 & 3 & \vdots & a \\ 2 & 3 & 8 & \vdots & b \\ 3 & 1 & 5 & \vdots & c \end{bmatrix},$$

$$\begin{bmatrix} 1 & 1 & 3 & \vdots & a \\ 0 & 1 & 2 & \vdots & b - 2a \\ 0 & -2 & -4 & \vdots & c - 3a \end{bmatrix},$$

$$\begin{bmatrix} 1 & 0 & 1 & \vdots & 3a - b \\ 0 & 1 & 2 & \vdots & b - 2a \\ 0 & 0 & 0 & \vdots & c + 2b - 7a \end{bmatrix}.$$

In case $c + 2b - 7a$ is not zero, then there is no root. Thus, in general, the vector

$$\begin{bmatrix} a \\ b \\ c \end{bmatrix}$$

is not an image under the given linear transformation $\mathcal{T}$. However, if $c = 7a - 2b$, then the third row in the last augmented matrix is a zero row. In this case, the equation $\mathcal{T}(V) = D$ has many roots. In fact, for any number p, the vector

$$p \begin{bmatrix} -1 \\ -2 \\ 1 \end{bmatrix} + \begin{bmatrix} 3a - b \\ -2a + b \\ 0 \end{bmatrix}$$

is a root. Since p is arbitrary, there are infinitely many roots.

The preceding two examples suggest the following development. Let $\mathcal{T}$ be a linear transformation on a vector space. It is possible (depending on the chosen $\mathcal{T}$) that there is in the space a vector that is not an image under $\mathcal{T}$. That is, there may be a vector W such that the equation $\mathcal{T}(V) = W$ has no root V.

On the other hand, it is possible that every vector in the space is an image under $\mathcal{T}$. Furthermore, it may be that each vector appears as an image just once. In other words, it may be the case that, given any vector U in the space, the equation $\mathcal{T}(V) = U$ has exactly one root V. If so, we may associate the root V with the given vector U. This new method of assignment, which matches a vector with each vector in the space, is a linear transformation. This (new) linear transformation (call it $\mathcal{R}$) is a sort of reversal of the original

Figure 1

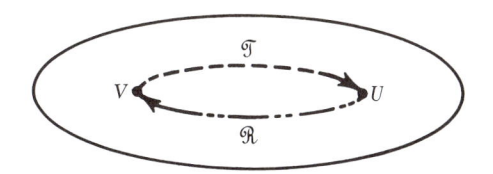

linear transformation $\mathcal{T}$. (See Fig. 1.) For if $\mathcal{T}$ associates U with V, then $\mathcal{R}$ associates V with U. If $\mathcal{T}(V) = U$, then $\mathcal{R}(U) = V$.

Definition Let $\mathcal{T}$ be a linear transformation with the property that every vector in the vector space is the image under $\mathcal{T}$ of exactly one vector. Then the *inverse of the linear transformation* $\mathcal{T}$ is the linear transformation $\mathcal{R}$ given by the condition that, for each vector U, $\mathcal{R}(U) = V$ if $\mathcal{T}(V) = U$.

In Example 7 we discussed a linear transformation that does not have an inverse. The linear transformation in Example 6 does have an inverse.

Let $\mathcal{T}$ be a linear transformation that has an inverse $\mathcal{R}$. Consider any vector V in the space. If $\mathcal{T}(V) = U$, then $\mathcal{R}(U) = V$. Hence,

$$\mathcal{R}(\mathcal{T}(V)) = \mathcal{R}(U) = V.$$

That is, the composite linear transformation $\mathcal{R}\mathcal{T}$ maps the vector V onto itself. The composite $\mathcal{R}\mathcal{T}$ is the identity transformation that matches every vector with itself.

Now consider any vector U in the space. If $\mathcal{R}(U) = V$, then $\mathcal{T}(V) = U$. Consequently,

$$\mathcal{T}(\mathcal{R}(U)) = \mathcal{T}(V) = U.$$

The composite linear transformation $\mathcal{T}\mathcal{R}$ is also the identity transformation. In particular, $\mathcal{R}\mathcal{T}$ and $\mathcal{T}\mathcal{R}$ are the same. (Recall that, generally speaking, two linear transformations $\mathcal{T}$ and $\mathcal{S}$ do not have the property that $\mathcal{S}\mathcal{T}$ and $\mathcal{T}\mathcal{S}$ are the same.)

EXERCISES D

1. Let $\mathcal{T}$ be the linear transformation given by the matrix $\begin{bmatrix} \frac{2}{3} & -\frac{5}{3} \\ -\frac{1}{3} & \frac{4}{3} \end{bmatrix}$.

    (a) If a and b are numbers, find the vector V whose image under $\mathcal{T}$ is $\begin{bmatrix} a \\ b \end{bmatrix}$.

    (b) Express V as the product of a 2×2 matrix and the matrix $\begin{bmatrix} a \\ b \end{bmatrix}$.

    (c) What matrix is associated with the inverse $\mathcal{R}$ of the linear transformation $\mathcal{T}$?
    (d) Calculate the product of the matrix associated with $\mathcal{T}$ and the matrix associated with $\mathcal{R}$.
    (e) Calculate the product of the matrix associated with $\mathcal{R}$ and the matrix associated with $\mathcal{T}$.

5 / Inverse of a linear transformation

2. Let the linear transformation $\mathcal{T}$ be given by the condition that if

$$U = \begin{bmatrix} x \\ y \\ z \end{bmatrix}, \quad \text{then} \quad \mathcal{T}(U) = \begin{bmatrix} x+y \\ x+y+z \\ y+z \end{bmatrix}.$$

(a) What matrix is associated with $\mathcal{T}$?

(b) Find the vector whose image under $\mathcal{T}$ is $\begin{bmatrix} a \\ b \\ c \end{bmatrix}$.

(c) Express the vector obtained in part (b) as the product of a 3×3 matrix and the matrix $\begin{bmatrix} a \\ b \\ c \end{bmatrix}$.

(d) What matrix is associated with the inverse of $\mathcal{T}$?

(e) Calculate the product of the matrices in parts (d) and (a).

3. Suppose that the matrix

$$A = \begin{bmatrix} -1 & 1 & 0 \\ 1 & -1 & 1 \\ 0 & 1 & -1 \end{bmatrix}$$

is associated with the linear transformation $\mathcal{T}$.

(a) Find the matrix B associated with the inverse $\mathcal{R}$ of the linear transformation $\mathcal{T}$.

(b) Find the matrix associated with the composite $\mathcal{T}\mathcal{T}$.

(c) Find the matrix associated with the composite $\mathcal{R}\mathcal{R}$.

(d) Find the matrix associated with the composite $\mathcal{R}\mathcal{T}$.

(e) Calculate the product of the matrices in parts (b) and (c).

4. Let the linear transformation $\mathcal{S}$ be given by

$$\mathcal{S}(V) = \begin{bmatrix} 2 & -1 & -3 \\ 6 & 4 & -2 \\ 8 & 17 & 9 \end{bmatrix} V.$$

(a) Show that the vector $\begin{bmatrix} 3 \\ 1 \\ 5 \end{bmatrix}$ is not an image under $\mathcal{S}$.

(b) Show that the vector $\begin{bmatrix} 1 \\ 2 \\ 1 \end{bmatrix}$ is an image under $\mathcal{S}$.

(c) Find all the vectors U such that $\mathcal{S}(U) = \begin{bmatrix} 1 \\ 2 \\ 1 \end{bmatrix}$.

(d) Show that the vector $\begin{bmatrix} a \\ b \\ c \end{bmatrix}$ is not an image under $\mathcal{S}$ unless $c = 3b - 5a$.

(e) In case $c = 3b - 5a$, show that the graph of the set of roots of the equation

$$\mathcal{S}(U) = \begin{bmatrix} a \\ b \\ c \end{bmatrix}$$

is a line on a three-dimensional coordinate system.

5. Let the linear transformation $\mathcal{T}$ be given by

$$\mathcal{T}(V) = \begin{bmatrix} 3 & 5 & -2 & 4 \\ -2 & -20 & 13 & -1 \\ 2 & 0 & 1 & 3 \\ 0 & 10 & -7 & -1 \end{bmatrix} V.$$

(a) Show that the vector $\begin{bmatrix} a \\ b \\ c \\ d \end{bmatrix}$ is not an image under $\mathcal{T}$ unless both $b = 5c - 4a$ and $d = 2a - 3c$.

(b) In case both $b = 5c - 4a$ and $d = 2a - 3c$, find the set of all vectors U such that

$$\mathcal{T}(U) = \begin{bmatrix} a \\ b \\ c \\ d \end{bmatrix}.$$

(c) Show that the set of roots U in part (b) may be expressed as the set of all products

$$\begin{bmatrix} 3 & -4 & -15 & 10 \\ 0 & 0 & 1 & 0 \\ 0 & 0 & 0 & 1 \\ -2 & 3 & 10 & -7 \end{bmatrix} \begin{bmatrix} a \\ c \\ p \\ q \end{bmatrix},$$

where p and q are arbitrary real numbers.

6 Inverse of a matrix

Suppose that $\mathcal{T}$ is a linear transformation that has an inverse $\mathcal{R}$. Let A be the matrix that determines $\mathcal{T}$. There is, of course, a matrix, say B, that is associated with the linear transformation $\mathcal{R}$. We wish to investigate the relationships between A and B.

The matrix that is associated with the identity transformation is the identity matrix I. From Section 4, we know that the product of A and B is the matrix that determines the composite of the linear transformations $\mathcal{T}$ and $\mathcal{R}$. Hence,

$A \cdot B = I$. Likewise, the product of B and A is associated with the composite $\mathcal{RJ}$. Therefore, $BA = I$.

We find that $AB = I$ and $BA = I$.

Definition Let A be an $n \times n$ matrix, and let I be the $n \times n$ identity matrix. If there is a matrix B such that $AB = I$ and $BA = I$, then the matrix B is called the *inverse of the matrix* A.

The relationship is, in fact, mutual: if B is the inverse of A, then A is the inverse of B.

Suppose that a given matrix A has an inverse B. If W is a vector in the space, then the root of the equation $AU = W$ is BW. This assertion is easily verified:

$$A(BW) = (AB)W = IW = W.$$

In particular, suppose that W is a column of the identity matrix I, say the kth. In the kth column

$$\begin{bmatrix} 0 \\ \vdots \\ 0 \\ 1 \\ 0 \\ \vdots \\ 0 \end{bmatrix}$$

of I, every element is 0 except the kth element, which is 1. In this case, the root of the equation $AU = W$, namely,

$$B \cdot \begin{bmatrix} 0 \\ \vdots \\ 1 \\ \vdots \\ 0 \end{bmatrix},$$

is[1] the kth column of the matrix B. This remark suggests a procedure for calculating the inverse of a matrix. We use an adaptation of the complete Gauss elimination method, in which we attempt to solve simultaneously the equations of the type $AU = W$, for every column W of I.

---

[1] Compare Exercise A-6.

To calculate[2] the inverse of an $n \times n$ matrix A, form the matrix $[A \mid I]$ with n rows and $2n$ columns. In the usual manner of the elimination method, manipulate the rows in an attempt to convert the matrix $[A \mid I]$ into a matrix with I to the left of the separation mark. When this goal has been achieved, the inverse of A will appear on the right of the separation: thus, the final matrix in the process is $[I \mid B]$, where B is the inverse of A.

Example 8 Find the inverse of the matrix $\begin{bmatrix} 3 & 7 \\ 1 & 2 \end{bmatrix}$.

We use the method outlined above.

$$\begin{bmatrix} 3 & 7 & \mid & 1 & 0 \\ 1 & 2 & \mid & 0 & 1 \end{bmatrix},$$

$$\begin{bmatrix} 1 & \frac{7}{3} & \mid & \frac{1}{3} & 0 \\ 0 & -\frac{1}{3} & \mid & -\frac{1}{3} & 1 \end{bmatrix},$$

$$\begin{bmatrix} 1 & 0 & \mid & -2 & 7 \\ 0 & 1 & \mid & 1 & -3 \end{bmatrix}.$$

The matrix $\begin{bmatrix} -2 & 7 \\ 1 & -3 \end{bmatrix}$ is the inverse of the given matrix $\begin{bmatrix} 3 & 7 \\ 1 & 2 \end{bmatrix}$. The reader should verify that the product of the given matrix and its inverse (as claimed) is I.

Example 9 Find an inverse (if any) for the matrix

$$D = \begin{bmatrix} 2 & 3 & 1 \\ -3 & 1 & -7 \\ 1 & 5 & -3 \end{bmatrix},$$

We begin by forming the 3×6 matrix $[D \mid I]$.

$$\begin{bmatrix} 2 & 3 & 1 & \mid & 1 & 0 & 0 \\ -3 & 1 & -7 & \mid & 0 & 1 & 0 \\ 1 & 5 & -3 & \mid & 0 & 0 & 1 \end{bmatrix},$$

$$\begin{bmatrix} 1 & \frac{3}{2} & \frac{1}{2} & \mid & \frac{1}{2} & 0 & 0 \\ 0 & \frac{11}{2} & -\frac{11}{2} & \mid & \frac{3}{2} & 1 & 0 \\ 0 & \frac{7}{2} & -\frac{7}{2} & \mid & -\frac{1}{2} & 0 & 1 \end{bmatrix},$$

---

[2] Compare Exercise B-5.

$$\begin{bmatrix} 1 & 0 & 2 & \frac{1}{11} & -\frac{3}{11} & 0 \\ 0 & 1 & -1 & \frac{3}{11} & \frac{2}{11} & 0 \\ 0 & 0 & 0 & -\frac{16}{11} & -\frac{7}{11} & 1 \end{bmatrix}.$$

The process has stopped without achieving our goal. This means that the matrix D has no inverse. Indeed, the first four columns of the matrices in the successive steps of the procedure represent a search for a root of the equation

$$D \cdot U = \begin{bmatrix} 1 \\ 0 \\ 0 \end{bmatrix}.$$

The final augmented matrix reveals that this equation has no solution. In other words, the vector

$$\begin{bmatrix} 1 \\ 0 \\ 0 \end{bmatrix}$$

is not an image under the linear transformation $\mathcal{S}$ determined by D. So this linear transformation does not have any inverse.

We may also observe that

$$\begin{bmatrix} 0 \\ 1 \\ 0 \end{bmatrix}$$

is not an image under $\mathcal{S}$. The first, second, third, and fifth columns of the above matrices were the steps in the search for a root of the equation

$$D \cdot U = \begin{bmatrix} 0 \\ 1 \\ 0 \end{bmatrix}.$$

The appearance of the final row $[0 \ \ 0 \ \ 0 \ \vdots \ -\frac{7}{11}]$ shows that this equation has no root.

A similar examination of the first, second, third, and sixth columns reveals that no vector has

$$\begin{bmatrix} 0 \\ 0 \\ 1 \end{bmatrix}$$

as image under $\mathcal{S}$.

Linear Transformations

In summary, the Gauss elimination method is a satisfactory method for investigating the inverse (if any) for a given matrix. If an $n \times n$ matrix A is given, form the $n \times (2n)$ matrix $[A \mid I]$. Apply the method to $[A \mid I]$. If successive row maneuvers lead to $[I \mid B]$, then B is the desired inverse of the given matrix A. In case the process does not yield I to the left of the separation mark, then the given matrix has no inverse. Thus, the elimination method not only produces the desired inverse whenever there is an inverse, but also reveals (in an efficient manner) the lack of an inverse whenever no inverse exists.

EXERCISES E

1. Let $\mathfrak{T}$ be the linear transformation given by the matrix

$$A = \begin{bmatrix} 2 & 1 \\ -3 & -1 \end{bmatrix}.$$

   (a) Find the vector whose image under $\mathfrak{T}$ is $\begin{bmatrix} 1 \\ 0 \end{bmatrix}$.

   (b) Find the vector whose image under $\mathfrak{T}$ is $\begin{bmatrix} 0 \\ 1 \end{bmatrix}$.

   (c) Give the inverse of the matrix A.

2. Let a, b, c be nonzero numbers, and let $\mathfrak{S}$ be the linear transformation given by the matrix

$$M = \begin{bmatrix} a & b \\ 0 & c \end{bmatrix}.$$

   (a) Find the root of the equation $\mathfrak{S}(V) = \begin{bmatrix} 1 \\ 0 \end{bmatrix}$.

   (b) Find the root of the equation $\mathfrak{S}(V) = \begin{bmatrix} 0 \\ 1 \end{bmatrix}$.

   (c) Give the inverse of the matrix M.

3. Calculate the inverse of each of the following matrices.

   (a) $A = \begin{bmatrix} 11 & 4 \\ 7 & 3 \end{bmatrix}.$

   (b) $B = \begin{bmatrix} 9 & 1 \\ 1 & 9 \end{bmatrix}.$

   (c) $C = \begin{bmatrix} -2 & 1 & 0 \\ 1 & -2 & 1 \\ 0 & 1 & -2 \end{bmatrix}.$

   (d) $D = \begin{bmatrix} 8 & 0 & 0 \\ 0 & -7 & 0 \\ 0 & 0 & 3 \end{bmatrix}.$

   (e) $E = \begin{bmatrix} 0 & 1 & -2 \\ -1 & 1 & 3 \\ 2 & -3 & 0 \end{bmatrix}.$

   (f) $F = \begin{bmatrix} 1 & 2 & -1 & 3 \\ 2 & 5 & 0 & 7 \\ 1 & 2 & 0 & 2 \\ 2 & 3 & -4 & 6 \end{bmatrix}.$

4. Suppose that each of the numbers a, b, c, d, e, f, g, h is different from zero. Find the inverse of the matrix

$$\begin{bmatrix} a & e & 0 & h \\ 0 & b & f & 0 \\ 0 & 0 & c & g \\ 0 & 0 & 0 & d \end{bmatrix}.$$

5. Let M be the matrix

$$\begin{bmatrix} 1 & 2 & -1 \\ 3 & -1 & -4 \\ 1 & 16 & 1 \end{bmatrix},$$

and let S be the linear transformation given by $S(V) = MV$.

(a) Find a vector U (if there is any) such that $S(U) = \begin{bmatrix} 0 \\ 7 \\ -14 \end{bmatrix}$.

(b) Find a nonzero vector U (if there is any) such that $S(U) = \begin{bmatrix} 0 \\ 0 \\ 0 \end{bmatrix}$.

(c) Find a vector U (if there is any) such that $S(U) = \begin{bmatrix} 1 \\ 0 \\ 0 \end{bmatrix}$.

(d) Explain why the matrix M does not have an inverse.

6. Use the Gauss elimination method in order to show that each of the following matrices does not have an inverse.

(a) $A = \begin{bmatrix} 5 & -3 & 1 \\ 8 & 1 & 4 \\ 1 & 11 & 5 \end{bmatrix}$.

(b) $B = \begin{bmatrix} 20 & -15 \\ -28 & 21 \end{bmatrix}$.

(c) $C = \begin{bmatrix} 0 & 1 & 2 & 3 \\ -1 & 0 & 1 & 2 \\ -2 & -1 & 0 & 1 \\ -3 & -2 & -1 & 0 \end{bmatrix}$.

7. Suppose that the $n \times n$ matrix A has an inverse B and that the $n \times n$ matrix E has an inverse F.
(a) What is the inverse of the product matrix AA?
(b) Find the product $(AE)(FB)$.

(c) Find the product $(FB)(AE)$.
(d) What is the inverse of the product matrix AE?
(e) What is the inverse of the matrix EA?
(f) What is the inverse of the matrix EAE?

8. Suppose that the $n \times n$ matrix A has an inverse C and that M is an $n \times p$ matrix.
   (a) Show that the product CM is a root of the equation $AX = M$. [HINT: Simplify $A(CM)$.]
   (b) Show that every root of the equation $AX = M$ is equal to CM.

9. Suppose that the $n \times n$ matrix A has an inverse C. Does C have an inverse; if so, what is it?

10. If a, b, c, d are nonzero numbers, find the inverse of the matrix
$$\begin{bmatrix} a & 0 & 0 & 0 \\ 0 & b & 0 & 0 \\ 0 & 0 & c & 0 \\ 0 & 0 & 0 & d \end{bmatrix}.$$

Linear Inequalities 4

1 Introduction

In the previous three chapters we have often encountered the problem of solving a single linear equation or a system of several linear equations. In many contemporary applications of mathematics, inequalities are fully as important as equations. In this chapter, we examine the problem of solving a single linear inequality or of solving a system of several linear inequalities. As the discussion is developed, be sure to compare and contrast the situation for inequalities with the familiar corresponding theory for equations. The chapter culminates in a brief glance at one of the most significant topics in contemporary mathematics, namely, linear programming. Throughout, we shall rely heavily upon geometry to aid us with a visual interpretation for our algebraic work.

2 Inequalities

We begin our study of inequalities with an investigation of the system of real numbers insofar as order and its relationship with the arithmetical operations are concerned. If a is a number and if b is a number, then we may compare them. They may be equal, or they may be different. In case they are not the same, we often prefer to be more specific by identifying which of them is greater.

We say that the number a is greater than the number b provided $a - b$ is positive. We write in symbols, $a > b$. Equivalently, we may say that b is less than a; in symbols, $b < a$. Geometrically speaking, the statement that $b < a$ means that, on the usual one-dimensional coordinate system, the point b is to the left of the point a. See Fig. 1.

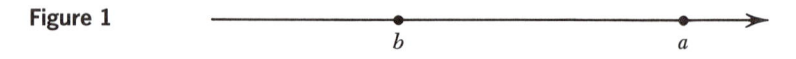

Figure 1

Example 1 Since $6 - 2 = 4$ is positive, $6 > 2$. Also $5 > -3$, because $(5) - (-3)$ is the positive number 8. Since $(-1) - (-7) = 6$ is positive, -1 is greater than -7.

On the coordinate system in Fig. 2, we note that 2 is to the left of 6; -3 is to the left of 5; and -1 is to the right of -7.

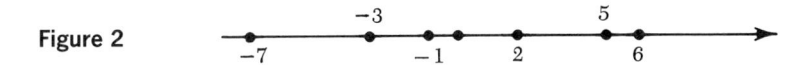

Figure 2

Example 2 On a one-dimensional coordinate system, sketch the graph of the inequality $x > 3$.

The statement that the root x of the inequality is greater than 3 means that the point x is to the right of the point 3. The graph (shown in Fig. 3) is the portion of the coordinate axis consisting of all points to the right of 3. Notice that the point 3 does not belong to the graph, for the number 3 is not greater than itself. The graph is a so-called open halfline.

Figure 3

Definition Let r be a real number. On a one-dimensional coordinate system, the graph of the inequality $x > r$ is a set of points called an *open halfline*. The graph of the inequality $x < r$ is also called an *open halfline*.

Example 3 The open halfline given by the inequality $x < \frac{3}{2}$ is sketched in Fig. 4.

Figure 4

In many situations a strict inequality is too restrictive. Suppose a and b are numbers. The statement that $a \geqq b$ means that either a is the same as b or else a is greater than b. In other words, $a \geqq b$ denies the possibility that a is less than b. Equivalently, $b \leqq a$, that is, b is equal to or less than a.

Definition Let r be a real number. On a one-dimensional coordinate system, the graph of the inequality $x \geqq r$ is a *closed halfline*. Likewise, the graph of $x \leqq r$ is a *closed halfline*.

The closed halfline given by $x \geqq r$ is distinguished from the graph of $x > r$ by the fact that the point r is included in the closed halfline but does not belong to the open halfline.

Example 4 In Fig. 5 the set consists of the point -2.5 and all points to the right of -2.5. What inequality has this set of points as its graph?

Figure 5 portrays the inequality $x \geq -2.5$. Note in particular that the number -2.5 satisfies the inequality $x \geq -2.5$.

Figure 5

Example 5 In Fig. 6 the set consists of the point 2, the point 6, and all points between them. If x is the coordinate of a point in the set, then x is at least as large as 2, that is, $2 \leq x$; on the other hand, x is no greater than 6, that is, $x \leq 6$. In order to indicate that x is a number between 2 and 6, or possibly 2, or 6, we write a chain inequality: $2 \leq x \leq 6$.

Figure 6

Let c and d be two real numbers such that $c < d$. The statement $c < x < d$ is an abbreviation for two statements: both $c < x$ and $x < d$; in other words, x is a number between c and d. The statement $c \leq x \leq d$ is a condensation of the statement that $c \leq x$ and $x \leq d$. Rephrased, x is a number between c and d, inclusive.

3 Inequalities and the arithmetical operations

Next, we summarize the interconnections between inequalities and the basic arithmetical operations of addition and multiplication.

The sum of any two numbers is less than the sum of two respectively greater numbers. Thus, if $a < b$ and if $c < d$, then the sum $a + c$ is less than the sum $b + d$. We may prove this result by a direct appeal to the definition of "less than." The hypothesis means that $b - a$ and $d - c$ are positive numbers. Since the sum of positive numbers is positive,
$$(b - a) + (d - c) = (b + d) - (a + c)$$
is a positive number. Consequently, again by the definition, $a + c < b + d$.

An extension of the preceding argument enables us to establish the result that if $a \leq b$ and $c \leq d$, then $a + c \leq b + d$.

122   Linear Inequalities

Example 6 In order to find the roots of the inequality $x - 2 \geq 0$, we may apply the preceding discussion to the given inequality and the statement $2 \geq 2$. We obtain $(x - 2) + 2 \geq 0 + 2$, that is, $x \geq 2$. Every number that is not less than 2 is a root.

Example 6 suggests an important special case of the above development: if $a \leq b$, then $a + c \leq b + c$. This is a particular application, since the inequality $c \leq c$ is certainly true. (Here is a simple illustration of the occasional desirability of being content with a weak statement, even though we have enough knowledge to assert a stronger statement: We know that $c = c$; hence, either $c = c$ or else $c < c$; by merely asserting that $c \leq c$, we are able to apply the above proposition directly.)

Example 7 If r is a given number, then the roots of the inequality $-x \geq r$ are the numbers x such that $x \leq -r$.

We obtain part of this conclusion by adding $x - r$ to each member of the given inequality. Thus,

$$(-x) + (x - r) \geq r + (x - r).$$

This statement can be simplified to $-r \geq x$, which is equivalent to

$$x \leq -r.$$

To complete the job, we must verify that any number d satisfying the condition $d \leq -r$ is a root of the given inequality. If $d \leq -r$, then $-r \geq d$ and

$$-r + (r - d) \geq d + (r - d),$$

that is

$$-d \geq r.$$

In other words, d does satisfy the given inequality; that is, replacement of the letter x in the inequality $-x \geq r$ by the number d yields the true statement $-d \geq r$.

The preceding example illustrates the important principle that the statements $a > b$ and $-a < -b$ have the same meaning. Indeed, the first means that $a - b$ is positive. The second inequality means that $(-b) - (-a)$ is positive; but $(-b) - (-a)$ is the same as $a - b$. In summary, we may reverse the sign of each member of an inequality if we also reverse the so-called *sense* of the inequality, that is, exchange the symbol $<$, or $>$, for the other one.

Inequalities behave very much like equations, insofar as adding is concerned. We manipulate inequalities in a natural manner, and we seldom go wrong. Much more caution is required, however, when multiplying is involved.

We have just seen that multiplying each member of an inequality by the number -1 requires that the sense of the inequality be reversed. On the other hand, multiplying each member of an inequality by a positive number preserves the sense of the inequality.

Specifically, if $a > b$ and if p is a positive number, then $pa > pb$. In order to prove this result, we again appeal to the definition of an inequality. By hypothesis, $a - b$ is positive. Since p is also positive (by hypothesis), their product $p \cdot (a - b)$ is positive. This product is $pa - pb$. Hence, pa is greater than pb.

A useful extension of this principle is that if $a \geq b$ and if p is positive, then $pa \geq pb$.

Example 8 In order to obtain the roots of the inequality $3x \geq 15$, we apply the above result by using the positive multiplier $\frac{1}{3}$. We obtain

$$\tfrac{1}{3} \cdot 3x \geq \tfrac{1}{3} \cdot 15 \quad \text{or} \quad x \geq 5.$$

We must now verify that each number equal to or greater than 5 is a root. To do so, we apply the above principle with the positive multiplier 3 and observe that if $x \geq 5$, then $3x \geq 15$.

Example 9 Sketch the graph of the inequality $1 - 2x < 5$.

First, we solve the inequality. We add -1 to each member of the inequality; the result is $-2x < 4$. We multiply both members of the new inequality by $\frac{1}{2}$; the result is $-x < 2$. We reverse the sign of each member and also reverse the sense of the inequality. The roots of the given inequality satisfy the condition $x > -2$. Furthermore, every number greater than -2 is a root of the given inequality (see Exercise 2 below). Therefore, the roots of the inequality $1 - 2x < 5$ are the numbers satisfying the condition $x > -2$.

Figure 7

Now we may draw the graph (Fig. 7). As a partial check, we observe that the diagram shows that 3 belongs to the graph; since $1 - 2 \cdot 3 = -5$ is less than 5, the number 3 is a root.

The effect of multiplication on an inequality may be summarized as follows. Let a, b, c be three numbers such that $a > b$. In case c is positive, then $ca > cb$. In case c is negative, then $ca < cb$. In case c is zero, then $ca = cb$. In case we do not know the character of the number c, then we do not compare ca and cb. (Disregard of the significance of the preceding sentence can be costly. For emphasis, we repeat. The knowledge that $a > b$ is not enough information to determine whether or not ca is greater than cb.)

4 Summary of properties

We collect together for reference the properties that will be most useful in the remainder of the chapter.

Theorem 1 Let a, b, c be real numbers, and let p be a positive number. Then:

(a) $a \leq a$.
(b) If a and b are different, then either $a \leq b$ or else $b \leq a$.
(c) If $a \leq b$ and $b \leq c$, then $a \leq c$.
(d) If $a \leq b$, then $a + c \leq b + c$.
(e) If $a \leq b$, then $-b \leq -a$.
(f) If $a \leq b$, then $ap \leq bp$.

EXERCISES A

1. A missing number is indicated by each of the following blanks. Fill in each blank with the best answer.
   (a) If x is greater than 3, then $x + 1$ is greater than _____.
   (b) If $2x$ is greater than 12, then x is greater than _____.
   (c) If $x > 3$ and $y > 5$, then $x + y$ is greater than _____.
   (d) If $3x$ is less than -18, then x is less than _____.
   (e) If $-3x$ is less than 18, then x is greater than _____.

2. Let x be a number greater than -2. Complete the argument in Example 9 by showing, in succession, each of the following: $-4 < 2x$; $-2x < 4$; $1 - 2x < 5$.

3. For each of the following, sketch the graph of the inequality (or system of inequalities) on a one-dimensional coordinate system.
   (a) $x \leq 0$.
   (b) $x + 3 > 0$.
   (c) $2x \geq 8$.
   (d) $1 - y > 0$.
   (e) $x > 3$, $x < 7$.
   (f) $x < 1$, $x < 2$.
   (g) $3t - 3 \leq 0$.
   (h) $-1 \leq x \leq 4$.
   (i) $0 < x < 1$.
   (j) $x \neq 3$.

4. Find the roots of each of the following inequalities.
   (a) $2x + 3a < 4$, where a is a given number.
   (b) $1 - 3y \geq b$, where b is a given number.
   (c) $2s + 5x \leq t$, where s and t are given numbers.

5. Consider the inequality $5x + 7y \geq 2$.
   (a) Which of the following ordered pairs (x, y) satisfy the inequality?
      $(1, 2)$; $(0, 0)$; $(-1, 1)$; $(2, -1)$; $(1, -2)$; $(\frac{5}{2}, -\frac{3}{2})$.
   (b) The given inequality may be expressed in the following form, where the blank indicates a missing number: _____ $x + y \geq \frac{2}{7}$. Fill the blank.
   (c) The same inequality may be expressed in the following form:
      $y \geq$ _____ $x +$ _____. Fill each blank with the appropriate number.

6. Consider the inequality $2x - 3y \leq 5$.
   (a) Which of the following ordered pairs (x, y) satisfy the inequality?
      $(4, 2)$; $(4, 0)$; $(4, 1)$; $(0, 0)$; $(1, -1)$; $(1, -2)$; $(1, 2)$.
   (b) The given inequality may be expressed in the form $2x +$ _____ $\leq 3y$. Fill the blank with the appropriate number.
   (c) The same inequality may be expressed as follows: _____ $x +$ _____ $\leq y$. Fill each blank with the appropriate number.

5 Halfplane

Example 10 On a two-dimensional coordinate system, sketch the graph of the system $x = 3$ and $y < 2x + 1$.

By the given equation, we restrict attention to points whose coordinates have the form $(3, y)$. Among such points we select those having a y-coordinate smaller than the number that is one more than double the x-coordinate. Since the x-coordinate is 3, the y-coordinate is less than $2 \cdot 3 + 1 = 7$. The graph is an open halfline, namely, the portion of the vertical line $x = 3$ "below" the point $(3, 7)$. (See Fig. 8.)

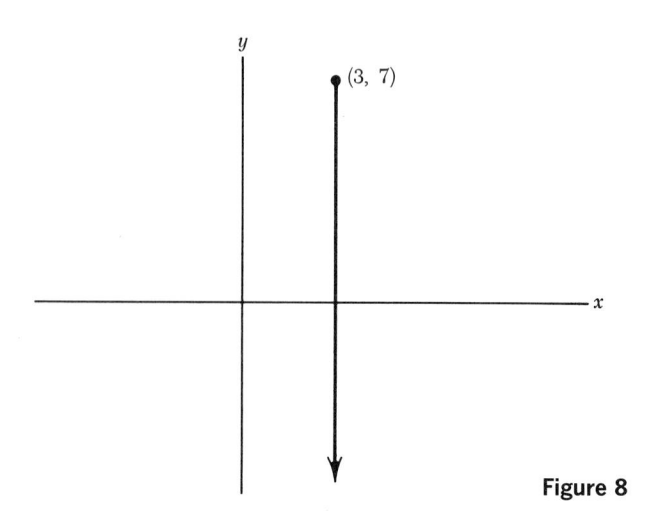

Figure 8

Example 11 On a two-dimensional coordinate system, sketch the graph of the inequality $y < 2x + 1$.

For each number c, the point (c, y) belongs to the graph if $y < 2c + 1$. For each c, the points (c, y) for which $y < 2c + 1$ form a vertical halfline below the point $(c, 2c + 1)$. The totality of these halflines for all possible numbers c is the desired graph (see Fig. 9). The graph consists of all points in the plane on one side ("below") the line whose equation is $y = 2x + 1$.

Example 12 Sketch the graph of the relation $y \geq -\frac{1}{2}x + 1$.

For each number c, a point with x-coordinate c belongs to the graph if its y-coordinate satisfies the restriction $y \geq -\frac{1}{2}c + 1$. For any particular c, the collection of such points is a closed halfline with endpoint $\left(c, 1 - \frac{c}{2}\right)$ and extending vertically upward. The desired graph (shown in Fig. 10) is the totality of these halflines for all possible choices of c.

126 Linear Inequalities

Figure 9

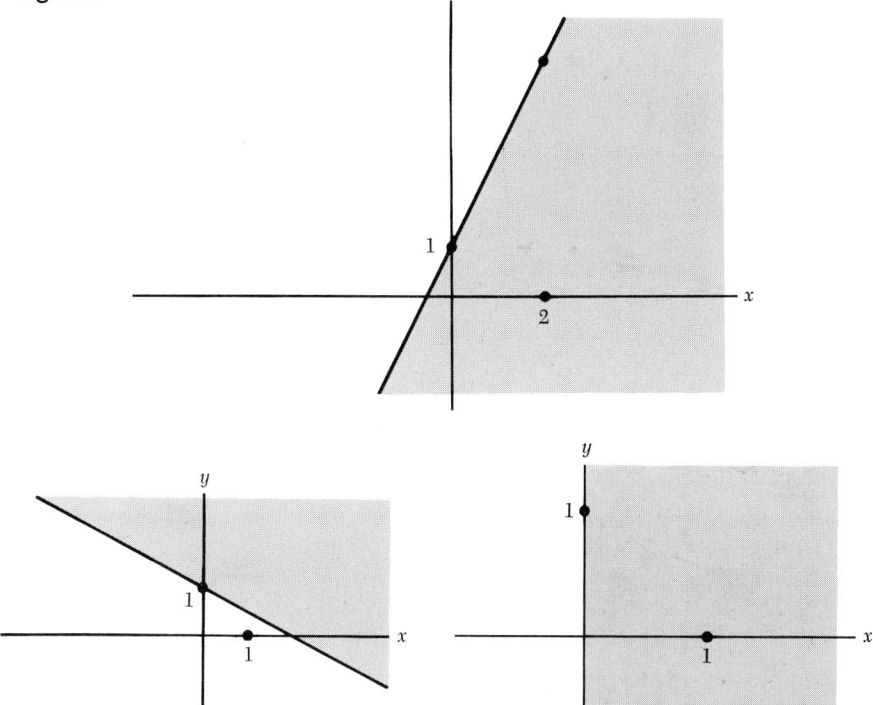

Figure 10

Figure 11

Example 13 On a two-dimensional coordinate system, the graph of the inequality $x \geq 0$ is a closed halfplane, the right coordinate halfplane. (See Fig. 11.)

Definition Let $[a \ \ b]$ be a nonzero vector, and let c be a number. On a two-dimensional coordinate system, the graph of the inequality

$$[a \ \ b]\begin{bmatrix} x \\ y \end{bmatrix} \geq c$$

is a *closed halfplane*. Likewise, the graph of the inequality

$$[a \ \ b]\begin{bmatrix} x \\ y \end{bmatrix} \leq c$$

is a *closed halfplane*. These two halfplanes have a common *boundary*, namely, the line whose equation is

$$[a \ \ b]\begin{bmatrix} x \\ y \end{bmatrix} = c.$$

5 / Halfplane

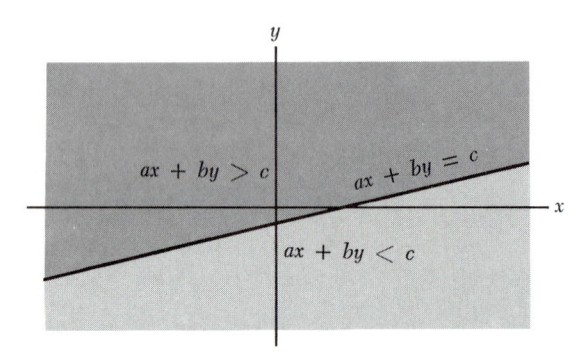

Figure 12

In Fig. 12, $b > 0$. The plane is partitioned into three pieces by the line. One set is the line itself, given by the equation $ax + by = c$. Another set is the portion of the plane "above" the line; its points satisfy the condition $ax + by > c$. The third set consists of the points in the plane "below" the line, points whose coordinates are roots of the inequality $ax + by < c$. A closed halfplane consists of the points on the line together with those on one side of the line. The line is the boundary, or edge, of the halfplane.

EXERCISES B

1. Consider the inequality $2x + y \leq 3$.
   (a) On a two-dimensional coordinate system, sketch the line that is the graph of the associated equation $2x + y = 3$.
   (b) Choose three different numbers. After you have made your selection, label one of the numbers you have picked by the letter f, another by the letter g, and the remaining one by h.
   [NOTE: Make specific choices. For example, you could choose the three numbers $4, -2, \frac{1}{3}$; afterward you could assign labels $f = \frac{1}{3}, g = 4, h = -2$. As another example, you could choose the three numbers $-15, \sqrt{2}/5, 0$, and you could assign labels $f = \sqrt{2}/5, g = 0, h = -15$. Do not choose the same numbers as in these illustrations—instead, pick three specific numbers that are your own favorites.]
   (c) Find the point on the line in part (a) whose x-coordinate is the number you picked and labeled f in part (b).
   (d) Choose any four other points, each of which has the same x-coordinate f as the point in part (c). For each of these points decide whether its coordinates satisfy the inequality, and observe whether the point is "above" or "below" the line.
   [NOTE: Again, make specific choices. Pick definite numbers (not letters!) as the coordinates of your points.]
   (e) Find the point on the line whose x-coordinate is the number you picked and labeled g in part (b).
   (f) Find three other points, each of which has x-coordinate equal to your chosen g and each of which is "above" the line. Which of these points has coordinates that satisfy the inequality?
   (g) Find the point on the line whose x-coordinate is the number you selected and called h in part (b).
   (h) Find two other points, each of which is "below" the line and has the same x-coordinate as the point in part (g). Which of the points you have found has coordinates that satisfy the inequality?

2. Answer all parts of Exercise 1 as applied to the inequality $x - 2y \leq 3$ and its associated equation $x - 2y = 3$.

3. For each of the following, sketch the graph of the inequality on a two-dimensional coordinate system.
   (a) $3x + 2y \leq 6$.
   (b) $2x + 3y \geq 6$.
   (c) $x - 5y \leq 3$.
   (d) $y - 5x \leq 3$.
   (e) $-x - 2y \geq 4$.
   (f) $4x - 3y \geq 8$.

4. Let P, Q, R be the points with respective coordinates $(4, 1), (-2, -1), (11, 3)$. The line containing P and Q is the boundary of a closed halfplane containing R. Find an inequality whose graph is this halfplane.

5. Repeat Exercise 4, with the following modification: the point P has coordinates $(5, 1)$ instead of $(4, 1)$.

6. Two halfplanes are given by the respective inequalities $3x + y \geq 7$ and $-x + 3y \leq 11$. Find the point of intersection of the boundaries for these halfplanes.

7. Find the coordinates of the point of intersection of the lines that are the boundaries of the halfplanes given by $-x + y \geq -2$ and $x + 7y \geq 6$, respectively.

8. For each of the following, give a word description of the graph of the inequality on a three-dimensional coordinate system.
   (a) $z \geq 1$.
   (b) $y \geq -4$.
   (c) $2x + 1 \leq 0$.
   (d) $y \neq 0$.
   (e) $-1 \leq z \leq 3$.

9. Give a word description of the set of all points (x, y, z) on a three-dimensional coordinate system that satisfy all three of the following conditions: $x \geq 0, y \geq 0, z \geq 0$. [REMARK: On a coordinate system, the set of all points, each of whose coordinates is not negative, may be called the *nonnegative orthant.*]

6 System of linear inequalities

Example 14 Sketch the graph of the system of two simultaneous inequalities

$$\begin{cases} x + 2y \geq 5, \\ 3x - 2y \geq 7. \end{cases}$$

The graph of the inequality $x + 2y \geq 5$ is a closed halfplane with the line $x + 2y = 5$ as boundary (see Fig. 13a).

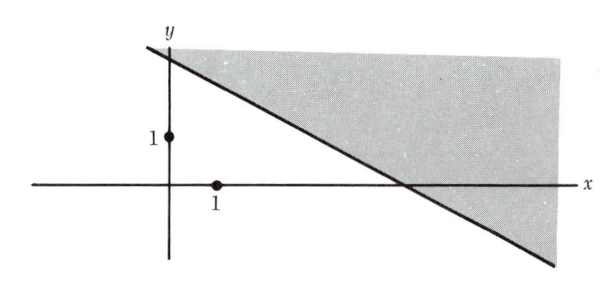

Figure 13a

Figure 13b

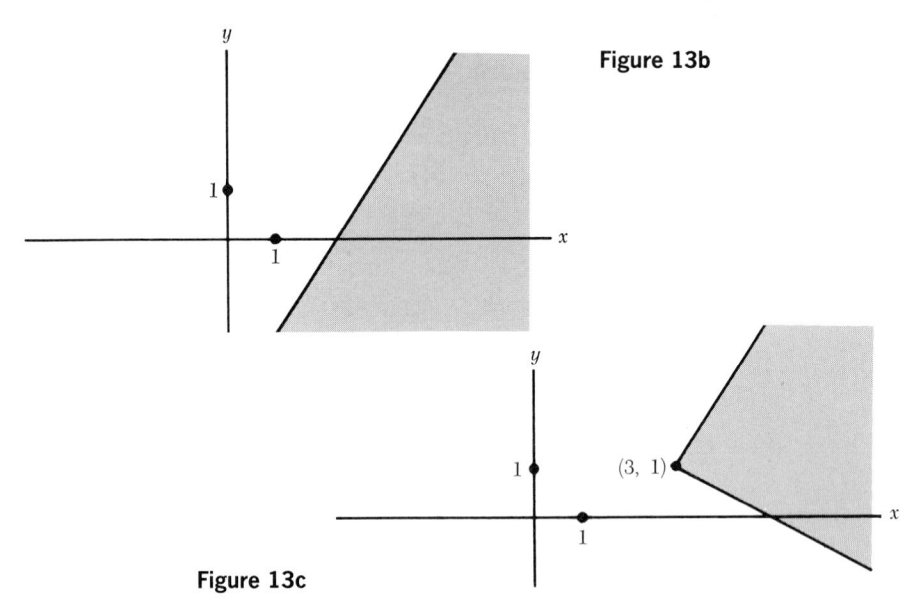

Figure 13c

The closed halfplane $3x - 2y \geq 7$ has as its boundary the line $3x - 2y = 7$ (see Fig. 13b).

The graph of the system of inequalities consists of all points that satisfy both of the inequalities. A point whose coordinates satisfy each inequality belongs to each halfplane. Consequently, the desired graph is the intersection of the two halfplanes (see Fig. 13c). The set of points common to the two halfplanes is an angular (closed) region. The vertex of the angle is the point (3, 1). We find this vertex by observing that it is the point of intersection of the boundaries of the two halfplanes. In other words, the vertex is the point whose coordinates satisfy both of the equations $x + 2y = 5$ and $3x - 2y = 7$. The simultaneous solution of these equations yields the common root (3, 1).

Example 15 Discuss and sketch the graph of the system of inequalities

$$\begin{cases} -x + 7y \leq 9, \\ 4x - 3y \leq 14, \\ -3x - y \leq 5, \\ -y \leq 2. \end{cases}$$

Each of the four inequalities has a closed halfplane as its graph. The graph of the system is the set of all points that belong to all four halfplanes. In other words, the graph is the intersection of the four closed halfplanes.

For each halfplane, we sketch its boundary (given by the corresponding equation) and indicate which side of the line is in the halfplane (see Fig. 14a).

Figure 14

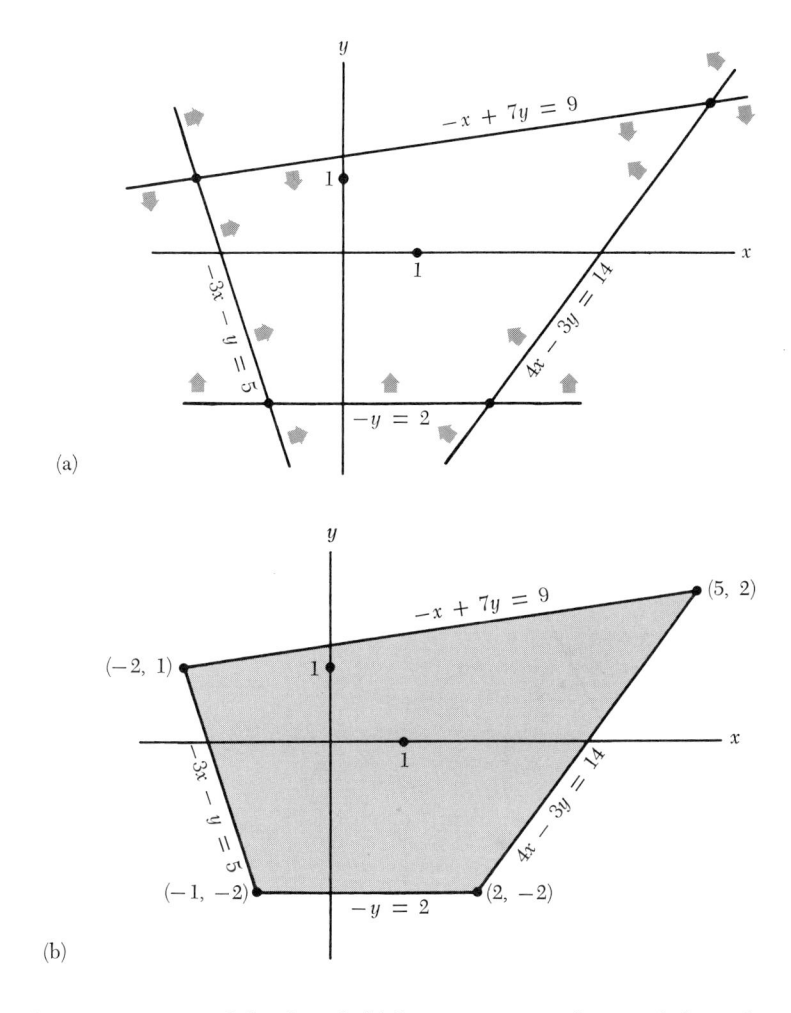

(a)

(b)

The intersection of the four halfplanes consists of a quadrilateral and its interior (see Fig. 14b).

Each vertex of the quadrilateral is the point at which the boundaries of two of the halfplanes intersect. We may find the coordinates of a vertex by solving simultaneously the equations for the appropriate two boundaries. The vertices are $(5, 2)$, $(2, -2)$, $(-1, -2)$, $(-2, 1)$.

We recall that a system of simultaneous linear equations may be expressed as a single equation if we adopt matrix notation. An analogous maneuver enables us to convert the above system of simultaneous linear inequalities into a single inequality. Instead of writing

$$\begin{cases} -x + 7y \leq 9, \\ 4x - 3y \leq 14, \\ -3x - y \leq 5, \\ -y \leq 2, \end{cases}$$

6 / System of linear inequalities   131

we may write
$$\begin{bmatrix} -x+7y \\ 4x-3y \\ -3x-y \\ -y \end{bmatrix} \leqq \begin{bmatrix} 9 \\ 14 \\ 5 \\ 2 \end{bmatrix}.$$

The left member of the preceding inequality is a product of matrices. Thus, our system of inequalities can be expressed as
$$\begin{bmatrix} -1 & 7 \\ 4 & -3 \\ -3 & -1 \\ 0 & -1 \end{bmatrix} \begin{bmatrix} x \\ y \end{bmatrix} \leqq \begin{bmatrix} 9 \\ 14 \\ 5 \\ 2 \end{bmatrix}.$$

The preceding examples lead us to introduce several new ideas. We begin by discussing inequalities for matrices.

Let A and E be two matrices with the same number of rows and with the same number of columns. The statement $A \leqq E$ means that every element in A is equal to or less than the element in E occupying the same position in the array. Equivalently, we may write $E \geqq A$.

Example 16

$$\begin{bmatrix} 1 & -2 \\ 3 & 5 \\ 2 & -1 \end{bmatrix} \leqq \begin{bmatrix} 4 & 0 \\ 3 & 8 \\ 9 & -1 \end{bmatrix};$$

but the following statement is false: $[5 \quad 1 \quad 3] \leqq [6 \quad 1 \quad 2]$.

A system of simultaneous linear (nonstrict) inequalities may be expressed, in a condensed fashion, as $AX \leqq B$. Here A is an $m \times n$ matrix whose elements are known numbers, B is an $m \times 1$ matrix of known numbers, while X is the $n \times 1$ matrix whose elements are to be found.

Notice that if a given relation employs the symbol $\geqq$, it may be converted to the desired form by sign reversal. The system
$$\begin{cases} ax + by \leqq c, \\ dx + ey \geqq f, \\ gx + hy \leqq k, \end{cases}$$
in which $a, b, c, d, e, f, g, h, k$ are known, may be converted to the system
$$\begin{cases} ax + by \leqq c, \\ -dx - ey \leqq -f, \\ gx + hy \leqq k, \end{cases}$$

132 Linear Inequalities

which is equivalent to

$$\begin{bmatrix} a & b \\ -d & -e \\ g & h \end{bmatrix} \begin{bmatrix} x \\ y \end{bmatrix} \leq \begin{bmatrix} c \\ -f \\ k \end{bmatrix}.$$

In our applications in the immediate future, we will not be concerned with linear (strict) inequalities involving the symbol $<$ or the symbol $>$.

EXERCISES C

1. For each of the following, sketch the graph of the inequality.

   (a) $[3 \quad 4] \begin{bmatrix} x \\ y \end{bmatrix} \geq 12.$ (b) $[-3 \quad 1] \begin{bmatrix} x \\ y \end{bmatrix} \geq 5.$ (c) $[4 \quad -3] \begin{bmatrix} x \\ y \end{bmatrix} \geq 12.$

2. On a two-dimensional coordinate system, show the points that belong to both the halfplane given by $3x - y \geq 0$ and the halfplane given by $-2x + y \geq 1$.

3. Sketch the set of vectors that are roots of the inequality

   $$\begin{bmatrix} 1 & 0 \\ 0 & 1 \\ -3 & 2 \\ 1 & -5 \end{bmatrix} \begin{bmatrix} x \\ y \end{bmatrix} \geq \begin{bmatrix} 0 \\ 0 \\ -6 \\ -10 \end{bmatrix}.$$

4. Express each of the following systems of inequalities in the form $AX \geq B$, where A, X, B are matrices such that the elements of A and B are numbers.
   (a) $5x + y \geq 3$, $-x - 2y \geq 1$, $-7x + 3y \geq -18$.
   (b) $2x - y \geq 4$, $-x + 3y \geq -1$, $5x - 2y \leq 12$.
   (c) $x + 3y \leq -5$, $-2x + y \geq 3$, $-2x + 3y \leq 14$.

5. The vertices of a triangle are the points $E = (1, 4)$, $D = (7, 2)$, $C = (9, 5)$.
   (a) Find an inequality whose graph is the closed halfplane that contains C and has as boundary the line containing E and D.
   (b) Find the system of three inequalities whose graph is the set of points that are on or inside the triangle.
   (c) Express this system of inequalities in the form $AX \leq B$, where A, X, B are matrices.

6. Repeat Exercise 5 for the triangle whose vertices E, D, C are the points $(0, 6)$, $(8, 2)$, $(-1, -1)$, respectively.

7. Give a word description of the graph of each of the following inequalities.

   (a) $[0 \quad 1] \begin{bmatrix} x \\ y \end{bmatrix} \geq 2.$

   (b) $[0 \quad 1 \quad 0] \begin{bmatrix} x \\ y \\ z \end{bmatrix} \geq 2.$ [HINT: The dimensionality is two in part (a), but is three in part (b).]

8. Give a word description of the graph of each of the following inequalities.

(a) $\begin{bmatrix} 1 & 0 \\ 0 & 1 \end{bmatrix} \begin{bmatrix} x \\ y \end{bmatrix} \geq \begin{bmatrix} 3 \\ 2 \end{bmatrix}$.

(b) $\begin{bmatrix} 1 & 0 & 0 \\ 0 & 1 & 0 \end{bmatrix} \begin{bmatrix} x \\ y \\ z \end{bmatrix} \geq \begin{bmatrix} 3 \\ 2 \end{bmatrix}$.

(c) $\begin{bmatrix} 1 & 0 & 0 \\ 0 & 1 & 0 \\ 0 & 0 & 1 \end{bmatrix} \begin{bmatrix} x \\ y \\ z \end{bmatrix} \geq \begin{bmatrix} 3 \\ 2 \\ 1 \end{bmatrix}$.

7 Convex polyhedral set

Let n be a natural number, let $R = [a_1 \ a_2 \ \cdots \ a_n]$ be a $1 \times n$ matrix of numbers, and let k be a number. If

$$X = \begin{bmatrix} x_1 \\ \vdots \\ x_n \end{bmatrix}$$

is an $n \times 1$ matrix, then the inequality $RX \leq k$ represents the single condition

$$a_1x_1 + a_2x_2 + \cdots + a_nx_n \leq k.$$

We suppose that R is not the zero matrix. We discuss the graph of the inequality $RX \leq k$ on an n-dimensional coordinate system.

In case $n = 1$, the inequality is $a_1x_1 \leq k$ and is equivalent to either $x \leq k/a_1$ or $x \geq k/a_1$, according as a_1 is positive or negative. We have called the graph a closed *halfline*. It consists of a boundary point and all points of the coordinate line that are on one particular side of the boundary. The boundary point is identified by the equation $RX = k$ (that is, $a_1x_1 = k$).

In case $n = 2$, the inequality is $a_1x_1 + a_2x_2 \leq k$. We have called the graph a closed *halfplane*. It consists of a boundary line and all points of the coordinate plane that are on a certain side of the boundary. The boundary line is given by the equation $RX = k$ (that is, $a_1x_1 + a_2x_2 = k$).

In case $n = 3$, the inequality is $a_1x_1 + a_2x_2 + a_3x_3 \leq k$. We may call the graph a closed *halfspace*. It consists of a boundary plane and all points of the space on one side of the boundary. The boundary plane is given by the equation $RX = k$ (that is, $a_1x_1 + a_2x_2 + a_3x_3 = k$).

Even though our ability to draw (and interpret) meaningful pictures in higher dimensions is meager, the language of geometry is both convenient and useful. For any value of n, the graph of the inequality $RX \leq k$ may be called a closed *halfspace*. The graph consists of a boundary hyperplane and all points of the n-dimensional space "on one side" of the boundary. The hyperplane is the graph of the associated equation $RX = k$.

134   Linear Inequalities

Example 17 On a three-dimensional coordinate system, the graph of the equation

$$2x + y + 3z = 6$$

is a plane. The plane separates the space (in the manner of a slanting wall). We may think of the region "above" the plane and the region "below" the plane. The closed halfspace consisting of the plane and all points below is the graph of

$$2x + y + 3z \leq 6.$$

The closed halfspace consisting of the plane and all points above is the graph of the inequality

$$2x + y + 3z \geq 6$$

(or equivalently, $-2x - y - 3z \leq -6$). (See Fig. 15.)

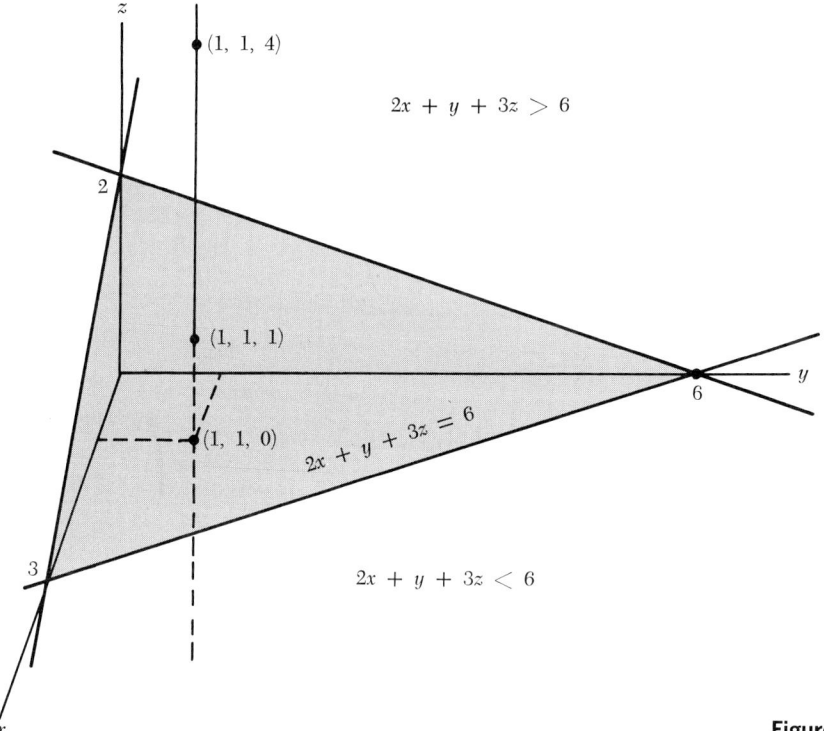

Figure 15

The point $(1, 1, 1)$ lies on the plane, because its coordinates satisfy the equation $2x + y + 3z = 6$. The point $(1, 1, 0)$ is located one unit "below" $(1, 1, 1)$. We note that the coordinates of the point $(1, 1, 0)$ satisfy the inequality $2x + y + 3z < 6$. By contrast, the point

7 / Convex polyhedral set   135

(1, 1, 4), which is located three units "above" (1, 1, 1), has coordinates satisfying the inequality $2x + y + 3z > 6$. The vertical line shown in Fig. 15 consists of all points with x-coordinate 1 and y-coordinate 1. The line pierces the plane at (1, 1, 1). The coordinates of another point on the line satisfy either $2x + y + 3z < 6$ or $2x + y + 3z > 6$, according as the point is below or above the plane. A similar discussion applies to every other line parallel to the z-axis and aids in our description of the halfspaces determined by the plane.

Example 18 The geometric interpretation of the inequality

$$5w - \tfrac{1}{2}x + 31y + 8z \leq -14$$

is a halfspace on a four-dimensional coordinate system. The point $(2, 6, -1, 1)$ belongs to the halfspace, since $5(2) - \tfrac{1}{2}(6) + 31(-1) + 8(1) = -16$ and $-16 \leq -14$. The origin $(0, 0, 0, 0)$ does not belong to the halfspace. The boundary of the halfspace is the hyperplane whose equation is

$$5w - \tfrac{1}{2}x + 31y + 8z = -14.$$

The point $(-1, 2, 0, -1)$ is on the hyperplane and belongs to the closed halfspace.

Now consider several halfspaces in n-dimensional space. Their intersection is the set of all points that belong to each of them. Each of them is described algebraically by a linear inequality. The intersection consists of all points whose coordinates satisfy each of the inequalities. In other words, the intersection is the graph of the system of simultaneous inequalities.

Let the matrix A have m rows and n columns. Denote its rows by $R_1, \ldots, R_m$; thus

$$A = \begin{bmatrix} R_1 \\ R_2 \\ \vdots \\ R_m \end{bmatrix}.$$

We suppose that no row of A is the zero matrix. Let B be the $m \times 1$ matrix

$$\begin{bmatrix} b_1 \\ b_2 \\ \vdots \\ b_m \end{bmatrix}.$$

On an n-dimensional coordinate system, the graph of the inequality $AX \leqq B$ is the intersection of the m halfspaces given by the respective linear inequalities $R_1 X \leqq b_1, R_2 X \leqq b_2, \ldots, R_m X \leqq b_m$.

In Example 15 (in Section 6) we examined a situation in two dimensions. The graph in that case is a certain polygon and its interior. In three dimensions an intersection of halfspaces frequently is a polyhedron and its interior. In every case, the intersection has characteristics that remind us of polyhedra. Furthermore, the intersection is a set that is convex, because it contains every segment whose endpoints belong to the set. These notions suggest the following name.

> **Definition** In n-dimensional space, an intersection of finitely many closed halfspaces is called a *convex polyhedral set*.

The above discussion has established the following result.

> **Theorem 2** If A is an $m \times n$ constant matrix (not the zero matrix) and B is an $m \times 1$ constant matrix, then the graph of the inequality $AX \leqq B$ is a convex polyhedral set in n-dimensional space.

EXERCISES D

1. On a three-dimensional coordinate system, a plane contains the points $(2, 0, 0)$, $(0, -5, 0)$, $(0, 0, 1)$. A closed halfspace consists of all points on or "above" this plane. Find an inequality whose graph is the halfspace.
2. Give a word description of the graph of the inequality $4x + 3y + 2z \geqq 12$.
3. Using geometrical language, give a word description of the set of roots of the inequality

$$\begin{bmatrix} 4 & 3 & 2 \\ 1 & -2 & 4 \end{bmatrix} \begin{bmatrix} x \\ y \\ z \end{bmatrix} \leqq \begin{bmatrix} 12 \\ 8 \end{bmatrix}.$$

4. Describe the set of roots of the inequality

$$\begin{bmatrix} 1 & 0 & 0 \\ 0 & 1 & 0 \\ 0 & 0 & 1 \\ -1 & 0 & 0 \\ 0 & -1 & 0 \\ 0 & 0 & -1 \end{bmatrix} \begin{bmatrix} x \\ y \\ z \end{bmatrix} \leqq \begin{bmatrix} 6 \\ 5 \\ 1 \\ -2 \\ -4 \\ 1 \end{bmatrix}.$$

What type of geometrical object is formed by the points that are roots?

7 / Convex polyhedral set

Figure 16

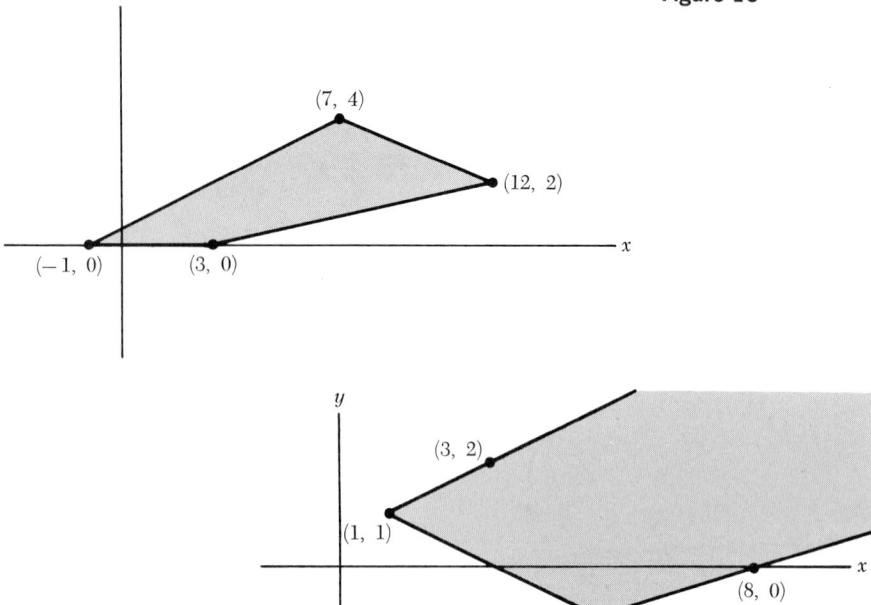

(7, 4)

(12, 2)

(−1, 0)   (3, 0)

Figure 17

(3, 2)

(1, 1)

(8, 0)

(5, −1)

5. Find a 4×2 matrix A and a 4×1 matrix B such that the convex polyhedral set shown in Fig. 16 is the set of roots of the inequality $AX \leq B$.

6. Find a 3×2 matrix A and a 3×1 matrix B such that the convex polyhedral set shown in Fig. 17 is the graph of the inequality $AX \geq B$.

7. A certain convex polyhedral set consists of a pentagon and its interior. The vertices of the pentagon are $(3, -2), (-1, 0), (1, 8), (2, 7), (4, 4)$.
   (a) Sketch the convex polyhedral set.
   (b) Find matrices A and B such that the convex polyhedral set is the graph of the inequality $AX \geq B$.

8. On a three-dimensional diagram, show the set of roots of the inequality

$$\begin{bmatrix} -1 & 0 & 0 \\ 0 & -1 & 0 \\ 0 & 0 & -1 \\ 4 & 2 & 3 \end{bmatrix} \begin{bmatrix} x \\ y \\ z \end{bmatrix} \leq \begin{bmatrix} 0 \\ 0 \\ 0 \\ 12 \end{bmatrix}.$$

8 Extreme point

A re-examination of Example 15 (in Section 6) reveals how important the vertices of the quadrilateral can be in describing the convex polyhedral set. In general, it is of considerable help in studying a convex polyhedral set to know its vertices (corner points) or, as they are commonly called, its *extreme points*.

138   Linear Inequalities

In the two-dimensional situation encountered in Example 15, each vertex is the intersection of two of the boundary lines. On the other hand, observe that not every pair of boundary lines intersect at an extreme point because in some instances the point of intersection does not belong to the convex polyhedral set.

A method for finding an extreme point of a convex polyhedral set is the following. Suppose the convex polyhedral set is given as an intersection of m halfspaces in n-dimensional space. Choose n of the boundary hyperplanes for the respective halfspaces; find the intersection of these hyperplanes; if there is only one point of intersection, determine whether it belongs to the convex polyhedral set; if so, the point is an extreme point.

By investigating every possible choice in the above method, all extreme points may be located.

Example 19 Find all extreme points of the convex polyhedral set given by

$$\begin{bmatrix} 1 & 0 \\ 0 & 1 \\ 1 & 3 \\ 3 & 2 \\ 2 & -1 \end{bmatrix} \begin{bmatrix} x \\ y \end{bmatrix} \geq \begin{bmatrix} 0 \\ 0 \\ 6 \\ 11 \\ -2 \end{bmatrix}.$$

The drawing of a good diagram often pays handsome dividends.[1] We sketch the boundary of each of the five halfspaces and indicate which side of the line is in the halfplane. An inspection of Fig. 18a

Figure 18a

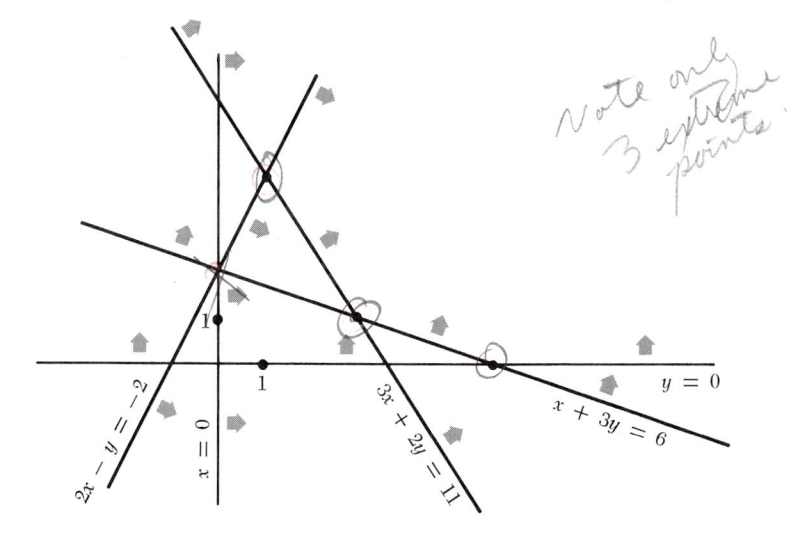

---
[1] Not every diagram is good, and the drawing of a bad diagram can be deceiving and costly.

8 / Extreme point 139

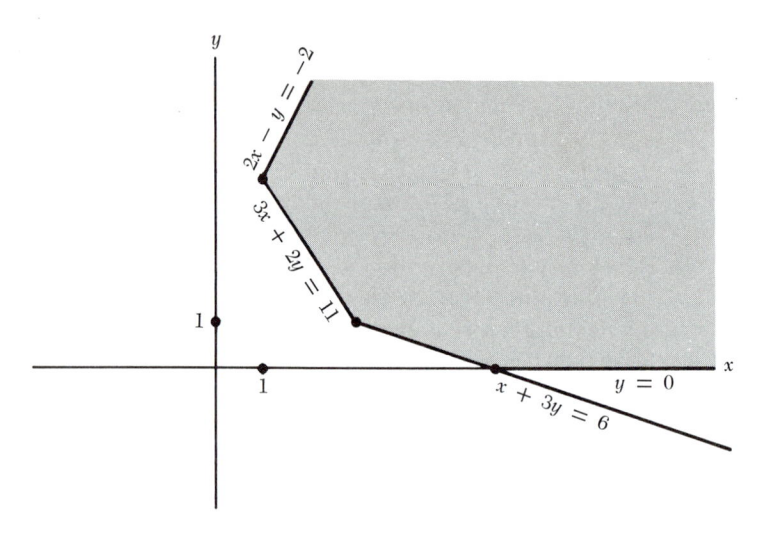

Figure 18b

shows that the boundary of the convex polyhedral set is composed of two segments and two halflines. The requirement that $x \geq 0$ appears to be unnecessary, since it imposes no further restriction besides the restrictions which the other four inequalities give. There are only three extreme points to be identified. See Fig. 18b. One extreme point is the root of the system

$$\begin{bmatrix} 0 & 1 \\ 1 & 3 \end{bmatrix} X = \begin{bmatrix} 0 \\ 6 \end{bmatrix},$$

namely, the point (6, 0). Another is the point of intersection of the two lines $x + 3y = 6$ and $3x + 2y = 11$; its coordinates are (3, 1). The third extreme point (1, 4) is obtained by solving the equation

$$\begin{bmatrix} 3 & 2 \\ 2 & -1 \end{bmatrix} X = \begin{bmatrix} 11 \\ -2 \end{bmatrix}.$$

Notice how the diagram is useful in helping us choose which pairs of boundary lines have an extreme point as their intersection. Without a guiding picture we would likely need to test several other possibilities. For example, consider the third and fifth inequalities in the description of the convex polyhedral set. The boundary lines for these halfplanes have the respective equations $x + 3y = 6$ and $2x - y = -2$. The point of intersection of the two lines is (0, 2). (Note this point in Fig. 18a.) This candidate for an extreme point fails to qualify because it does not belong to the convex polyhedral set; we discover this by observing that (0, 2) does not satisfy the inequality $3x + 2y \geq 11$.

140   Linear Inequalities

We remark that a convex polyhedral set in two dimensions does not necessarily have a polygon as its boundary, nor does a convex polyhedral set in three dimensions necessarily have a polyhedron as its boundary.

Our primary reason for interest in the extreme points of a convex polyhedral set is the following: *The greatest value, if any, assumed by a linear function defined on the set is assumed at an extreme point.* The same statement, modified by replacing "greatest" by "least," is also true.

Example 20 On a two-dimensional coordinate system, consider the convex polyhedral set shown in Fig. 19a. The vertices of the triangle, $(-1, 2)$, $(1, -1)$, $(4, 1)$, are the extreme points of the convex polyhedral set.

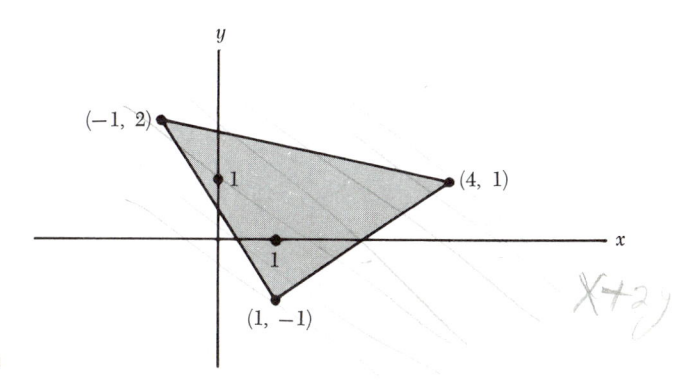

Figure 19a

Consider the linear function

$$[1 \quad 2]\begin{bmatrix} x \\ y \end{bmatrix}, \quad \text{or} \quad x + 2y.$$

At each point of the triangular set, we may evaluate $x + 2y$. At $(3, 1)$, say, the value is $3 + 2 \cdot 1 = 5$; at $(1, \frac{1}{2})$, the value is $1 + 2 \cdot \frac{1}{2} = 2$. At what point of the convex polyhedral set does $x + 2y$ take on its greatest value? At what point does $x + 2y$ assume its least value? The answer to each of these questions is an extreme point. At the vertex $(-1, 2)$, the value is 3; at $(1, -1)$, the value is -1; at $(4, 1)$, the value is 6. Hence, 6 is the greatest value $x + 2y$ assumes in the convex polyhedral set, and -1 is the least.

In order to understand why the "extreme" (that is, greatest or least) value is assumed at an extreme point, consider the equations $x + 2y = d$ for various choices of the number d. Each such equation represents a line. The line has slope $-\frac{1}{2}$. The lines corresponding to two different choices of d are parallel to each other. For all the possible values of d, we obtain all the lines with slope $-\frac{1}{2}$, that is, a family of mutually parallel lines.

8 / Extreme point

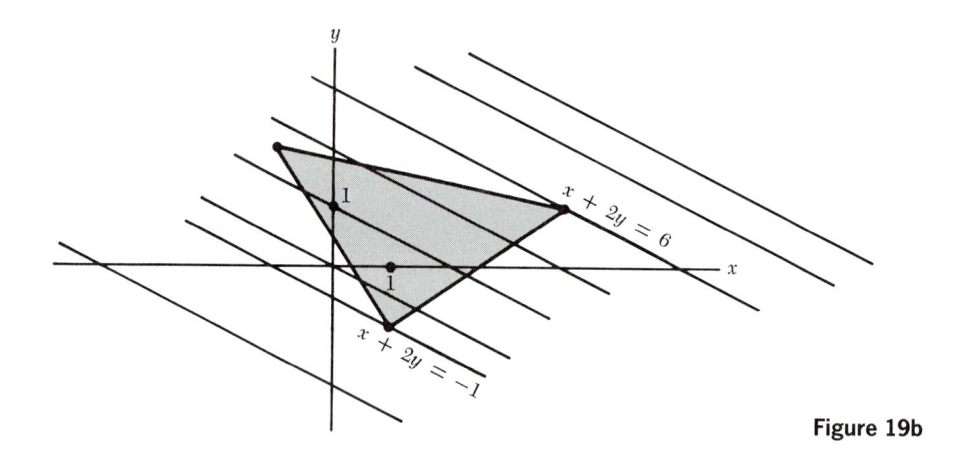

Figure 19b

Figure 19b shows several members of the family. Along each line the value of $x + 2y$ remains the same. The farther right, or the higher up, a line is, the greater the corresponding value of $x + 2y$. The picture suggests that an extreme value of $x + 2y$, for points (x, y) in the convex polyhedral set, is attained on the boundary—specifically, can be attained at a vertex, or extreme point.

The situation illustrated in the preceding example is typical. We will examine a few applications of these principles in the next section.

EXERCISES E

1. A certain convex polyhedral set is the graph of the inequality

$$\begin{bmatrix} 0 & -1 \\ -3 & 2 \\ 2 & 1 \\ 2 & 3 \end{bmatrix} \begin{bmatrix} x \\ y \end{bmatrix} \leqq \begin{bmatrix} 0 \\ 0 \\ 22 \\ 26 \end{bmatrix}.$$

   (a) Sketch the set.
   (b) How many extreme points (or vertices) does the set have?
   (c) Find each extreme point.
   (d) Name two points that belong to the graph but that are not on the boundary of the convex polyhedral set.
   (e) Find three points on the boundary of the set such that no one of them is an extreme point and such that no two of them belong to the same edge of the boundary.

2. For each of the following inequalities, find all of the extreme points of the graph of the inequality.

   (a) $\begin{bmatrix} 2 & -3 \\ -11 & 3 \\ 1 & 3 \\ 5 & -3 \end{bmatrix} \begin{bmatrix} x \\ y \end{bmatrix} \leqq \begin{bmatrix} 10 \\ -1 \\ 23 \\ 25 \end{bmatrix}.$

   (b) $\begin{bmatrix} 2 & -1 \\ -1 & 3 \\ 1 & 0 \\ 0 & 1 \end{bmatrix} X \geqq \begin{bmatrix} -5 \\ -11 \\ 1 \\ -2 \end{bmatrix}.$

142   Linear Inequalities

(c) $\begin{bmatrix} -1 & 0 & 0 \\ 0 & -1 & 0 \\ 0 & 0 & -1 \\ 2 & 4 & 1 \\ 3 & 5 & 15 \end{bmatrix} \begin{bmatrix} x \\ y \\ z \end{bmatrix} \leq \begin{bmatrix} 0 \\ 0 \\ 0 \\ 4 \\ 15 \end{bmatrix}.$ (d) $\begin{bmatrix} 0 & 1 & 1 \\ -1 & 0 & -1 \\ -1 & -1 & 0 \\ 4 & 1 & 1 \end{bmatrix} X \geq \begin{bmatrix} 1 \\ -3 \\ -2 \\ 5 \end{bmatrix}.$

(e) $\begin{bmatrix} -1 & 0 \\ 0 & -1 \\ 1 & 1 \\ 1 & 2 \\ 1 & -1 \end{bmatrix} X \leq \begin{bmatrix} 0 \\ 0 \\ 9 \\ 14 \\ 5 \end{bmatrix}.$

3. A certain convex polyhedral set is the graph of the inequality

$$\begin{bmatrix} 1 & 0 \\ 0 & 1 \\ 4 & 1 \\ 1 & 1 \\ 1 & 3 \end{bmatrix} \begin{bmatrix} x \\ y \end{bmatrix} \geq \begin{bmatrix} 0 \\ 0 \\ 8 \\ 5 \\ 9 \end{bmatrix}.$$

(a) Sketch the set.
(b) Find each extreme point of the set.
(c) Calculate the value, at each extreme point, of the function given by the formula $2x + 3y$.
(d) At what point of the convex polyhedral set does $2x + 3y$ have its least value?

4. Among all the roots of the inequality in Exercise 1, which one has the maximum image under the function given by the formula $5x + 2y$?

5. Find the least value of the function $-3x + 5y$ on the set given in Exercise 2a.

6. Consider the values of the function $-x + 3y + 2z$ at points belonging to the set given in Exercise 2c.
    (a) What is the maximum value?
    (b) At what point (or points) is the maximum value assumed?
    (c) What is the minimum value?
    (d) At what point (or points) does the function take on its minimum value?

7. At what point (or points) of the convex polyhedral set shown in Fig. 16 (see Exercise D-5) does the linear function $y - 4x$ have its greatest value?

8. At what point (or points) of the convex polyhedral set given in Exercise 2e does the linear function $x + 2y$ have its greatest value?

9. The graph of the system of inequalities

$$x \geq 0, \quad y \geq 0, \quad x + y \geq 3$$

is a convex polyhedral set. Show that on this set the linear function $x + 2y$ does not have a greatest value.

8 / Extreme point

9 Linear programming

Among the comparatively recent developments in mathematics is mathematical programming, a subject with rapidly increasing applications. Here, we briefly discuss a special case of mathematical programming known as linear programming. With the development in the past couple of decades of computing machinery capable of handling massive problems quickly and a breakthrough in the methods adaptable to machine use, we can now handle problems that formerly were too difficult. Among these problems are situations demanding a decision—specifically, a decision of how best to use what is available or needed. A businessman wishes to use his resources (products, labor, capital) to make as large a profit as possible. A space traveler wishes to select his equipment to occupy as little space as possible—or perhaps to have as little weight as possible—or perhaps to cost as little as possible. If there are many alternative ways in which some activity can be accomplished, which should be selected in order to achieve a certain goal in the best possible manner?

Example 21 A firm manufactures three products whose trade names are JIJ, JIK, and JIL. One plant can produce one JIJ, four JIKs, and four JILs in an hour at a cost of $30. Another plant can produce three JIJs, five JIKs, and one JIL in an hour at a cost of $20. The firm has an order for 80 JIJs, 230 JIKs, and 110 JILs. How much time should each plant run so that the order can be filled at the least cost?

The problem asks a typical query in mathematical programming: how can a certain required situation be achieved with minimum cost? Two unknowns are involved in this example, namely, the length of time each of the two plants is in operation. We proceed to formulate the requirements and the cost in terms of these unknown numbers.

Let the number of hours that the first plant runs be denoted by x, and let the second plant run y hours. Consider the product JIJ. The first plant produces one per hour, and in x hours produces x JIJs. The second plant, producing three each hour, produces $3y$ JIJs during its run. The total number produced by both plants is $x + 3y$. Since 80 JIJs have been ordered, together the plants must produce at least that many. In other words,

$$x + 3y \geq 80.$$

We next analyze the requirement for the product JIK. Running for x hours and producing four each hour, the first plant produces $4x$ JIKs. The second plant, producing five per hour, produces $5y$ JIKs. The total number of JIKs produced is $4x + 5y$. Since 230 JIKs are ordered, we require that

$$4x + 5y \geq 230.$$

A similar discussion for the product JIL shows that the total number produced by the two plants is $4x + y$ and that

$$4x + y \geq 110.$$

It is evident (and almost too trivial to mention) that the numbers x and y, representing the lengths of time of operation for the plants, are not negative. Hence, $x \geq 0$ and $y \geq 0$. We summarize the requirements by the following system of inequalities:

$$\begin{cases} x & \geq 0, \\ y \geq 0, \\ x + 3y \geq 80, \\ 4x + 5y \geq 230, \\ 4x + y \geq 110. \end{cases}$$

The graph of this system of linear inequalities is the convex polyhedral set shown in Fig. 20.

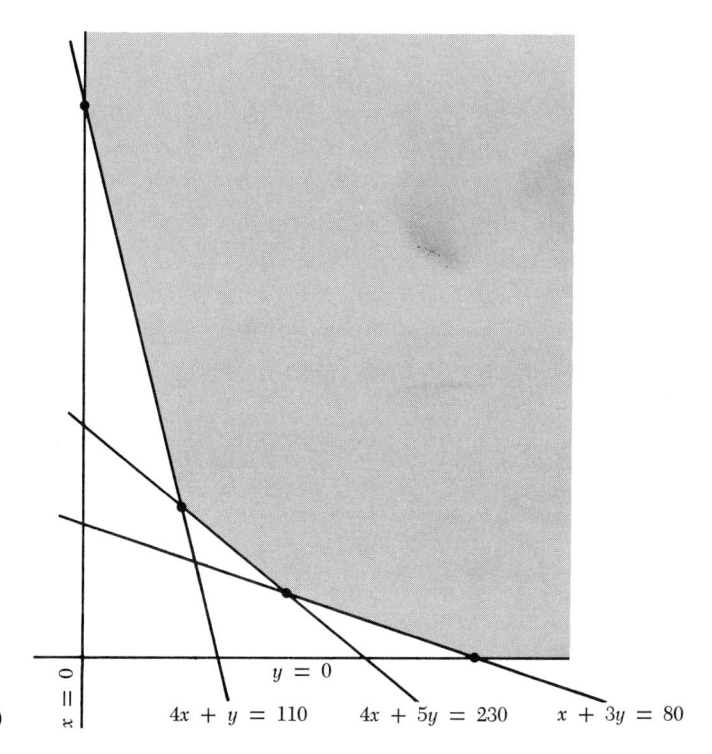

Figure 20

$x = 0$, $y = 0$, $4x + y = 110$, $4x + 5y = 230$, $x + 3y = 80$

There are four extreme points: $(80, 0)$, $(\frac{290}{7}, \frac{90}{7})$, $(20, 30)$, $(0, 110)$.

The cost of running the plants x and y hours, respectively, is $30x + 20y$ dollars. We wish to find the point (x, y) in the convex polyhedral set at which the function $30x + 20y$ assumes its least value. We have noted in Section 8 the general principle that we should

9 / Linear programming 145

examine the values of $30x + 20y$ at the four extreme points. The value at $(80, 0)$ is 2400, at $(\frac{230}{7}, \frac{90}{7})$ it is 1500, at $(20, 30)$ it is 1200, and at $(0, 110)$ it is 2200. The smallest of these values is 1200.

The firm should run the first plant 20 hours and the second plant 30 hours. The minimal cost is $1200.

We may note that this management decision means that the required JIKS and JILS will be produced and that a total of 110 JIJS will be produced, giving a surplus of 30 above the requirements. Any attempt to avoid this surplus would result either in failure to meet all other demands or in an increased cost. Since the three lines whose equations are

$$x + 3y = 80, \qquad 4x + 5y = 230, \qquad 4x + y = 110$$

do not have a common point, it is not possible for the firm to produce exactly the number of each product ordered.

EXERCISES F

1. Perform the following partial checks that $1200 is the least cost for the firm in Example 21.
   (a) Show that the point $(40, 15)$ is in the convex polyhedral set, and then evaluate the function $30x + 20y$ at this point.
   (b) Repeat part (a) for the point $(25, 30)$.
   (c) Repeat part (a) for the point $(20, 35)$.
   (d) Select two other points in the convex polyhedral set (not extreme points), and evaluate the cost function at each of them.
   (e) In each of parts (a)–(d), observe that the functional value exceeds 1200.

2. A refinery produces two fuel blends. A low-grade blend is 50% crude oil number 1, 40% crude oil number 2, and 10% additives; it nets the company a profit of $3 per barrel. A high-grade blend is 20% crude oil number 1, 60% crude oil number 2, and 20% additives; it nets a profit of $5 per barrel. The refinery has on hand 3000 barrels of crude oil number 1 and 3200 barrels of crude oil number 2 and 1000 barrels of additives.
   (a) Suppose that the refinery produces x barrels of the low-grade blend and y barrels of the high-grade blend. Explain why $\frac{5}{10}x + \frac{2}{10}y \leq 3000$.
   (b) Find the inequality that expresses the restriction that 3200 barrels of crude oil number 2 is available.
   (c) Find the inequality which the vector (x, y) must satisfy and which pertains to the supply of additives on hand.
   (d) Express by inequalities the restrictions on the individual components of the vector (x, y).
   (e) Sketch the convex polyhedral set that is the graph of the system of inequalities obtained in the preceding parts.
   (f) Find each extreme point of the set.
   (g) Express the total profit as a function of x and y.
   (h) What is the maximum profit that the refinery can achieve?
   (i) How many barrels of each blend should the refinery produce in order to obtain the maximum profit?

3. A factory makes three products that we may call A, B, C. An older machine turns out three items of product A, one of product B, and two of product C in a day at a cost of $300. A newer machine turns out two items of product A, four of product B, and five of product C in a day at a cost of $400. The manufacturer has an order for 26 items of A, 21 of B, and 32 of C.
   (a) How many days should each machine run in order that the demand be filled at the least cost?
   (b) What is the minimum cost?

4. Suppose that each of two foods supplies three nutrients not otherwise available. Each pound of Slix gives two units of nutrient E, four units of nutrient F, and one unit of nutrient G; it costs $4. Each pound of Krix gives three units of nutrient E, one unit of nutrient F, and two units of nutrient G; it costs $6. If a person's diet needs one unit of each of the three nutrients, how many pounds of each food should he eat in order to achieve the minimum cost?

5. A chemist is making certain solutions by mixing. Each liter of one solution requires two grams of chemical J and three grams of chemical L. Each liter of a second solution requires four grams of chemical J and one gram of chemical L. Each liter of a third solution requires one gram of chemical J and two grams of chemical L. The stockroom has available four grams of chemical J and six grams of chemical L. How many liters of each solution should the chemist make if he wishes to have the largest possible total number of liters of the three solutions?

6. A producer makes two kinds of alloy. The chart shows, for each of three essential substances, the quantities required in the alloys and the quantity available.

	Number of kilograms of substance needed in each metric ton of durable alloy	Number of kilograms of substance needed in each metric ton of strong alloy	Number of kilograms of substance available to producer
SUBSTANCE A	60	75	1260
SUBSTANCE D	10	15	240
SUBSTANCE G	30	20	560

The producer can easily sell all the alloy he makes. Each metric ton of the durable alloy nets $30 profit, and each metric ton of the strong alloy nets $36 profit.
   (a) How much of each alloy should be produced to attain maximum profit?
   (b) What is the maximum possible profit?
   (c) If the price of the strong alloy can be adjusted so that the profit on it is increased by $4 per metric ton, then what is the maximum possible total profit for the producer?

PART II
Probability

Sets and Logic

Counting

Powers and Sequences

Probability

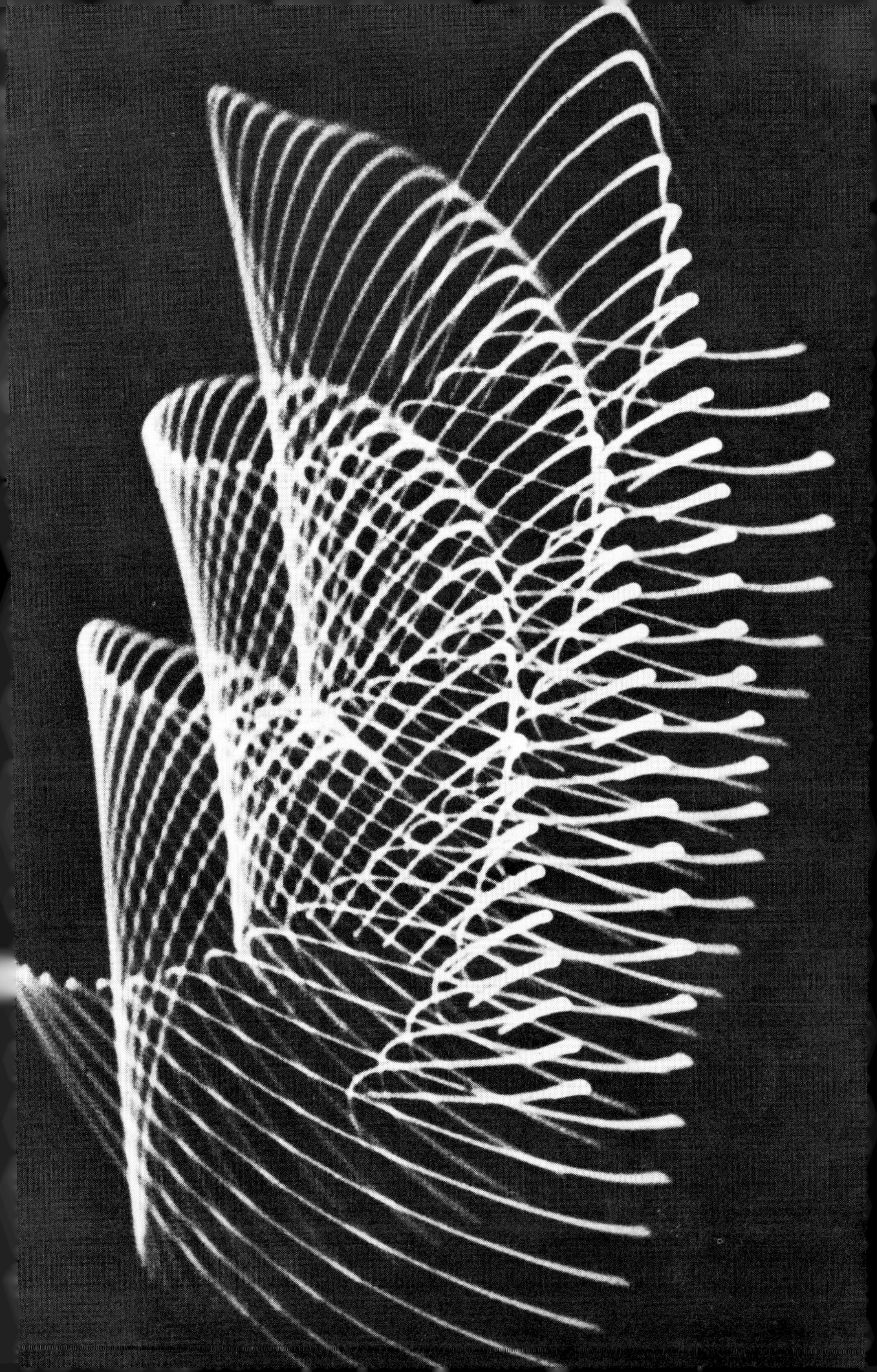

Sets and Logic

1 Introduction

We have, in a variety of situations, encountered the notion of a set. We have talked about a set of points, about a set of numbers, about a set of vectors, and about other sets. The graph of a linear equation on a two-dimensional coordinate system is a line, and a line is a collection, or set, of points. The roots of such an inequality as $x > 0$ form the set of all positive numbers. In a three-dimensional vector space we may consider the set of all vectors whose second component is zero.

If we have a system of linear inequalities in x and y, the graph of each inequality is a halfplane. The graph of the system is the intersection of the various halfplanes, that is, the set of points that belong to each and every one of the halfplanes. Since each halfplane is a set of points, we find here a way to describe a new set in terms of several given sets.

Much of mathematics—like much of science—is dependent, either explicitly or implicitly, upon the basic notions of set theory. We wish now to investigate these fundamental ideas.

2 Set, member, belongs to

The concept of a set itself is usually considered to be too primitive to be defined. In everyday language there are a number of words that are approximately synonymous with "set," such as "collection," "group," "aggregate," "bunch," "pile," etc. Very often one of these words is followed by the preposition "of." We speak of a pile of books, a collection of coins, a group of people. In general, a set is composed of "things," perhaps animate, perhaps inanimate, and in fact not necessarily tangible. Each of these things may be called a *member* of the set, and is said to *belong to* the set. An alternative word for *member* in this context is *element*.

Example 1 The number 7 is a member of the set of all positive numbers. The linear function $3x - 2$ belongs to the set of all linear functions whose graphs have slope 3. The column matrix $\begin{bmatrix} 1 \\ 6 \end{bmatrix}$ is an element of the set of roots of the inequality

$$\begin{bmatrix} 2 & 1 \\ 7 & -1 \end{bmatrix} X \geq \begin{bmatrix} 5 \\ 1 \end{bmatrix}.$$

Iowa belongs to the set of American states.

Very often a set is denoted by a capital letter, while a member of the set is denoted by a lower-case letter. Thus, one of the members of the set S may be the element b. To symbolize the statement that b belongs to S, we write $b \in S$. The symbol $\in$ may be read "belongs to," or "is a member of," or some equivalent phrase; sometimes we simply say "is in."

If a set has only a few elements, it may be easily identified by giving the entire membership list. For this purpose, braces, { }, are used to denote the set whose elements are listed within the braces.

Example 2 The set that has the three elements 0, 1, -1 is denoted by $\{0, 1, -1\}$.

Example 3 The letter e belongs to the set of English vowels; we may write $e \in \{a, e, i, o, u\}$.

When a set has many members, a complete membership list may not be convenient.

Often a set is described by naming a property that characterizes its members. Every element of the set has the property, and conversely every possessor of the property is an element of the set. Thus, possession of the property is the qualification for membership in the set.

Example 4 We may speak of the set of all integers. In this case the property is "being an integer." The integers are the numbers 0, 1, -1, 2, -2, 3, -3, 4, -4, etc. An attempt to make a complete listing would be an unending process. Nevertheless, we may consider the collection of all of them. Every member of the collection is an integer, and every integer belongs to the set.

As another illustration, in describing a three-dimensional vector space, we considered the set of all vectors that have three components. The membership qualification in this case is being a vector with three components.

3 Subset

Example 5 In the set of all integers, some members have the additional property of being positive. The positive integers (namely, 1, 2, 3, . . .) form a set by themselves. The set of positive integers may be called a subset of the set of integers.

Definition Suppose that S is a set and that T is a set. If each element that belongs to S also is a member of T, then S is called a *subset* of T.

We may write $S \subset T$; this statement, which means that S is a subset of T, is sometimes in pictorial language read as "S is contained in T."

Example 6 $\{a, i, y\} \subset \{a, e, i, o, u, x, y, z\}$. The set consisting of the elements a, i, y is contained in the set whose members are a, e, i, o, u, x, y, z.

Suppose that T is a set, and suppose we have a certain property that is meaningful when applied to the elements of T. The elements of T that possess the property form a subset of T. (Review Example 5, where we considered the set of integers and the property of being positive. For each integer, it is meaningful to discuss whether the integer is positive.)

The subset of T, consisting of precisely those elements with the property P, may be denoted by the notation $\{x \mid x \text{ has property } P\}$. Notice the use of the braces again to name a set. We read the vertical bar in this context by the phrase "such that"; thus the entire symbol $\{x \mid x \text{ has property } P\}$ may be read "the set of all x such that x has property P." (Sometimes a colon is used instead of the vertical bar.)

Example 7 In the set of integers, $\{x \mid x > 0\}$ represents the subset of positive integers. We may read $\{x \mid x > 0\}$ as "the set of all (integers) x such that x is greater than 0." The fact that x in this situation refers to an integer is known from the context. The use of the particular letter x is immaterial. The same subset of positive integers is equally well described by the notation $\{y \mid y > 0\}$ or by $\{k \mid k > 0\}$.

Example 8 On a two-dimensional coordinate system,

$$\{(x, y) \mid x + 2y \geq 0\}$$

is a closed halfplane—namely, the closed halfplane consisting of all points on or above the boundary $x + 2y = 0$, the line containing the origin and having slope $-\frac{1}{2}$. This halfplane, considered as a set of points, is a subset of the coordinate plane. The fact that x and y in this situation refer to real numbers is suggested by the context; any

Figure 1

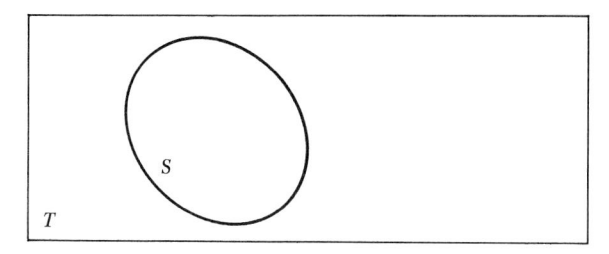

possible doubt is removed by the introductory phrase mentioning a "two-dimensional coordinate system."

Throughout this book, wherever such notation as $\{x \mid \ldots\}$ or $\{(x, y) \mid \ldots\}$ occurs, the elements x (or the elements x and y) are real numbers, unless the contrary is clear from the context.

The relationship between a set and a subset may be schematically pictured by a diagram (often called a *Venn diagram*). If, in Fig. 1, we imagine that the rectangle fences in the elements of the set T, and the oval encircles the members of the set S, then S is a subset of T, for a point within the oval is surely also inside the rectangle.

EXERCISES A

1. Let S be the set $\{a, b, c, d\}$. One of the subsets of S with two members is $\{c, b\}$. List all the other subsets of S that have two elements each. [HINT: There are five of them.]

2. Let T be the set $\{f, g, h\}$.
   (a) List all the subsets of T that have two elements each.
   (b) List all the subsets of T that have just one member each.
   (c) What subset of T has three elements?

3. Let W be the set $\{5, 8, -1\}$, and let Z be the set $\{8, 3, \frac{1}{2}\}$. Which (if any) of the following statements are true?
   (a) $5 \in W$.
   (b) $3 \in W$.
   (c) $3 \in Z$.
   (d) W and Z have the same number of elements.
   (e) W and Z have the same elements.
   (f) $\{-1, 5\}$ is a subset of W.
   (g) $\{\frac{1}{2}\}$ is a subset of Z.
   (h) $\{-1\} \subset W$.

4. Let M be the set $\{1, -1, 2, -2, 3\}$. List all the subsets of M that have four members each.

5. Suppose that a certain set has seven elements.
   (a) How many subsets of this set have six members each?
   (b) How many subsets of this set have one member each?
   [HINT: If necessary, make up an example of a set with seven elements, list the desired subsets, and count. Try to discover the pattern, and then decide whether your answer would be applicable to every set with seven elements.]

6. Let k be a natural number greater than 1, and let the set G have k members.
   (a) How many subsets of G have $k-1$ elements each?
   (b) How many subsets of G have one element each?

7. For each of the following, name the geometrical figure that is the graph of the given set.
   (a) $\{x \mid x \geq 2\}$.
   (b) $\{(x, y) \mid 8x + 11y = 6\}$.
   (c) $\{(x, y, z) \mid 2x - 3y - 5z = 1\}$.
   (d) $\{(x, y) \mid y = 0\}$.
   (e) $\{y \mid y = 0\}$.
   (f) $\left\{ \begin{bmatrix} x \\ y \end{bmatrix} \middle| \begin{bmatrix} 1 & 3 \\ 2 & 5 \end{bmatrix} \begin{bmatrix} x \\ y \end{bmatrix} = \begin{bmatrix} 4 \\ 7 \end{bmatrix} \right\}$.
   (g) $\left\{ \begin{bmatrix} x \\ y \end{bmatrix} \middle| \begin{bmatrix} -1 & -5 \\ 2 & 3 \\ -1 & 4 \end{bmatrix} \begin{bmatrix} x \\ y \end{bmatrix} \geq \begin{bmatrix} -17 \\ 9 \\ 1 \end{bmatrix} \right\}$.
   (h) $\{(5, -1, 2)\}$.
   (i) $\{5, -1, 2\}$.
   (j) $\{(x, y) \mid x + y \geq 2\}$.

8. Which, if any, of the following statements are true?
   (a) $(4, 1) \in \{(x, y) \mid x + y \geq 2\}$.
   (b) $9 \in \{t \mid t < 20\}$.
   (c) $(3, 2) \in \{(x, y) \mid 11y - 8x + 2 = 0\}$.
   (d) $\{2, -3, 4\} \subset \{v \mid v > 0\}$.
   (e) $\left\{ \begin{bmatrix} x \\ y \end{bmatrix} \middle| \begin{bmatrix} -1 & 2 \\ 3 & -1 \end{bmatrix} \begin{bmatrix} x \\ y \end{bmatrix} = \begin{bmatrix} 4 \\ 13 \end{bmatrix} \right\} = \left\{ \begin{bmatrix} 6 \\ 5 \end{bmatrix} \right\}$.
   (f) $\{(x, y) \mid x + y \geq 2\} \subset \{(x, y) \mid 2x + y \geq 3\}$.
   (g) If M is the set of all matrices each of which has an inverse, then
   $$\begin{bmatrix} 3 & 1 & 2 \\ 5 & 6 & 4 \end{bmatrix} \in M.$$

9. Make your own example of each of the following.
   (a) A set with exactly 50 members.
   (b) A set such that the number of its members is between 10,000 and 20,000.
   (c) A set with exactly one member.

10. For each of the sets you selected in Exercise 9, give a subset of the set.

4 Intersection, union, complement

Example 9 On a two-dimensional coordinate system, the lines whose respective equations are $-3x + 2y = 7$ and $x + y = 1$ intersect at the point $(-1, 2)$ (see Fig. 2). Each of the lines is a subset of the plane. The set $\{(-1, 2)\}$, consisting of the single point $(-1, 2)$, is a subset of the plane.

Example 10 On a two-dimensional coordinate system, each of the halfplanes $\{(x, y) \mid x + 3y \geq 5\}$ and $\{(x, y) \mid 2x - y \leq 1\}$ is a

154 Sets and Logic

Figure 2

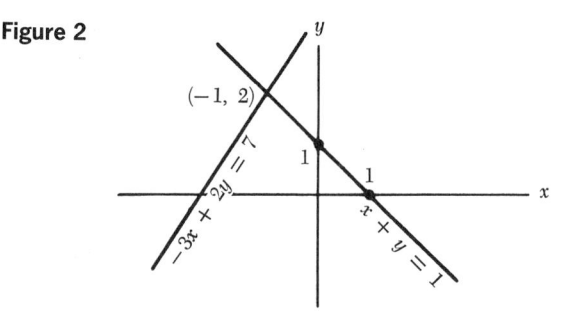

Figure 3

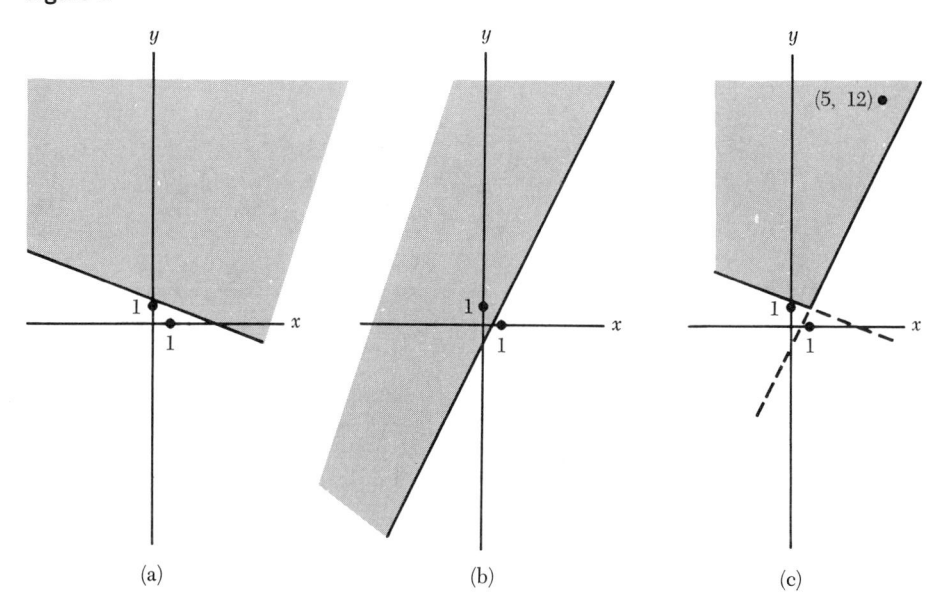

(a) (b) (c)

subset of the plane (see Figs. 3a and b). There are some points which belong to both halfplanes, for example, (5, 12) (see Fig. 3c). The set of all points that are members of both halfplanes is the intersection of the halfplanes.

The language of geometry suggests the following.

Definition Suppose that R, S, T are sets such that each of the sets R and S is a subset of T. The intersection of R and S is the subset of T consisting of all elements that belong to each of the sets R and S.

The intersection of R and S is denoted by $R \cap S$.

More generally, if we have several subsets of a set T, their intersection is the subset of T consisting of all elements that belong to each of the given subsets.

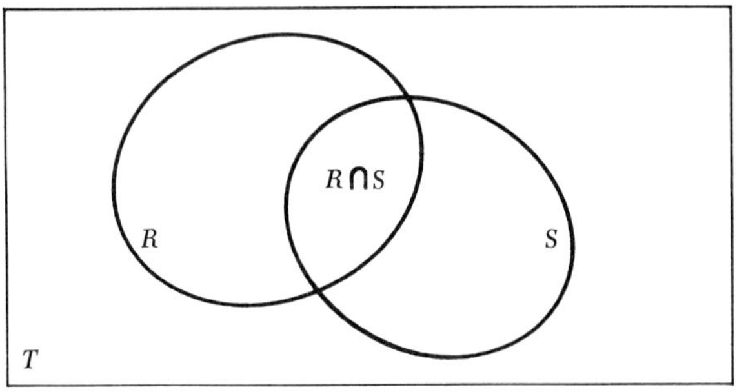

Figure 4

On the Venn diagram in Fig. 4, the intersection of the subsets R and S is represented by the lunar-shaped region inside both ovals.

Example 11 Let a and b be real numbers such that $a < b$. On a one-dimensional coordinate system, the intersection of the halfline given by the inequality $x \geq a$ and the halfline $x \leq b$ is the graph of the chain inequality $a \leq x \leq b$. (See Fig. 5.) Using our new symbolism, we may write

$$\{x \mid x \geq a\} \cap \{x \mid x \leq b\} = \{x \mid a \leq x \leq b\}.$$

Figure 5

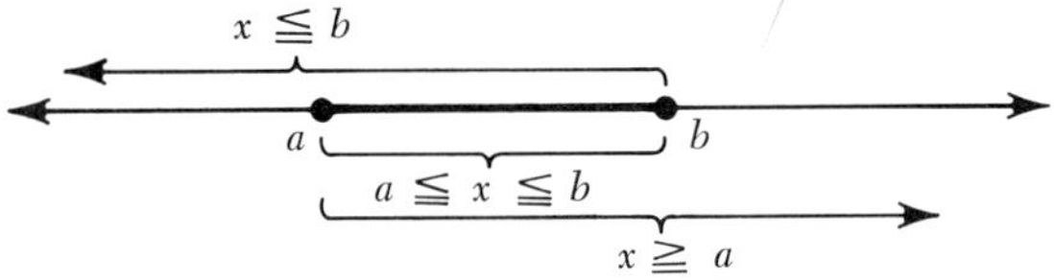

The operation of forming the intersection of given subsets is a method of constructing a new subset in terms of the given ones. Another important method involves the operation of forming the union.

Definition Suppose that R, S, T are sets such that each of the sets R and S is a subset of T. The union of R and S is the subset of T consisting of all elements that belong to R or belong to S or belong to both.

In other words, an element of T is a member of the union provided it belongs to at least one of the given subsets. The union of R and S is denoted by $R \cup S$.

More generally, if we have several subsets of a set T, their union is the subset of T consisting of all elements that belong to one or more of the given subsets.

Figure 6

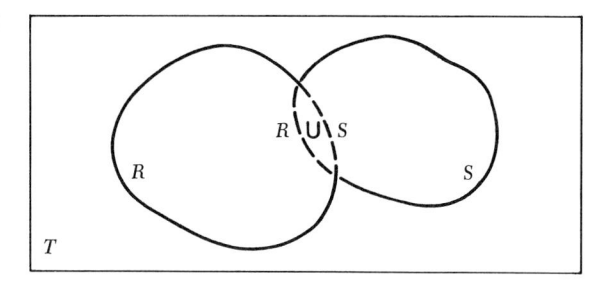

[Venn diagram showing two overlapping ovals R and S within rectangle T, with R ∪ S labeled]

On the Venn diagram in Fig. 6, the union of the subsets R and S is represented by the double-lobed region inside either (or both) ovals.

Don't count twice

Example 12 $\{a, b, c, d, e\} \cup \{a, e, i, o, u\} = \{a, b, c, d, e, i, o, u\}$. The union of the five-member set $\{a, b, c, d, e\}$ and the five-member set $\{a, e, i, o, u\}$ is a set with eight members. Each of the two elements a and e in the intersection $\{a, b, c, d, e\} \cap \{a, e, i, o, u\}$ is a member of the union, but is considered as a member of the union only once. Thus, we may obtain a collection of eight objects by putting together a collection of five objects and another collection of five objects.

Example 13 Again consider the set of (lower-case) letters of the English alphabet. If R is the subset $\{a, b, c, d\}$ and S is the subset $\{m, t, x\}$, then $R \cup S = \{a, m, b, t, d, x, c\}$. In this case, the union of R with four elements and S with three elements is a set with $4 + 3$ members. The reason that the number of members in $R \cup S$ is the sum of the number of elements in R and the number in S is that no letter belongs to both R and S. In other words, R and S do not overlap. In technical language, R and S are *disjoint* subsets.

In this case, what do we mean by $R \cap S$? The intersection of two subsets has been described as consisting of all elements that belong to both. Here, there is no such element. At first glance it may seem reasonable to deny that R and S have an intersection. In the long run, however, our work progresses more smoothly if we agree that two disjoint subsets do have an intersection and that the intersection is a subset with no members.

By special agreement, there is a set that has no member. It may be called the *empty set*. *or null set*

If any two subsets of a set have no element in common, their intersection is the empty set. An important illustration of two subsets of a set that do not have a common member is given by a subset and its complement.

4 / Intersection, union, complement   157

Definition Let R and T be sets such that R is a subset of T. The *complement* of R is the subset of T consisting of all elements that do not belong to R.

We may denote the complement of R by $\tilde{R}$.

Example 14 In the set of integers, the complement of the set of negative integers is the set of integers which are nonnegative, that is, which are positive or zero. If $N = \{x \mid x < 0\}$, then $\tilde{N} = \{x \mid x \geq 0\}$.

Example 15 In the set $\{a, b, c, d, e, f, g\}$, the complement of the subset $\{b, e\}$ is the subset $\{a, c, d, f, g\}$. In a set with seven elements, the complement of a subset with two members has $7 - 2$ members.

If R is a subset of a set T, then $R \cap \tilde{R}$ is the empty set; indeed no element of T can both belong to R and not belong to R. Also, $R \cup \tilde{R} = T$; indeed, every element of T either belongs to R or else does not belong to R.

EXERCISES B

1. Let $T = \{a, b, c, d, e, f, g, h, i, j, k\}$. Using the brace notation, express each of the following subsets.
   (a) The union of $\{a, f, d\}$ and $\{f, d, g, j\}$.
   (b) The complement of $\{d, f, i, k, b\}$.
   (c) $\{b, c, e, h, i, j\} \cap \{a, b, e, g, i, k\}$.
   (d) $\{d, j, f\} \cap \{e, a, f, k, b\}$.
   (e) $\tilde{R}$, where $R = \{c, e, g, k\} \cup \{a, f, i, k\}$.

2. Which, if any, of the following statements are true?
   (a) The complement of $\{x \mid x \geq 0\}$ is $\{x \mid x \leq 0\}$.
   (b) The union of $\{x \mid x \geq 3\}$ and $\{x \mid x \geq 1\}$ is $\{x \mid x \geq 1\}$.
   (c) The complement of $\{(x, y) \mid (x, y) > (0, 0)\}$ is $\{(x, y) \mid (x, y) \leq (0, 0)\}$.
   (d) $\{x \mid x \leq \frac{11}{3}\} \cap \{x \mid x \geq \frac{1}{4}\} = \{x \mid \frac{1}{4} \leq x \leq \frac{11}{3}\}$.
   (e) $\left\{ \begin{bmatrix} x \\ y \end{bmatrix} \Big| [2 \ 3] \begin{bmatrix} x \\ y \end{bmatrix} \geq 9 \right\} \cap \left\{ \begin{bmatrix} x \\ y \end{bmatrix} \Big| [-1 \ 4] \begin{bmatrix} x \\ y \end{bmatrix} \geq 1 \right\}$
   $= \left\{ \begin{bmatrix} x \\ y \end{bmatrix} \Big| \begin{bmatrix} 2 & 3 \\ -1 & 4 \end{bmatrix} \begin{bmatrix} x \\ y \end{bmatrix} \geq \begin{bmatrix} 9 \\ 1 \end{bmatrix} \right\}$.

3. Let a, b, c, d, e, f, g, h be nonzero numbers. Show that
$$\left\{ \begin{bmatrix} x \\ y \\ z \end{bmatrix} \Big| ax + by + cz = d \right\} \cap \left\{ \begin{bmatrix} x \\ y \\ z \end{bmatrix} \Big| ex + fy + gz = h \right\}$$

$$= \left\{ \begin{bmatrix} x \\ y \\ z \end{bmatrix} \middle| \begin{bmatrix} a & b & c \\ e & f & g \end{bmatrix} \begin{bmatrix} x \\ y \\ z \end{bmatrix} = \begin{bmatrix} d \\ h \end{bmatrix} \right\}.$$

4. Let P and Q be subsets of a set T. In each of the following Venn diagrams a subset of T is shaded. Express the shaded subset in terms of P and/or Q.

(a)

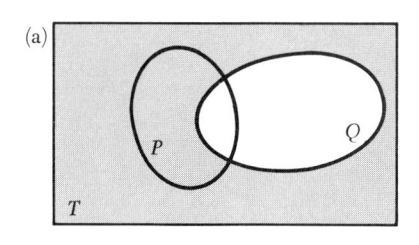

(b)

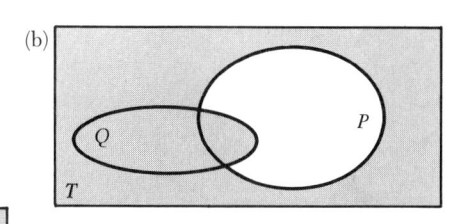

(c)

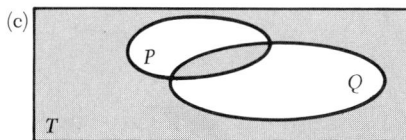

5. Let P, Q, R be subsets of a set T. Using the following figure as a pattern, represent each of the following subsets of T on a Venn diagram by shading the portion of the picture that depicts the subset.
   (a) $(P \cap \tilde{R}) \cup Q$.
   (b) $(\tilde{P} \cup \tilde{Q}) \cap (Q \cup R)$.
   (c) $Q \cup \tilde{T}$.
   (d) The complement of $P \cap \tilde{Q} \cap R$.
   (e) The complement of $\tilde{P} \cup \tilde{Q} \cup \tilde{R}$.

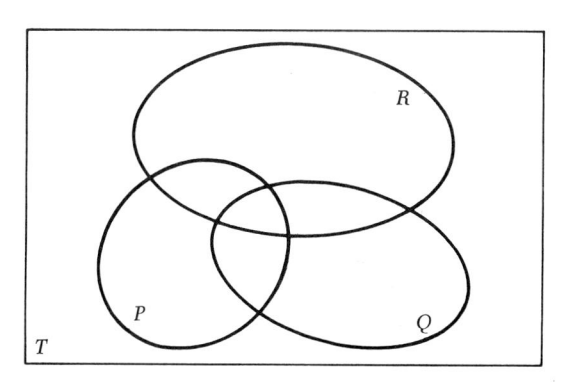

4 / Intersection, union, complement

6. Let P, Q, R be subsets of a set T. In each of the following Venn diagrams a subset of T is shaded. Express the shaded subset in terms of P and/or Q and/or R.

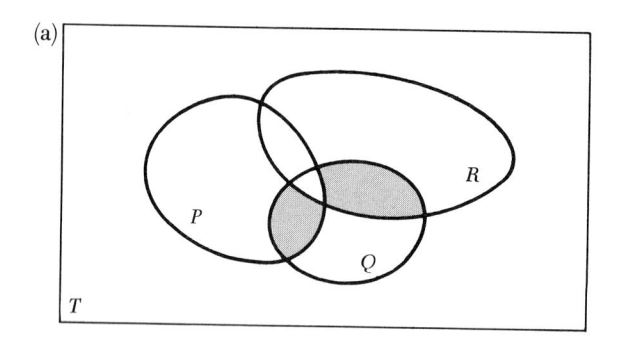

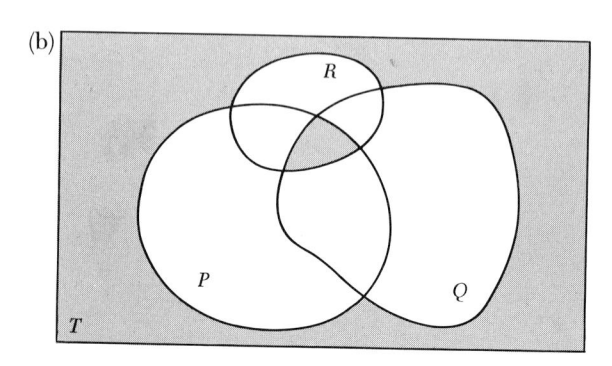

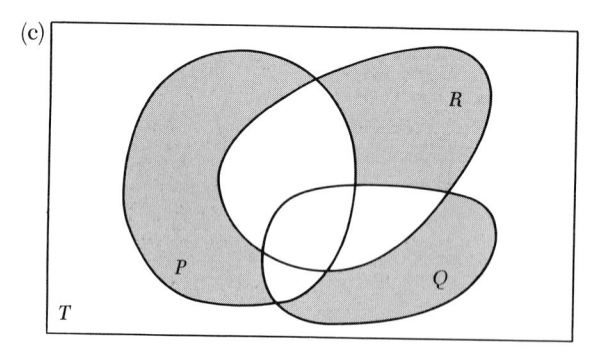

(a)

(b)

(c)

7. Let R and S be subsets of a set T. [HINT: In one or more parts of this problem, a diagram may be revealing.]
   (a) Find a simpler name for the subset $R \cap T$.
   (b) Simplify $S \cup T$.
   (c) Explain why $R \cap S$ is the same subset as $S \cap R$.
   (d) Find a simpler name for the subset $S \cup S$.
   (e) Simplify $R \cap R$.
   (f) Simplify $S \cup \tilde{T}$.
   (g) Simplify $\tilde{R} \cap \tilde{T}$.
   (h) Simplify $R \cup (R \cap S)$.
   (i) Simplify $S \cap (R \cup S)$.

160   Sets and Logic

(j) Simplify $R \cup (\tilde{R} \cap S)$.
(k) Explain why $(R \cap S) \cup (R \cap \tilde{S})$ is the same subset as R.
(l) Simplify $(S \cup R) \cap (S \cup \tilde{R})$.

8. Let R and S be subsets of a set T. If $R \subset \tilde{S}$, show that the intersection $R \cap S$ is the empty set.

9. Let P and Q be subsets of a set T. If P and Q are disjoint, show that $P \subset \tilde{Q}$ and that $Q \subset \tilde{P}$.

10. Let I be the set $\{\square, \star, \triangle\}$. Let A, B, C be the subsets of I given by $A = \{\star, \triangle\}$, $B = \{\triangle, \square\}$, $C = \{\square, \star\}$. Let R, S, T be the singleton subsets of I given by $R = \{\square\}, S = \{\star\}, T = \{\triangle\}$. Let Q denote the empty subset of I.
(a) Construct a table showing the intersection of each ordered pair of subsets of I.
[HINT: The pattern appears in Table 1. The intersection symbol in the upper left corner identifies the operation. Each subset appears as a column heading and as a row heading. In the body of the table, the entry in row A and column B is T because $A \cap B = \{\star, \triangle\} \cap \{\triangle, \square\} = \{\triangle\} = T$. The entry in row C and column S is S because $C \cap S = S$. Copy the pattern and fill in all the missing entries in the table.]
(b) Construct a table showing the union of each ordered pair of subsets of I.
(c) Construct a table showing the complement of each subset of I.
[HINT: The pattern appears in Table 2. Each subset appears as a row heading. Opposite it, in the column indicated by the complementation symbol, is its complement. For example, B appears opposite S, because $\tilde{S} = B$, that is, the complement of $\{\star\}$ is $\{\triangle, \square\}$.]

Table 1

$\cap$	Q	I	R	A	S	B	T	C
Q								
I								
R	Q	R						
A					T			
S								
B								
T								
C				S				

Table 2

	$\sim$
Q	
I	
R	
A	
S	B
B	
T	
C	

5 Number of elements

If R is a subset of a finite set T, then the number of elements in $\tilde{R}$ is the difference between the number of members of T and the number of elements in R. Examples 12 and 13 suggest the following result.

> **Theorem** If R and S are subsets of a finite set T, then the sum of the number of elements in R and the number in S is the same as the sum of the number of elements in $R \cup S$ and the number in $R \cap S$.

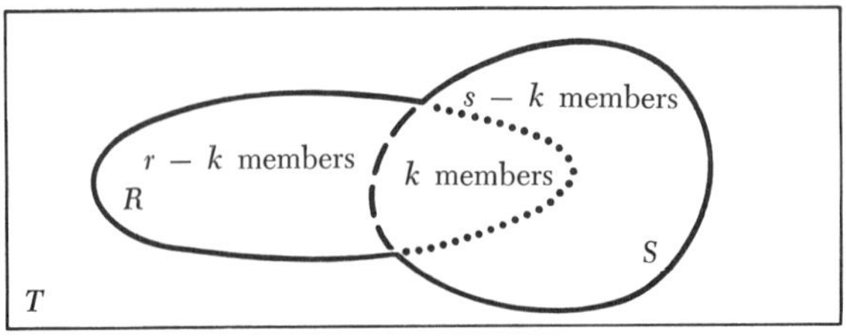

Figure 7

PROOF Suppose there are r elements in R, and s elements in S, and k elements in $R \cap S$, and u elements in $R \cup S$. In this notation, we are claiming that $r + s = u + k$. Figure 7 shows how $R \cup S$ is decomposed into three nonoverlapping subsets. There are k members in $R \cap S$. Since R has r elements, the remaining $r - k$ elements lie in the left-hand region. Similarly, the right-hand region has $s - k$ members. Summing, we obtain

$$u = (r - k) + k + (s - k).$$

That is, $u = r + s - k$, or

$$r + s = u + k;$$

thus, our assertion is established.

In Chapter 6 we will pursue much more extensively the notion of the number of elements in a set.

EXERCISES C

1. Let A and B be subsets of a set T. If $A \subset B$, show that $\tilde{B} \subset \tilde{A}$.

2. Suppose that T is a set with 40 members, and suppose that A and B are disjoint subsets of T with 23 and 17 elements, respectively.
   (a) Give a simpler name for the subset $A \cup B$.
   (b) How many elements belong to the set $\tilde{A}$?
   (c) How many members does the set $\tilde{B}$ have?
   (d) Explain why $\tilde{A}$ and $\tilde{B}$ are disjoint subsets.

3. Suppose that T is a set with 40 members, and suppose that C and D are disjoint subsets of T with 16 elements and 11 elements, respectively. Decide whether $\tilde{C}$ and $\tilde{D}$ are disjoint subsets; justify your decision.

4. Suppose that T is a set with 40 members. Suppose that A and B are subsets of T with 20 elements and 30 elements, respectively. Which, if any, of the following statements are incompatible (that is, inconsistent) with the preceding information?
   (a) There are fewer than eight elements in $A \cap B$.
   (b) $A \subset \tilde{B}$.
   (c) $A \subset B$.
   (d) The subset $\tilde{A} \cap \tilde{B}$ consists of five members.
   (e) The number of elements in $A \cap \tilde{B}$ is 12.

5. Suppose that T is a set with 60 elements. Suppose that P and Q are subsets of T such that 22, 34, and 15 are the numbers of members in P, in Q, and in $P \cap Q$, respectively. How many members does each of the following subsets have?
   (a) $\tilde{Q}$.   (b) $P \cup Q$.   (c) $P \cap \tilde{Q}$.   (d) $\tilde{P} \cap Q$.   (e) $\tilde{P} \cap \tilde{Q}$.   (f) $\tilde{P} \cup \tilde{Q}$.

6. Suppose that P and R are subsets of a set S. Suppose that the numbers of elements in the sets $P \cap R$, $\tilde{P} \cap \tilde{R}$, $P \cap \tilde{R}$, and R are, respectively, 33, 65, 45, and 90. Calculate the number of members in each of the following subsets.
   (a) $\tilde{P} \cap R$.   (b) P.   (c) $P \cup R$.   (d) S.   (e) $R \cup (R \cap \tilde{P})$.   (f) $R \cap \tilde{R}$.

6 And, not, or

Our discussion of the subsets of a set has relied upon our understanding of the simple words AND, OR, NOT. Suppose that R and S are subsets of a set T. Then $R \cap S$ is described as the set $\{x \mid x \in R \text{ and } x \in S\}$; a key word is AND. The union is given by $R \cup S = \{x \mid x \in R \text{ or } x \in S\}$; a crucial word is OR. The complement of R is identified as $\tilde{R} = \{x \mid x \text{ does not belong to } R\}$; a decisive word is NOT. These three words are vital in describing subsets in terms of given subsets. They are the basic words in forming compound properties that serve as the membership qualifications for subsets.

We wish to study the role of these three words in mathematics. Each of the words, AND, OR, is used to connect two simpler sentences into a single compound sentence; the word NOT also changes a simpler structure into a more complex one. It is convenient to refer to each of the three as a connective, or more specifically, as a logical connective. Our problem is to investigate how a compound statement formed by using one or more of these connectives on simple statements is related to the individual simple statements. The relationship involves the notion of *true* as applied to statements.

> **Definition**   A sentence that is either true or else false may be called a *statement*.

> **Example 17**   An example of a true statement is $5 > 3$. An example of a false statement is $1 = 0$.

> **Example 18**   A sentence that (standing alone) is not a statement is $x > 2$. This sentence is neither true nor false; it has no truth value. Instead, it hints at a question. It suggests the question: What number (or numbers) can be substituted for x in order to make the resulting sentence a true statement? Substitution of 7 for x yields "$7 > 2$," a statement which is true. On the other hand, replacement of x by 1 yields the false statement $1 > 2$. A number which, upon substitution for x, converts $x > 2$ into a true statement is called, as we know, a root of the inequality.

The general problem of solving an equation or solving an inequality is the problem of identifying what substitutions in a sentence, which itself is neither true nor false, will convert the sentence into a true statement.

A statement has a truth value. Either the statement is true or else it is false. We often denote a statement by a lower-case letter such as p or q or r.

Suppose we have two statements. We may combine them into a single statement with the connective AND. How is the truth value of the compound statement related to the truth values of the individual statements? Reflection concerning the common usage of the connective AND suggests that the compound statement is true in case both of the simple statements are true. On the other hand, in case one, or perhaps both, of the individual statements are false, then the statement obtained by joining them with AND is itself false.

Example 19 Each of the two following statements is true:

"Boston is a seaport";
"The Empire State Building is in New York City."

The following compound statement is also true:

"Boston is a seaport and the Empire State Building is in New York City."

However, the following compound statement is false:

"Boston is a seaport and the Empire State Building is in Boston."

Likewise, the following statement, formed from two false statements, is false:

"Boston is a port on the Gulf of Mexico and the Empire State Building is in Boston."

Definition Let p be a statement, and let q be a statement. The compound statement formed from the two given statements with the logical connective AND, namely, the statement "p AND q," is called the *conjunction* of p and q.

We sometimes use the special symbol $\wedge$ for the logical connective AND. Thus, the conjunction of p and q may be written $p \wedge q$.

A table is a convenient way to summarize our remarks about the truth value of a conjunction of statements. We list in Table 3a, on different rows, all the possibilities for the truth values of the statements in the pair (p, q). (The usual initials T, F refer to true, false, respectively.) We note there are

Table 3a

p	q	$p \wedge q$
T	T	
T	F	
F	T	
F	F	

Sets and Logic

four possible cases for the ordered pair (p, q). We now complete the table by giving, opposite each possibility, the corresponding truth value of $p \wedge q$. The finished product is Table 3b.

Table 3b

p	q	$p \wedge q$
T	T	T
T	F	F
F	T	F
F	F	F

An extremely important fact to be noted is that the truth value of the conjunction, $p \wedge q$, depends only on the truth value of p and the truth value of q. It does not depend directly upon what the individual statements p, q are, that is, upon their content. The subject matter they discuss is irrelevant.

Example 20 Let p be the statement:

"Boston is a seaport."

The negation, or denial, of p is formed by using the connective NOT. The new statement may be called NOT-p. However, in English sentences, the word "not" more commonly appears near the verb:

"Boston is not a seaport."

Consider the statement

"The Empire State Building is in Boston";

its negation is

"The Empire State Building is not in Boston."

Definition Let p be a statement. The statement formed from p by using the logical connective NOT, namely, the statement "NOT-p," is called the *negation* of p.

Ordinary usage of the word "not" leads us to decide that the negation of a true statement is false, while the negation of a false statement is true. The negation of a statement p may be written $\sim p$, which we read as NOT-p. The table for the logical connective NOT is Table 4. In brief, a statement and its negation have opposite truth values.

Table 4

p	$\sim p$
T	F
F	T

The connective AND and the connective NOT have rather clear-cut meanings in ordinary language, and we have been able easily to construct appropriate truth tables. By contrast, the connective OR does not have a uniform connotation in everyday language. Let p, q be statements. What can we say about the truth value of the compound statement p OR q? Certainly, p OR q is a false statement in case each of the given statements is false. Furthermore, p OR q is a true statement in case exactly one of the given statements is true. Thus, we have no difficulty in filling in most of the entries (see Table 5a) in a truth table for the statement p OR q.

Table 5a

p	q	
T	T	
T	F	T
F	T	T
F	F	F

The uncertainty arises in deciding the truth value for the compound statement p OR q, in case (p, q) is a pair of true statements. This important case is very often excluded from consideration in the everyday use of the word "or." The exclusion is commonly tacit and is based on the content of the particular statements p, q in the sentence, namely, in case the particular p and the particular q are inconsistent, that is, cannot both be true.

We need a truth value for the compound statement p OR q in case (p, q) is a pair of true statements. A convenient choice is to agree that the compound statement is true in this case. Our use of the logical connective OR then has the same meaning as the legal phrase "and/or." When we say that p OR q is true, we mean that p is true, or q is true, or both p and q are true; in other words, we mean that at least one of the simple statements is true.

Definition Let p be a statement, and let q be a statement. The compound statement formed from the given statements with the connective OR, namely, the statement "p OR q" (subject to the above agreement on the meaning of the connective), is called the *disjunction* of p and q.

A special symbol for this logical connective is $\vee$. Thus, the disjunction of p and q may be written $p \vee q$.

Table 5b for the disjunction emphasizes our agreement about the meaning of OR. We stress that the truth value of a disjunction of statements depends only upon the truth values of the individual statements and not specifically upon the content of the statements.

Table 5b

p	q	$p \vee q$
T	T	T
T	F	T
F	T	T
F	F	F

166   Sets and Logic

EXERCISES D

1. Let p, q be statements.
   (a) Show that, no matter what the truth value of p is and regardless of what the truth value for q is, the truth value of the conjunction $p \wedge q$ is the same as the truth value for the conjunction $q \wedge p$.
   (b) Show that the truth value for the disjunction $p \vee q$ is the same as the truth value for the disjunction $q \vee p$ in every possible case.

2. Let p be a statement.
   (a) Write the truth table for the compound statement $p \wedge p$. [HINT: Two rows in the table suffice.]
   (b) Show that, in every possible case, the conjunction $p \wedge p$ has the same truth value as p.
   (c) Write the truth table for the compound statement $p \vee p$.
   (d) Show that, in every possible case, the disjunction $p \vee p$ has the same truth value as p.

3. Let p be a statement, and let u be a true statement.
   (a) Write the truth table for the compound statement $p \wedge u$.
   (b) What simple statement has, in every case, the same truth value as the conjunction $p \wedge u$?
   (c) Write the truth table for $p \vee u$.
   (d) What simple statement has, in every case, the same truth value as $p \vee u$?

4. Let w be a false statement, and let q be a statement. For each of the following, what simple statement has, in every case, the same truth value as the given compound statement?
   (a) $w \vee q$. (b) $w \wedge q$. (c) $w \vee w$.

5. Let the (true) statement

   The earth is a planet

   be abbreviated by p, and let the (false) statement

   Water is a chemical element

   be abbreviated by q.
   (a) Write, in English prose, the conjunction of p and q.
   (b) Write, in English prose, the disjunction of $\sim p$ and q.
   (c) What is the truth value for $p \vee q$?
   (d) Give the truth value for $(\sim p) \wedge q$.
   (e) Give the truth value for $p \wedge (\sim q)$.
   (f) Write, in English prose, the negation of the disjunction of p and $\sim q$.

7 Compound statements

Many compound statements may be formed from simple statements by successive use of our three basic connectives.

Example 21 Let p, q, r be statements. Then $p \wedge q$ is a statement. Also, $\sim r$ is a statement. Hence, $(p \wedge q) \vee (\sim r)$ is a statement. Furthermore, $[(p \wedge q) \vee (\sim r)] \wedge [\sim(p \vee r)]$ is a statement.

Example 22 Write a table showing the truth value of the compound statement $p \wedge (\sim q)$ for each of the four possible cases of truth values associated with the ordered pair (p, q).

We first handle the negation and afterward consider the conjunction. In Table 6a each entry in the column headed $\sim q$ is the

Table 6a

p	q	$\sim q$	$p \wedge (\sim q)$
T	T	F	
T	F	T	
F	T	F	
F	F	T	

opposite of the corresponding entry in the column headed q. Now, in order to evaluate the entries in the next column, we need to compare the entries in the column headed p and the column headed $\sim q$. We recall that a conjunction of two statements is true in case each of the statements is true, but it is false in every other case. An examination of the first and third columns reveals that each of p and $\sim q$ is true in the second row but in no other row. Thus, in the fourth column we enter T in the second row and F in every other row. The completed final column is shown in Table 6b.

Table 6b

p	q	$\sim q$	$p \wedge (\sim q)$
T	T	F	F
T	F	T	T
F	T	F	F
F	F	T	F

With the understanding that the symbol $\sim$ applies to the next following symbol, we customarily suppress the parentheses in $p \wedge (\sim q)$ and write $p \wedge \sim q$.

Example 23 Write a table showing how the truth value of the compound statement $\sim p \vee q$ is determined by the truth values of p and of q.

According to the above convention, $\sim p \vee q$ means $(\sim p) \vee q$. Our task is a two-stage process. First we enter the truth values for $\sim p$, obtaining Table 7a. The disjunction of two statements is false in case each individual statement is false, but it is true in every other case. A comparison of the third and second columns shows that each of the statements $\sim p$ and q is false in the second row but in no other row. Thus, the completed table is Table 7b.

Table 7a

p	q	$\sim p$	$\sim p \vee q$
T	T	F	
T	F	F	
F	T	T	
F	F	T	

Table 7b

p	q	$\sim p$	$\sim p \vee q$
T	T	F	T
T	F	F	F
F	T	T	T
F	F	T	T

We remark that the column headed $\sim p$ is a useful aid in attaining the goal, but is not an essential part of the answer. The answer to the problem consists of the first, the second, and the final columns of Table 7b.

Example 24 If p, q, r are statements, write a truth table for the compound statement $(p \wedge q) \vee \sim r$.

Since there are two possible truth values for each of the statements p, q, r, there are eight possible cases to be considered for the ordered triple (p, q, r). Each possibility is listed in one of the eight rows in the body of Table 8a. We fill in the column headed $p \wedge q$ by comparing

Table 8a

p	q	r	$p \wedge q$	$\sim r$	$(p \wedge q) \vee \sim r$
T	T	T			
T	F	T			
F	T	T			
F	F	T			
T	T	F			
T	F	F			
F	T	F			
F	F	F			

the first two columns. Each entry in the fifth column is the opposite of the corresponding entry in the column headed r. When this portion of the work has been finished, the uncompleted assignment looks like Table 8b.

Table 8b

p	q	r	$p \wedge q$	$\sim r$	$(p \wedge q) \vee \sim r$
T	T	T	T	F	
T	F	T	F	F	
F	T	T	F	F	
F	F	T	F	F	
T	T	F	T	T	
T	F	F	F	T	
F	T	F	F	T	
F	F	F	F	T	

Examination of the auxiliary columns for $p \wedge q$ and $\sim r$ enables us to complete the table. We enter F wherever both $p \wedge q$ and $\sim r$ are false; otherwise, we enter T. The desired result (with the auxiliary columns omitted) is Table 8c.

Table 8c

p	q	r	$(p \wedge q) \vee \sim r$
T	T	T	T
T	F	T	F
F	T	T	F
F	F	T	F
T	T	F	T
T	F	F	T
F	T	F	T
F	F	F	T

Example 25 If p and q are statements, an important compound statement is the following: $(p \wedge q) \vee (\sim p \wedge \sim q)$. Construction of its truth table is accomplished in five steps. First, we evaluate $p \wedge q$; second, $\sim p$; third, $\sim q$. After these steps, we have Table 9a. The

Table 9a

p	q	$p \wedge q$	$\sim p$	$\sim q$	$\sim p \wedge \sim q$	$(p \wedge q) \vee (\sim p \wedge \sim q)$
T	T	T	F	F		
T	F	F	F	T		
F	T	F	T	F		
F	F	F	T	T		

entries in the column headed $\sim p \wedge \sim q$ are obtained from the preceding two columns. Recall that a conjunction is true in case each simple statement is true, but false otherwise. The final column is completed by examining the third and sixth columns. Recall that a disjunction is false in case each statement is false, but true otherwise.

From Table 9b, we observe that the compound statement $(p \wedge q) \vee (\sim p \wedge \sim q)$ has the following characteristic: it is true in case the truth values of p and q are the same, but it is false in case p and q have opposite truth values.

Table 9b

p	q	$p \wedge q$	$\sim p$	$\sim q$	$\sim p \wedge \sim q$	$(p \wedge q) \vee (\sim p \wedge \sim q)$
T	T	T	F	F	F	T
T	F	F	F	T	F	F
F	T	F	T	F	F	F
F	F	F	T	T	T	T

EXERCISES E

1. Let p be a statement, and let q be a statement. Write the truth table for each of the following compound statements.
   (a) $p \vee \sim q$.
   (b) $\sim(p \vee q)$.
   (c) $\sim p \wedge \sim q$.
   (d) $(p \vee q) \wedge (\sim p \vee \sim q)$.
   (e) $(p \vee q) \vee \sim q$.

2. Let p and q be statements. Show that, in every possible case, the truth value of $\sim(p \wedge q)$ is the same as the truth value of $\sim p \vee \sim q$.

3. At a certain place and time some possible weather forecasts for the next 12 hours are: it will be sunny; it will be rainy. Denote these respective statements by s and r. Translate each of the following forecasts into symbolic statements, using the symbols r, s, $\vee$, $\wedge$, $\sim$, appropriately.
   (a) "Either the sun will shine or it will rain."
   (b) "It will be sunny and no rain will fall."
   (c) "It will be neither sunny nor rainy."
   (d) "Either there will be sunshine but no rain, or there will be rain and no sunshine."

4. Let p, q, r be statements.
   (a) Write the truth table for the statement $p \vee (q \wedge r)$.
   (b) Write the truth table for the statement $(p \vee q) \wedge (p \vee r)$.
   (c) Show that the two compound statements in parts (a) and (b) have the same truth value in every possible case.

5. Let p, q, r be statements.
   (a) Write the truth table for $p \wedge (\sim q \vee r)$.
   (b) Write the truth table for $\sim p \vee (q \wedge \sim r)$.
   (c) Compare the truth tables in parts (a) and (b).

6. Let p be a statement.
   (a) Show that the statement $p \vee \sim p$ is true, no matter what the statement p is.
   (b) Show that, in every case, the statement $p \wedge \sim p$ is false.

7. Let p be a statement.
   (a) Write the truth table for $\sim(\sim p)$.
   (b) Show that, in every case, the truth value for $\sim(\sim p)$ is the same as the truth value for p.

8. Let p, q be statements.
   (a) Write the truth table for $p \vee (p \wedge q)$.
   (b) What simple statement has, in every possible case, the same truth value as $p \vee (p \wedge q)$?
   (c) Write the truth table for $(p \vee q) \wedge q$.
   (d) What simple statement agrees in truth value with $(p \vee q) \wedge q$ in every case?

9. Let q, r, s be statements. For each of the following write, in English prose, an argument that justifies the claim. (In this problem, inspection of truth tables is not an acceptable argument.)
   (a) There is one and only one case for which the compound statement $q \wedge (r \wedge s)$ is true; it is the case in which each of the individual statements q, r, s is true.
   (b) There is exactly one case for which $(q \wedge r) \wedge s$ is true, namely, the case in which each of the given statements q, r, s is true.
   (c) The two compound statements $q \wedge (r \wedge s)$ and $(q \wedge r) \wedge s$ have the same truth value in all possible cases.
   (d) The compound statement $q \vee (r \vee s)$ is true in every case except one, namely, the case in which each of the statements q, r, s individually is false.

(e) The statements $q \lor (r \lor s)$ and $(q \lor r) \lor s$ agree in their truth value in all possible cases.

10. Let p, q, r, s be statements.
    (a) Insofar as truth value is concerned, how many possible cases are there for the ordered quadruple (p, q, r, s)?
    (b) Write the truth table for the compound statement $[\sim p \lor \sim q] \land [q \lor (r \lor s)]$.
    (c) Write the truth table for the compound statement $[(q \land r) \land s] \lor [\sim(p \land r)]$.

8 If . . . , then . . . ; if and only if

Another significant logical connective is IF . . . , THEN . . . . Let p be a statement, and let q be a statement. Then the sentence IF p, THEN q is a statement. In case each of the given statements is true, we readily agree that the compound statement is true. Common usage of the word "if" also dictates that the compound statement is false in case p is true but q is false. Table 10a summarizes these remarks and contains space for two more entries.

Table 10a

p	q	
T	T	T
T	F	F
F	T	
F	F	

When a statement of the type "if p, then q" is made in everyday language, the primary concern is vested in the case that p is true. The proper interpretation in the case that p is false is more obscure. Thus, common knowledge does not appear to prescribe the two missing entries in Table 10a. We again adopt the charitable viewpoint of assigning "true" as the truth value in any case left in doubt by common sense. Fortunately, this choice turns out to be a conveniently useful decision in our later development.

The statement IF p, THEN q is symbolized by $p \to q$ and is called a *conditional* statement. The completed table for the conditional is Table 10b. In accordance with our agreement, the conditional $p \to q$ is true in every case except the case in which p is true and q is false.

A comparison of Table 10b with Table 7b in Example 23 reveals that the two compound statements, $p \to q$ and $\sim p \lor q$, agree in their truth values in

Table 10b

p	q	$p \to q$
T	T	T
T	F	F
F	T	T
F	F	T

172   Sets and Logic

every possible case. The two compound sentences may be considered to have the same meaning.

Example 26 Let A be a certain triangle, and let B be a particular triangle in the same plane as A. Let p be the statement that

A and B are congruent;

let q be the statement that

A and B are similar;

let r be the conditional statement that

if A and B are congruent, then A and B are similar.

In other words, r is an abbreviation for $p \to q$. It is possible that each of the statements p, q is true; in this case, r is also true. It is possible that A and B are not congruent but are similar; in this case, p is false, q is true, and (in accordance with our agreement) $p \to q$ is true. It is also possible that A and B are neither congruent nor similar; in this case, p is false, q is false, but r is true.

The appropriate geometric definitions reveal that, in this example, the conditional statement $p \to q$ is true in every case. Thus, it is not possible for r to be false; in other words, the following case is not possible: p is true and q is false. The fact that r is true for every pair (A, B) of coplanar triangles is expressed by calling the statement

if A and B are congruent, then A and B are similar

a *theorem* in plane geometry.

The fifth logical connective we wish to discuss is IF AND ONLY IF. We describe this connective in terms of those we have already studied. Let p be a statement, and let q be a statement. The sentence p IF AND ONLY IF q is a statement that means the same as the statement IF q, THEN p; AND IF p, THEN q. Thus, the new connective is an abbreviation for the conjunction of a pair of conditionals; we call the statement p IF AND ONLY IF q the *biconditional* of p and q. The symbol which we use is a double arrow $\leftrightarrow$.

From the definition, the biconditional statement $p \leftrightarrow q$ has the same truth table as the conjunction of the individual conditional statements. We apply our methods for constructing the truth table for $(q \to p) \wedge (p \to q)$. After two steps, we have Table 11a. We use the third and fourth columns to fill in the entries in the final column. Since the heading of the final column in Table 11a

Table 11a

p	q	$q \to p$	$p \to q$	$(q \to p) \wedge (p \to q)$
T	T	T	T	
T	F	T	F	
F	T	F	T	
F	F	T	T	

Table 11b

p	q	$p \leftrightarrow q$
T	T	T
T	F	F
F	T	F
F	F	T

has the same meaning as the biconditional, we obtain Table 11b (in which the auxiliary columns of Table 11a have been suppressed).

Note that the biconditional statement $p \leftrightarrow q$ is characterized as follows: it is true in case the truth values of p and q are the same, but it is false in case p and q have opposite truth values. This observation recalls the final remark in Example 25. The two compound statements, $p \leftrightarrow q$ and $(p \wedge q) \vee (\sim p \wedge \sim q)$, have the same truth tables. In other words, they agree in their truth values in every possible case.

We may analyze this particular agreement in truth value further. The conjunction $\sim p \wedge \sim q$ is true in case each of the statements $\sim p$, $\sim q$ is true, that is, in case each of the statements p, q is false. The conjunction $p \wedge q$ is of course true in case each of the statements p, q is true. Hence, the disjunction $(p \wedge q) \vee (\sim p \wedge \sim q)$ is true either in case p and q are both true or in case they are both false. More briefly, the disjunction is true in case p, q have the same truth value.

In a compound statement involving several connectives, it is convenient to have conventions that reduce the number of parentheses needed for clarity. We agree that the connective $\sim$ is applied first, that the connectives $\wedge$, $\vee$ are applied next, and that the connectives $\rightarrow$, $\leftrightarrow$ are applied last.

Example 27 By our agreement, $p \wedge q \leftrightarrow q \vee r$ means $(p \wedge q) \leftrightarrow (q \vee r)$.

Also, $\sim p \rightarrow (p \leftrightarrow q)$ means $(\sim p) \rightarrow (p \leftrightarrow q)$; note that

$$\sim p \rightarrow p \leftrightarrow q$$

does not have a meaning, according to the convention.

Likewise, $p \wedge q \vee r$ does not have a meaning; we use parentheses to distinguish between $(p \wedge q) \vee r$ and $p \wedge (q \vee r)$.

EXERCISES F

1. Let p be a statement, and let q be a statement. Write the truth table for each of the following compound statements.
   (a) $\sim(p \rightarrow q)$.   (b) $\sim p \rightarrow q$.   (c) $\sim p \leftrightarrow \sim q$.
   (d) $(\sim q \vee p) \rightarrow (q \rightarrow p)$.   (e) $(p \wedge q) \rightarrow p$.
   (f) $(p \wedge (p \rightarrow q)) \rightarrow q$.   (g) $(p \vee \sim p) \leftrightarrow (q \vee \sim q)$.

2. At a certain place and time some possible weather forecasts for the next 12 hours are: it will be sunny; it will be rainy; it will be hot. Denote these respective statements by s, r, h. Translate each of the following forecasts into symbolic statements.

(a) "If it will be sunny, it will be hot."
(b) "It will not be hot if and only if it will be rainy."
(c) "The sun will not shine in case it rains, and it will not rain in case the sun shines."
(d) "It will be neither hot nor sunny if and only if it will be rainy."

3. Show that each of the following compound statements is true in every possible case.
   (a) $\sim(\sim p) \leftrightarrow p$.
   (b) $(p \wedge \sim q) \to p$.
   (c) $p \to (p \vee q)$.
   (d) $p \wedge (p \vee q) \leftrightarrow p$.
   (e) $p \wedge (q \vee r) \leftrightarrow (p \wedge q) \vee (p \wedge r)$.
   (f) $q \vee r \leftrightarrow q \vee (\sim q \wedge r)$.
   (g) $[(p \to q) \wedge (q \to r)] \to [p \to r]$.
   (h) $(p \vee r) \wedge (\sim p \vee q) \to (p \wedge q) \vee (\sim p \wedge r)$.

9 Sentences pertaining to the outcome of an activity

In a typical application, we are interested in a sentence that pertains to the outcome (or result) of some activity. For one or more possible outcomes, the sentence may be true, while for other possibilities, the sentence may be false. Thus, although the sentence by itself may not have a truth value and therefore may not be a statement, it becomes a statement whenever the outcome is specified.

Example 28 Roger and Susan are the nominees for club president. A sentence pertaining to the election is, "A girl was elected president." This sentence is a true statement in case Susan is elected; it is a false statement in case Roger is elected. When the outcome of the election is specified, the sentence has a truth value.

Example 29 Bill and Carl are competing against each other in a match. Each game has a winner. The player who first wins two games wins the match. A possible outcome for the games in the match may be given by the ordered triple (b, c, c); by this we mean that Bill won the first game, Carl the second, and Carl the third. The initial of the winner of each game is the corresponding component of the triple. Another possible outcome is given by the ordered pair (b, b); a third game is not played after Bill has won the first two games. Altogether there are six possible cases, as itemized in Table 12. (The order of listing is immaterial.)

Table 12

CASE 1	(b, c, b)
CASE 2	(b, c, c)
CASE 3	(c, b, c)
CASE 4	(c, b, b)
CASE 5	(b, b)
CASE 6	(c, c)

Now consider the simple sentence, "Bill wins the match after losing one game." In each of Cases 1 and 4, this sentence is a true statement; in each of the other four cases, the sentence is a false statement.

As a more complicated situation, consider the sentence, "Carl wins the first game if and only if the number of games in the match is three." Suppose we let p denote the sentence that Carl wins the first game; suppose we let q denote the sentence that the number of games in the match is three. For each of the six cases, p and q are statements, and we can find the truth value for the biconditional statement $p \leftrightarrow q$. We tabulate the results in Table 13. The entries in

Table 13

CASE	p	q	$p \leftrightarrow q$
(b, c, b)	F	T	F
(b, c, c)	F	T	F
(c, b, c)	T	T	T
(c, b, b)	T	T	T
(b, b)	F	F	T
(c, c)	T	F	F

the column headed p are determined, in each case, by noting whether Carl does, or does not, win the first game. Likewise, the entries in the column headed q are determined by the number of games in the match, for each case. In the final column the entries are found by a comparison of the corresponding entries in the second and third columns. (It is also possible to obtain the entries in the fourth column in the same manner as in the second column; the alternative method serves as a check on accuracy.)

It may be of interest to note Case 5, in which the biconditional statement is true because each of p and q is false.

The preceding examples have illustrated an analysis of a sentence pertaining to the outcome of an activity. The activity may be an election, a contest, a market survey, a campaign, a scientific experiment, etc. We suppose that the activity yields one of several outcomes, or results, or outputs. Each of these we may call a logically possible case, or a possible outcome. The sentence pertaining to the outcome becomes, in each possible case, a statement with a truth value.

In order to make an effective analysis of such a sentence, we should be sure that we have an appropriate set of logical possibilities. Every conceivable situation should be accounted for, exactly once, in the list of possible outcomes. No possibility should be overlooked, and no possibility need be handled more than once.

Example 30 An urn contains one red ball and five indistinguishable green balls. Consider the following activity: one ball is withdrawn at random from the urn and not replaced, a second ball is drawn and not replaced, a third ball is drawn. The color of each ball drawn is noted. If we use the letters r and g to denote the drawing of a red or green ball, respectively, then an outcome is an ordered triple such as (g, r, g). There are four logically possible cases altogether, namely, (r, g, g), (g, r, g), (g, g, r), (g, g, g).

Consider the sentence, "If the first ball drawn is red, then the second ball drawn is green." From common sense, this sentence seems a reasonable assertion. Let us analyze the sentence to see why it appears reasonable. For abbreviation, denote by p the sentence that "the first ball drawn is red," and denote by q the sentence that "the second ball drawn is green." For each of the logically possible cases, p and q are statements, and we are interested in the conditional statement $p \to q$. We construct a table. In the first column we list all possible cases; in the second column we give the corresponding truth value for p; in the third column, the truth value for q; in the fourth column we combine the entries in the second and third columns to give the truth values for $p \to q$ (see Table 14). We note that in every possible outcome of the activity described, the statement $p \to q$ is true, which is why common sense accepts the assertion, "If the first ball drawn is red, then the second ball drawn is green."

Table 14

CASE	p	q	$p \to q$
(r, g, g)	T	T	T
(g, r, g)	F	F	T
(g, g, r)	F	T	T
(g, g, g)	F	T	T

Definition If a sentence that pertains to the outcome of an activity is true for every possible outcome, then we may call the sentence a *logically true* statement. Likewise, a sentence that is false in every possible case may be called a *logically false* statement.

Definition A pair of sentences pertaining to the outcome of an activity may be called a *logically equivalent pair* of statements provided, for every possible outcome, the two statements agree in their truth value.

Thus, for a logically equivalent pair of statements, each is true in every case in which the other is also true. Another way to describe a logically equivalent pair is the following: the statements are sentences such that their biconditional is a logically true statement.

In contrast to logical equivalence, we also have the notion of logical contradiction.

Definition A pair of sentences pertaining to the outcome of an activity may be called a *logically contradictory pair* of statements provided the biconditional of the statements is logically false.

In other words, a logically contradictory pair has the characteristic that, in every possible case, the statements have opposite truth values—one is true, but the other is false.

Example 31 For any sentences p, q pertaining to the outcome of an activity, show that the conditional statements $\sim p \to q$ and $\sim q \to p$ form a logically equivalent pair.

One method is to construct the truth tables for the respective conditional statements. Since a comparison reveals that the fourth and sixth columns in Table 15 are alike, the conditional statements have the same truth value in every case and therefore form a logically equivalent pair.

An alternative method is to complete the seventh column in Table 15, giving the respective truth values for the biconditional $(\sim p \to q) \leftrightarrow (\sim q \to p)$. Since the biconditional is true in every possible case, we conclude as before that the two conditional statements form a logically equivalent pair.

Table 15

p	q	$\sim p$	$\sim p \to q$	$\sim q$	$\sim q \to p$	$(\sim p \to q) \leftrightarrow (\sim q \to p)$
T	T	F	T	F	T	
T	F	F	T	T	T	
F	T	T	T	F	T	
F	F	T	F	T	F	

EXERCISES G

1. Four cards numbered 1, 2, 3, 4, respectively, form a deck. Bob draws a card from the deck and does not replace it. Fred draws a card from the reduced deck. Let the numbers on Bob's card and Fred's card be denoted by b and f, respectively.
   (a) List all the possible outcomes (b, f).
   (b) For which of the possible outcomes is the statement $b < f$ true?
   (c) For which of the possible outcomes is the following statement true? "The number b is even or the number f is odd."
   (d) For which of the cases is the following statement true? "Exactly one of the components of the vector (b, f) is odd if and only if $b + f = 5$."

2. Three cards numbered 1, 2, 3, respectively, form a deck. Harry draws a card from the deck and replaces it. Gary draws a card from the deck. Let the numbers on Harry's and Gary's cards be denoted by h and g, respectively.

(a) List all the possible outcomes (h, g).
(b) Which of the outcomes are roots of the equation $h + g = 4$?
(c) For which of the outcomes is the following statement true? "If $h = 2$, then $g \leq 2$."

3. An urn contains two red balls, one green ball, and one blue ball. Two balls are withdrawn together, and the colors of the balls are observed.
   (a) List all the possible outcomes for the colors of the balls withdrawn. Disregard any order for the colors.
   (b) For which of the outcomes is the following statement true? "If at least one of the balls is green or red, then at least one of the balls is blue."
   (c) For which of the outcomes is the following statement true? "If both of the balls are green, then at least one of the balls is not red."
   (d) For which of the outcomes is the following statement true? "At least one of the balls is blue or green if and only if both balls are red."
   (e) Among the following nine statements pertaining to the outcome of the activity, identify two such that each of them is logically true, identify one that is logically false, identify two that form a logically equivalent pair, and identify two that form a logically contradictory pair.
   p_1: "At least one of the balls is green or red."
   p_2: "At least one of the balls is blue."
   p_3: "Both of the balls are green."
   p_4: "At least one of the balls is not red."
   p_5: "At least one of the balls is blue or green."
   p_6: "Both balls are red."
   p_7: "If at least one of the balls is green or red, then at least one of the balls is blue."
   p_8: "If both of the balls are green, then at least one of the balls is not red."
   p_9: "At least one of the balls is blue or green if and only if both balls are red."

10 Truth set

Definition Suppose we have an activity, and let T denote the set of all possible outcomes. Let p be a sentence about the outcome such that, in every possible case, p is a statement. The subset of T consisting of all outcomes for which p is a true statement is called the *truth set for the statement p*.

Example 32 In an ordinary deck of 52 playing cards, each suit consists of 13 cards: the ace, the king, the queen, the jack, and nine cards bearing the respective numbers 10, 9, 8, 7, 6, 5, 4, 3, 2. Each card that is a king or a queen or a jack is a face card. In certain games an honor card is a card having any one of the following five denominations: ace, king, queen, jack, or ten. A deuce and a trey are cards whose denominations are 2 and 3, respectively.

There are four suits. The spade and club suits are black; the heart and diamond suits are red. Let us denote a card of a suit by the corresponding initial letter s, c, h, d. From the deck two cards are drawn together (without replacement). Disregarding any order for the two cards, we list the possibilities concerning the suits of the selected cards. They are: $\{s, h\}$, $\{s, d\}$, $\{s, c\}$, $\{h, d\}$, $\{h, c\}$, $\{d, c\}$, $\{s, s'\}$, $\{h, h'\}$, $\{d, d'\}$, $\{c, c'\}$. By the notation $\{s, d\}$ we mean one of the two chosen cards is a spade and the other is a diamond; by $\{h, h'\}$ we mean that both cards are hearts.

Let p be the sentence that at least one of the cards drawn is a heart. The truth set for this statement consists of the four outcomes $\{s, h\}$, $\{h, d\}$, $\{h, c\}$, $\{h, h'\}$. Let q be the sentence that both cards drawn are red. The truth set for q is the subset $\{\{h, d\}, \{h, h'\}, \{d, d'\}\}$. The truth set for the conjunction $p \wedge q$ has only two members, namely, $\{h, d\}$ and $\{h, h'\}$; we note that this truth set is the intersection of the truth set for p and the truth set for q.

There is a very close relationship between our study of subsets of a set and our study of logical connectives. Consider the set T of all possible outcomes for a given activity. Let P and Q be the respective truth sets for sentences p and q pertaining to the outcome. The following remarks typify the relationship.

1. P and Q are the same set precisely in case (p, q) is a logically equivalent pair of statements.
2. The truth set for the conjunction $p \wedge q$ is the intersection $P \cap Q$.
3. The truth set for the disjunction $p \vee q$ is the union $P \cup Q$.
4. The complement $\tilde{P}$ is the truth set for the negation $\sim p$.

As byproducts, we may take advantage of the fact that the conditional and biconditional have, respectively, the same meaning as certain compound statements involving the basic connectives. Thus, since the statements $p \to q$ and $\sim p \vee q$ form a logically equivalent pair, the truth set for $p \to q$ is $\tilde{P} \cup Q$. Also, since $p \leftrightarrow q$ and $(p \wedge q) \vee (\sim p \wedge \sim q)$ form a logically equivalent pair, the truth set for the biconditional is $(P \cap Q) \cup (\tilde{P} \cap \tilde{Q})$.

The second and third remarks above help to explain the similar appearance of the symbols $\wedge$ and $\cap$ for conjunction and intersection, and the strong resemblance between the symbols $\vee$ and $\cup$.

EXERCISES H

1. Suppose that p and q are sentences pertaining to the outcome of a certain activity. Suppose that T is the set of all possible outcomes and that P and Q are the respective truth sets for p and q. Express the truth set for each of the following compound statements.
   (a) $(p \vee \sim q) \wedge (\sim p \vee q)$.
   (b) $p \vee q \to p$.
   (c) $p \vee (p \wedge q) \leftrightarrow p$.
   (d) $\sim(p \to p \vee \sim p)$.

2. From an ordinary deck of playing cards, three cards are drawn together. In each of the following parts, tell whether the drawing of the three named cards belongs to the truth set of the statement that appears after the subset.
   (a) {diamond king, club jack, heart jack}. "Each card drawn is a face card."

180   Sets and Logic

(b) {spade 3, heart 9, diamond ace}. "A face card or a deuce is drawn."
(c) {club 7, diamond jack, heart 4}. "If at least one spade is drawn, then at least one red card is drawn."
(d) {club queen, club 6, spade 8}. "A diamond is drawn if and only if a heart is drawn."
(e) {heart queen, heart king, heart jack}. "At least one of the cards drawn is a heart or a face card."

3. A positive integer x is chosen. What is the truth set for the statement $x < 6$?

4. Let L be the linear function given by $L(x) = mx + b$, and let d be a number. If an activity consists of selecting a number t, what is the truth set for the statement $L(t) = d$?

5. Show that an equation of the type

$$[a \ b]\begin{bmatrix}x\\y\end{bmatrix} = c,$$

where a, b, c are nonzero numbers, may be interpreted as a sentence about the vector $\begin{bmatrix}x\\y\end{bmatrix}$. With this interpretation, show that the set of roots of the equation is the truth set for the statement.

6. Find the truth set for each of the following.
(a) $5x + 4 = 2x - 2$.
(b) $\begin{bmatrix}5 & 7\\1 & 3\end{bmatrix}\begin{bmatrix}x\\y\end{bmatrix} = \begin{bmatrix}3\\-1\end{bmatrix}$.

7. Let A be an $m \times n$ matrix and B be an $m \times 1$ matrix. Discuss what is meant by the truth set for the inequality $AX \leq B$.

11 Boolean algebra

In logic, we have emphasized three basic connectives. Each of these connectives enables us to form a new statement from one or two given statements. In set theory, we have emphasized three basic operations. Each of them gives us a way of describing a new subset in terms of one or two given subsets. We have remarked on the fact that these two systems have many features in common. With the aid of the notion of truth set for a statement, any observation or result in one system seems to have a corresponding fact in the other system. When a scientist discovers such a marked resemblance between two seemingly different investigations, he curiously explores further in an attempt to gain a deeper understanding.

Several properties of our two systems are similar to the familiar rules for handling ordinary numbers. In arithmetic, there are two fundamental ways to combine a pair of numbers, namely, addition and multiplication. The statement that $P \cup Q = Q \cup P$ for all subsets P, Q of a set T looks like the statement that $x + y = y + x$ for all numbers x, y. The fact that $p \wedge (q \wedge r)$ and $(p \wedge q) \wedge r$ form a logically equivalent pair of statements is reminiscent of the fact that $x \cdot (y \cdot z)$ and $(x \cdot y) \cdot z$ are equal numbers. On the other hand, some other characteristics of our two systems do not have a counterpart in the system of ordinary numbers.

We will abstract the essential features that characterize our study of logic and our study of set theory and formalize them into a mathematical system. This system is a different kind of algebra from the algebra of ordinary numbers. In honor of George Boole, an English mathematician who early recognized the algebraic aspects of logic, the system is called a *Boolean algebra*.

A definition of a Boolean algebra can be given in a variety of ways. We choose a way that is suitable for our purposes. The definition is a rather lengthy description. First, we give the definition, using several technical phrases. An appreciation of these special words makes the overall concept of a Boolean algebra easier to grasp. Since the reader may not know all of these words, a fully elaborated definition will be given later.

> **Definition** A Boolean algebra is a mathematical system, consisting of a set of two or more elements, together with three operations (called addition, multiplication, and complementation) on the set, having the following properties: addition is a binary operation that has the associative, the commutative, the identity, and the idempotent properties; multiplication is a binary operation that has the associative, the commutative, the identity, and the idempotent properties; multiplication is distributive over addition; complementation is a unary operation; each element and its complement have the additive identity as their product and the multiplicative identity as their sum.

In the elaborated definition we introduce appropriate notation. Often in mathematics, symbolism can aid in the understanding of unfamiliar words.

> **Definition** A Boolean algebra is a mathematical system, consisting of a set of two or more elements, together with three operations customarily called addition, multiplication, and complementation, having the following properties.
>
> [POSTULATES CONCERNING ADDITION ONLY]
>
> 1. Addition is a closed binary operation on the set.
>   In other words, to every ordered pair of elements in the set, there corresponds a uniquely determined element in the set called the sum of the elements in the given pair. The sum of the elements in the pair (a, b) may be denoted by the symbol $a + b$.
> 2. Addition is an associative operation.
>   In symbols, for every element a, every element b, every element c in the set,
> $$(a + b) + c = a + (b + c).$$

3. Addition is a commutative operation.
   In symbols, for every element a and every element b in the set,
   $$a + b = b + a.$$
4. There is in the set an identity element for addition.
   This identity element is customarily called zero and denoted by the symbol 0. In symbols, the element 0 has the property that, for every element a in the set,
   $$a + 0 = a.$$
5. Each element in the set is idempotent under addition.
   In symbols, for every element a in the set,
   $$a + a = a.$$

[POSTULATES CONCERNING MULTIPLICATION ONLY]

6. Multiplication is a closed binary operation on the set.
   In other words, to every ordered pair of elements in the set, there corresponds a uniquely determined element in the set called the product of the elements in the given pair. The product of the elements in the pair (a, b) may be denoted by the symbol $a \cdot b$.
7. Multiplication is an associative operation.
   In symbols, for every element a, every element b, every element c in the set,
   $$(a \cdot b) \cdot c = a \cdot (b \cdot c).$$
8. Multiplication is a commutative operation.
   In symbols, for every element a and every element b in the set,
   $$a \cdot b = b \cdot a.$$
9. There is in the set an identity element for multiplication.
   This identity element is customarily called one and denoted by the symbol 1. In symbols, the element 1 has the property that, for every element a in the set,
   $$a \cdot 1 = a.$$
10. Each element in the set is idempotent under multiplication.
    In symbols, for every element a in the set,
    $$a \cdot a = a.$$

[POSTULATE CONCERNING COMPLEMENTATION ONLY]

11. Complementation is a closed unary operation on the set.
    In other words, to every element in the set, there corresponds a uniquely determined element in the set called the complement

of the given element. The complement of the element a may be denoted by the symbol a'.

[POSTULATES INTERCONNECTING THE OPERATIONS]

12. The product of each element and its complement is the additive identity.
    In symbols, for every element a in the set,
    $$a \cdot (a') = 0.$$

13. The sum of each element and its complement is the multiplicative identity.
    In symbols, for every element a in the set,
    $$a + (a') = 1.$$

14. Multiplication is distributive with respect to addition.
    In symbols, for every element a, every element b, every element c in the set,
    $$a \cdot (b + c) = (a \cdot b) + (a \cdot c).$$

We remark that the special elements 0 and 1 in a Boolean algebra are not the ordinary numbers zero and one. We use the same names and symbols because these elements have properties—in the Boolean algebra—that are in many respects similar to the properties of the familiar numbers.

In ordinary algebra we learn to interpret $1 + 2x$ as meaning that first 2 and x are multiplied and afterward 1 and $2x$ are added: in brief, multiplications are performed before additions. A corresponding convention in our Boolean algebra will enable us to avoid excessive use of parentheses. In an expression involving more than one operation, we agree that the complementations are applied first, the multiplications next, and the additions last. Furthermore, in case no ambiguity can occur, we sometimes suppress the multiplication symbol $\cdot$, and write simply ab rather than $a \cdot b$.

In order to specify a Boolean algebra, we need to give the set of elements and we need to tell what the three operations are. If the number of elements in the set is small, a convenient means for describing an operation is by means of a table.

Example 33 Consider a set of eight elements. As convenient names for these elements we will use 0, 1, 2, 3, 4, 5, 6, 7. A word of warning may be appropriate: usage of these names does *not* mean that the elements are ordinary numbers. The three operations on the set $\{0, 1, 2, 3, 4, 5, 6, 7\}$ are given by the respective tables for addition, multiplication, and complementation presented in Table 16.

To find the sum $a + b$, look for a in the left-hand column, and for b in the top row, of the addition table; at the intersection of the

Table 16

+	0	1	2	3	4	5	6	7
0	0	1	2	3	4	5	6	7
1	1	1	1	1	1	1	1	1
2	2	1	2	1	7	5	5	7
3	3	1	1	3	3	1	3	1
4	4	1	7	3	4	1	3	7
5	5	1	5	1	1	5	5	1
6	6	1	5	3	3	5	6	1
7	7	1	7	1	7	1	1	7

·	0	1	2	3	4	5	6	7
0	0	0	0	0	0	0	0	0
1	0	1	2	3	4	5	6	7
2	0	2	2	0	0	2	0	2
3	0	3	0	3	4	6	6	4
4	0	4	0	4	4	0	0	4
5	0	5	2	6	0	5	6	2
6	0	6	0	6	0	6	6	0
7	0	7	2	4	4	2	0	7

	′
0	1
1	0
2	3
3	2
4	5
5	4
6	7
7	6

row headed a and the column headed b we find the sum $a + b$. For example, $3 + 5 = 1$ and $6 + 2 = 5$. From the multiplication table, we find, by the same method, that $3 \cdot 5 = 6$ and $6 \cdot 2 = 0$. To find the complement a', look for a in the left-hand column of the complementation table; opposite it in the right-hand column is a'. Thus, $2' = 3$ and $7' = 6$.

The equation $5 + x = 5$ has four solutions. From the row headed 5 in the addition table, we find that each of the elements 0, 2, 5, 6 satisfies the equation. On the other hand, the equation $5 + y = 6$ has no solution, since the row headed 5 does not contain the entry 6.

The equation $3x = 4$ has two roots, namely, 4 and 7, while there is no root for the equation $5x = 4$.

An inspection of the complementation table reveals that the eight elements split into four pairs so that, within any pair, each of the two elements is the complement of the other. Thus, 4 and 5 are complements of one another; so are 6 and 7; so also $2' = 3$ and $3' = 2$.

Example 34 The simplest, and one of the more important, illustrations of a Boolean algebra is a system with just two elements. Consider the set $\{0, 1\}$ and the operations given by the tables in Table 17.

Table 17

+	0	1
0	0	1
1	1	1

·	0	1
0	0	0
1	0	1

	′
0	1
1	0

In each of Examples 33 and 34, we have not actually verified that the system is a Boolean algebra. In order to do this, it would be necessary to check that all the parts of the definition of a Boolean algebra are satisfied by the system. Such a check can be a tedious task. Suffice it to say, the verification can be done, but we will not do so.[1]

---

[1] To the reader who wonders where the tables in Example 33 came from, we remark that they are the same as the tables the reader constructed in Exercise B-10, except for the change in notation. The letters Q, I, R, A, S, B, T, C and the operational symbols $\cup$, $\cap$, $\sim$ have been replaced by the symbols 0, 1, 2, 3, 4, 5, 6, 7, and $+$, $\cdot$, $'$, respectively.

EXERCISES I

1. Calculate each of the following by using the Boolean algebra tables in Example 33.
   (a) $2 + 4$.
   (b) $2 \cdot 4$.
   (c) $4 \cdot (5 + 6)$.
   (d) $6' + 2'$.
   (e) $6'(3 + 7)'$.
   (f) $5 + 5'$.

2. Consider the Boolean algebra given in Example 33.
   (a) Find each element x (if there is any) such that $4 + x = 3$.
   (b) Find each root of the equation $5x = 2$.
   (c) Find each element x (if there is any) such that $x' = 3'$.
   (d) Solve the equation $5 + y = 5y$.
   (e) What is the truth set for the equation $3 + 3x = 3$?
   (f) How many roots does the equation $5 + 5x = 1$ have?

3. Verify that the system in Example 34 satisfies both part 12 and part 13 of the definition for a Boolean algebra.

4. Consider the Boolean algebra in Example 34.
   (a) Verify that $(a + b)(a + c) = a + bc$ in the case that $a = 0, b = 1, c = 0$.
   (b) Verify that $a + a'b = a + b$ in the case that $(a, b) = (0, 0)$.
   (c) Verify that $a + ab = a$ in the case that $(a, b) = (0, 1)$.

5. Consider the Boolean algebra in Example 33.
   (a) Find every ordered pair (x, y) such that $x + y = 0$.
   (b) Find the truth set for the equation $xy = 1$.

12 Proofs in Boolean algebra

Example 35 Let T be a nonempty set. Consider the collection of all subsets of T. These subsets of T may be considered as the elements of a Boolean algebra. The "sum" of two elements P and Q is the union $P \cup Q$ of the subsets. The "product" of the elements P and Q is the intersection $P \cap Q$ of the subsets. The "complement" of an element P is the subset $\tilde{P}$. All the various properties required by the definition of a Boolean algebra are satisfied in this model. For example, condition 5 is met because every subset P of T has the property that $P \cup P = P$. The additive identity element demanded in condition 4 is the empty subset, and the element "one" in condition 9 is the subset T. Note that $P \cap T = P$ for every subset P.

Example 36 Consider the collection of all statements pertaining to the outcome of a certain activity. We obtain a Boolean algebra by making the following interpretations: interpret logical equivalence of statements as equality in the Boolean algebra sense, interpret the disjunction of statements as the sum of elements in the Boolean algebra, interpret the conjunction of statements as the product of elements, and interpret the negation of a statement as the complement of an element.

In this illustration, a logically true statement plays the role of the element 1, and a logically false statement plays the role of 0. In the lengthy task of verifying that all the parts of the definition of a Boolean algebra are satisfied, we encounter (for example) condition 13. Since every statement p has the property that $p \lor {\sim}p$ is logically true, we check that part 13 is satisfied. Other parts may be verified similarly.

From the various properties of a Boolean algebra embodied in the definition, many other facts can be deduced by logical argument. Much of mathematics is concerned with the pleasant task of deriving consequences from certain chosen assumptions. Our work with Boolean algebra is intended to give us some appreciation of this vital aspect of mathematics.

The information we have to work with (that is, our chosen assumption) is the definition of a Boolean algebra. Each general consequence of this information may be called a theorem. Once we have established that a particular theorem is a true statement, then we may use it to shorten the proof of later theorems. In the following presentation we shall give a number of theorems. For some of them we write the proof in detail; for others, we outline the proof, leaving the details to the reader; in still other cases, the entire proof is left as an exercise.

Notice that each step in a proof requires a reason.

In order to have readily available the assumed properties of a Boolean algebra, we repeat here the various parts of the definition, in a highly condensed form.

1. $+$,
2. $(a + b) + c = a + (b + c)$,
3. $a + b = b + a$,
4. $a + 0 = a$,
5. $a + a = a$,
6. $\cdot$,
7. $(ab)c = a(bc)$,
8. $ab = ba$,
9. $a \cdot 1 = a$,
10. $aa = a$,
11. $'$,
12. $aa' = 0$,
13. $a + a' = 1$,
14. $a(b + c) = ab + ac$.

To avoid repetition in the following statements, we shall agree that we have an arbitrary Boolean algebra and that all elements discussed are members of it.

Theorem 1a The complement of one is zero.
This assertion, in symbols, is $1' = 0$.

12 / Proofs in Boolean algebra   187

PROOF $1' = 1' \cdot 1$ [because 1 is the multiplicative identity; see condition 9]
$ = 1 \cdot 1'$ [because multiplication is commutative; see condition 8]
$ = 0$ [because the product of any element and its complement is zero; see condition 12].

Theorem 1b   The complement of zero is one.
In symbols, $0' = 1$.

The proof is requested in Exercise 1 below.

Theorem 2a   The sum of any element and one is one.
In symbols, for every element a, $a + 1 = 1$.

PROOF $a + 1 = a + (a + a')$ [because the sum of any element and its complement is one; see condition 13]
$ = (a + a) + a'$ [because addition is associative; see condition 2]
$ = a + a'$ [because each element is idempotent under addition; see 5]
$ = 1$ [because the sum of any element and its complement is one; see 13].

Theorem 2b   The product of any element and zero is zero.
In symbols, for every element a, $a \cdot 0 = 0$.

The proof is left to Exercise 2.

Theorem 3a   For every element a and every element b, $a + (a \cdot b) = a$.

PROOF $a + ab = a \cdot 1 + ab$ [because 1 is the multiplicative identity; 9]
$ = a(1 + b)$ [because multiplication is distributive with respect to addition; 14]
$ = a(b + 1)$ [because addition is commutative; 3]
$ = a \cdot 1$ [by Theorem 2a]
$ = a$ [because 1 is the multiplicative identity].

Theorem 3b For every element a and every element b, $a \cdot (a + b) = a$.

The proof is left to Exercise 3.

The results announced in Theorems 3a and 3b may appear quite strange to a person accustomed to ordinary algebra. These unusual features are two of the most distinctive characteristics in Boolean algebra, and they are particularly useful in deducing further important information.

Compare Theorem 3a with the analogous result in set theory, namely, that $P \cup (P \cap Q)$ is the same as the subset P, no matter what Q may be. See Fig. 8. Since $P \cap Q$ is a subset of P itself, the union of $P \cap Q$ and P is simply P.

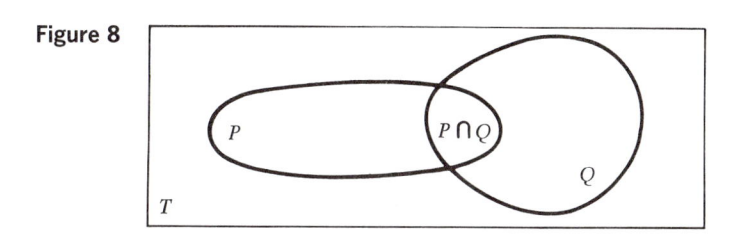

Figure 8

Also notice the interpretation of Theorem 3a in logic (see Example 36). Examination of the appropriate truth table shows that the statement $p \vee (p \wedge q)$ and the statement p form a logically equivalent pair.

Example 37 As an application of the theory studied thus far, we prove the following proposition: if c and d are elements of a Boolean algebra such that $c + d = d$, then $cd = c$. Each step in the proof requires a reason, which we give in brackets. We seek to show that cd is the same element as c. We, of course, use the given hypothesis at some stage in the discussion, and we also freely use any parts of the definition of a Boolean algebra or any theorem that we have already proved. The proof is as follows.

$$\begin{aligned} cd &= c(c + d) \quad \text{[by hypothesis, } d = c + d\text{]} \\ &= cc + cd \quad \text{[because multiplication is distributive with respect to addition]} \\ &= c + cd \quad \text{[because each element is idempotent under multiplication]} \\ &= c \quad \text{[by Theorem 3a].} \end{aligned}$$

Example 38 As a second application of the theory studied thus far, we sketch the proof of the following proposition: if c and d are elements of a Boolean algebra such that $c + d = d$, then $cd' = 0$.

12 / Proofs in Boolean algebra

The reasons are omitted below, because the reader is requested in Exercise 8 to supply the reasons.

$$\begin{aligned} cd' &= cd' + 0 \\ &= cd' + dd' \\ &= d'c + d'd \\ &= d'(c + d) \\ &= d'd \\ &= dd' \\ &= 0. \end{aligned}$$

We can present a briefer proof if we accept Theorem 2b and if we exploit the proposition established in Example 37. The shorter proof is as follows:

Since $\quad c + d = d \quad$ [by hypothesis],
$\quad\quad\quad\quad cd = c \quad$ [by the proposition in Example 37].
Therefore, $cd' = (cd)d' \quad$ [by the previous step]
$\quad\quad\quad\quad = c(dd') \quad$ [because multiplication is associative]
$\quad\quad\quad\quad = c \cdot 0 \quad$ [because the product of any element
$\quad\quad\quad\quad\quad\quad\quad\quad$ and its complement is zero]
$\quad\quad\quad\quad = 0 \quad$ [by Theorem 2b].

EXERCISES J

1. Prove Theorem 1b. [HINT: The proof resembles the proof of Theorem 1a, with the roles of addition and multiplication interchanged.]

2. Prove Theorem 2b.

3. Prove Theorem 3b. [HINT: Use Theorem 3a.]

4. Write the proof for each of the following propositions. Be sure to give the reason (or reasons) for each step in your proof.
   (a) For every element a, $0 + a = a$.
   (b) For every element a, $a'a = 0$.
   (c) For every pair (a, b) of elements, $(ab)a = ab$.
   (d) For every quadruple (c, d, e, f) of elements, $(c + d)(e + f) = (c + d)e + (c + d)f$.
   (e) For every triple (u, v, w) of elements, $(u + v)w = uw + vw$.
   (f) For every pair (e, f) of elements, $ef + ef = ef$.
   (g) For every triple (a, b, c) of elements, $(a + b + c)(a + b + c) = a + b + c$.
   (h) For every triple (a, b, c) of elements, $a + a(b + c) = a$.
   (i) For every quadruple (a, b, c, d) of elements, $c(a + b' + c + d) = c$.

5. Simplify each of the following elements.
   (a) $a + (b + a)$.
   (b) $(a + a')'$.
   (c) $c[a' + (b + a)]$.
   (d) $d' + (d')'$.
   (e) $(a + a) + (a + a)$.
   (f) $[(c + d) \cdot (c + d)']'$.
   (g) $(bb + bb) + bb$.
   (h) $(bb + cc)(bb)$.
   (i) $(f' + g)(g + f)$.

6. The interpretation of Theorem 1a in Example 35 is the statement that the complement of T is the empty subset. Interpret each of the following propositions in the Boolean algebra discussed in Example 35.
   (a) Theorem 1b.
   (b) Theorem 2a.
   (c) Theorem 3b.
   (d) Exercise 4e.

7. Consider the Boolean algebra discussed in Example 36. The interpretation of Theorem 1b in this example is the assertion that the negation of a false statement is true. Interpret each of the following propositions in this Boolean algebra.
   (a) Theorem 1a.
   (b) Theorem 2b.
   (c) Theorem 3b.
   (d) Exercise 4a.

8. Write in full the proof sketched in Example 38 by copying the given steps and supplying the reason for each of the seven steps.

9. Prove each of the following propositions. Be sure to give the reason (or reasons) for each step in your proof.
   (a) If c and d are elements such that $cd = c$, then $c + d = d$.
   (b) If c and d are elements such that $cd = c$, then $cd' = 0$.

10. Consider the Boolean algebra described in Example 33.
    (a) Find a pair (c, d) of distinct elements in the algebra such that c is not 0, d is not 1, and $c + d = d$. For the pair you find, verify that $cd = c$.
    (b) Find a different pair (c, d) of distinct elements [that is, not the same pair as you selected in part (a)] such that $c \neq 0$, $d \neq 1$, and $cd = c$. For the pair you choose, verify that $c + d = d$.

11. Prove the following proposition: if the sum of the elements in a pair is the same as their product, then the elements are the same. [HINT: Suppose $b + c = bc$. Then $b = b + bc = b + b + c = b + c$. Supply the missing reasons for the preceding steps, and finish the task of showing that $b = c$.]

13 Distributivity in a Boolean algebra

Condition 14 in the definition of a Boolean algebra is familiar because this interconnection between multiplication and addition is true in ordinary algebra. We say that multiplication is distributive over addition. In Boolean algebra there is another type of distributive property. If we interchange the roles of the two operations, we obtain the condition that addition is distributive over multiplication. This result is the next theorem. Note that the occurrences of $+$ and $\cdot$ are interchanged from the formula in condition 14.

Theorem 4 Addition is distributive with respect to multiplication.

In symbols, for every element a, every element b, every element c, $a + (b \cdot c) = (a + b) \cdot (a + c)$.

OUTLINE OF PROOF

$$\begin{aligned}
a + bc &= a(a + b + c) + bc && \text{[by Theorem 3b]} \\
&= a(a + b) + ac + bc && \text{[because multiplication is distributive over addition]} \\
&= (a + b)a + (a + b)c && \text{[because multiplication is commutative and is distributive over addition]} \\
&= (a + b)(a + c) && \text{[because multiplication is distributive over addition].}
\end{aligned}$$

A word of elaboration about the application of Theorem 3b may be helpful. According to Theorem 3b, no matter what the element x is, a is the same as $a(a + x)$. In particular then, using the element $b + c$ in place of x, we obtain $a = a(a + b + c)$.

The preceding proof exploited the associativity of addition (condition 2) several times without explicit mention. As we gain more familiarity with a theory, we begin to take more and more of the basic parts for granted when we offer reasons in a proof. The omissions prevent the proof from becoming excessively long. Considerable experience is the best way to learn what steps in a proof should be discussed in detail and what should not. Whenever in doubt, give the details. Above all, be sure that you can give reasons if you are challenged!

Theorem 5a For every element a and every element b, $a + a'b = a + b$.

PROOF
$$\begin{aligned}
a + a'b &= (a + a')(a + b) && \text{[because addition is distributive over multiplication; see Theorem 4]} \\
&= 1 \cdot (a + b) && \text{[because the sum of an element and its complement is 1]} \\
&= a + b && \text{[because multiplication is commutative and has the identity element 1].}
\end{aligned}$$

Theorem 5b For every element a and every element b, $a \cdot (a' + b) = ab$.

The proof is left to Exercise 1 below.

Theorem 6 Given any element a, there is exactly one element that is a root of the system of simultaneous equations:

$$\begin{aligned}
a + x &= 1, \\
a \cdot x &= 0;
\end{aligned}$$

the root is a'.

PROOF According to parts 12 and 13 of the definition for a Boolean algebra, a' is one root of the system. It remains to show that a' is the only root. Suppose that s is any root of the system; then

$$\begin{aligned}
s &= s \cdot 1 &&\text{[because 1 is the multiplicative identity]} \\
&= s \cdot (a + a') &&\text{[because } a + a' = 1\text{]} \\
&= sa + sa' &&\text{[because multiplication is distributive over addition]} \\
&= 0 + sa' &&\text{[since } s \text{ is a root of } ax = 0, \, sa = as = 0\text{]} \\
&= aa' + sa' &&\text{[because } aa' = 0\text{]} \\
&= (a + s)a' &&\text{[because multiplication is commutative and is distributive over addition]} \\
&= 1 \cdot a' &&\text{[since } s \text{ is a root of } a + x = 1\text{]} \\
&= a' &&\text{[because multiplication is commutative and has identity element 1].}
\end{aligned}$$

Since every root s is the same as a', the only root is a'.

Observe what Theorem 6 means in terms of set theory. Given a subset P of the set T, then the complement $\tilde{P}$ is the one and only subset X of T having

Figure 9

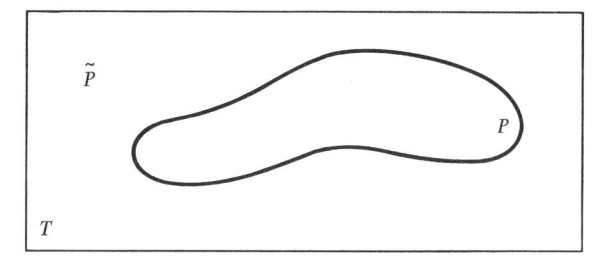

the properties that $P \cup X = T$ and $P \cap X$ is empty. See Fig. 9. Any subset X has the property that $P \cap X$ is empty if and only if $X \subset \tilde{P}$. On the other hand, X has the property that $P \cup X$ is the entire set T if and only if $\tilde{P} \subset X$. Thus, X has both properties if and only if $X = \tilde{P}$.

Generally speaking, if a is an element of a Boolean algebra, there may be many solutions to the equation $a + x = 1$ and there may be many roots for the equation $a \cdot x = 0$. This is in marked contrast to the situation in ordinary algebra.

EXERCISES K

1. Prove Theorem 5b.

2. Write the proof for each of the following propositions.
   (a) If the sum of two elements is 0, then each of the elements is 0.
   (b) If the product of two elements is 1, then each of the elements is 1.

3. For the Boolean algebra discussed in Example 35, interpret each of the following propositions.
   (a) Theorem 4. (b) Theorem 5b. (c) Exercise 2b.

4. For the Boolean algebra discussed in Example 36, interpret each of the following propositions.
   (a) Theorem 4.   (b) Theorem 5a.   (c) Exercise 2a.
5. Let p, q, r be sentences pertaining to the outcome of an activity. Show that each two of the three following compound statements form a logically equivalent pair: $(p \vee q) \wedge (p \rightarrow r)$; $(p \wedge r) \vee (\sim p \wedge q) \vee (q \wedge r)$; $(p \wedge r) \vee (\sim p \wedge q)$.
6. The logical equivalences in Exercise 5 suggest that the corresponding proposition may be proved in any Boolean algebra.
   (a) State this proposition. [HINT: The proposition has the following form: for every element a, every element b, every element c, three particular elements expressed in terms of a, b, c are equal to one another.]
   (b) Prove the proposition.
7. Simplify each of the following elements. Give a reason for each step in your simplification.
   (a) $(a + b)(a' + b)$.   (b) $ab(a + b)$.   (c) $a + b + a'b'$.
8. Construct an example of a Boolean algebra with four elements. That is, give the elements and give the three operations (say, by giving tables for the operations, as in Example 33).
9. Show that there is no Boolean algebra with exactly three elements.

14 De Morgan's laws

The following are some important applications of Theorem 6.

Theorem 7 The complement of the complement of any element is the element itself.

In symbols, for every element a, $(a')' = a$.

PROOF Consider the system of equations $a' + x = 1$, $a' \cdot x = 0$. Since addition and multiplication are commutative operations, the element a is a root of the system. However, by Theorem 6, the only root of the system is the complement of a', namely, $(a')'$. Thus, $(a')'$ and a are the same element.

Theorem 8a The complement of the product of any two elements is the sum of the complements of the individual elements.

In symbols, for every element a and every element b, $(ab)' = a' + b'$.

PROOF We show that $a' + b'$ is a root of the system of equations $ab + x = 1$, $ab \cdot x = 0$. First,

$$ab + (a' + b') = (a + a' + b')(b + a' + b') \quad \text{[an application of Theorem 4]}$$
$$= (1 + b')(1 + a')$$
$$= 1 \cdot 1 \quad \text{[two applications of Theorem 2a]}$$
$$= 1.$$

194   Sets and Logic

Second, $ab \cdot (a' + b') = ab \cdot a' + ab \cdot b'$
$= b \cdot 0 + a \cdot 0$
$= 0.$

By Theorem 6, the only root of the system $ab + x = 1$, $ab \cdot x = 0$ is $(ab)'$; thus, $(ab)' = a' + b'$.

Theorem 8b The complement of the sum of any two elements is the product of the complements of the individual elements.
In symbols, for every element a and every element b, $(a + b)' = a'b'$.

The proof is left to Exercise 1 below.

Theorems 8a and 8b are commonly known as *de Morgan's laws* in commemoration of Augustus de Morgan, who shared with George Boole the lead in the nineteenth-century development of logic. It is instructive to examine the meaning of these two theorems in our two familiar examples of Boolean algebra. Suppose that P and Q are subsets of a set T. Then, according to Theorem 8a, the complement of $P \cap Q$ is $\tilde{P} \cup \tilde{Q}$. Also, $P \cup Q$ and $\tilde{P} \cap \tilde{Q}$ are, by Theorem 8b, complements of each other.

Suppose that p and q are sentences pertaining to the outcome of an activity. What is the negation of their conjunction? What is the negation of their disjunction? The interpretation of Theorem 8a is that the statement $\sim(p \wedge q)$ and the statement $\sim p \vee \sim q$ form a logically equivalent pair. In other words, the negation of the conjunction of two statements has the same meaning as the disjunction of the negations of the individual statements. We interpret Theorem 8b as asserting that $\sim(p \vee q)$ and $\sim p \wedge \sim q$ form a logically equivalent pair of statements.

Example 39 Consider the statement from ordinary language, "On my European trip I am going to neither Spain nor Italy." The speaker is negating the statement that he is going to either Spain or Italy. This negation of a disjunction means that he is not going to Spain and he is not going to Italy. The connective AND shows that the negation of a disjunction of statements is the conjunction of their negations. The phrase "neither this nor that" means "NOT (this OR that)," which is equivalent to "(NOT this) AND (NOT that)."

Example 40 An experiment consists in selecting one card from an ordinary deck of playing cards. Consider the following statement about the outcome, "The card drawn is not a heart or not an ace." This statement has the same meaning as the statement that the card drawn is not the ace of hearts. Observe the illustration of the logical equivalence between $\sim p \vee \sim q$ and $\sim(p \wedge q)$.

Example 41 For every ordered triple of elements (a, b, c),
$$(a + b)(a' + c) = ac + a'b + bc = ac + a'b.$$

PROOF

$$\begin{aligned}
(a + b)(a' + c) &= aa' + ac + a'b + bc \\
&= ac + a'b + bc & &\text{[because } aa' = 0 \text{ is the additive identity]} \\
&= ac + a'b + (a + a')bc & &\text{[because } a + a' = 1 \text{ is the multiplicative identity]} \\
&= (ac + acb) + (a'b + a'bc) \\
&= ac + a'b & &\text{[two applications of Theorem 3a].}
\end{aligned}$$

A key step in the preceding argument depends on the fact that the element bc is expressible as the sum of two elements, one with the factor a and the other with the factor a'; thus, $bc = a \cdot bc + a' \cdot bc$.

Example 42 Let x, y be elements of a Boolean algebra. Show that $x + y$ is a factor of the element xy'.
Indeed, $xy' = xy' + 0 = xy' + yy' = (x + y) \cdot y'$.

Example 43 Let f, g, h be any elements in a Boolean algebra. Simplify $hf' + (h' + fg)' + g$.
The following sequence of steps reveals that the given element is simply $h + g$.

$$\begin{aligned}
hf' + (h' + fg)' + g &= hf' + (h')'(fg)' + g & &\text{[by Theorem 8b]} \\
&= hf' + h(fg)' + g & &\text{[by Theorem 7]} \\
&= hf' + h(f' + g') + g & &\text{[by Theorem 8a]} \\
&= hf' + hf' + hg' + g & &\text{[by the distributive property]} \\
&= hf' + hg' + g & &\text{[by the idempotent property for addition]} \\
&= hf' + h + g & &\text{[by Theorem 5a]} \\
&= h + g & &\text{[by Theorem 3a].}
\end{aligned}$$

One of the most striking features about Boolean algebra is the dual relationship between addition and multiplication. Any true statement involving the operations in a Boolean algebra remains true if the roles of addition and multiplication are interchanged. This symmetry is strongly noticeable in the definition: parts 6 through 10 parallel parts 1 through 5; parts 12 and 13 are duals of each other. Part 11 does not mention the binary operations. Only part 14 does not seem to have a counterpart, but this apparent deficiency is remedied in Theorem 4. Most of the theorems in our development have occurred in pairs, distinguished by the labels "a" and "b." In each case, either theorem in the pair is obtainable from the other by interchanging the roles of addition and multiplication. This interchange can be effected symbolically by substituting $+$ for $\cdot$, substituting $\cdot$ for $+$, replacing 0 by 1, and replacing 1 by 0, wherever the various symbols occur. In practice, this principle of duality reduces our work considerably. We do not need to give separate discussions for a pair of dual results. As soon as we establish one of them, the other is also considered proved.

EXERCISES L

1. Prove Theorem 8b. (Give a proof not based on a duality argument.)
2. State the dual of the proposition in Example 41, and then prove it. (Give a proof not based on the duality argument.)
3. Simplify each of the following elements. Give a reason for each step.
   (a) $(a + b + c + d)(a + b + d)(a + c)$.
   (b) $abc + ab'c + ab'c' + abc'$.
   (c) $abc + ab'c + a'b'c$.
4. Show that the sum $xy' + x'y$ can be expressed as a product of two factors, one of which is $x + y$.
5. Show that the sum $xy + x'y'$ can be expressed as a product of two factors.
6. Show that the sum $xy'w + x'yz$ can be expressed as a product of two factors, one of which is $x' + y'$.
7. Consider the elements $xy' + x'y$ and $xy + x'y'$.
   (a) Show that these two elements are complements of each other.
   (b) For the Boolean algebra discussed in Example 36, show that the interpretation of $xy + x'y'$ is the biconditional.
   (c) Hence, conclude that the interpretation for $xy' + x'y$ is the negation of the biconditional.
   (d) Let p and q be sentences pertaining to the outcome of an activity, as in Example 36. Deduce from part (c) that a statement that is true precisely in case p, q have opposite truth values is the compound statement $(p \wedge {\sim}q) \vee ({\sim}p \wedge q)$.
   (e) Deduce from Exercise 4 that another statement that is true precisely in case p, q have opposite truth values is the conjunction $(p \vee q) \wedge ({\sim}p \vee {\sim}q)$.
8. Let a, b, c be elements in a Boolean algebra. Let p, q, r be sentences pertaining to the outcome of an activity, as in Example 36.
   (a) As an application of de Morgan's laws, show that
   $$a' + (b' + c) = (ab)' + c = b' + (a' + c).$$
   (b) Interpret part (a) with reference to the three conditional statements
   $p \to (q \to r)$, $(p \wedge q) \to r$, $q \to (p \to r)$.
9. Simplify each of the following elements. Give a reason for each step.
   (a) $xy' + z(x' + y + w)$.
   (b) $[x + y][(xy)' + x'z]' + z$.
   (c) $(uv)'(uv + uw)$.

Counting 6

1 Counting

An important application of our work in the preceding chapter will be in the study of probability. In order to have an adequate preparation for solving problems in probability, we need to be able to find the number of elements in a set. The following is a typical situation: we are given one or more sets and we know how many members each of them has; another set is described, using the given sets; the question is to find how many elements the described set has.

Example 1 Suppose the set $S = \{0, 1, 2\}$ with three elements is given. Consider the set T of all ordered pairs of elements of S, each of which has two different components. The elements in a membership list for T are $(0, 1), (0, 2), (1, 0), (1, 2), (2, 0), (2, 1)$. By counting we find that T has six members.

As illustrated by the example, a straightforward approach to a counting problem is to list all the members individually and then to count. This method may require painstaking care to ensure that the list is both complete and free of duplication. Since we are often concerned with sets having many members, this method is seldom efficient. We wish to discover general principles of counting which will help us.

2 Partition

Example 2 Suppose we have a set T of current American coins. The set T may be separated into subsets according to any of various classifications. We may consider a two-way classification into the set S of coins with silver content and the set $\tilde{S}$ of coins that do not contain any silver. Or we may consider the classification according to monetary value. The latter classification is a partition of T into six subsets: the subset P of pennies, the subset N of nickels, the subset

D of dimes, the subset Q of quarters, the subset H of halves, and the subset W of silver dollars. These six subsets have the property that every coin in T belongs to exactly one of the subsets. The union of the six subsets is T, and the intersection of any two of the subsets is empty.

The elements of T also may be classified according to the year of minting. Here again we have a separation of the coins into several categories. The corresponding subsets have T as their union, and each two of them are disjoint.

A significant part of science is concerned with classification. Much of the technical language in a science is introduced to name a property of an object or a category of objects. Clear-cut classifications are most satisfying, namely, those in which each object under investigation belongs to exactly one of the categories. These ideas lead us to introduce the notion of a partition.

Definition Let T be a nonempty set. A *partition* of T is a collection of nonempty subsets of T such that each element of T belongs to exactly one of the subsets in the collection.

The requirement that every member of T belongs to at least one of the subsets in the collection means that the union of the subsets is T itself. The requirement that every member of T belongs to at most one of the subsets means that the intersection of any two of the subsets in the collection is empty. Thus, an alternative description of a partition is as follows:

Alternative definition A *partition* of T is a collection of nonempty subsets of T such that every two of them have an empty intersection and the union of all of them is T.

The subsets of T that form a partition may be called the *cells* of the partition. Figuratively, we may imagine the elements of T as being separated into compartments or cells. In fact, a worthwhile pictorial notion of a partition is that of taking the elements of T, one at a time, and tossing them into boxes or cells. (See Fig. 1.) Each element, of course, falls into exactly one cell.

We are most often interested in a partition consisting of two cells. Such a partition corresponds to splitting the set T into two nonempty subsets. Since their union is T and their intersection is empty, each of them is the complement of the other. (Compare this statement with the remark following Theorem 6 in Section 13 of Chapter 5.)

Figure 1

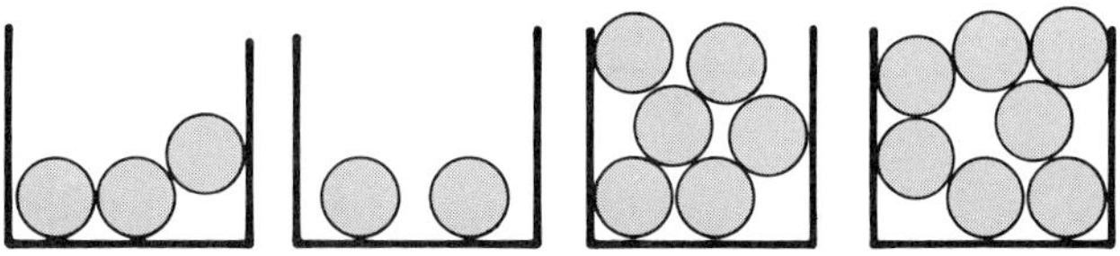

Example 3 Suppose that a set Q has a partition consisting of four cells. Suppose that the numbers of elements in the respective cells are 20, 40, 35, 15. How many members does the set Q have?

We may answer this question by simply adding the numbers of elements in the several cells. There are $20 + 40 + 35 + 15 = 110$ members in Q.

This general principle is important enough to be formulated as a theorem.

Theorem 1 For any partition of a finite set T, the number of elements in T is the sum of the numbers of elements in all the cells of the partition.

Example 4 Consider a dot and dash type of code, similar to the Morse code. Suppose that in the particular code under consideration each character is transmitted by no more than four signals, where each signal is either a dot or a dash. A character may be a letter of the alphabet, a punctuation mark, a numeral, a word, or some other coded information. Let T be the set of transmittable characters. One method of classifying a character is according to the number of signals required. This classification procedure gives a partition of T consisting of four cells: the subset P consisting of one-signal characters, the subset Q consisting of two-signal characters, the subset R of three-signal characters, and the subset S of four-signal characters. If we find that P has 2 members, Q has 4 members, R has 8, and S has 16, then we may conveniently discover that the number of characters in T is the sum $2 + 4 + 8 + 16 = 30$. Such a system is adequate for the English alphabet, but does not take care of the alphabet plus the digits, 0, 1, 2, ..., 9. For this reason, the Morse code uses four signals for alphabet characters and five signals for the digits, 0, 1, 2, ..., 9.

3 Number of elements in a union

In Section 5 of Chapter 5 we discussed another principle concerning counting.

Theorem 2 Let A and B be finite subsets of a set T. Then the sum of the numbers of elements in A and in B is the same as the sum of the numbers of elements in $A \cup B$ and $A \cap B$.

If A, B, $A \cap B$, $A \cup B$ have r, s, k, u members, respectively, then

$$r + s = u + k.$$

Figure 2

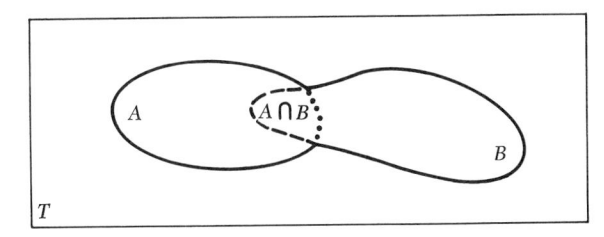

We review the proof of this important theorem. The set A is the union of the two disjoint subsets $A \cap B$ and $A \cap \tilde{B}$. Since A has r elements and $A \cap B$ has k elements, we conclude that $A \cap \tilde{B}$ has $r - k$ elements. (See Fig. 2.) The set B is the union of two disjoint subsets $A \cap B$ and $\tilde{A} \cap B$. Since B and $A \cap B$ have s and k members, respectively, $\tilde{A} \cap B$ has $s - k$ members. The union of the mutually disjoint subsets $A \cap \tilde{B}$, $A \cap B$, $\tilde{A} \cap B$ is $A \cup B$. Thus, $(r - k) + k + (s - k) = u$. This equation can be simplified to $r + s = u + k$.

In many cases the three subsets $A \cap \tilde{B}$, $A \cap B$, $\tilde{A} \cap B$ form a partition of the set $A \cup B$. This is certainly the situation depicted in Fig. 2. In our proof, however, we have not used the language of partitions because it is possible that one or more of the three subsets $A \cap \tilde{B}$, $A \cap B$, $\tilde{A} \cap B$ may be empty. (See Figs. 3a, b.) Note that this possibility does not affect the validity of our argument, for the number of elements in the empty set is simply 0.

Figure 3a illustrates the case where $s - k = 0$; then $u = (r - k) + k + 0 = (r - k) + k + (s - k)$, which agrees with the general result. Figure 3b

Figure 3

(a) $s - k = 0$; then,
$u = (r - k) + k + 0$
$= (r - k) + k + (s - k)$.

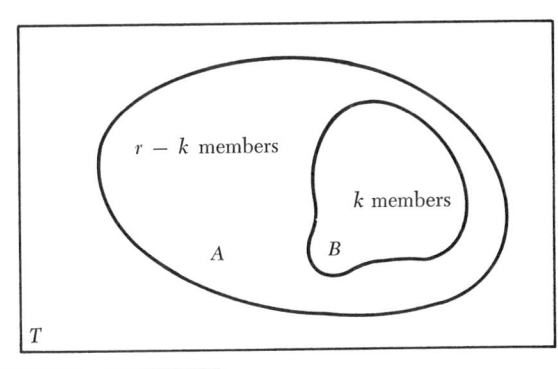

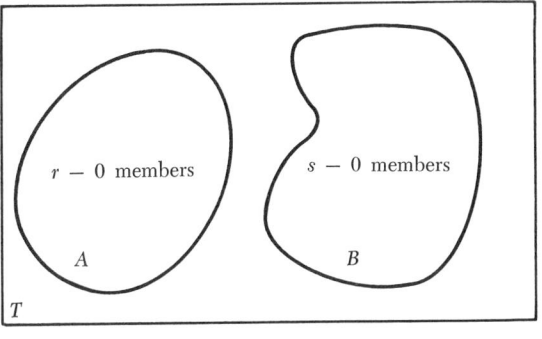

(b) $k = 0$; then,
$u = r + s$
$= (r - k) + k + (s - k)$.

3 / Number of elements in a union

portrays the case where $k = 0$; then $u = r + s = (r - k) + k + (s - k)$, which again agrees with the conclusion in the general case.

EXERCISES A

1. The automobiles in a certain parking lot at a certain time may be classified according to make. They may also be classified according to year of model. Cite several other ways in which they may be classified, and in each case describe the corresponding partition of the set of automobiles.

2. The performers in a college marching band at a certain football game may be classified according to the number of semesters of membership in the band. Give several other ways in which they may be classified, and in each case describe the corresponding partition of the band.

3. If a marching band has three piccolo players, 24 clarinetists, ten saxophonists, 12 cornetists, six French horn players, five baritone players, nine trombonists, five sousaphonists, and seven in the percussion section, how many players belong to the band?

4. How many students are enrolled in a certain language course that has 18 sections with 23 students in each section?

5. How many fellows live in four barracks if each building has three wings with 16 bunks in each wing?

6. How many problems does a student solve during a semester when he successfully completes 36 assignments with 15 exercises in each assignment?

7. Let h and k be natural numbers. A partition of a certain set consists of h subsets, each of which has k elements. How many members belong to the set?

8. A certain set has 120 members.
   (a) If a partition of the set has ten cells and each cell has the same number of elements, what can be deduced about the number of elements in each cell?
   (b) If a partition of the set has 30 cells, what can be deduced about how many elements belong to each cell?
   (c) If a partition of the set has 103 cells, give whatever information you can about the number of elements in the various cells.

9. Suppose that the set W has 60 members. Suppose that $\{A, B\}$ is a partition of W such that the subset A of W has 26 members. Suppose that $\{P, Q\}$ is a partition of W and that there are 38 elements in the subset P. Suppose that the subset $A \cup P$ has 52 elements.
   (a) Find the number of elements in the intersection $A \cap P$.
   (b) Show that the two subsets $A \cap P$ and $A \cap Q$ are the two cells of a partition of the set A.
   (c) Find the number of elements in $A \cap Q$.
   (d) Show that $\{A \cap P, B \cap P\}$ is a partition of P.
   (e) Find the number of elements in $B \cap P$.
   (f) How many elements belong to $B \cap Q$?
   (g) Show that $\{A \cap P, B \cap P, A \cap Q, B \cap Q\}$ is a partition of W consisting of four cells.

10. Let A, B, C be subsets of a certain set. Suppose that A has 90 elements, B has 80 elements, C has 70 elements, $A \cap B$ has 50 elements, $A \cap C$ has 40 elements,

$B \cap C$ has 30 elements, and $A \cap B \cap C$ has ten elements. Using the given information, find the number of elements in each of the following subsets.
(a) $A \cup B$.
(b) $(A \cap C) \cap (B \cap C)$.
(c) $(A \cap C) \cup (B \cap C)$.
(d) $(A \cup B) \cap C$.
(e) $(A \cup B) \cup C$.

11. Let A, B, C be finite subsets of a certain set. Suppose that A has r elements, B has s elements, and C has t elements. Suppose that $A \cap B$ has f members, $A \cap C$ has g members, and $B \cap C$ has h members. Suppose that there are k elements in the intersection $A \cap B \cap C$ and u members in the union $A \cup B \cup C$. Apply Theorem 2 several times in order to answer the following questions in terms of the given information. [HINT: The answer to one part of the exercise may be useful in solving a later part.]
(a) Find the number of elements in the set $A \cup B$.
(b) How many members belong to the set $(A \cap C) \cap (B \cap C)$?
(c) Find the number of elements in the set $(A \cap C) \cup (B \cap C)$.
(d) How many members belong to the set $(A \cup B) \cap C$?
(e) Express the number of elements in the set $(A \cup B) \cup C$ in terms of your answers to parts (a) and (d).
(f) By simplifying the result in part (e), show that
$$u = r + s + t - f - g - h + k.$$

12. A store has a set V of vases on display. The vases may be classified according to color: some are white, the others are blue. Hence, the color classification gives a partition of the set V with two cells. The vases may be classified according to height: tall, medium, or short. From this classification V has a partition with three cells. If each intersection of a color cell and a height cell has five vases, how many vases are there in the set V?

13. Suppose that a set of books is packed for shipment in five cartons.
(a) If each carton contains t books, how many books are in the set?
(b) If the cartons are labeled 1, 2, 3, 4, 5, respectively, and if each carton contains twice as many books as its label, how many books are in the set?

4 Application

Example 5 The manufacturer of the products Ak, Bu, Ci authorizes his advertising agency to conduct a survey. The agency interviews 400 people and reports the following information:

150 interviewees use Ak,
100 interviewees use Bu,
120 interviewees use Ci,
40 use both Ak and Bu,
50 use both Ak and Ci,
20 use Bu and Ci but not Ak,
10 use all three products.

Three questions are posed. How many interviewees use none of the three products? How many use Ak but not Ci? How many either use both Bu and Ci or use neither Bu nor Ci?

In the set I of interviewees, let A, B, C be the subsets consisting of the people who use Ak, Bu, Ci, respectively. A diagram may be

Figure 4a

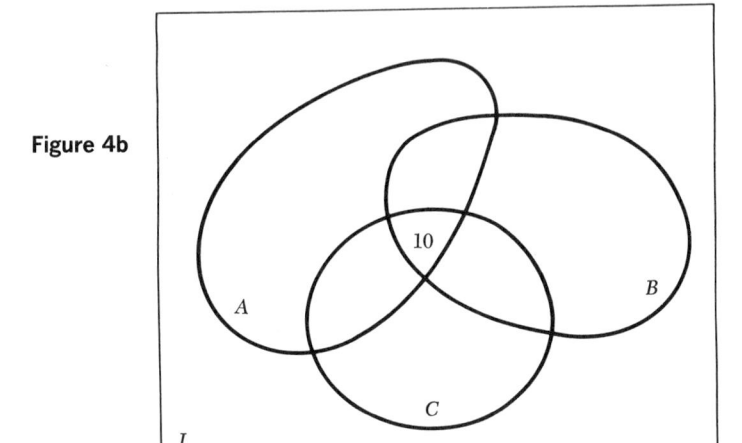

Figure 4b

helpful. The three ovals in Fig. 4a separate the region into eight subsets which are mutually disjoint. We wish to find how many interviewees belong to each of these eight subsets. These calculations we make in a stepwise fashion, working, in a sense, "from the inside out." We begin by noting the given information that ten people belong to the intersection $A \cap B \cap C$. We may mark this number on the diagram in the proper compartment (Fig. 4b).

The 20 people who use Bu and use Ci but do not use Ak are the members of the set $B \cap C \cap \tilde{A}$. The corresponding region is inside the oval encircling B, inside the oval encircling C, but outside the oval surrounding A. We may mark the number 20 in the proper compartment on the diagram.

The set $A \cap C$ has 50 members. Of these, ten are also in the set B. The remaining people, $50 - 10 = 40$ in number, belong to A and to C but not to B. We may mark 40 in the corresponding compartment.

204   Counting

Figure 4c

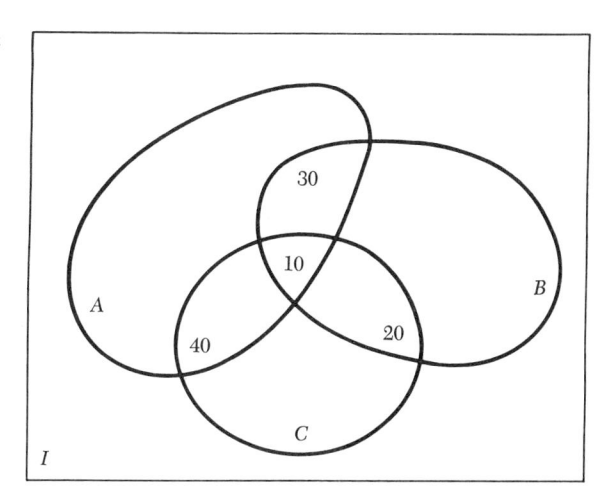

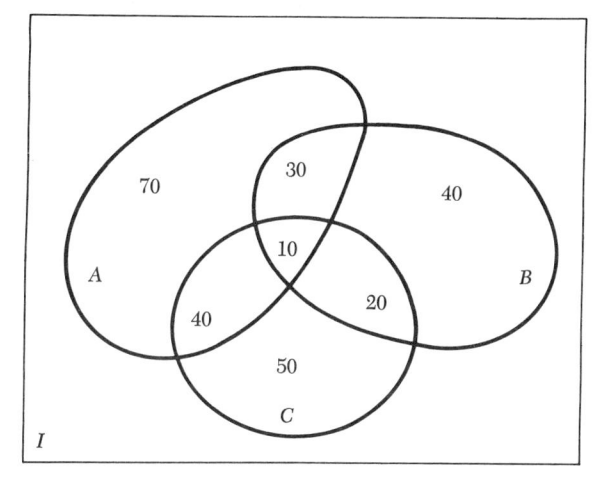

Figure 4d

According to the given information, $A \cap B$ has 40 members. Since ten of these are included in the set $A \cap B \cap C$, the remaining 30 are not in the set C. After marking 30 at the appropriate place, we obtain Fig. 4c.

Inside the oval corresponding to C, we have thus far accounted for $40 + 10 + 20 = 70$ members; since C has a total membership of 120 people, there are 50 who are elements of neither A nor B.

We apply a similar discussion to the set B; of the 100 members, $30 + 10 + 20 = 60$ are accounted for already; the remaining 40 belong to neither A nor C.

Finally, in the set A, $30 + 10 + 40 = 80$ members are accounted for, and $150 - 80 = 70$ members belong to neither B nor C. We may mark in the respective compartments on the diagram these three numbers: 50, 40, 70 (see Fig. 4d).

The sum of the numbers recorded thus far is $10 + 20 + 40 + 30 + 50 + 40 + 70 = 260$, which is the number of elements in the

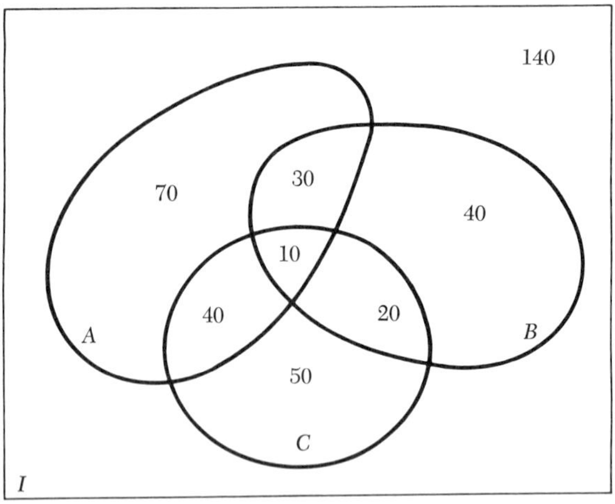

Figure 4e

[Venn diagram with three overlapping sets A, B, C inside universe I. Regions labeled: 140 (outside all sets), 70 (A only), 30 (A∩B only), 40 (B only), 10 (A∩B∩C), 40 (A∩C only), 20 (B∩C only), 50 (C only).]

union $A \cup B \cup C$. Since I has 400 members, there are $400 - 260 = 140$ in the complement of $A \cup B \cup C$. We are now ready to answer the questions asked.

The people who use none of the three products belong to neither A nor B nor C. They are the members of the complement of $A \cup B \cup C$. There are 140 of them (see Fig. 4e).

The people who use Ak but not Ci are the members of the set $A \cap \tilde{C}$. They belong to A and do not belong to C. From Fig. 4e we find $70 + 30 = 100$ of them. The number 30 tells how many of the people under consideration also use Bu, the number 70 tells how many do not.

The subset $B \cap C$ has $10 + 20 = 30$ members. The subset $\tilde{B} \cap \tilde{C}$ has $70 + 140 = 210$ elements. Their union has $30 + 210 = 240$ members. Hence, there are 240 people who either belong to $B \cap C$ or belong to $\tilde{B} \cap \tilde{C}$, that is, people who either use both Bu and Ci or use neither Bu nor Ci. We may note that the union is expressed symbolically by $(B \cap C) \cup (\tilde{B} \cap \tilde{C})$. An alternative description of this set is the set of all people each of whom uses Bu if and only if he uses Ci.

EXERCISES B

1. Suppose that P, Q, R are subsets of a set T. The following data tell the number of elements belonging to various subsets of T: P, 19; R, 17; $P \cap Q$, 9; $P \cup Q$, 26; $\tilde{P} \cap Q \cap R$, 4; $Q \cap R$, 11; $P \cap \tilde{R}$, 11; T, 31. Calculate the number of members in each of the following subsets.
    (a) Q.   (b) $Q \cup R$.   (c) $P \cup R$.   (d) $\tilde{P} \cap \tilde{Q} \cap \tilde{R}$.

2. Suppose that A, B, C are subsets of a set S. The following data tell how many members there are in each of various subsets of S: C, 19; B, 28; $A \cap B$, 14; $\tilde{A} \cap B \cap C$, 5; $A \cap \tilde{B} \cap C$, 1; $A \cap \tilde{C}$, 12; $A \cap B \cap C$, 6; S, 50. Find the number of elements belonging to each of the following subsets.
    (a) A.   (b) $B \cup C$.   (c) $A \cup \tilde{C}$.
    (d) $\tilde{A} \cap \tilde{B} \cap \tilde{C}$.   (e) $(A \cup C) \cap \tilde{B}$.   (f) $(A \cap C) \cup (B \cap C)$.

3. Suppose that P, Q, R are subsets of a set T. The following table gives the number of elements contained in each of various subsets of T.

Subset	$Q \cap R$	$P \cap Q$	$P \cap \tilde{R}$	$Q \cap \tilde{R}$	$P \cap \tilde{Q} \cap R$	$P \cup R$	$\tilde{Q} \cup R$	$\tilde{P} \cap \tilde{Q} \cap \tilde{R}$
NUMBER OF MEMBERS	11	12	10	8	8	33	36	6

Determine the number of elements in each of the following subsets of T.
(a) Q. (b) R. (c) P. (d) T. (e) $P \cup Q$.

4. Suppose that 1000 families are asked what type of vacation they like. The report of this poll states that 678 families enjoy a vacation in a big city, 646 families enjoy a vacation at the seashore, 603 families enjoy a vacation in the mountains, 347 like both the big city and the mountains, 388 like both the big city and the seashore, 291 like both the seashore and the mountains, 106 enjoy all three types. What significant deductions can you make from the report?

5 Tree diagrams

Example 6 A businessman plans a trip from New York to Los Angeles. He wishes to visit the branch office in one of the four cities, Detroit, Cleveland, Pittsburgh, or Nashville; afterward he wants to stop at the branch office in one of the two cities, Minneapolis or St. Louis. How many possible itineraries are available to him?

Two successive activities are involved in this problem. The businessman must select one of the eastern cities; there are four possible choices. He must also pick one of the central cities, and there are two available opportunities. Each possible itinerary is essentially an ordered pair of cities. Suppose we write a list of all possible itineraries, using an initial letter to designate each city. We have the ordered pairs (d, m), (d, s), (c, m), (c, s), (p, m), (p, s), (n, m), (n, s). In the language of Chapter 5, we may consider each itinerary as an outcome of the decision process. With four choices for the first component and two choices for the second component, we find there are eight possible outcomes for our sequence of activities.

A treelike diagram (see Fig. 5) pictures the possible outcomes in terms of the two successive choices. Each ordered pair is represented by a path upward through the tree from the bottom to the top.

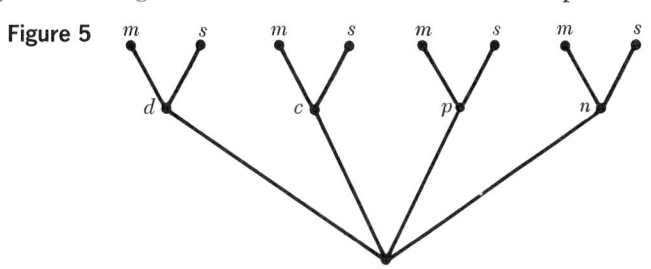

Figure 5

Example 7 A committee has five members. The committee must select one of its members as chairman and another as secretary. How many possible official pairs are available to the committee?

Suppose we identify the members of the committee by the names a, b, c, d, e. An official pair may be considered as an ordered pair of members such that the first component is the chairman and the second component is the secretary. Any of the five members is eligible to be chairman. Any member is eligible to be secretary. However, the same member cannot serve in both official capacities. Thus, the set of ordered pairs we wish to count has the following characterization: any member of the committee may be the first component, any member may be the second component, but the components of any given pair are different.

We make a list of the ordered pairs as described:

$$(a, b), \quad (a, c), \quad (a, d), \quad (a, e);$$
$$(b, a), \quad (b, c), \quad (b, d), \quad (b, e);$$
$$(c, a), \quad (c, b), \quad (c, d), \quad (c, e);$$
$$(d, a), \quad (d, b), \quad (d, c), \quad (d, e);$$
$$(e, a), \quad (e, b), \quad (e, c), \quad (e, d).$$

There are 20 pairs. In the listing they have been arranged in five batches of four each. Each batch corresponds to the selection of a chairman, since the ordered pairs in a batch have the same first component. Each batch has four pairs, since there are, after the selection of a chairman, only four remaining members eligible for the secretaryship.

The twenty possible outcomes for this sequence of activities are pictured in the tree diagram shown in Fig. 6. Each ordered pair is represented by an upward path through the tree.

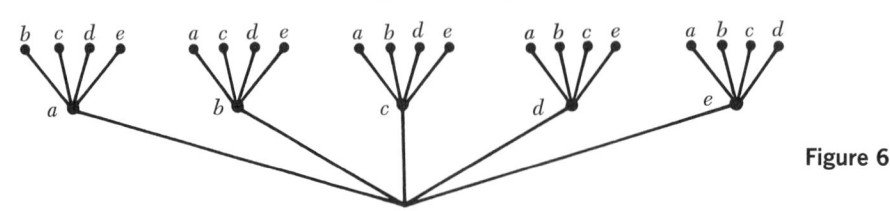

Figure 6

6 Number of permutations

Example 8 Let S be the set $\{a, b, c\}$. Find the number of ordered triples of elements of S such that each ordered triple has every element of S as a component.

Let T be the set of ordered triples as described. We may classify the members of T according to the first component. Since S has three elements and each is eligible to be the first component, T has a partition consisting of three cells: the subset A consisting of ordered triples with first component a, the subset B which consists of ordered triples with first component b, and the subset C composed of ordered triples with left component c. An element of A has the form $(a, \_, \_)$, where one blank is to be filled by b and the other by c. There are only two possibilities, namely, (a, b, c) and (a, c, b). Thus, A has two members. A similar argument shows that B has two members: indeed, an element of B has the form $(b, \_, \_)$, where the first blank may be filled by either a or c and the second blank may be filled by the remaining member of S; thus, $B = \{(b, a, c), (b, c, a)\}$. Likewise, C is a subset with two elements: $C = \{(c, a, b), (c, b, a)\}$. Since each cell of the partition has two members and there are three cells, T contains $2 \cdot 3 = 6$ elements.

The selection of an ordered triple belonging to T is accomplished by a sequence of three activities: first, the choice of a left component; second, the choice of a middle component; third, the choice of a right component. This viewpoint is illustrated in the tree diagram in Fig. 7. In summary, the six ordered triples are (a, b, c), (a, c, b), (b, a, c), (b, c, a), (c, a, b), (c, b, a). Each of these ordered triples gives a permutation of the set S. Thus, S has six permutations.

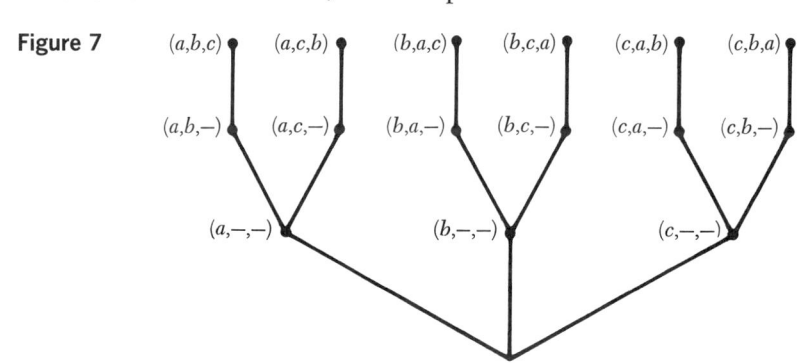

Figure 7

Definition Let n be a natural number, and let S be a set with n members. An ordered n-tuple (a vector with n components) such that each element of S is a component of the n-tuple describes a permutation of the set S. The number of such n-tuples is the *number of permutations* of S.

Example 9 Let R be the set $\{x, y, z\}$. Find the number of permutations of R.

We list the ordered triples that describe the various permutations; they are (x, y, z), (x, z, y), (y, x, z), (y, z, x), (z, x, y), (z, y, x). The number of permutations is six.

We compare this result with Example 8. Each of the sets S and R has three members. A comparison of the list of ordered triples given in Example 8 and the list in the present example indicates a natural way of pairing the permutations of S and the permutations of R. It seems reasonable then to expect that any set with three elements has six permutations.

The remarks in Example 9 suggest the following important observation. The number of permutations of a set depends only on how many members the set itself has and is not influenced by what the individual elements actually are. For this reason, we may speak meaningfully about "the number of permutations of a set with m elements," without specifying the actual set.

EXERCISES C

1. A restaurant offers a choice of four entrees and three desserts. The entrees are beef, fish, lamb, turkey; the desserts are cake, pie, sherbet. Suppose that each customer orders both an entree and a dessert.
   (a) How many different dinners are available to the customer?
   (b) Sketch a tree diagram picturing the possible outcomes for the activity of ordering dinner.

2. A man invites his wife out to dinner and the theater. Five favorite restaurants specialize in American, Chinese, French, German, Mexican food, respectively. Four theaters offer attractive billings. The husband has learned from experience that dining at certain restaurants does not allow enough time to reach all the theaters before the performance begins. They cannot eat American food and go to the Palace; they cannot eat Chinese food and go to the Arena; if they eat German food, they can go only to the Bijou; if they eat Mexican food, they must go to either the Garden or the Palace.
   (a) How many different evenings out are available to the couple?
   (b) Sketch a tree diagram picturing the possible outcomes for the decision process.

3. Let h and k be natural numbers. Suppose a tree diagram has h branches at the bottom, and suppose that each of these separates into k branches at the top. How many upward paths through the tree are there?

4. A traveler from Boston to Honolulu may visit exactly one of the two cities, New York or Washington; then exactly one of Chicago or St. Louis; then exactly one of Los Angeles or San Francisco or Seattle.
   (a) How many routes are offered to the traveler?
   (b) Sketch a tree diagram for this sequence of three decisions.

5. From an ordinary deck of playing cards one card is drawn and not replaced, and then a second card is drawn. For each of the following, find the number of possible drawings that satisfy the given condition.
   (a) The first card drawn is a spade, and the second is a heart.
   (b) A black card is drawn, and then a heart is drawn.
   (c) The respective drawings give a spade and a black card.
   (d) A black and a spade card are drawn in that order.

6. From an ordinary deck of playing cards two cards are drawn in succession. The first card is not replaced before the second card is drawn. For each of the following, find

the number of possible drawings that belong to the truth set of the given statement.
(a) The first card is a king, and the second card is an ace.
(b) The drawing of a king is first, and the drawing of a face card is second. [NOTE: See Example 32 in Section 10 of Chapter 5.]
(c) A face card and a king appear in that order.

7. From an ordinary deck of playing cards two cards are picked successively, without replacing the first before choosing the second. Find the number of members in the truth set of each of the following:
   (a) First a spade is drawn, and then a red ace is drawn.
   (b) A spade is drawn, and then an ace is drawn.
   (c) An ace is picked first and is followed by a spade.

8. (a) Make a list of all the ordered pairs of elements from the set $\{c, d\}$ such that the two components of each pair are different.
   (b) Make a list of all the ordered pairs of elements from the set $\{x, y\}$ such that each ordered pair has every member of $\{x, y\}$ as a component.
   (c) How many permutations does a set with two members have?

9. Let T be the set $\{a, b, c, d\}$.
   (a) Make a list of all the ordered pairs of elements from T such that the two components of each pair are different.
   (b) How many ordered pairs of elements of T have different components?
   (c) Find how many ordered pairs of elements of T there are.

10. Let $T = \{a, b, c, d\}$. Find the number of ordered triples of elements of T such that, in each ordered triple, the second component is different from the first component and the third component is the same as the second component.

7 Summary of counting principles

We are now ready to summarize the counting ideas that have been illustrated by the examples in this chapter.

1. To find the number of elements in a finite set, the basic technique is to list the members and count them, one by one.

This procedure may be tedious, and more efficient methods are desirable in many cases. A more efficient method often demands more careful analytical thinking. Thus, a price may be paid for efficiency. Fortunately, the price is low in comparison with value received.

2. To find the number of elements in a finite set, we may select a partition of the set, find the number of elements in each cell of the partition, and compute the sum of the numbers for the various cells.

In this method, analytical thinking is important in choosing a worthwhile partition and in verifying that the selection actually is a partition.

An important special case arises if all the cells have the same number of elements; then the number of members in the set is the product of the number of cells and the number of elements in each cell.

Suppose that the members of a set T are ordered pairs of elements. Suppose that there are m elements that are first components of the pairs. Suppose that each of these elements is the first component of n different ordered pairs. Then the number of ordered pairs in T is mn. Indeed, we may classify the members of T according to their first components. This classification yields a partition of T with m cells, each cell having n members. By the special case, T has mn members.

3. Suppose we have a sequence of activities. If the first activity has m outcomes, and if, for each of these, the second activity has n possible outcomes, then the sequence of two activities has $m \cdot n$ outcomes.

A pictorial representation of this type of situation is given in Figs. 5 and 6. A sequence of more than two activities is often encountered.

Theorem 3 Let a sequence of finitely many activities be given. Suppose that there are m possible outcomes for the first activity; suppose that, for each of these, there are k possible outcomes for the second activity; suppose that, for each of the pairs of outcomes for the preceding activities, there are h possible outcomes for the third activity; etc. Then the number of outcomes for the entire sequence of activities is the product of the numbers m, k, h, etc.

The number of outcomes for a sequence of n activities is the same as the number of ordered n-tuples in which the first component corresponds to the outcome of the first activity, the second component to the outcome of the second activity, etc.

Example 10 Suppose that a company leases three suites in an office building. Six employees are to be assigned working space, two in each suite. In how many ways may the assignments be made?

One method for doing the assignment job is via the following sequence of activities:

(a) Choose a suite for the person with highest seniority.
(b) Choose an office mate for him.
(c) Choose a suite for the person with highest seniority among those not yet assigned.
(d) Choose an office mate for him.
(e) Assign the remaining two employees to the remaining unoccupied suite.

If we adopt this analysis of the counting problem, then we may find the total number of possible assignments by multiplying the numbers of outcomes for the five successive activities.

Activity (a) has 3 outcomes, since there are three suites.

Activity (b) has 5 outcomes, since there are five employees who have not been assigned an office after activity (a) has been finished.

Activity (c) has 2 outcomes, since two suites remain unoccupied.

Activity (d) has 3 outcomes, since there remain three unassigned employees after the first three activities are done.

Activity (e) has only 1 outcome, since there remain only one vacant suite and two employees.

Consequently, the number of possible assignments is
$$3 \cdot 5 \cdot 2 \cdot 3 \cdot 1 = 90.$$

EXERCISES D

1. From an ordinary deck of playing cards one card is drawn; without replacement of the first card, a second card is drawn; without replacement of either of the first two cards, a third card is drawn. For each of the following, find the number of possible sequences of drawings that yield cards of the prescribed categories in the designated order.
   (a) First a spade, second a heart, third a club.
   (b) A spade, then a heart, then a spade.
   (c) A spade, a spade, a spade.
   (d) A spade, a spade, lastly a black card.
   (e) A spade, then a black card, third a spade.
   (f) A black card, second a spade, third a club.
   (g) First a red ace, second the club king, third a black queen.
   (h) A red ace, then a black king, finally a club.
   (i) A queen, a spade, a jack.

2. The six faces of a die have 1, 2, 3, 4, 5, 6 spots, respectively. Suppose that a player rolls two dice, of which one is red and the other blue. Let the outcome of a roll be denoted by the ordered pair (r, b) of numbers, in which the first component r is the number of spots showing on the red die and the second component is the number on the blue die.
   (a) Find the number of possible outcomes.
   (b) Find the number of members of the truth set for the statement $r = b$.
   (c) Find the number of outcomes that satisfy the condition that $r < b$.
   (d) List the ordered pairs (r, b) such that $r < b$.
   (e) For how many possible outcomes is the number on the blue die smaller than the number on the red die?
   (f) How many outcomes belong to the truth set of the disjunction that either $r - b$ or $b - r$ is equal to one?

3. Let S be a set with five members, and let T be the set of ordered triples of elements from S.
   (a) Find the number of elements in T.
   (b) Find the number of ordered triples belonging to T such that each ordered triple has three different components.
   (c) How many ordered triples in T have the property that all three components of the triple are the same?
   (d) How many members of T belong to the truth set of the following statement? Two of the components of the triple are the same but are not equal to the other component.

7 / Summary of counting principles

(a,b,c,d) (a,b,d,c) (a,c,b,d) (a,c,d,b) (a,d,b,c) (a,d,c,b)   (b,a,c,d) (b,a,d,c) (b,c,a,d) (b,c,d,a) (b,d,a,c) (b,d,c,a)

(a,b,c,—) (a,b,d,—) (a,c,b,—) (a,c,d,—) (a,d,b,—) (a,d,c,—)   (b,a,c,—) (b,a,d,—) (b,c,a,—) (b,c,d,—) (b,d,a,—) (b,d,c,—)

(a,b,—,—)   (a,c,—,—)   (a,d,—,—)   (b,a,—,—)   (b,c,—,—)   (b,d,—,—)

(a,—,—,—)   (b,—,—,—)

Figure 8

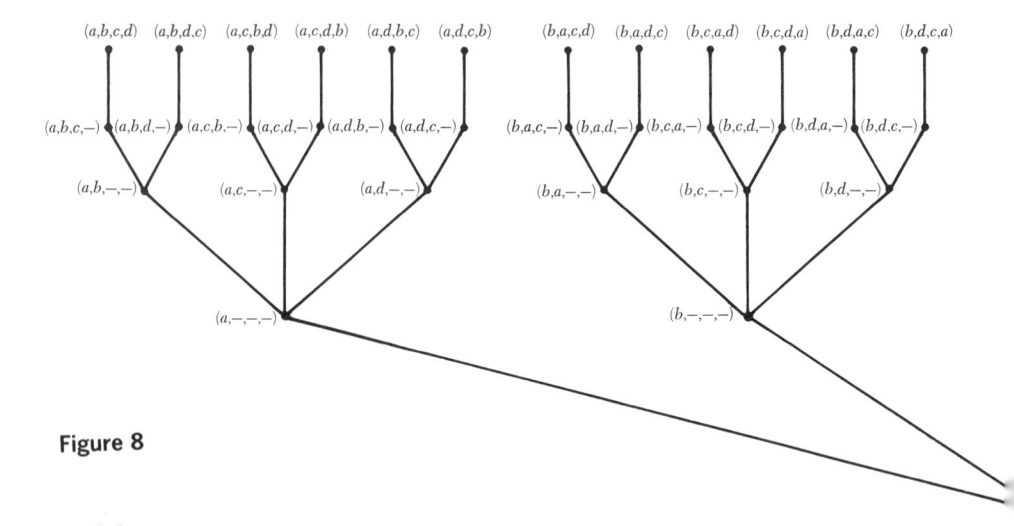

4. Eight basketball teams in different cities form a league. During the season each team is host on its own court to every other team in the league twice. How many games are played in the league season?

5. Six violinists are competing in a contest. Each violinist performs before the panel of judges. In how many ways can the contest officials schedule the sequence of performances by the respective contestants?

6. Three pianists are entered in a contest. Each contestant performs twice before the panel of judges. In his first appearance he plays classical compositions; in his second, contemporary music. In arranging the schedule of auditions, the contest committee makes certain that no pianist immediately follows himself on the program, or in other words, that between the two appearances of any contestant at least one of the other players is heard. Find the number of possible ways in which the sequence of auditions before the panel can be scheduled.

8 Number of permutations of a finite set

The number of permutations of a set with four elements is 24. Indeed, suppose we have such a set, say $S = \{a, b, c, d\}$. Consider the set T of ordered quadruples that describe permutations of S. Each ordered quadruple belonging to T has four different components. An ordered quadruple may be classified according to its first component. Thus, T has a partition consisting of four cells: one cell contains the quadruples with first component a, another the quadruples with first component b, and so on. (See Fig. 8.) Typical among these cells is $\{(c, a, b, d), (c, a, d, b), (c, b, a, d), (c, b, d, a), (c, d, a, b), (c, d, b, a)\}$. We note that the last three components of each quadruple in this cell form an ordered triple which describes a permutation of $\{a, b, d\}$. Furthermore, all such permutations are accounted for. The number of permutations of a set with three elements is six (see Example 9). Each cell in the partition of T has six members. Thus, the number of elements in T is $4 \cdot 6 = 24$.

214   Counting

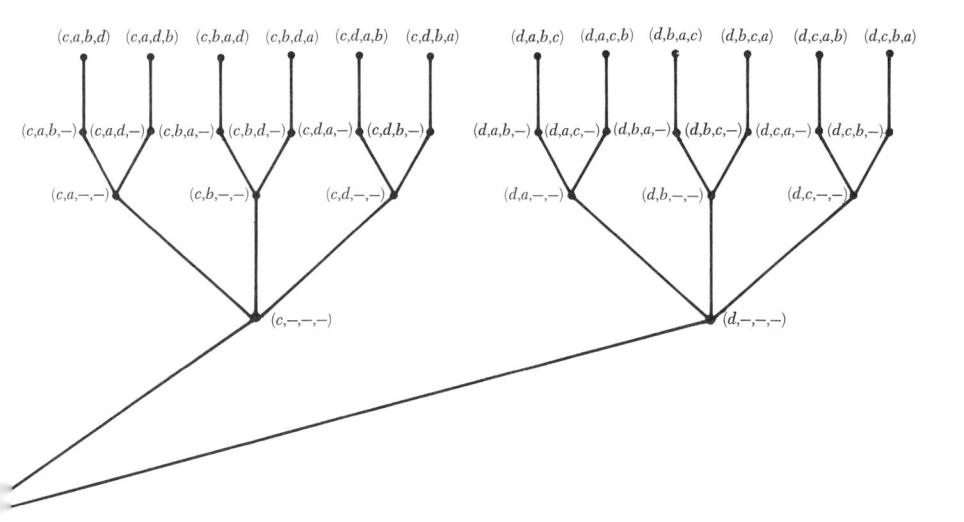

The type of argument in the preceding paragraph enables us to find the number of permutations of a set with five elements. The corresponding set of ordered 5-tuples has a partition consisting of five cells such that each cell has as many members as there are permutations of a set with four elements, namely, 24. Hence, the number of permutations of a set with five elements is 5 · 24 = 120.

The incomplete tree in Fig. 9 suggests a sequence of five activities: the picking of the five components, one at a time, from left to right. Each branching in the picture looks like Fig. 8 and corresponds to filling in the last four components of the ordered 5-tuple from the subset of elements not already picked as the left component.

Consider now the permutations of a set with six elements. A partition of the set of ordered 6-tuples, classified according to first component, has six cells, and each cell has as many members as there are permutations of a set with five elements, namely, 120. Thus, there are 6 · 120 = 720 permutations of a set with six elements.

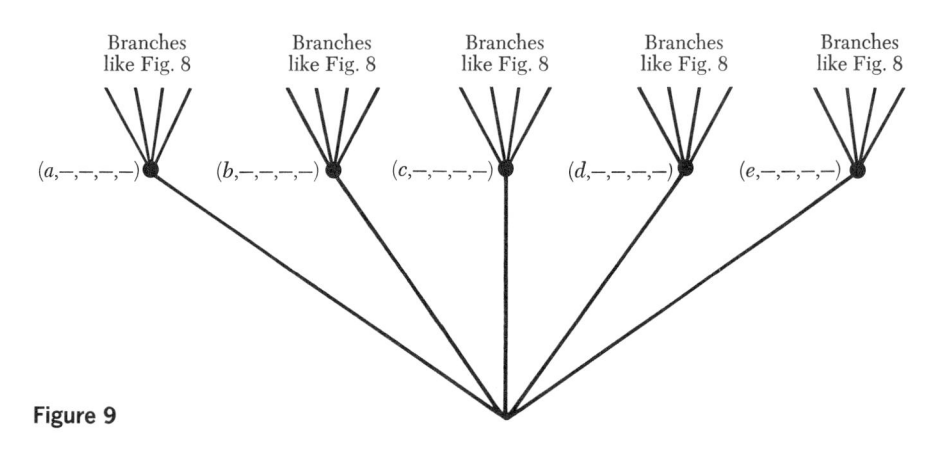

Figure 9

8 / Number of permutations of a finite set

Our results are summarized in Table 1. The first row lists the number n of elements in a set (for selected values of n) and the second row tells the number of permutations of a set with n elements.

Table 1

n	3	4	5	6
	$6 = 3 \cdot 2$	$24 = 4 \cdot 6 = 4 \cdot 3 \cdot 2$	$120 = 5 \cdot 24 = 5 \cdot 4 \cdot 3 \cdot 2$	$720 = 6 \cdot 120 = 6 \cdot 5 \cdot 4 \cdot 3 \cdot 2$

The second row of the table reveals a pattern for the numbers in factored form. Corresponding to the number n we have the product of all the natural numbers from 2 to n inclusive. Since a factor of 1 does not alter a product, it is simpler to describe the entry in the second row corresponding to n as the product of all the natural numbers from 1 to n inclusive. Thus, as examples, $24 = 1 \cdot 2 \cdot 3 \cdot 4$ and $720 = 1 \cdot 2 \cdot 3 \cdot 4 \cdot 5 \cdot 6$.

The discussion in Section 6 and the present section makes plausible the following general result.

Theorem 4 For every natural number n, the number of permutations of a set with n elements is the product of all the natural numbers from 1 to n inclusive.

It is desirable to have a convenient name and notation for this product. The name assigned to the product of all natural numbers between 1 and n inclusive is: *n factorial*. We may use the notation $1 \cdot 2 \cdot 3 \cdot \ldots \cdot n$ for this product: the initial factors 1, 2, 3 set the pattern of successive natural numbers beginning with 1; the terminal factor n indicates how far to carry the pattern; the three dots denote the possibility of intermediate factors not explicitly written. A much briefer notation for n factorial is the symbol $n!$.

Example 11
$7! = 1 \cdot 2 \cdot 3 \cdot 4 \cdot 5 \cdot 6 \cdot 7 = 5040;$
$8! = (7!)8 = 40320;$
$9! = (8!)9 = 362880;$
$2! = 1 \cdot 2 = 2;$
$1! = 1.$

Theorem 4 may now be restated as follows. For each natural number n, the number of permutations of a set with n elements is $n!$.

EXERCISES E

1. Calculate each of the following numbers.
   (a) $2(5!)$.
   (b) $10!$.
   (c) $5! + 3!$.
   (d) $\dfrac{6!}{3!}$.
   (e) $(4!)(6!)$.
   (f) $\dfrac{7!}{4!}$.
   (g) $\dfrac{7!}{4!3!}$.
   (h) $\dfrac{7!}{3!2!2!}$.

2. Suppose you have calculated 25!, or perhaps you have found its value printed in a reference table. How can you easily compute 26!?

3. If you have calculated 100!, how can you evaluate 101!?

4. Let n be a natural number. Simplify the product $(n!) \cdot (n + 1)$.

5. Let n be a natural number greater than 1. Simplify each of the following.

   (a) $(n - 1)!(n)$. (b) $\dfrac{n!}{(n - 1)!}$. (c) $\dfrac{(n + 1)!}{(n - 1)!}$.

6. In how many ways may the cards in an ordinary deck be stacked in a pile?

7. Five black pieces and four white pieces are picked from a chess set; the black pieces are the king, the queen, a bishop, a rook, and a pawn; the white pieces are the king, the queen, a knight, and a pawn. The nine pieces are placed on a chessboard, filling nine of the 64 squares. How many different outcomes are there for the activity of arranging the pieces?

8. Let the natural numbers r, s, t be given. Consider a sequence of three activities. Suppose that the number of possible outcomes for the sequence of activities is t. Suppose that for each of the possible outcomes for the first activity, there are r possible outcomes for the second. Suppose that for each of the outcomes for the first pair of activities, there are s possible outcomes for the third activity. How many possible outcomes are there for the first activity?

9. Give a specific example of the following situation. A sequence of two activities has 30 possible outcomes, the first activity has six possible outcomes, the second activity has six possible outcomes.

10. Seven distinguished guests of equivalent rank are to be assigned the seven chairs of honor in a row on a dais before the audience. How many different seating assignments are possible?

11. Five men and four women are seated in a row so that no man sits next to another man. How many seating arrangements of the prescribed type are possible for these nine people?

12. At a reception there are ten distinguished co-hosts eligible to serve on the receiving line. The committee plans the receiving line to consist of at least three but not more than six of the eligible hosts. In how many different ways may the members of the receiving line be chosen and arranged?

9 Ordered partition with two cells

Let n and k be natural numbers such that $k < n$. Let S be a set with n elements. Then S has a partition consisting of two cells such that the first cell has k members. Since the two cells are complements of one another, the second cell has of course the remaining $n - k$ elements of S. Since we are interested in distinguishing between the first and second cells, we may speak of the partition as being *ordered*. How many different ordered partitions of this type does S have?

In an attempt to obtain clues to the solution, we look at two numerical illustrations.

Example 12 Let T be a set with four members, say $T = \{a, b, c, d\}$. Consider the ordered partitions of T such that each partition has two cells, the first containing two elements of T and the second containing the remaining two elements of T. How many such ordered partitions does T have?

We may think pictorially of separating the elements of T into two compartments, with two elements in the first compartment and two in the second (Fig. 10).

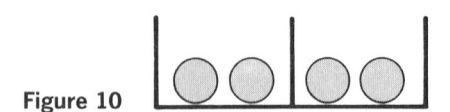

Figure 10

There are $4! = 24$ permutations of the set T. That is, there are 24 ways of arranging the members of T to form an ordered quadruple. In other words, there are 24 ways in which the four elements of T may be substituted for the objects in Fig. 10. The results of these replacements are given in Table 2.

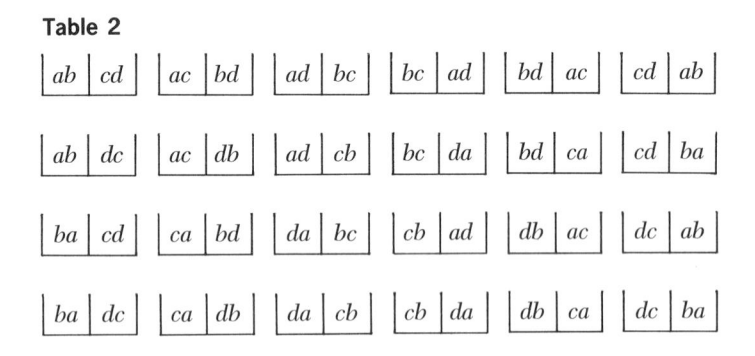

Table 2

ab	cd		ac	bd		ad	bc		bc	ad		bd	ac		cd	ab

| ab | dc | | ac | db | | ad | cb | | bc | da | | bd | ca | | cd | ba |

| ba | cd | | ca | bd | | da | bc | | cb | ad | | db | ac | | dc | ab |

| ba | dc | | ca | db | | da | cb | | cb | da | | db | ca | | dc | ba |

Examination of the table leads to several observations. The arrangement of elements within a cell is immaterial; therefore, in each column of the table, all the ordered partitions listed in the column are the same. On the other hand, an ordered partition in one column is different from an ordered partition in another column. Thus, the number of partitions in which we are interested is the same as the number of columns in the table, namely, six.

Note further that each column in the table has the same number of rows. The number of entries in the table (namely, 24) is the product of the number of columns (namely, six) and the number of rows (namely, four) in each column. In order to gain clues for the general problem, we analyze in more detail why each column has four rows.

In any particular column, the top two entries differ only in the arrangement of elements in the second cell. Likewise, the bottom two entries agree except for an interchange of elements in the second cell. Since the second cell has two members, there are 2! = 2 different ways of ordering the members within the cell.

In any particular column, the top pair of entries and the bottom pair of entries are alike except for the arrangement of elements in the first cell. Since the first cell has two members, they may be arranged within the cell in 2! = 2 different ways.

In any particular column, we obtain the entries by considering all ways of arranging the elements in the first cell and all ways of ordering the members of the second cell. Thus, the number of entries in the column is the product of the number of permutations of the first cell and the number of permutations of the second cell.

Example 13 Let W be a set with seven members. Consider the ordered partitions of W such that each partition has two cells, the first containing four elements and the second containing the remaining three elements. How many such ordered partitions does W have?

We may adopt the same attack as in Example 12. Pictorially a partition may be represented as shown in Fig. 11.

Figure 11

Since W has 7! = 5040 permutations, a table such as Table 2 should be contemplated rather than written. All the entries in any one column describe the same ordered partition, while entries in different columns give different ordered partitions. Hence, the number of columns is the same as the number we want.

In any particular column the various entries correspond to all possible ways of arranging the four elements in the first cell and all possible ways of ordering the three elements in the second cell. Since there are 4! permutations of the first cell and 3! permutations of the second cell, the number of possible outcomes for this sequence of two activities is the product (4!)(3!).

Every column contains the same number of rows, namely, (4!)(3!). Since there are 7! entries in the table and they are arranged in a rectangular fashion with (4!)(3!) rows, the number of columns is $\dfrac{7!}{4!3!}$.

This is the number of ordered partitions of W with two cells of which the first contains four elements.

We now return to the general case. Suppose that n and k are natural numbers such that $k < n$. If the set S has n members, find the number of ordered partitions of S with two cells such that the first cell has k elements. So frequently do we meet this type of problem that special notations have been introduced for the desired number. The notation we adopt, which emphasizes that the number depends upon n and k, is $\binom{n}{k}$. Except for the condition that the number of elements in S is n, the nature of S is irrelevant. For a reason to be developed later (see Section 3 in Chapter 7), the symbol $\binom{n}{k}$ may be read "binomial coefficient, n, k."

Example 14 The number requested in Example 12 may be expressed as $\binom{4}{2}$; thus

$$\binom{4}{2} = 6.$$

The number sought in Example 13 is $\binom{7}{4}$; thus

$$\binom{7}{4} = \frac{7!}{(4!)(3!)}.$$

Since $7! = 5040$, $4! = 24$, $3! = 6$, we find that

$$\binom{7}{4} = \frac{5040}{24 \cdot 6} = 35.$$

Consider the following sequence of three activities:

(a) For the set S with n elements, select an ordered partition with two cells such that there are k elements in the first cell.
(b) Arrange the members of the first cell to form an ordered k-tuple.
(c) Arrange the $n - k$ elements of the second cell to form an ordered $(n - k)$-tuple.

Effectively, this sequence of activities yields an ordered n-tuple such that every element of S is a component. In other words, a permutation of S is described.

Now we apply the counting principle concerning the number of possible outcomes for a sequence of activities. On the one hand, the number of outcomes in the present situation is the number of permutations of S, namely, $n!$. On the other hand, the number of outcomes for activity (a) is the number we have denoted by $\binom{n}{k}$; the number of outcomes for activity (b) is $k!$, since there are $k!$ permutations of a cell with k members; for activity (c) the number of outcomes is $(n - k)!$, by the same type of reasoning. Therefore

$$n! = \binom{n}{k} k!(n - k)!.$$

We solve this equation for the binomial coefficient to obtain

$$\binom{n}{k} = \frac{n!}{k!(n-k)!}.$$

Example 15 Evaluate $\binom{8}{3}$.

The product of the natural numbers from 1 to 8 inclusive is also the product of the natural numbers from 8 descending to 1 inclusive. Thus,

$$\binom{8}{3} = \frac{8!}{(3!)(5!)} = \frac{8 \cdot 7 \cdot 6 \cdot (5 \cdot 4 \cdot 3 \cdot 2 \cdot 1)}{1 \cdot 2 \cdot 3 \cdot (5 \cdot 4 \cdot 3 \cdot 2 \cdot 1)}.$$

Rather than evaluate the factorials, we ordinarily simplify the fractional expression. Suppressing the common factor of 5! in the numerator and denominator, we obtain

$$\frac{8 \cdot 7 \cdot 6}{1 \cdot 2 \cdot 3}.$$

Thus, $\binom{8}{3} = 56$.

We remark further that, in the form

$$\frac{8 \cdot 7 \cdot 6}{1 \cdot 2 \cdot 3},$$

the numerator and denominator have the same number of factors, namely, three. In the numerator we begin with the factor 8 and successively descend by one; in the denominator we begin with the factor 1 and successively ascend.

Example 16 Evaluate $\binom{9}{2}$.

According to the pattern observed in Example 15,

$$\binom{9}{2} = \frac{9 \cdot 8}{1 \cdot 2} = 36.$$

In the form

$$\frac{9 \cdot 8}{1 \cdot 2},$$

the denominator is the factorial of 2; and the numerator has the same number of factors as the denominator, beginning with 9 and successively descending.

We check our result by noting that

$$\binom{9}{2} = \frac{9 \cdot 8 \cdot (7 \cdot 6 \cdot 5 \cdot 4 \cdot 3 \cdot 2 \cdot 1)}{1 \cdot 2 \cdot (7 \cdot 6 \cdot 5 \cdot 4 \cdot 3 \cdot 2 \cdot 1)} = \frac{9 \cdot 8}{1 \cdot 2}.$$

In general, we may express $\binom{n}{k}$ as a fraction of the form

$$\frac{\overbrace{n(n-1)\cdots}^{k \text{ factors}}}{k!},$$

in which the numerator has k factors, beginning with n and successively decreasing by one.

EXERCISES F

1. Evaluate each of the following binomial coefficients.

   (a) $\binom{6}{4}$. (b) $\binom{8}{2}$. (c) $\binom{12}{3}$. (d) $\binom{100}{98}$. (e) $\binom{11}{4}$. (f) $\binom{20}{15}$.

2. Calculate each of the following differences.

   (a) $\binom{7}{2} - \binom{7}{5}$. (b) $\binom{9}{6} - \binom{9}{3}$. (c) $\binom{75}{40} - \binom{75}{35}$.

3. Find the number of partitions of an ordinary deck of playing cards such that each partition has two cells, of which one is a bridge hand with 13 cards.

4. Suppose that each of the 12 boys on a basketball squad is capable of playing any position. Find the number of partitions of the squad each of which consists of the team in action on the floor and the subset of seven players on the bench.

10 Number of subsets with a specified number of elements

Consider an ordered partition of a set T with two cells. The first cell C is a subset of T. The second cell is the complement of the first cell. The second cell is completely determined by T and C. Thus, the ordered partition essentially selects the subset C of T.

On the other hand, consider any subset of T that is neither T nor empty. This subset corresponds to an ordered partition of T with two cells, namely, the partition having the given subset as first cell and its complement as second cell.

Suppose that n and k are natural numbers such that $k < n$. Let T be a set with n members. On the basis of the two preceding paragraphs, the selection of an ordered partition of T with two cells such that the first cell has k elements is equivalent to the selection of a subset of T with k elements. Consequently, the number of subsets of T with k members each is the same as the number of ordered partitions of the type described, namely, $\binom{n}{k}$.

Theorem 5 A set with n members has $\binom{n}{k}$ subsets with k elements each.

Example 17 Consider a case in which $(n, k) = (4, 3)$. The set $\{e, f, g, h\}$ with four members has four subsets with three elements each, namely, $\{e, f, g\}, \{e, f, h\}, \{e, g, h\}, \{f, g, h\}$. Note that $\binom{4}{3} = 4$.

Example 18 A club has 20 members. How many committees with five members each may be formed in the club?

A committee is a subset of the club. The number of committees is $\binom{20}{5}$. If desired, we may evaluate the binomial coefficient:

$$\binom{20}{5} = \frac{20 \cdot 19 \cdot 18 \cdot 17 \cdot 16}{1 \cdot 2 \cdot 3 \cdot 4 \cdot 5} = 19 \cdot 3 \cdot 17 \cdot 16 = 15{,}504.$$

Example 19 From an ordinary deck of playing cards, a poker hand is a subset with five members. Find the number of poker hands that consist of two hearts and three black cards.

A hand of the prescribed type is the outcome of a sequence of two activities: (a) the selection of two cards from the heart suit, (b) the choice of three cards from the black suits. Activity (a) has $\binom{13}{2}$ possible outcomes, since a set with 13 members has $\binom{13}{2}$ subsets with two elements each. The number of possible outcomes for activity (b) is $\binom{26}{3}$. Consequently, the number of hands with two hearts and three black cards is $\binom{13}{2} \cdot \binom{26}{3}$. In case the computed result is desired, we obtain

$$\binom{13}{2} \cdot \binom{26}{3} = \frac{13 \cdot 12}{2} \cdot \frac{26 \cdot 25 \cdot 24}{6} = 202{,}800.$$

Example 20 A commission has nine items of unfinished business pending. Two items are policy matters and seven are administrative questions. Four items will be selected and arranged in order as the agenda for the next meeting. By-laws prescribe that, at any meeting, all policy matters for discussion must precede any administrative question on the agenda. Find the number of possible agendas for the next meeting.

We classify an agenda according to the number of policy matters on it. Thus, the set of possible agendas has a partition, consisting of the subset A of agendas with two policy matters, the subset B of agendas with one policy matter, and the subset C of agendas with zero policy matters.

An agenda in set A is an ordered quadruple of items such that the first two components are policy matters and the last two components are administrative questions. The selection of an ordered

10 / Subsets with a specified number of elements

quadruple is a sequence of four activities: (a) the choice of first component, (b) the choice of second component, (c) the choice of third component, (d) the choice of fourth component. The number of possible outcomes for these four activities are 2, 1, 7, 6, respectively. Thus, set A has $2 \cdot 1 \cdot 7 \cdot 6 = 84$ members.

An agenda in set B is an ordered quadruple of items such that the first component is a policy matter and the remaining three are administrative questions. An analysis of the sequence of four activities, as in the preceding paragraph, reveals that B has $2 \cdot 7 \cdot 6 \cdot 5 = 420$ members.

A similar discussion applied to set C shows that C has $7 \cdot 6 \cdot 5 \cdot 4 = 840$ members.

The total number of possible agendas is $84 + 420 + 840 = 1344$.

EXERCISES G

1. Find the number of different subsets contained in a set with 18 elements such that each subset has six elements.

2. Consider a set with five members.
   (a) How many subsets of the set have two elements apiece?
   (b) Verify your response to (a) by choosing a set with five elements, making a list of all the subsets with two members, and counting them.

3. A bridge hand consists of 13 cards from an ordinary deck. Find the number of possible bridge hands of each of the following types.
   (a) The hand contains four spades, four hearts, one diamond, four clubs.
   (b) The hand contains one card from one suit and four cards from each of the other suits (that is, the hand has a so-called 4-4-4-1 distribution).
   (c) The hand contains six hearts, five diamonds, two clubs.
   (d) The hand contains six cards from one of the red suits, five cards from one of the black suits, and two cards from the other black suit.
   (e) The hand contains six cards from one suit, five cards from another suit, and two cards from a third suit (that is, the distribution is 6-5-2-0).
   (f) The hand contains one card of each denomination, that is, one ace, one king, one queen, etc.
   (g) Every card in the hand is an honor card. [NOTE: See Example 32 in Section 10 of Chapter 5.]
   (h) No card in the hand is an honor card. (The hand is a Yarborough.)
   (i) The hand contains exactly two red cards and exactly three kings.
   (j) The hand contains exactly two red cards but no kings.

4. Consider a time when the United States Senate has a full membership of 100. Suppose that the Senate rules prescribe that each committee shall consist of 15 members, no two of which represent the same state. Find the total number of possible committees of the type described.

5. A box of 1200 manufactured parts contains 18 defective items. A sample of 30 parts from the box is picked for inspection.
   (a) How many different samples of the prescribed type are possible?
   (b) How many of these samples belong to the truth set of the statement that there are three defective items in the sample?

6. A small company has 84 employees of whom three are women. An advertising agency plans to select a sample (that is, a subset) of seven employees for its campaign. (In each part of this exercise you may express your answer in simple form, without performing extensive arithmetical computation.)
   (a) Find the number of possible samples.
   (b) Find the number of samples that are composed entirely of men.
   (c) How many of the samples include all three of the women employees?
   (d) How many samples satisfy the requirement that the sample contains exactly one woman?
   (e) How many samples belong to the truth set of the statement that there are exactly two women in the sample?
   (f) Find the sum of the four numbers in parts (b), (c), (d), (e), and show that the sum is the same as the number in part (a).

7. Thirty students are enrolled in a class. Each student is required to go on exactly one field trip. There are accommodations for 13 in October, 11 in November, and six in December.
   (a) Find the number of possible outcomes for the activity of selecting students for the October trip.
   (b) Supposing the October trip is finished, find the number of possible selections of students for the November trip.
   (c) Supposing the November trip is over, how many possible ways are there for choosing the students to go on the December trip?
   (d) Find the number of possible outcomes for the activity of assigning students to the three field trips.
   (e) If a set has 30 members, find the number of ordered partitions of the set with three cells, such that the first, second, and third cells have 13, 11, and six elements, respectively.

8. A certain set has ten members.
   (a) Find the number of ordered partitions of the set with three cells such that the first cell has two elements, the second has five, and the third has three.
   (b) Find the number of ordered partitions of the set with three cells such that the first, second, and third cells have three, two, and five elements, respectively.

9. Let a, b, c, n be natural numbers such that $a + b + c = n$. Consider a set with n members.
   (a) Find the number of ordered partitions of the set with three cells such that the first cell, the second cell, and the third cell have a, b, c elements, respectively.
   (b) Show that the number in part (a) may be expressed as $\dfrac{n!}{a!b!c!}$.

11 Properties of the binomial coefficients

Let n be a natural number, and consider a set with n elements. We have interpreted $\binom{n}{k}$ as the number of subsets of the set, each having k elements. Until now we have supposed that k is a natural number less than n. We may extend the notion by allowing $k = n$ or $k = 0$. Since the given set has only one subset with n elements, namely itself, we agree that $\binom{n}{n}$ is the number 1.

Since the given set has only one subset with 0 elements, that is, only one empty subset, we agree that $\binom{n}{0}$ is the number 1.

Consider a set with n members. To each subset with k elements there corresponds naturally a complementary subset with $n - k$ elements, and vice versa. Thus, there are precisely as many subsets with k members as there are subsets with $n - k$ members each. Since the number of subsets with k elements each is $\binom{n}{k}$ and the number of subsets with $n - k$ elements each is $\binom{n}{n - k}$, we have proved the following result.

Theorem 6 For every natural number n and every integer k such that $0 \leq k \leq n$,
$$\binom{n}{k} = \binom{n}{n - k}.$$

An alternate proof of the theorem may be given by comparing the number
$$\binom{n}{k} = \frac{n!}{k!(n - k)!}$$

and the number
$$\binom{n}{n - k} = \frac{n!}{(n - k)![n - (n - k)]!} = \frac{n!}{(n - k)!k!}.$$

Let n and k be natural numbers such that $k < n$. Then $\binom{n}{k + 1}$ may be expressed as a fraction of the form

$$\underbrace{\frac{\overbrace{n(n - 1) \cdots}^{k + 1 \text{ factors}}}{\underbrace{1 \cdot 2 \cdot \ldots}_{k + 1 \text{ factors}}}},$$

in which the numerator and denominator each have $k + 1$ factors. In the denominator the final factor is $k + 1$; since corresponding factors in numerator and denominator have sum $n + 1$, the final factor in the numerator is $(n + 1) - (k + 1) = n - k$. If we isolate the final factors, we have

$$\underbrace{\frac{\overbrace{n(n - 1) \cdots}^{k \text{ factors}}}{\underbrace{1 \cdot 2 \cdot \ldots}_{k \text{ factors}}}} \cdot \frac{n - k}{k + 1},$$

226 Counting

in which the numerator and denominator of the first fraction have k factors each. Thus, the first fraction is simply $\binom{n}{k}$. We have established the following theorem, except in the case where $k = 0$.

Theorem 7 For every natural number n and every integer k such that $0 \leq k < n$,

$$\binom{n}{k+1} = \binom{n}{k} \cdot \frac{n-k}{k+1}.$$

To verify the equation for the case $k = 0$, we note that the left member is

$$\binom{n}{0+1} = \binom{n}{1} = n,$$

while the right member is the product of

$$\binom{n}{0} = 1 \quad \text{and} \quad \frac{n-0}{0+1} = n;$$

since $n = 1 \cdot n$, the equation is satisfied in case $k = 0$.

Example 10 (revisited) With more tools at our disposal, we may now give a shorter solution to the question in Example 10. The assignment job may be accomplished via the following sequence of activities:

(a) From the six employees, select two and assign them to the suite with the lowest room number.
(b) From the remaining four employees, choose two and place them in the suite with the next smallest room number.
(c) Put the remaining two workers in the one vacant suite.

The number of possible outcomes for activity (a) is $\binom{6}{2}$, for activity (b) is $\binom{4}{2}$, and for activity (c) is 1. Thus, the number of possible assignments is the product

$$\binom{6}{2} \cdot \binom{4}{2} \cdot 1 = 15 \cdot 6 \cdot 1 = 90.$$

Note that the sequence of activities in the alternative solution is quite different from the sequence of activities in the previous method. Very often a counting problem may be acceptably analyzed in more than one manner. It is essential that the set of possible outcomes for the sequence of activities in any analysis is the set of outcomes wanted in the problem. Furthermore, it is important that the activities in the sequence be given a clear description in writing.

11 / Properties of the binomial coefficients

EXERCISES H

1. First calculate $\binom{7}{2}$ and $\binom{7}{3}$; then compute $\binom{7}{2} \cdot \frac{7-2}{2+1}$ and compare it with $\binom{7}{3}$.

2. Calculate $\binom{8}{4}$ and $\binom{8}{5}$; then compute $\binom{8}{4} \cdot \frac{8-4}{4+1}$ and compare it with $\binom{8}{5}$.

3. Show each of the following inequalities:
$$\binom{6}{0} < \binom{6}{1} < \binom{6}{2} < \binom{6}{3} \quad \text{and} \quad \binom{6}{3} > \binom{6}{4} > \binom{6}{5} > \binom{6}{6}.$$

4. Show each of the following:
$$\binom{7}{0} < \binom{7}{1} < \binom{7}{2} < \binom{7}{3} \quad \text{and} \quad \binom{7}{3} = \binom{7}{4} \quad \text{and} \quad \binom{7}{4} > \binom{7}{5} > \binom{7}{6} > \binom{7}{7}.$$

5. Let n be a natural number, and let k be either zero or a natural number less than n.

   (a) Show that if $\frac{n-k}{k+1} > 1$, then $k < \frac{1}{2}(n-1)$.

   (b) Show that if $k < \frac{1}{2}(n-1)$, then $\frac{n-k}{k+1} > 1$.

   (c) Show that $\binom{n}{k+1} > \binom{n}{k}$ if and only if $k < \frac{1}{2}(n-1)$.

   (d) Show that $\binom{n}{k+1} = \binom{n}{k}$ if and only if n is an odd number and $k = \frac{n-1}{2}$.

   (e) Show that $\binom{n}{k+1} < \binom{n}{k}$ if and only if $k > \frac{1}{2}(n-1)$.

   (f) Consider the sequence of binomial coefficients
$$\binom{n}{0}, \binom{n}{1}, \binom{n}{2}, \ldots, \binom{n}{n}.$$
   Show the following. In the first part of the sequence each coefficient is larger than its predecessors. In the latter part the coefficients are successively smaller. The greatest of the coefficients is in the middle.

12 Summary

The two most important principles concerning counting involve addition and multiplication, respectively. According to Theorem 1, the number of elements in a finite set is the sum of the numbers of elements in the various cells of any partition of the set. The number of outcomes for a finite sequence of activities is expressible, by Theorem 3, as a product in many situations—namely, in case at each stage in the sequence the number of possible outcomes for the activity under consideration is not dependent on the result of prior activities in the sequence.

EXERCISES I

1. There are 18 contestants in a talent search. Two of the contestants are full-blooded American Indians; they are Miss Iroquois and Mr. Cheyenne. A television sponsor plans to select five of the contestants and arrange their acts in succession for a guest program.
   (a) How many different program arrangements are possible?
   (b) How many of these program arrangements do not include Mr. Cheyenne's act?
   (c) How many of these program arrangements include Miss Iroquois as a performer?
   (d) In how many of these arrangements does Mr. Cheyenne appear among the last three performers?
   (e) In how many of the programs do the two American Indian performers appear in consecutive spots on the program?
   (f) In how many of the programs do both of the American Indian performers appear but not consecutively?

2. A three-digit number is formed from the digits 2, 3, 4, 5, 6, 7, 8, 9. The middle digit is greater than the left digit, and the right digit is greater than the middle digit. How many three-digit numbers satisfy the requirements?

3. For each of the following, find the number of possible poker hands of the given type.
   (a) The hand contains four aces.
   (b) The hand contains four cards of the same denomination (that is, four of a kind).
   (c) The hand contains three kings and a pair of nines.
   (d) The hand is a full house (that is, contains three cards of one denomination and two cards of another denomination).
   (e) The hand contains two jacks, two sixes, and a four.
   (f) The hand contains a pair of jacks, two cards of another denomination, and one card of a third denomination.
   (g) The hand contains two cards of each of two denominations and one card of a third denomination (that is, two pairs).
   (h) The hand is a subset of the heart suit.
   (i) All cards in the hand belong to the same suit.
   (j) The hand contains cards of five different denominations.
   (k) The hand contains a king, a queen, a jack, a ten, and a nine which are not all in the same suit (that is, a straight with king high).

4. Let n be a natural number. We assert that the following equation is satisfied for every number k such that each of the binomial coefficients is meaningful.
$$\binom{n}{k} + \binom{n}{k+1} = \binom{n+1}{k+1}.$$
   (a) By expressing the binomial coefficients in terms of factorials, show that the equation holds for any natural number k such that $k + 1 < n$.
   (b) Show that the equation holds in case $k = 0$.
   (c) Show that the equation is satisfied when $k + 1 = n$.
   (d) By an entirely different method, namely, by exploiting Theorem 5 (in Section 10), show that the equation is satisfied for every number k such that the binomial coefficients are meaningful.

Powers and Sequences

1 Introduction

The desire for a simple representation of a product of several equal factors led to the seventeenth-century development of our exponent notation. We begin the present chapter by studying properties of powers in which the exponents are natural numbers. Later we shall extend the notion of a power by introducing exponents that are rational numbers. A further extension will be made in Chapter 12.

In the previous two chapters we have considered sequences of activities. The latter sections of the present chapter are devoted to sequences of numbers, with special emphasis on arithmetic sequences and geometric sequences.

A third topic that we shall introduce is the convenient summation notation. For our major results throughout the chapter we shall offer arguments designed to make them plausible rather than give formal proofs of the theorems.

2 The binomial theorem

In order to study a power of a sum, we exploit the binomial coefficients introduced in Section 9 of Chapter 6.

Example 1 Expand $(x + y)^6$.

We recall that the exponent 6 means that $(x + y)^6$ is the product of six factors, each of which is $x + y$. That is,

$$(x + y)^6 = (x + y)(x + y)(x + y)(x + y)(x + y)(x + y).$$

By straightforward, but tedious, multiplication, we find that

$$(x + y)^6 = x^6 + 6x^5y + 15x^4y^2 + 20x^3y^3 + 15x^2y^4 + 6xy^5 + y^6.$$

Fortunately, a more efficient method is offered by the binomial theorem.

Our problem is to study products of the type $(a + b)^n$, where n is a natural number. In particular, we want to develop the expanded form of such a product. Since multiplication is distributive with respect to addition (in ordinary algebra),

$$(a + b)(x + y) = ax + bx + ay + by.$$

The product of the binomial $a + b$ and the binomial $x + y$ is the sum of the four products ax, bx, ay, by. These are all the products obtainable by selecting one term from each binomial and multiplying them. In particular,

$$(a + b)(a + b) = aa + ba + ab + bb,$$

or

$$(a + b)^2 = a^2 + 2ab + b^2.$$

The third power of $a + b$, namely, $(a + b)^3 = (a + b)(a + b)(a + b)$, is also equal to a sum of products; specifically, all the products obtainable by selecting one term from each binomial and multiplying them. There are $2 \cdot 2 \cdot 2 = 8$ products of this type:

$$aaa, \quad baa, \quad aba, \quad bba, \quad aab, \quad bab, \quad abb, \quad bbb.$$

Their sum is

$$a^3 + 3a^2b + 3ab^2 + b^3 = (a + b)^3.$$

As a typical situation, we investigate $(a + b)^5$ with the intention of finding the general pattern for expansions. Now $(a + b)^5$ is the product

$$(a + b)(a + b)(a + b)(a + b)(a + b).$$

In expanded form $(a + b)^5$ is a sum of many products. Specifically, it is the sum of all the products we can obtain if we select one term from each of the five binomials $a + b$ and multiply the chosen terms. Since each binomial offers two choices, there are $2 \cdot 2 \cdot 2 \cdot 2 \cdot 2 = 32$ products altogether. Each of these products has five factors; each factor is either a or b. In a particular product the number of factors b is 0, 1, 2, 3, 4, or 5; the remaining 5, 4, 3, 2, 1, or 0 factors, respectively, are each equal to a. Thus, there are six possible (distinct) products, namely,

$$a^5, \quad a^4b, \quad a^3b^2, \quad a^2b^3, \quad ab^4, \quad b^5.$$

Several of these possibilities occur more than once, for the total number of products (including repetitions) is, as we have noted, 32. An interesting portion of our task is to analyze how many of each of the different products appear in the expansion.

Consider the product a^2b^3. It arises in case we select the term b from each of exactly three binomials. Then from the other two binomials the chosen term is a. To find the number of occurrences of the product a^2b^3 we ask how many selections of three binomials from a set of five binomials are possible.

The number is $\binom{5}{3}$. Alternatively, we may classify the five binomials according to the term that is selected; the subset of three binomials from which b is chosen and the subset of two binomials from which a is chosen are the first and second cells of an ordered partition of the set of five binomials. The number of such ordered partitions is $\binom{5}{3}$. From either viewpoint, we find that the product a^2b^3 appears ten times in the expansion of $(a + b)^5$.

The product a^4b arises in case we select the term b from exactly one binomial. In the set of five binomials the number of subsets consisting of one binomial is $\binom{5}{1}$. Thus, the numerical coefficient of a^4b in the expansion of $(a + b)^5$ is 5.

The argument above, when applied to each of the distinct products, shows that the numbers of appearances of the products

$$a^5, \quad a^4b, \quad a^3b^2, \quad a^2b^3, \quad ab^4, \quad b^5$$

are

$$\binom{5}{0}, \quad \binom{5}{1}, \quad \binom{5}{2}, \quad \binom{5}{3}, \quad \binom{5}{4}, \quad \binom{5}{5},$$

respectively. These numbers are 1, 5, 10, 10, 5, 1, respectively. Therefore,

$$(a + b)^5 = a^5 + 5a^4b + 10a^3b^2 + 10a^2b^3 + 5ab^4 + b^5.$$

The discussion of the preceding special cases provides the clues for the pattern in the general situation. Let n be a natural number, and consider $(a + b)^n$. This product of n factors each equal to $a + b$ is, in expanded form, a sum of many products. It is the sum of all possible products obtainable if we select one term from each of the n binomials $a + b$ and multiply the chosen terms. Each product has n factors; each factor is either a or b. A product that has b as a factor k times has $n - k$ factors a. The number of appearances of this product is the same as the number of ways of selecting k binomials from the set of n binomials, which is the number of subsets, with k members each, in the set of n binomials. It is $\binom{n}{k}$. Thus, for each natural number k less than n, there are $\binom{n}{k}$ appearances of the product $a^{n-k}b^k$ in the expansion of $(a + b)^n$. We may now state our result.

Theorem 1 (The Binomial Theorem) Let n be a natural number, and let a and b be numbers. Then,

$$(a + b)^n = \binom{n}{0}a^n + \binom{n}{1}a^{n-1}b + \binom{n}{2}a^{n-2}b^2$$
$$+ \cdots + \binom{n}{k}a^{n-k}b^k + \cdots + \binom{n}{n}b^n.$$

Powers and Sequences

Example 2 Expand and simplify $(x + 2y)^4$.

We let x, $2y$, 4 play the respective roles of a, b, n in the binomial theorem. Then,

$$(x + 2y)^4 = \binom{4}{0}x^4 + \binom{4}{1}x^3(2y) + \binom{4}{2}x^2(2y)^2$$
$$+ \binom{4}{3}x(2y)^3 + \binom{4}{4}(2y)^4.$$

We simplify each term in the expansion, evaluating the binomial coefficients and the powers of $2y$. We obtain

$$(x + 2y)^4 = 1x^4 + 4x^3(2y) + 6x^2(4y^2) + 4x(8y^3) + 1(16y^4)$$
$$= x^4 + 8x^3y + 24x^2y^2 + 32xy^3 + 16y^4.$$

Example 3 Expand and simplify $(1 - 4x)^3$.

Since $1 - 4x = 1 + (-4x)$, we let 1, $-4x$, 3 play the respective roles of a, b, n in the binomial theorem. Then,

$$(1 - 4x)^3 = 1^3 + \binom{3}{1}1^2(-4x) + \binom{3}{2}1(-4x)^2 + (-4x)^3$$
$$= 1 + 3(-4x) + 3(16x^2) + (-64x^3)$$
$$= 1 - 12x + 48x^2 - 64x^3.$$

Since the successive terms in an expansion involve successively ascending powers of b, the expanded form of $(a - b)^n$ is the same as the expansion of $(a + b)^n$ except for the alternation of signs from each term to the next.

Example 4 Express the sum $81r^4 + 108r^3s + 54r^2s^2 + 12rs^3 + s^4$ in factored form.

The sum resembles the expansion of a power of a binomial. If we can fit the sum into a proper pattern, we may recognize the desired factorization. Now $81r^4 = (3r)^4$. Thus, we are led to link 3 and r together. Doing so, we obtain

$$81r^4 + 108r^3s + 54r^2s^2 + 12rs^3 + s^4$$
$$= (3r)^4 + 4(3r)^3s + 6(3r)^2s^2 + 4(3r)s^3 + s^4.$$

The respective coefficients 1, 4, 6, 4, 1 are the binomial coefficients

$$\binom{4}{0}, \binom{4}{1}, \binom{4}{2}, \binom{4}{3}, \binom{4}{4}.$$

Consequently, the given sum is $(3r + s)^4$.

The binomial theorem states that a certain power and a certain sum are the same. Thus, if we are given either one, we may express it in the other form whenever the other form is more suitable.

Example 5 Let w be a positive real number. Show that $1 + w^2 < (1 + w)^2$.

Since w is positive by hypothesis, $2w > 0$. Hence,
$$(1 + w)^2 = 1 + 2w + w^2 = (1 + w^2) + 2w > 1 + w^2.$$

EXERCISES A

1. Use the binomial theorem to expand each of the following and then simplify.
   (a) $(a + x)^2$. (b) $(a + 2y)^2$. (c) $(3c + 5d)^2$.
   (d) $(\tfrac{1}{4}r + \tfrac{1}{2}q)^2$. (e) $\left(\dfrac{a}{b} + \dfrac{c}{d}\right)^2$. (f) $(v - w)^2$.
   (g) $\left(\dfrac{u}{4} - \dfrac{x}{5}\right)^2$. (h) $(1 - 5x^2y^2)^2$. (i) $\left(1 - \dfrac{t + t^2}{2}\right)^2$.

2. Using the binomial theorem, expand each of the following and then simplify.
   (a) $(a + x)^3$. (b) $(b - w)^3$. (c) $(c + 5d)^3$.
   (d) $(2a - 3b)^3$. (e) $(b + y)^4$. (f) $(k + g)^5$.

3. In the binomial expansion of $(5a + 4b)^7$, one term has the form $\binom{x}{4}(5a)^y(4b)^4$. Find x and y.

4. In the expansion of $(1 - t)^{15}$, one term has the form $\binom{x}{y}t^8$. Find x and y.

5. In the expansion of $(1 + t^2)^{80}$, one term has the form $\binom{x}{16}t^y$. Find x and y.

6. How many different products appear in the expansion of each of the following?
   (a) $(a + b)^4$. (b) $(2 + 7x)^{19}$. (c) $(1 + t)^k$.

7. In the binomial expansion of $(a + b)^{60}$ there are 61 terms. The "middle" term is the thirty-first. It has the form $\binom{60}{x}a^y b^z$. Find x and y.

8. Give the middle term in the expansion of $(x + 5z)^{72}$.

9. There are two "middle" terms in the expansion of $(3x - 2y)^{45}$. What are they?

10. Show that $1 + 4x + 4x^2$ is the expansion of a power of a binomial.

11. Show that $1 - 6x + 12x^2 - 8x^3$ is the third power of a binomial.

12. Express each of the following sums as a power of a binomial.
    (a) $9c^2 - 6c + 1$.
    (b) $36u^2 + 60uv + 25v^2$.
    (c) $1 + 12y + 48y^2 + 64y^3$.
    (d) $8a^3 - 36a^2b + 54ab^2 - 27b^3$.

13. Evaluate in simplest form:
    $p^5 + 5p^4(1 - p) + 10p^3(1 - p)^2 + 10p^2(1 - p)^3 + 5p(1 - p)^4 + (1 - p)^5$.

14. Let x, y, z be positive real numbers.
    (a) If $x + y = z$, show that $x^2 + y^2 < z^2$.
    (b) If $x^2 + y^2 = z^2$, show that $x + y > z$.

3 Binomial coefficients and the binomial theorem

The numbers $\binom{n}{k}$ introduced in the preceding chapter serve as the numerical coefficients in the expansion of a power of a binomial, which explains the name "binomial coefficients." The properties of these numbers may be used to reduce the computational labor in an expansion.

According to Theorem 6 in Chapter 6, if n is a natural number and if k is an integer such that $0 \leqq k \leqq n$, then

$$\binom{n}{k} = \binom{n}{n-k}.$$

Applied to the binomial theorem, this property means that the sequence of successive binomial coefficients in each expansion is the same, whether ordered from left to right, or from right to left. In other words, there is a symmetry about the middle. In a lengthy expansion, when we reach the middle, no further computation is needed to find the remaining coefficients.

Example 6 Note the symmetry in the following sequences. The binomial coefficients corresponding to the exponent $n = 8$ are, respectively, 1, 8, 28, 56, 70, 56, 28, 8, 1. If $n = 9$, the respective binomial coefficients are 1, 9, 36, 84, 126, 126, 84, 36, 9, 1.

Another important property of the binomial coefficients developed in the preceding chapter (see Theorem 7) is summarized by the equation

$$\binom{n}{k+1} = \binom{n}{k} \cdot \frac{n-k}{k+1}.$$

Applied to the binomial theorem, this property offers a stepwise procedure for using each coefficient in turn to compute the next. In the sequence of successive coefficients appearing in the expansion of a power with exponent n, the coefficient $\binom{n}{k}$ is followed by the coefficient $\binom{n}{k+1}$, for each $k < n$. Knowing the number $\binom{n}{k}$, we obtain the next coefficient if we multiply by $n - k$ and divide by $k + 1$. Since $\binom{n}{k}$ is the coefficient in the $(k + 1)$th term $\binom{n}{k} a^{n-k} b^k$ of the expansion of $(a + b)^n$, a convenient memory device is the following: to obtain the binomial coefficient in the next term, multiply the coefficient in the present term by the exponent on a in the present term and divide by the number of the present term.

Example 7 Expand $(f+g)^7$.
The expansion begins
$$f^7.$$
The binomial coefficient in the next term is the exponent 7 of the power; hence, the expansion begins
$$f^7 + 7f^6g.$$
To obtain the binomial coefficient in the next term, we use the term $7f^6g$ as follows: multiply the coefficient 7 by the exponent 6 on f and divide by the number of the term, namely, 2 (since it is the second term); $7 \cdot 6/2 = 21$. The expansion begins
$$f^7 + 7f^6g + 21f^5g^2.$$
To obtain the binomial coefficient in the next term, we multiply the coefficient 21 by the exponent 5 and divide by the number 3 of the (third) term; $21 \cdot 5/3 = 35$. The expansion begins
$$f^7 + 7f^6g + 21f^5g^2 + 35f^4g^3.$$
The symmetry property enables us to complete the expansion:
$$(f+g)^7 = f^7 + 7f^6g + 21f^5g^2 + 35f^4g^3$$
$$+ 35f^3g^4 + 21f^2g^5 + 7fg^6 + g^7.$$

Example 8 Suppose that money is invested in a savings account. Suppose that at the end of each regular period of time, interest is paid on the amount in the account during the time period. If an investment of A dollars is deposited and if the rate of interest is r each period, what is the value of the account after m periods?

For any particular period, suppose B dollars are in the account at the beginning of the period. The interest paid at the end of the period is rB dollars. When this interest is credited to the account, the value of the account becomes $B + rB$ dollars. Now $B + rB = B(1 + r)$. Thus, to find the change in the value from the beginning of any period to the beginning of the following period, we may multiply by the number $1 + r$. We apply this result repeatedly in the next sentence.

The initial investment of A dollars increases to $A(1 + r)$ dollars at the end of one period, to
$$[A(1 + r)](1 + r) = A(1 + r)^2$$
dollars at the end of two periods, to
$$[A(1 + r)^2](1 + r) = A(1 + r)^3$$
dollars at the end of three periods, and so on. At the end of m interest-paying periods, the investment is worth approximately $A(1 + r)^m$ dollars. We say "approximately" because financial transactions seldom involve a fraction of a cent. The accumulated effects of rounding off

interest payments each paying period may produce a slight discrepancy between $A(1 + r)^m$ and the value in practice.

The total increase in the value of the investment is the difference between the final value and the initial value; in dollars, it is $A(1 + r)^m - A = A[(1 + r)^m - 1]$. In case the investment is continued for at least two periods, then $m \geq 2$ and

$$(1 + r)^m - 1 = [1 + mr + \binom{m}{2}r^2 + \cdots] - 1,$$

where the three dots represent the unwritten terms in the expansion of $(1 + r)^m$. Each unwritten term (if there is any) is of the form $\binom{m}{k}r^k$ and is certainly positive since $\binom{m}{k}$ is a natural number and the interest rate r is positive. Hence,

$$(1 + r)^m - 1 \geq 1 + mr + \binom{m}{2}r^2 - 1$$

$$= mr + \binom{m}{2}r^2$$

$$> mr.$$

Unless m and r are relatively small, $(1 + r)^m - 1$ is not only greater than mr but considerably greater. Consequently, $A(1 + r)^m - A$ is, in many practical situations, much greater than Amr. Since Amr is the number of dollars earned by simple interest and $A(1 + r)^m - A$ is the number of dollars earned by compound interest, we have a comparison between the two types of interest.

EXERCISES B

1. Use the binomial theorem to expand each of the following. Simplify.
   (a) $(c - w)^4$. (b) $(1 - \tfrac{5}{2}x)^3$. (c) $(2m + 5r)^4$. (d) $(1 + 3x)^5$.
   (e) $(k + p)^6$. (f) $(h + q^2)^7$. (g) $(3a + 2b)^5$. (h) $(3a - 2b)^6$.

2. Using the binomial theorem, expand each of the following and then simplify.
   (a) $(1 + 0.4x)^3$. (b) $(-1 + 0.3u)^4$. (c) $(1 - 0.01)^5$. (d) $(1 + 0.1)^4$.
   (e) $(\sqrt{2} + \sqrt{3})^4$. (f) $(\sqrt{5} - 1)^3$. (g) $(1 - 0.02)^4$.

3. Express each of the following sums as a power of a binomial.
   (a) $1 + 12x + 54x^2 + 108x^3 + 81x^4$.
   (b) $1 - 10y + 40y^2 - 80y^3 + 80y^4 - 32y^5$.

4. Show that $4p^3(1 - p) + 6p^2(1 - p)^2 + 4p(1 - p)^3 + (1 - p)^4$ is the same as $1 - p^4$.

5. Let p and q be real numbers such that $p + q = 1$. Simplify the sum
$$\binom{6}{1}p^5q + \binom{6}{2}p^4q^2 + \binom{6}{3}p^3q^3 + \binom{6}{4}p^2q^4 + \binom{6}{5}pq^5 + q^6.$$

6. Compute (with minimal computational effort) the following sum:
$$5(0.2)^4(0.8) + 10(0.2)^3(0.8)^2 + 10(0.2)^2(0.8)^3 + 5(0.2)(0.8)^4 + (0.8)^5.$$

7. Suppose that $2000 is deposited in a savings account paying interest annually at the rate of 3%.
   (a) With the aid of the binomial theorem, find the (approximate) value of the account five years later.
   (b) Calculate the exact value of the account at the end of the five years. To do this, determine the interest credited to the account at the end of each year. (Whenever the calculated interest payment involves a fraction of a cent, round off to the nearest penny.)

8. Suppose that $1000 is invested in a savings account paying interest at the quoted annual rate of 4%, compounded quarterly. (This means that interest is paid at the rate of 1% each three months.)
   (a) Find the (approximate) value of the account three years later.
   (b) Calculate the exact value of the account at the end of the three years. To do this, determine the interest credited to the account at the end of each three months. (Whenever the calculated interest payment involves a fraction of a cent, round off to the nearest penny.)

9. Suppose that $4000 is invested to earn interest at the rate of 2% each six months. Find the (approximate) value of the investment three years later.

10. Suppose that $7000 is invested to earn interest at the rate of $\frac{1}{2}$% each month. Find the (approximate) value of the investment one year later.

11. Suppose that $12,000 is deposited in a savings account. During the next six years interest is credited to the account at the rate of $1\frac{1}{2}$% semiannually. The rate is then increased to 1% quarterly. Express the (approximate) value of the account two and one-half years after the rate increase. (Computation is not requested.)

12. Let m be a natural number greater than one.
    (a) Explain why: if x is positive, then $(1 + x)^m$ is greater than $1 + mx$.
    (b) If x is positive and if $m = 2$, can you deduce that $(1 - x)^m > 1 - mx$? Why?
    (c) Decide whether the following statement is true, and give the reason for your decision: if $x > 0$, then $(1 - x)^m > 1 - mx$.
    (d) If $x < 0$, can you deduce that $(1 + x)^m > 1 + mx$? Why?

4 Powers

The binomial theorem is concerned with a power of a sum. A much simpler situation occurs in the consideration of a power of a product. We now review the basic notions and properties of powers, and afterward we will extend the concept as we introduce rational exponents. A number that is a quotient of two integers is called a *rational* number.

> **Definition** Let b, r, p be numbers such that $b^r = p$. We call b^r a *power* of b, more specifically the rth power of b. The numbers b and r are called the *base* of the power and the *exponent* of the power, respectively.

> **Definition** If b is a real number and n is a natural number, then b^n denotes the product of n factors, each of which is b. In the special case that $n = 1$, the power is simply b itself.

Example 9 The third power of 4 is 64, and we write $4^3 = 64$. The second power of $-\frac{3}{4}$ is $\frac{9}{16}$, and we write $(-\frac{3}{4})^2 = \frac{9}{16}$. Note that $-3^2 = -9$, since the notation -3^2 means the negative of the second power of 3; the exponent 2 applies to the base 3.

Example 10 A condensed representation for the number 29,800,000,000,000 is 2.98×10^{13}.

The basic properties of powers involving exponents that are natural numbers are summarized in the following formulas, which are applicable for any real number b, any real number a, any natural number n, any natural number k (unless further restricted).

(i) $b^n \cdot b^k = b^{n+k}$.

(ii) If $b \neq 0$ and if $n > k$, then $\dfrac{b^n}{b^k} = b^{n-k}$.

(iii) $(b^n)^k = b^{nk}$.

(iv) $(ab)^n = a^n b^n$.

(v) If $b \neq 0$, then $\left(\dfrac{a}{b}\right)^n = \dfrac{a^n}{b^n}$.

Example 11 For each of the five properties above, we give below an immediate application and follow it with a justification for the particular situation that makes plausible the general principle.

First, $x^3 \cdot x^5 = x^8$, because x^3 means xxx and x^5 means $xxxxx$ and $x^3 x^5 = (xxx)(xxxxx) = x^8$.

Second,
$$\frac{(-2)^5}{(-2)^2} = (-2)^3 = -8,$$
because
$$\frac{(-2)^5}{(-2)^2} = \frac{(-2)(-2)(-2)(-2)(-2)}{(-2)(-2)} = (-2)(-2)(-2) = (-2)^3.$$

Third, $(u^4)^3 = u^{12}$, because
$$(u^4)^3 = u^4 u^4 u^4 = (uuuu)(uuuu)(uuuu) = u^{12}.$$

Fourth, $(am)^3 = a^3 m^3$, because
$$(am)^3 = amamam = (aaa)(mmm) = a^3 m^3.$$

Fifth,
$$\left(\frac{5}{4}\right)^2 = \frac{5^2}{4^2} = \frac{25}{16},$$
because
$$\left(\frac{5}{4}\right)^2 = \frac{5}{4} \cdot \frac{5}{4} = \frac{5^2}{4^2}.$$

Example 12 As an application of properties (iv), (iii), (ii), we have:
$$\frac{(2x^3y^2)^4}{x^5} = \frac{2^4(x^3)^4(y^2)^4}{x^5} = \frac{2^4 x^{12} y^8}{x^5} = 16x^7 y^8.$$

Example 13 Let a and c be natural numbers such that $a < c$. Show that $10^a < 10^c$.

Since $a < c$, the difference $c - a$ is a natural number. Hence, 10^{c-a}, the product of several tens, is greater than one. Thus
$$10^c = 10^{a+(c-a)} = 10^a \cdot 10^{c-a} > 10^a \cdot 1 = 10^a.$$

EXERCISES C

1. Let a, b, c be nonzero numbers. Find each of the following.

   (a) $a^8 \cdot a^{15}$.   (b) c^{19}/c^{12}.   (c) $(a^3)^8$.   (d) $\dfrac{(a^4)^5}{a^9}$.

   (e) $(a^2 b)^{14}$.   (f) $\left(\dfrac{ab^2}{c^3}\right)^9$.   (g) $\dfrac{(a^3 b^4)^6}{(a^2 b)^5}$.

2. Let a, b, c, d be positive numbers. For each of the following equations, find one root.

   (a) $x^3 = a^9$.   (b) $x^5 = \dfrac{b^{20}}{c^{35} d^5}$.   (c) $(d^4)^x = d^{12}$.

   (d) $(-a^x)^3 + a^{12} = 0$.   (e) $b^{3+x} = b^{7-x}$.

3. Evaluate each of the following.

   (a) $\dfrac{(-2)^9}{2^8}$.   (b) $\dfrac{-2^9}{(-2)^8}$.   (c) $\dfrac{-2^8}{(-2)^9}$.

4. Express $(1 + x^2 y^3)^4$ as a sum that does not involve parentheses.

5. Let k, m, n be natural numbers such that $n > m$. Find each of the following.

   (a) $\dfrac{2^{nk}}{2^{mk}}$.   (b) $(2^{nk})^k$.

6. Evaluate $(0.1)^{12}$.

7. If $(0.01)^{371}$ is written in customary decimal form, how many zeros follow the decimal point before the digit 1?

8. Express each of the following products in customary decimal form.
   (a) 4.36×10^9.
   (b) 1.008×10^{14}.
   (c) 2.51×10^{20}.

9. Let r, s, t be natural numbers such that $r > s$.
   (a) Explain why: if $b > 1$, then $b^t > 1$.
   (b) Explain why: if $0 < b < 1$, then $0 < b^t < 1$.
   (c) Explain why: if $b > 1$, then $b^r > b^s$.
   (d) Explain why: if $0 < b < 1$, then $b^r < b^s$.
   (e) Explain why: if $b \in \{0, 1\}$, then $b^r = b^s$.
   (f) What may be said about whether b^r is greater than, less than, or equal to b^s, in case b is negative?

5 Rational exponents

Property (i) in Section 4 states that the product of powers of the same base is also a power of the same base in which the exponent is the sum of the exponents of the given powers. Property (iii) states that a power of a power of a base is a power of that base in which the exponent is the product of the exponents of the given powers. Property (iv) states that a power of a product of numbers is the product of the powers of the individual factors with the same exponent as the given power.

The notion of a power may be extended by introducing exponents that are not natural numbers. We treat certain powers in each of which the exponent is a rational number. The meanings that we adopt in the new situations are dictated by the desire that the basic properties in Section 4 be applicable to all exponents. Fortunately, this desire can be achieved. Although we shall make no attempt to carry out the tedious task of showing that the extension exists and preserves the basic properties, we shall freely apply the properties whenever convenient. Furthermore, with the extended notion, we may remove the restriction in property (ii) that $n > k$.

The following definitions give the respective meanings for certain powers.

Definition If b is a real number different from zero, then b^0 is the number 1.

Definition If b is a real number that is not negative, and if q and r are natural numbers, then $b^{q/r}$ is the nonnegative real number whose rth power is b^q.

In the special case that $q = 1$, then $b^{1/r}$ is the nonnegative number whose rth power is b itself.

Example 14 Since 4 is the nonnegative number whose third power is 64, we find that $64^{1/3} = 4$. Likewise, $64^{1/2} = 8$, because $8^2 = 64$ and $8 \geq 0$. Moreover $8^{2/3} = (8^2)^{1/3} = 64^{1/3} = 4$. Also, $0^{1/5} = 0$. Lastly, $(\frac{1}{2})^0 = 1$.

Definition If b and r are numbers such that b^r is a real number different from zero, then b^{-r} is $\dfrac{1}{b^r}$.

Example 15 Since $2^3 = 8$ is not zero, $2^{-3} = \frac{1}{8}$. If $a + x \neq 0$, then $(a + x)^{-4} = 1/(a + x)^4$. Also $4^{-1/2} = \dfrac{1}{4^{1/2}} = \dfrac{1}{2}$.

Example 16 Simplify each of the following, where a and b are nonzero numbers: $(a^{-3}b^2)^{-4}$; $a^{-3}b^4 \cdot a^2 b^{-7}$.

First,
$$(a^{-3}b^2)^{-4} = (a^{-3})^{-4}(b^2)^{-4} = a^{12}b^{-8}.$$
We have applied properties (iv) and (iii). Note that $(-3)(-4) = 12$ and $2(-4) = -8$. An experienced evaluator may perform both of the steps together and write only the final result. If there is a specific reason for doing so, the result $a^{12}b^{-8}$ may be expressed without a negative exponent as $\dfrac{a^{12}}{b^8}$.

Second,
$$a^{-3}b^4 \cdot a^2b^{-7} = a^{-1}b^{-3}.$$
We have applied property (i) twice: the sum of the exponents -3 and 2 is -1, and $4 + (-7) = -3$. If there is a specific reason for preferring a fractional form, an alternative answer is $\dfrac{1}{ab^3}$.

Example 17 Let a, b, c be positive numbers. Simplify each of the following: $(a^{1/2})(a^{-3/2}); \dfrac{b^{1/2}}{b^{1/3}}; (c^{-2/3})^{-9/4}$.

First, $a^{1/2}a^{-3/2} = a^{-1}$, in accordance with property (i), for $\frac{1}{2} + (-\frac{3}{2}) = -1$.

Second, $\dfrac{b^{1/2}}{b^{1/3}} = b^{1/6}$, since $\frac{1}{2} - \frac{1}{3} = \frac{1}{6}$.

Third, we apply property (iii) to obtain $(c^{-2/3})^{-9/4} = c^{3/2}$.

Example 18 Simplify each of the following: $(3^2 + 4^2)^{1/2}$; $(a^2 + b^2)^{1/2}$.

First, $(3^2 + 4^2)^{1/2} = (9 + 16)^{1/2} = 25^{1/2} = 5$.

Second, $(a^2 + b^2)^{1/2}$ is itself in an acceptably simple form.

Example 19 Expressed in scientific notation, the number $0.000\ 000\ 074$ is 7.4×10^{-8}.

Theorem 2 Let b be a real number greater than 1, and let t be a rational number. If $t > 0$, then $b^t > 1$; if $t < 0$, then $b^t < 1$.

PROOF If t is a natural number, then b^t is the product of t factors, each of which is greater than 1; therefore, $b^t > 1$ in this case.

If $t = q/r$, where q and r are natural numbers, then $b^{q/r}$ is (by definition) the positive number whose rth power is b^q. That is, $b^{q/r}$ is the positive root x of the equation
$$x^r = b^q.$$
By the previous paragraph, $b^q > 1$. Hence, $x^r > 1$. We show that x itself is greater than 1 by observing that each of the alternatives

$x = 1$ or $x < 1$ leads to a contradiction: if $x = 1$, then $x^r = 1$ and we have a contradiction; if $x < 1$, then x^r, the product of r factors each less than 1, is less than 1; again we have a contradiction. In summary, $b^t > 1$ in this case.

If $t < 0$, then $t = -p$, where p is a positive rational number. We have just shown that $b^p > 1$. Hence, $b^t = b^{-p} = \dfrac{1}{b^p} < 1$.

Example 20 Find the rational numbers, if any, that are roots of the inequality $2^x \geq \tfrac{1}{8}$.

By Section 4 of Chapter 4, the desired numbers are the same as the roots of the inequality $8 \cdot 2^x \geq 1$. Now, $8 = 2^3$. Thus, our inequality becomes $2^{x+3} \geq 1$, which (by the preceding theorem) is satisfied if and only if $x + 3 \geq 0$. Thus, the desired roots are the rational numbers x such that $x \geq -3$.

EXERCISES D

1. Express each of the following in a simple form not involving a negative exponent.
   (a) $16^{1/2}$.   (b) 16^{-1}.   (c) -6^{-1}.   (d) -6^{-2}.
   (e) $8^{5/3}$.   (f) $4^{3/2} + 9^{3/2}$.   (g) $9^{-1/2}$.   (h) $(\tfrac{1}{8})^{-1/3}$.
   (i) $(0.0381)^0$.   (j) $0^{1/2}$.

2. Express each of the following products in customary decimal form.
   (a) 8.13×10^{-7}.   (b) 1.2×10^{-4}.   (c) 2.01×10^{-16}.

3. Let a, b, c, d, g, h be positive numbers. Simplify each of the following.
   (a) $a^{1/2} \cdot a^{3/2}$.   (b) $\dfrac{b^{5/4}}{b^{3/4}}$.   (c) $\dfrac{d^{-5/4}}{d^{-3/4}}$.   (d) $\dfrac{1}{c^{-2}}$.
   (e) $(ac^{-2})^3$.   (f) $\dfrac{4g^{1/2}}{9h^{-1/2}}$.   (g) $\dfrac{27g^{1/3}}{(cd)^0}$.   (h) $(a+b+c)^0$.
   (i) $(a^3 b^2)^{-4}$.   (j) $(b^{1/2} c^{-3/2})^{-6}$.   (k) $(a^3 b^4 / d^8)^{1/2}$.

4. In each of the following, expand by the use of the binomial theorem, and then simplify.
   (a) $(1 + x^{1/2})^4$.   (b) $(1 - 2x^{1/3})^3$.

5. Let p be a number such that $0 < p < \tfrac{1}{2}$. Express each of the following in simple form.
   (a) $(1 + 2p^2 + p^4)^{1/2}$.   (b) $\dfrac{(1+p^2)^{1/2}}{1+p}$.   (c) $(1 - 6p + 12p^2 - 8p^3)^{1/3}$.

6. Calculate the number
$$\left[2^6 + \binom{6}{1} \cdot 2^5 \cdot 3 + \binom{6}{2} \cdot 2^4 \cdot 3^2 + \binom{6}{3} \cdot 2^3 \cdot 3^3 + \binom{6}{4} \cdot 2^2 \cdot 3^4 + \binom{6}{5} \cdot 2 \cdot 3^5 + 3^6 \right]^{1/3}.$$

7. Show that, if $a > 0$, then $a^{-1/2} = \dfrac{a^{1/2}}{a}$.

8. Approximately, $2^{1/2}$ is 1.414, $3^{1/2}$ is 1.732, $5^{1/2}$ is 2.236, and $7^{1/2}$ is 2.645. Using this information to best advantage, find a decimal approximation for each of the following.
   (a) $2^{3/2}$.   (b) $3^{-1/2}$.   (c) $7^{-1/2}$.   (d) $5^{-3/2}$.

9. Find the roots, if any, for the inequality $1 + x^2 \leq (1 + x)^2$.

10. Let t be a positive number. Find the truth set for the statement $(1 + t)^{1/2} < 1 + t^{1/2}$.

11. Let q and r be natural numbers.
    (a) Show that $3^{1/r} > 1$.
    (b) Show that $3^{q/r} > 1$.
    (c) Show that 3^{-q} is between 0 and 1.
    (d) Show that $0 < 3^{-q/r} < 1$.

12. Let x belong to the set of rational numbers. Find the truth set for the inequality $5^x \geq 1$.

13. Find the rational numbers, if any, that are roots of the inequality $4^x < 8$.

14. Let x be a rational number. Solve each of the following inequalities.
    (a) $2^x > \frac{1}{32}$.   (b) $(\frac{1}{2})^x > 32$.   (c) $(\frac{1}{4})^x \leq 32$.

15. Let r and s be rational numbers such that $r < s$.
    (a) Explain why: $2^r < 2^s$.
    (b) Explain why: $3^r < 3^s$.
    (c) Explain why: $(\frac{1}{2})^r > (\frac{1}{2})^s$.

16. Let r and s be rational numbers.
    (a) Explain why: if $2^r < 2^s$, then $r < s$.
    (b) Explain why: if $(\frac{1}{2})^r < (\frac{1}{2})^s$, then $r > s$.

17. Let x belong to the set of rational numbers. Find the truth set for each of the following statements.
    (a) $2^x \geq (\frac{1}{2})^x$.   (b) $2^x = 4^{x-1}$.   (c) $2^{3x} < 4^{x-6}$.

6 Sequence

The integers are the members of the set $\{0, 1, -1, 2, -2, 3, -3, 4, -4, \ldots\}$. One partition of the set of integers consists of the set of positive integers, $\{1, 2, 3, 4, 5, \ldots\}$, the set $\{0\}$, and the set of negative integers, $\{-1, -2, -3, \ldots\}$. In order to discuss sequences, we need to consider certain subsets of the set of integers.

Let J be a set of two or more integers with the following properties: J has a least member (that is, there belongs to J an integer that is less than every other member in J); if a and b are members of J, then every integer between a and b belongs to J.

Example 21 Each of the following four sets is an illustration of a set that has the properties prescribed above: $\{1, 2, 3, 4, 5, 6\}$; the set of positive integers, $\{1, 2, 3, \ldots\}$; the set of nonnegative integers, $\{0, 1, 2, 3, \ldots\}$; and $\{0, 1, 2, \ldots, 1000\}$. The latter set consists of all integers between 0 and 1000, inclusive.

Each of the following three sets does not satisfy the conditions above: the set of all integers; $\{2, 4, 6, 8, 10\}$; $\{1\}$. The set of all integers does not have a least member; the set {2, 4, 6, 8, 10} contains 4 and 10 but not the integer 5, which is between 4 and 10; the set {1} fails to have at least two members.

Definition Let J be a set of integers as described above. A correspondence that assigns to every member of J a number is a function on J and is often called a *sequence* on J.

Very frequently the correspondence is specified by a formula. (Recall the use of a formula as a convenient description of a linear function.)

Example 22 Let J be the set {1, 2, 3, 4, 5}. The sequence f given by the formula $f(x) = x^2$ assigns to each member of J its second power. Thus, $f(1) = 1$, $f(2) = 2^2 = 4$, $f(3) = 3^2 = 9$, $f(4) = 16$, $f(5) = 25$. We sometimes arrange these numbers in order and simply write the sequence 1, 4, 9, 16, 25, in which case we understand that the first number listed corresponds to 1, the second number in the list corresponds to 2, the third to 3, etc.

In general, whenever we specify a sequence on a set J by writing a list of numbers in order, we shall agree that the first number in the list corresponds to the least member of J, the second number listed corresponds to the next smallest member of J, etc.

Example 23 Consider the sequence 2, 4, 8, 16, 32, 64. In this situation we may interpret the set J to be {1, 2, 3, 4, 5, 6} and the sequence to be the correspondence f given by the following table:

n	1	2	3	4	5	6
$f(n)$	2	4	8	16	32	64

A suitable formula for f is $f(n) = 2^n$.

7 Arithmetic sequence

Two important types of sequences are the arithmetic sequence and the geometric sequence.

Definition An *arithmetic sequence* is a sequence that may be given by a formula of the type $f(n) = dn + c$, where d and c are constants.

Example 24 The sequence 5, 8, 11, 14, 17, 20 is an arithmetic sequence on the set $\{1, 2, \ldots, 6\}$ given by the formula $f(n) = 3n + 2$. In this case, $(d, c) = (3, 2)$. We may verify the assertion by noting that

$$3 \cdot 1 + 2 = 5,$$
$$3 \cdot 2 + 2 = 8,$$
$$3 \cdot 3 + 2 = 11,$$
$$3 \cdot 4 + 2 = 14,$$
$$3 \cdot 5 + 2 = 17,$$
$$3 \cdot 6 + 2 = 20.$$

Example 25 The sequence $\frac{1}{2}$, 1, $\frac{3}{2}$, 2, ... is an arithmetic sequence on the set of positive integers; a formula for the sequence is $\frac{1}{2}n$. Here, $(d, c) = (\frac{1}{2}, 0)$.

The formula for an arithmetic sequence recalls the formula for a linear function. When we consider the linear function given by $mx + b$, we have an image for every real number x. By contrast, when we consider a sequence given by $dx + c$, the only numbers x that have images are the members of the set J on which the sequence is defined.

EXERCISES E

1. Let J be the set of positive integers. Let the arithmetic sequence f on J be given by the formula $f(n) = 2n + 1$.
   (a) Find $f(7)$. (b) Find $f(3 + 5)$. (c) Find $f(3) + f(5)$.
   (d) Find the positive integer m such that $f(m) = 59$.
   (e) Show that the sum $f(1) + f(100)$ and the sum $f(2) + f(99)$ and the sum $f(3) + f(98)$ are the same number.

2. Consider the arithmetic sequence f on the set of nonnegative integers given by the formula $f(n) = 100 - 6n$.
   (a) Find $f(10)$. (b) Find $f(1 - 2 + 3)$. (c) Find $f(1) - f(2) + f(3)$.
   (d) Find the root, if any, of the equation $f(k) = -50$.
   (e) How many roots does the inequality $f(n) > 0$ have?
   (f) How many roots does the equation $f(n) = 0$ have?
   (g) Show that the following three sums are the same number: $f(0) + f(80), f(1) + f(79), f(2) + f(78)$.

3. A sequence on the set $J = \{0, 1, 2, \ldots, 8\}$ is given by the formula $h(n) = \binom{8}{n}$.
   (a) Write the numbers of the sequence in the usual order.
   (b) Find the sum of the images under h of all the numbers in J.

4. A sequence on the set $\{1, 2, 3, 4, 5, 6\}$ is given by the formula
$$f(n) = \frac{n(n-1)(n-2)(n-3)(n-4)(n-5)}{6!}.$$
Find each of the images: $f(1), f(2), f(3), f(4), f(5), f(6)$.

5. A sequence t on the set J of nonnegative integers is given by the formula
$$t(n) = \frac{1 + (-1)^n}{2}.$$
Find the image of each of the first seven numbers in J.

6. The sequence 1, 0, 7, 6, 9, 0, 7, 4, 6, 1, 9, 5, 0, 9, 9, 4, 3, 7, 5, 3 was obtained by successively cutting a deck of numbered cards.
   (a) Find the sum of the numbers in the sequence.
   (b) Find the average of the numbers in the sequence.
   (c) Try to find a formula that gives the respective images under this sequence.

8 Geometric sequence

Definition A *geometric sequence* is a sequence that may be given by a formula of the type $g(n) = ar^n$, where a and r are constants different from zero.

Example 26 The sequence $\frac{1}{3}$, 2, 12, 72, 432 is a geometric sequence on the set $\{0, 1, 2, 3, 4\}$ given by the formula $g(n) = 3^{-1} \cdot 6^n$. In this case, $(a, r) = (\frac{1}{3}, 6)$. We check our claim by computing the following respective images under the function g and observing that they agree with the successive numbers in the given sequence:

$$g(0) = 3^{-1} \cdot 6^0 = (\tfrac{1}{3}) \cdot 1 = \tfrac{1}{3},$$
$$g(1) = 3^{-1} \cdot 6 = \tfrac{1}{3} \cdot 6 = 2,$$
$$g(2) = 3^{-1} \cdot 6^2 = \tfrac{1}{3} \cdot 36 = 12,$$
$$g(3) = 3^{-1} \cdot 6^3 = 72,$$
$$g(4) = 3^{-1} \cdot 6^4 = 432.$$

Example 27 Consider the geometric sequence on the set of positive integers given by the formula $g(n) = ar^n$, in which $a = 1$ and $r = \frac{1}{2}$. Then $g(n) = 1 \cdot (\frac{1}{2})^n = (\frac{1}{2})^n = 2^{-n}$. The sequence may be expressed as $\frac{1}{2}, \frac{1}{4}, \frac{1}{8}, \frac{1}{16}, \ldots$.

Example 28 The sequence 1, 4, 9, 16, 25 is neither an arithmetic sequence nor a geometric sequence.

Example 29 The sequence 7, 7, 7, 7 is both an arithmetic sequence and a geometric sequence. On the set $\{1, 2, 3, 4\}$, this sequence is given by the formula $h(n) = 7$. This is of the type $dn + c$, in which $(d, c) = (0, 7)$; consequently, the sequence is arithmetic. The formula is also of the type ar^n, in which $(a, r) = (7, 1)$; therefore, the sequence is also geometric.

9 Formulas for sequences

Example 30 Find a formula for the arithmetic sequence 1, 7, 13, ..., 97.

Since we are not given full information concerning the set on which the sequence is defined, we have some freedom of choice; as we shall see, the formula we obtain depends upon our choice. Consequently, the choice must be explicitly written.

(a) Suppose we choose a set J whose least element is 1. The desired formula has the form $f(n) = dn + c$, where the numbers d and c remain to be found. We find from the given information that $f(1) = 1$ and $f(2) = 7$. Thus,

$$d \cdot 1 + c = 1$$

and

$$d \cdot 2 + c = 7.$$

This system of equations has exactly one root, namely, $(d, c) = (6, -5)$. The desired formula therefore is

$$6n - 5.$$

As a check, we observe that $f(3) = 6 \cdot 3 - 5 = 18 - 5 = 13$ agrees with the given information. Finally, if $6n - 5 = 97$, then $n = 17$. This means that the largest number in J is 17. Hence, $J = \{1, 2, \ldots, 17\}$. There are exactly 17 numbers that have images under this sequence.

(b) An alternative selection is to choose a set whose least element is 0. In this case the desired formula may be expressed in the form $h(n) = an + b$. Since $h(0) = 1$ and $h(1) = 7$, we find that

$$a \cdot 0 + b = 1$$

and

$$a \cdot 1 + b = 7.$$

The root of this system of equations is $(a, b) = (6, 1)$. The formula in this case is

$$6n + 1,$$

and the sequence is defined on the set $\{0, 1, \ldots, 16\}$. The set in this case has largest element 16, since the root of the equation $6n + 1 = 97$ is 16.

(c) If we preferred to consider the given sequence as defined on the set $\{-8, -7, \ldots, 8\}$, then the appropriate formula is $6n + 49$. We obtain this result by the same method we used in the preceding two situations.

Powers and Sequences

In each of the formulas obtained, the coefficient of n is 6. Furthermore, 6 is the difference between the second image 7 and the first image 1 in the sequence. It is also the difference between 13 and 7.

Example 31 Find a formula for the geometric sequence 81, 27, 9, ..., $\frac{1}{729}$.

(a) Suppose the sequence is defined on a set J with least element 1. We seek a formula of the type ar^n. If $n = 1$, then $ar = 81$; if $n = 2$, then $ar^2 = 27$. We note that ar^2 is the product of ar and r; hence,

$$r = \frac{ar^2}{ar} = \frac{27}{81} = \frac{1}{3}.$$

Thus, $a \cdot \frac{1}{3} = 81$, or $a = 243$. The desired formula is

$$g(n) = 243(\tfrac{1}{3})^n.$$

Next, we seek the value of n such that $g(n) = \frac{1}{729}$. Since

$$g(n) = 243(\tfrac{1}{3})^n = 3^5 \cdot 3^{-n} = 3^{5-n}$$

and since $\frac{1}{729} = \frac{1}{3^6} = 3^{-6}$, the equation becomes $3^{5-n} = 3^{-6}$. Hence, $5 - n = -6$ or $n = 11$. Thus, the formula $g(n) = 243(\tfrac{1}{3})^n$, or $g(n) = 3^{5-n}$, is applicable in case the sequence is defined on the set $J = \{1, 2, \ldots, 11\}$. The sequence consists of 11 numbers.

(b) If we prefer to consider the sequence as defined on the set $\{0, 1, \ldots, 10\}$, then the same type of analysis shows that the formula is $81(\tfrac{1}{3})^n$, or 3^{4-n}.

EXERCISES F

1. Let J be the set of positive integers. Let the geometric sequence g on J be given by the formula $g(n) = 2 \cdot 3^n$.
   (a) Find $g(4)$. (b) Find $g(40)$. (c) Find the root of the equation $g(m) = 486$.
   (d) Which of the following two numbers is greater, $g(3 + 4)$ or the sum $g(3) + g(4)$? Justify your answer.
   (e) Which of the following two numbers is greater, $g(3 + 4)$ or the product $g(3) \cdot g(4)$? Give the reason.
   (f) Which of the following is greater, the product $g(3) \cdot g(4)$, or $g(3 \cdot 4)$? Justify your answer.

2. A geometric sequence is given by the formula $g(n) = 2^{4-n}$. This formula is of the type ar^n. Find the ordered pair of numbers (a, r).

3. Consider the arithmetic sequence 1, 4, ..., 250. This sequence can be interpreted as a function on any one of the following sets. In each case, find a formula that gives the sequence.
   (a) $\{1, 2, \ldots, 84\}$. (b) $\{0, 1, \ldots, 83\}$.
   (c) A set in which the least element is 18.

4. Consider the geometric sequence $3 \cdot 10$, $3 \cdot 10^4$, ..., $3 \cdot 10^{250}$. This sequence may be interpreted as a function on any of various sets.

(a) If the sequence is considered to be defined on the set $\{1, 2, \ldots, 84\}$, show that a formula for the sequence is $3 \cdot 10^{3n-2}$.
(b) Compare the result in part (a) with Exercise 3a.
(c) If we interpret the geometric sequence as a function on $\{0, 1, \ldots, 83\}$, find a formula for the sequence.
(d) Compare part (c) with Exercise 3b.

5. Consider the arithmetic sequence $17, 11, \ldots, -85$. This sequence may be interpreted as a function on a set J whose least element is 1.
   (a) Find a formula for the sequence on J.
   (b) Find the largest number in J.

6. Consider the arithmetic sequence $-\frac{31}{4}, -7, \ldots, \frac{121}{2}$ to be a function on a set J whose least number is 0.
   (a) Find a formula for the sequence on J.
   (b) Find the largest number in J.
   (c) How many members are in the set J?

7. Suppose that the sequence $20, \ldots, \frac{5}{2}$ on the set $\{0, 1, 2, 3\}$ is a geometric sequence.
   (a) Find a formula for the sequence.
   (b) Find the two numbers in the sequence that are not explicitly written above.

8. Suppose that the sequence $7, \ldots, 28$ on the set $\{0, 1, 2, 3, 4\}$ is a geometric sequence.
   (a) Find a formula for the sequence.
   (b) Find the respective images of 1, 2, 3 under the sequence.

9. Consider the geometric sequence $10^{-8}, 10^{-6}, \ldots, 10^{42}$. This sequence may be interpreted as a function on a set J whose least element is 1.
   (a) Find a formula for the sequence on J.
   (b) How many members are in the set J?

10. Let p and q be positive numbers. Consider the sequence p, q.
    (a) Show that the sequence p, q is an arithmetic sequence on the set $\{1, 2\}$, and find an appropriate formula for the sequence.
    (b) Show that the same sequence is an arithmetic sequence on the set $\{0, 1\}$, and find the corresponding formula.
    (c) How are the formulas in parts (a) and (b) alike, and how are they different?
    (d) Which of the two formulas do you consider simpler?
    (e) Show that the sequence p, q is a geometric sequence on the set $\{1, 2\}$, and give a formula for the geometric sequence.
    (f) Show that the same sequence is a geometric sequence on the set $\{0, 1\}$, and give the corresponding formula.
    (g) Compare the formulas in parts (e) and (f).

11. Suppose that u, v, w are real numbers such that the sequence u, v, w is an arithmetic sequence on the set $\{0, 1, 2\}$.
    (a) Find a formula for the sequence, such that the formula does not involve w.
    (b) Use the formula in part (a) and the fact that w is the image of 2 in order to express w in terms of u and v.
    (c) Find another formula for the sequence, by using the information that v and w are the images of 1 and 2, respectively.
    (d) Find a formula for the sequence such that the formula does not involve v.
    (e) Use the formula in part (d) to express v in terms of u and w.

12. Suppose that u, v, w are positive numbers such that the sequence u, v, w is a geometric sequence on the set $\{0, 1, 2\}$.
    (a) Using the data that u and v are the respective images of 0 and 1, find a formula for the sequence.

(b) Obtain another formula for the sequence, by using the images of 1 and 2.
(c) Find, for the sequence, a formula not involving v.
(d) Use the formula in part (a) to express w in terms of u and v.
(e) Use the formula in part (c) to express v in terms of u and w.

10 The summation notation

A highly useful symbol in many mathematical studies is the capital Greek letter "sigma" Σ used in summation notation.

Let f be a sequence defined on a set J. Let a and b be two members of J such that $a < b$. Each integer in J between a and b inclusive has an image under the function f. The sum of all these images, namely,

$$f(a) + f(a+1) + f(a+2) + \cdots + f(b),$$

is compactly denoted by the notation $\sum_{k=a}^{b} f(k)$.

Example 32 Let f be a sequence defined on a set J such that 0 and 4 are members of J. Then $\sum_{k=0}^{4} f(k)$ means the sum

$$f(0) + f(1) + f(2) + f(3) + f(4).$$

Example 33 Consider the sequence h on the set of positive integers given by the formula $h(n) = n^2$. Then,

$$\sum_{k=1}^{6} h(k) = \sum_{k=1}^{6} k^2$$
$$= 1^2 + 2^2 + 3^2 + 4^2 + 5^2 + 6^2$$
$$= 1 + 4 + 9 + 16 + 25 + 36$$
$$= 91.$$

Example 34 The number $\sum_{j=5}^{8} 2^j$ is the sum of the images 2^5, 2^6, 2^7, 2^8. Here, the function is given by the formula 2^n, and the sequence is defined on a set to which 5 and 8 belong. We calculate:

$$\sum_{j=5}^{8} 2^j = 2^5 + 2^6 + 2^7 + 2^8 = 32 + 64 + 128 + 256 = 480.$$

Note that $\sum_{k=5}^{8} 2^k$ also represents the sum $2^5 + 2^6 + 2^7 + 2^8 = 480$ and thus is the same number as $\sum_{j=5}^{8} 2^j$; the use of the letter j or the letter k is immaterial.

In the notational convention above, there is no requirement that the letter k be employed in the symbol $\sum_{k=a}^{b} f(k)$. Any other unused letter, such as j or i or n, may be substituted without affecting the meaning.

Example 35 The binomial theorem expresses a power of a binomial as a sum of products. In symbols,

$$(a+b)^n = a^n + \binom{n}{1}a^{n-1}b + \binom{n}{2}a^{n-2}b^2 + \cdots$$
$$+ \binom{n}{n-1}ab^{n-1} + b^n.$$

The typical term in the expansion appears to be of the type $\binom{n}{k}a^{n-k}b^k$. With our agreement that $b^0 = 1$ and $\binom{n}{0} = 1$, the first term a^n also fits the pattern, for $a^n = \binom{n}{0}a^n b^0$. Likewise, the last term b^n fits the pattern, since $a^0 = 1$ and $\binom{n}{n} = 1$; thus, $b^n = \binom{n}{n}a^0 b^n$.

Hence, $(a+b)^n$ is the sum of terms of the form $\binom{n}{k}a^{n-k}b^k$ for all integers k belonging to the set $\{0, 1, \ldots, n\}$. Use of the summation notation enables us to express the binomial expansion as follows:

$$(a+b)^n = \sum_{k=0}^{n} \binom{n}{k}a^{n-k}b^k.$$

EXERCISES G

1. Calculate each of the following sums.

    (a) $\sum_{k=0}^{6}(-1)^k.$
    (b) $\sum_{j=-3}^{3} j^2.$
    (c) $\sum_{x=0}^{4} \frac{1+x^2}{1+x}.$
    (d) $\sum_{n=1}^{5} n^{-1}.$
    (e) $\sum_{m=0}^{30} \frac{1+(-1)^m}{2}.$
    (f) $\sum_{k=1}^{1776} 1.$
    (g) $\sum_{k=19}^{84} 2.$

2. Find each of the following sums.

    (a) $\sum_{k=0}^{6} \binom{6}{k} 5^{6-k}(-3)^k.$
    (b) $\sum_{j=0}^{8} \binom{8}{j}(0.3)^{8-j}(0.7)^j.$
    (c) $\sum_{k=1}^{9} \binom{9}{k}(0.1)^{9-k}(0.9)^k.$

3. Let n be a natural number, and let x be a number different from zero. Evaluate each of the following sums.

    (a) $\sum_{k=0}^{n} \binom{n}{k} \cdot 1^{n-k} \cdot 1^k.$
    (b) $\sum_{k=0}^{n} \binom{n}{k}.$
    (c) $\sum_{k=0}^{n} \binom{n}{k}(-1)^k.$
    (d) $\sum_{k=0}^{n} \binom{n}{k} x^k.$

4. Let p be a number such that $0 < p < 1$.

    (a) Show that $\sum_{k=0}^{2} k\binom{2}{k}(1-p)^{2-k}p^k$ is $2p$.

(b) Show that $\sum_{k=0}^{3} k\binom{3}{k}(1-p)^{3-k}p^k$ is $3p$.

(c) First guess what the value of the following sum is, and then verify your guess:
$$\sum_{k=0}^{4} k\binom{4}{k}(1-p)^{4-k}p^k.$$

5. Let p be a number such that $0 < p < 1$, and let n be a natural number. We shall find later (see Section 25 of Chapter 8) that
$$\sum_{k=0}^{n} k^2 \binom{n}{k}(1-p)^{n-k}p^k = np(np - p + 1).$$

Using this information, evaluate each of the following.

(a) $\sum_{k=0}^{3} k^2 \binom{3}{k}\left(\frac{1}{3}\right)^{3-k}\left(\frac{2}{3}\right)^k$.   (b) $\sum_{k=0}^{100} k^2 \binom{100}{k}(0.97)^{100-k}(0.03)^k$.

(c) $\sum_{k=1}^{5} k^2 \binom{5}{k}\left(\frac{1}{2}\right)^5$.   (d) $\sum_{k=1}^{8} k^2 \binom{8}{k}$.

11 Sum of numbers in a sequence

Let m be a natural number. We wish to show that the sum of the natural numbers from 1 to m inclusive is $\frac{1}{2}m(m+1)$. To do this, let us denote the sum by S. Then,
$$S = 1 + 2 + 3 + \cdots + (m-2) + (m-1) + m.$$
The sum is not altered by arranging the terms in reverse order. Therefore,
$$S = m + (m-1) + (m-2) + \cdots + 3 + 2 + 1.$$
Adding corresponding members of these two equations, term by term, we obtain
$$2S = (m+1) + (m+1) + (m+1) + \cdots + (m+1) + (m+1) + (m+1)$$
$$= m(m+1).$$
Consequently, $S = \frac{1}{2}m(m+1)$.

Example 36 The sum of the first 200 natural numbers is
$$\sum_{k=1}^{200} k = \frac{1}{2}(200)(201) = 20{,}100.$$

Let f be an arithmetic sequence on the set of positive integers. There are constants d and c such that f is given by the formula $f(n) = dn + c$. Let m be a natural number. Then the sum $\sum_{k=1}^{m} f(k) = \sum_{k=1}^{m}(dk + c)$ is equal to $d \sum_{k=1}^{m} k + mc$, which we obtain as follows:

$$\sum_{k=1}^{m}(dk+c) = (d\cdot 1 + c) + (d\cdot 2 + c) + \cdots + (dm + c)$$
$$= (d\cdot 1 + d\cdot 2 + \cdots + dm) + (c + c + \cdots + c)$$
$$= d(1 + 2 + \cdots + m) + mc$$
$$= d\sum_{k=1}^{m} k + mc.$$

Example 37 Find the sum $5 + 8 + 11 + 14 + 17 + 20 + 23$.

One method is straightforward addition; the sum is 98. An alternative method is particularly useful in case a sum of many terms is involved. Applied to the present case, we begin by recognizing that 5, 8, 11, 14, 17, 20, 23 is an arithmetic sequence. Defined on the set $\{1, 2, \ldots, 7\}$, the sequence is given by the formula $3n + 2$. The desired sum is

$$\sum_{k=1}^{7}(3k+2) = 3\sum_{k=1}^{7} k + 7\cdot 2$$
$$= 3\cdot\frac{7\cdot 8}{2} + 14$$
$$= 98.$$

Example 38

$$\sum_{h=1}^{12}\left(\frac{1}{2}h - \frac{1}{3}\right) = \frac{1}{2}\sum_{h=1}^{12} h + 12\left(-\frac{1}{3}\right) = \frac{1}{2}\frac{12\cdot 13}{2} - 4 = 35.$$

We have been investigating a method for calculating easily the sum of the numbers in an arithmetic sequence. We now turn to the analogous problem for geometric sequences.

Let r be a real number that is neither 0 nor 1. Let m be a natural number. The sum of the powers of r from the 0th to the mth inclusive is $(1 - r^{m+1})/(1 - r)$. To show this, we proceed as follows. Let the sum be denoted by S. Thus,

$$S = 1 + r + r^2 + \cdots + r^{m-1} + r^m.$$

Multiplying each member of the equation by $1 - r$, we obtain

$$(1 - r)S = (1 - r)(1 + r + r^2 + \cdots + r^{m-1} + r^m)$$
$$= (1 - r) + (1 - r)r + (1 - r)r^2 + \cdots + (1 - r)r^{m-1} + (1 - r)r^m$$
$$= 1 - r + r - r^2 + r^2 - r^3 + \cdots + r^{m-1} - r^m + r^m - r^{m+1}$$
$$= 1 - r^{m+1}.$$

The simplification in the final step occurs because every power of r from the first to the mth appears with a minus sign and again with a plus sign. Solving for S, we obtain

$$S = \frac{1 - r^{m+1}}{1 - r}.$$

Example 39 The sum

$$\sum_{k=0}^{8} (\tfrac{1}{2})^k = 1 + 2^{-1} + 2^{-2} + \cdots + 2^{-8}$$

is

$$\frac{1 - (\tfrac{1}{2})^9}{1 - \tfrac{1}{2}} = \frac{1 - 2^{-9}}{\tfrac{1}{2}} = 2 - 2^{-8}.$$

Let g be a nonconstant geometric sequence on the set of nonnegative integers. There is a constant a, different from 0, and there is a constant r, which is neither 0 nor 1, such that g is given by the formula $g(n) = ar^n$. Let m be a natural number. Then the sum

$$\sum_{k=0}^{m} g(k) = \sum_{k=0}^{m} ar^k \text{ is equal to } \frac{a(1 - r^{m+1})}{1 - r}.$$

Indeed,

$$\sum_{k=0}^{m} ar^k = a + ar + ar^2 + \cdots + ar^m = a(1 + r + \cdots + r^m) = a \sum_{k=0}^{m} r^k.$$

Example 40 Let t be a positive number. Then

$$\sum_{j=0}^{99} A(1 + t)^j = A \cdot \frac{1 - (1 + t)^{100}}{1 - (1 + t)}.$$

In this application, $1 + t$ plays the role of r in the general situation. Algebraic simplification may be made. Note that $1 - (1 + t) = -t$. Hence,

$$A \frac{1 - (1 + t)^{100}}{1 - (1 + t)} = A \frac{1 - (1 + t)^{100}}{-t}$$

$$= \frac{A}{t}[-1 + (1 + t)^{100}]$$

$$= At^{-1}[(1 + t)^{100} - 1].$$

Example 41 The accumulation of money from a single investment at compound interest was discussed in Example 8 in Section 3. Suppose that an investor makes deposits at regular periods of time. Specifically, suppose that he invests A dollars in a savings account at the end of each interest-paying period. Suppose that the account pays interest at the rate of j per period. What is the accumulated value of the account at the end of 40 periods?

At the end of the 40th period the kth deposit, having been in the account $40 - k$ periods, is worth $A(1 + j)^{40-k}$ dollars. (Review Example 8.) Therefore, the total number of dollars in the account is the sum of the numbers $A(1 + j)^{40-k}$ for all members k of the set $\{1, 2, \ldots, 40\}$. This sum is

$$A + A(1 + j) + A(1 + j)^2 + \cdots + A(1 + j)^{39},$$

or in more compact notation,
$$\sum_{k=0}^{39} A(1+j)^k.$$

We evaluate the sum as
$$A\frac{1-(1+j)^{40}}{1-(1+j)} = Aj^{-1}[(1+j)^{40}-1].$$

As a numerical illustration, suppose that every three months the investor deposits $100 in an account. Suppose that interest is paid at the rate of one percent each quarter (a rate advertised as "4% compounded quarterly"). At the end of ten years, just after the 40th deposit has been made, the investment is worth $100(0.01)^{-1}[(1.01)^{40}-1]$ dollars, which is approximately $4888.

The principal results developed in this section are stated in the following two theorems.

Theorem 3 Let d, c be real numbers, and let m be a natural number. If f is the arithmetic sequence defined on the set of positive integers by the formula $f(n) = dn + c$, then
$$\sum_{k=1}^{m} f(k) = \frac{dm(m+1)}{2} + mc.$$

(See also the Remark in Exercise 21 below.)

Theorem 4 Let a, r be nonzero real numbers such that $r \neq 1$, and let m be a natural number. If g is the geometric sequence defined on the set of nonnegative integers by the formula $g(n) = ar^n$, then
$$\sum_{k=0}^{m} g(k) = a\frac{1-r^{m+1}}{1-r} = a\frac{r^{m+1}-1}{r-1}.$$

It is customary to use the form $a(r^{m+1}-1)/(r-1)$ in case $r > 1$, because then the denominator $r - 1$ is positive.

EXERCISES H

1. Find the sum of all the integers from 1 to 1,000,000 inclusive.
2. An arithmetic sequence f is defined on the set $\{1, 2, \ldots, 12\}$ by the formula $f(n) = -100n + 800$. Find the sum $f(1) + f(2) + \cdots + f(12)$.
3. A sequence f is defined on the set $\{1, 2, \ldots, 50\}$ by the formula $f(n) = \frac{3}{4}n - \frac{1}{2}$. Find $\sum_{k=1}^{50} f(k)$.

4. Evaluate each of the following sums.

   (a) $\sum_{k=1}^{15} (4k + 5)$.

   (b) $\sum_{j=1}^{60} (1 - \frac{1}{2}j)$.

5. Find the sum of all the even integers from 2 to 10,000 inclusive.

6. Find the sum $16 + 19 + \cdots + 253$, where $16, 19, \ldots, 253$ is an arithmetic sequence.

7. If $0.688, 0.681, \ldots, 0.135$ is an arithmetic sequence, calculate the sum $0.688 + 0.681 + \cdots + 0.135$.

8. Find the sum of all the powers of $\frac{2}{3}$ from the 0th to the 40th inclusive.

9. A geometric sequence g is defined on the set $\{0, 1, \ldots, 7\}$ by the formula $g(n) = \frac{1}{5} \cdot 3^n$. Find the sum $g(0) + g(1) + \cdots + g(7)$.

10. A sequence g is defined on the set $\{0, 1, \ldots, 5\}$ by the formula $g(n) = 5(0.2)^n$. Find
$$\sum_{k=0}^{5} g(k).$$

11. Evaluate the sum $\sum_{k=0}^{10} 2^{5-k}$.

12. If B and j are positive numbers, find the sum $\sum_{n=0}^{15} B(1 + j)^n$.

13. If q is a positive number, find the sum $\sum_{k=0}^{43} (1 + q)^{-k}$.

14. Express in customary decimal notation each of the following numbers.

    (a) $\sum_{k=1}^{12} 6(0.1)^k$.   (b) $\sum_{k=1}^{8} 37(0.01)^k$.

15. By division, show that $\frac{1}{3}$ is approximately $\sum_{k=1}^{1000} 3(0.1)^k$.

16. Show that $\frac{8}{33}$ is approximately $\sum_{k=1}^{200} 24(0.01)^k$.

17. Consider the sum $S = \sum_{k=1}^{473} (0.001)^k$.

    (a) Show that S is equal to $\sum_{k=0}^{472} (0.001)(0.001)^k$.

    (b) Using the information in part (a), evaluate the sum S.
    (c) Show that S is approximately $\frac{1}{999}$.

18. Let m be a large positive integer. Show that $\sum_{k=1}^{m} (0.01)^k$ is approximately $\frac{1}{99}$.

19. Let m be a large positive integer, and let q be an integer such that $0 < q < 100$. Show that $\sum_{k=1}^{m} q(0.01)^k$ is approximately $q/99$.

20. For each of the following, find a simple fraction that is very nearly equal to the given decimal.
    (a) 0.6262626262626262.   (b) 0.4848484848.   (c) 0.435435435435435435.

21. Let f be an arithmetic sequence defined on the set of positive integers. Suppose that the formula for f is $f(n) = dn + c$. Let m be a natural number.
    (a) Find $f(1)$.
    (b) Find $f(m)$.
    (c) Calculate $m \cdot \dfrac{f(1) + f(m)}{2}$.
    (d) Show that $\sum_{k=1}^{m} f(k)$ is the same as $m \cdot \dfrac{f(1) + f(m)}{2}$. [REMARK: The result in part (d) may be stated in words as follows. For an arithmetic sequence, the sum of the images of the numbers from 1 to m inclusive is m times the average of the images of 1 and m.]

22. Use the result of Exercise 21d to find each of the following sums, where the terms in each sum are an arithmetic sequence.
    (a) $1 + 2 + \cdots + 627$.
    (b) $1 + 3 + \cdots + 627$.
    (c) $(-8) + (-3) + \cdots + 132$.
    (d) $34 + 43 + \cdots + 907$.

Probability 8

1 Introduction

The subject of probability has a long and somewhat controversial history. Dating back some three centuries, it has undergone, in contemporary times, an extreme broadening of its significance, both in application and in theory.

The following is a typical problem of the elementary theory of probability. We have a certain experiment. There are several possible outcomes, or results, for the experiment. We have a certain sentence that pertains to the outcome of the experiment. For each possible outcome, the sentence is either true or else false (but not both) and consequently becomes a statement. We wish to assign to the statement a measure that, in some sense, suggests how much chance we think the statement has of turning out to be true. By "turning out to be true," we mean that the experiment yields an outcome for which the sentence is a true statement. This measure of chance or likelihood, this degree of our belief, for the statement we call the probability of the statement.

The measure is a number. There are two contrasting approaches to the task of how such a number may be assigned. One procedure is to establish a mathematical model, complete with an axiomatic system, and to apply common sense in the choice of numerical probabilities. An alternative method is to perform the experiment repeatedly, gathering data concerning the observed results, and from these data to estimate the probabilities for the outcomes of the experiment, or similar experiments, in the future. The former approach is emphasized by the theoretical probabilist, the latter, by the applied statistician. Both viewpoints are highly useful, and we shall try to blend them.

In its early history, probability was intimately linked with games of chance. The popular appeal for games and the standardization of their basic rules still make them fine illustrative material for an initial study of probability.

Statistical uses of probability are extremely numerous. Among the highly varied applications we mention only a few: in agricultural research, to determine the worth of a new chemical product for promoting growth; in manufacturing, to control the quality of the output in mass production; in transportation or communications, to expedite the smooth flow of traffic.

Our goal in this chapter is to develop some fundamental notions of probability that will serve as a basis for statistical training. In this task we shall exploit the recent chapters on set theory, logic, counting, and related topics.

2 Probability in the finite case

First, we turn our attention, in more detail, to the central idea of probability. Suppose we have a certain experiment or activity. The activity may perhaps be a compound situation consisting of a sequence of several simple activities. Further, suppose that the experiment has a nonempty set of possible outcomes. Although a significant portion of the theory of probability treats of other cases, we shall confine ourselves in this chapter to the case in which the set of possible outcomes is a finite set.

To each member of the set of logically possible outcomes we assign a measure. This measure is a number. We require that

each assigned measure be a positive number

and that

the sum of the measures assigned to all the outcomes be one.

Now consider any sentence pertaining to the outcome of the experiment under consideration, that is, any sentence which, for every possible outcome, has a truth value. The truth set for this statement is the subset consisting of those outcomes for which the sentence is a true statement. *The sum of the measures assigned to all the elements belonging to the truth set of the statement is called the probability of the statement.*

> **Example 1** One of the simplest illustrations involves the experiment of an ordinary tossing of a "true" coin. We agree that there are exactly two possible outcomes, namely, a head appears or a tail appears. We must now choose the measures to be assigned to these respective possibilities. The common-sense appeal to the trueness of the coin and the fairness of the toss as well as extensive past experience lead us to select the same measure m for each of the two outcomes. Since their sum $m + m$ is required to be 1, necessarily $m = \frac{1}{2}$. In other words, we assign the number $\frac{1}{2}$ as the measure of our belief that a head appears, and the number $\frac{1}{2}$ as the measure of the likelihood that a tail appears. We now consider a sentence pertaining to the outcome of the experiment, namely, "The toss does not result in a head." The truth set for this statement is the set[1] {TAIL} consisting

---

[1] We shall frequently name an outcome for an experiment by writing only one or two key words in capital letters. For example, here the capitalized word "TAIL" means "the tail appears on the toss of the coin." As another example, if an experiment consists in selecting one card from a deck, the outcome "the card drawn is the ace of spades" may be denoted by the phrase, "SPADE ACE."

of one outcome. Since the measure assigned to this single outcome in the truth set is $\frac{1}{2}$, the probability of the statement is $\frac{1}{2}$.

Example 2 An experiment consists in fairly drawing a single card from an ordinary deck of playing cards. If measures are assigned in a reasonable fashion to the various possible outcomes, what is the probability that the card drawn is a black ten?

Since the deck has 52 cards, there are 52 possible outcomes for the experiment. Since the drawing is unbiased, it seems most reasonable to assign the same measure to each of the 52 outcomes. If m is the number assigned to each outcome, then the sum $m + m + \cdots + m$, or $52m$ (that is, the total measure assigned to all the possible outcomes), is required to be 1. Hence, $m = \frac{1}{52}$.

We now consider the statement whose probability is requested; it is, "The card drawn is a black ten." The truth set for this statement has two members, the drawing of the ten of spades and the drawing of the ten of clubs. Since each of these two outcomes has been assigned the measure of $\frac{1}{52}$, the sum of their measures is $\frac{1}{52} + \frac{1}{52} = \frac{1}{26}$. The probability for the statement under consideration is $\frac{1}{26}$.

Example 3 In their manufacture, a pair of gambling dice may be specially weighted. Suppose the specifications for a certain die are that on an ordinary roll the likelihood of each face appearing is to be proportional to the number of spots on the face. On a single roll of this die, what is the probability of a deuce?

In this problem the activity is a single, ordinary roll of the weighted die. There are six possible outcomes, corresponding respectively to the six faces of the die. Let k be the constant of proportionality. Then Table 1a expresses the measures of the individual outcomes. The

Table 1a

Outcome (number of spots)	ONE	TWO	THREE	FOUR	FIVE	SIX
Measure	k	$2k$	$3k$	$4k$	$5k$	$6k$

entries in the table are based on the specification in the manufacture of the weighted die. The total of the six measures, $k + 2k + 3k + 4k + 5k + 6k = 21k$, must be 1. Consequently, the constant of proportionality k is $\frac{1}{21}$. The table now becomes that shown in Table 1b. The probability that a deuce appears is $\frac{2}{21}$.

Table 1b

Outcome (number of spots)	ONE	TWO	THREE	FOUR	FIVE	SIX
Measure	$\frac{1}{21}$	$\frac{2}{21}$	$\frac{3}{21}$	$\frac{4}{21}$	$\frac{5}{21}$	$\frac{6}{21}$

In summary, we are asking what is the probability of a certain statement. To answer the question,

1. We must identify the experiment to which the statement refers.
2. We must recognize all the logically possible outcomes.
3. We must assign in a reasonable manner a measure to each outcome.
4. We must determine the truth set for the statement.

Then we may calculate the desired probability by adding together the measures of all the outcomes that belong to the truth set.

EXERCISES A

1. An experiment consists in rolling one of a pair of balanced gambling dice.
   (a) List the possible outcomes.
   (b) If the rolling is done in a fair manner, what measure may be assigned to the outcome FIVE?
   (c) What is the probability that the number of spots showing is greater than four?

2. An experiment consists in a fair toss of a nickel and a dime. Denote each possible outcome by an ordered pair, in which the left component tells which face the nickel shows and the right component gives the face shown by the dime.
   (a) List all possible outcomes.
   (b) What measure may be assigned to each outcome?
   (c) What is the probability that at least one coin shows a tail?

3. An experiment consists in a fair toss of a nickel and a dime. Let Y be the number of heads that appear.
   (a) The possible outcomes are $Y = 0$, $Y = 1$, $Y = 2$. What measure may be reasonably assigned to each of these logical possibilities?
   (b) What is the probability that $Y > 0$?

4. An experiment consists in drawing one card from an ordinary deck of playing cards. For each of the following statements, find its probability.
   (a) The card drawn is a club.
   (b) The card drawn is a red card.
   (c) The card drawn is a deuce or a trey (that is, a two or a three).
   (d) The card drawn is not a face card.

5. A bowl contains seven tickets bearing respectively the numbers 1, 2, 3, 4, 5, 6, 7. An experiment consists in selecting by lot one ticket from the bowl.
   (a) How many possible outcomes are there?
   (b) What measure may be assigned to each outcome?
   (c) What is the probability that the number on the chosen ticket belongs to the set $\{1, 2, 4\}$?
   (d) What is the probability that the chosen ticket has an odd number?

6. A bowl contains four tickets of different colors: one red, one blue, one green, one white. An experiment consists in drawing a ticket from the bowl and afterward drawing another ticket from the bowl (without having replaced the first ticket drawn). Denote each possible outcome by an ordered pair, in which the left component tells the color of the first ticket drawn and the right component tells the color of the second ticket drawn.
   (a) List all the possible outcomes.

(b) What measure may be assigned to each outcome?
(c) Evaluate the probability that the first ticket drawn is blue.
(d) Evaluate the probability that the second ticket drawn is blue.
(e) What is the probability that neither ticket drawn is blue?
(f) What is the probability that the blue ticket is drawn?
(g) What is the probability that both tickets drawn are blue?

7. In a certain game a player draws, fairly, one card from an ordinary deck. In this game each red king and each trey are wild cards, scoring in favor of the player. The player wins if the card drawn belongs to the winning set or is wild. For each of the following winning sets, what is the probability that the player wins?
   (a) The winning set consists of the tens.
   (b) The winning set is the club suit.
   (c) The winning set consists of the face cards.
   (d) The winning set is the heart suit.
   (e) The winning set consists of the red cards.

8. A certain experiment has four possible outcomes. Three of them have been assigned measures, in the probability sense, of $\frac{1}{3}, \frac{1}{4}, \frac{2}{5}$, respectively. What measure is assigned to the other possible outcome?

9. A certain experiment has five possible outcomes. Three of them have been assigned probability measures of $\frac{1}{3}, \frac{1}{4}, \frac{1}{6}$, respectively. If the other two possible outcomes deserve equal measures, what is the measure for each of them?

10. In a certain state 100,000 citizens have an opinion, either for or against, a certain public issue. It is desired to estimate the probability of the statement that a citizen selected at random is in favor of the issue.
    (a) In a random sample of 1000 inquiries from an urban area in the state, 640 citizens favored the issue. On the basis of this information, what is the best estimate for the number of citizens in the state who are for the issue?
    (b) On the basis of the information in part (a), what is a reasonable estimate for the desired probability?
    (c) In a random sample of 1000 inquiries from a rural area in the state, 560 citizens favored the issue. On the basis of this information alone, what is a reasonable estimate for the desired probability?
    (d) Based on the information in both parts (a) and (c), what is a reasonable estimate for the desired probability?
    (e) Suppose that four times as many citizens in the state live in urban areas as in rural areas. Does this additional information affect your response to part (d)? If so, how?
    (f) Suppose that all of the citizens express their opinions in an election. Do a majority favor the issue? Justify your answer.

3 More examples

Example 4 In an ordinary roll of a pair of "true" dice, what is the probability that the total number of spots shown is six?

For identification purposes, suppose that the two dice are distinguishable by color, one being red and the other white. The experiment of rolling the pair of dice once consists of two simultaneous simple activities, the roll of the red die and the roll of the white die. The number of possible results for the activity of rolling the red die is six; for each of these, the number of possible faces for the white die to show is six; hence, by Theorem 3 in Section 7 of Chapter 6, there are $6 \cdot 6 = 36$ possible outcomes for the experiment. Each outcome for the experiment may be expressed as an ordered couple (r, w), in which the left component r is the number of spots shown on the red die and the right component w is the number of spots on the face of the white die. The 36 possible couples are listed in Table 2.

Table 2

(1, 1)	(1, 2)	(1, 3)	(1, 4)	(1, 5)	(1, 6)
(2, 1)	(2, 2)	(2, 3)	(2, 4)	(2, 5)	(2, 6)
(3, 1)	(3, 2)	(3, 3)	(3, 4)	(3, 5)	(3, 6)
(4, 1)	(4, 2)	(4, 3)	(4, 4)	(4, 5)	(4, 6)
(5, 1)	(5, 2)	(5, 3)	(5, 4)	(5, 5)	(5, 6)
(6, 1)	(6, 2)	(6, 3)	(6, 4)	(6, 5)	(6, 6)

Notice that (1, 2) and (2, 1) are different outcomes, since (1, 2) represents a roll in which one spot appears on the red die and two spots appear on the white die, while (2, 1) denotes the outcome of a deuce on the red die and an ace on the white die.

Since we have an unbiased roll of true dice in this experiment, we may reasonably assign the same measure to each of the 36 possible outcomes. With a total measure of 1, each outcome is then assigned the number $\frac{1}{36}$ as its measure.

The statement under consideration is, "The total number of spots shown on the pair of dice is six." The truth set for the statement is $\{(1, 5), (2, 4), (3, 3), (4, 2), (5, 1)\}$. Since each of the five members in the truth set has measure $\frac{1}{36}$, the desired probability is $\frac{5}{36}$.

Example 5 From an ordinary deck of playing cards, three cards are drawn in succession, without replacement. What is the probability that the first card drawn is a jack, the next is a six, and the last is an honor card?

By the phrase "without replacement," we mean that any card drawn is not returned to the deck before any subsequent drawing is made. Thus, the first card is selected from a set of 52 cards, the second from an incomplete deck of 51 cards, and the third from a set of only 50 cards. The number of possible outcomes for this sequence of three successive activities is $52 \cdot 51 \cdot 50 = 132600$. If we suppose that the drawings are done fairly, then each of the outcomes may be assigned the number $1/132600$ as its measure.

Next, we identify the truth set for the statement whose probability is requested. One member of the truth set is represented by the ordered triple (HEART JACK, DIAMOND SIX, CLUB ACE); another is (SPADE JACK, SPADE SIX, CLUB JACK). There are too many members to justify writing a complete roster. However, we may apply the principles discussed in Chapter 6 to find the number of members in the truth set. An ordered triple belonging to the truth set is obtained from the following sequence of three activities: first, the drawing of a jack as left component; next, the selection of a six as middle component; finally, the choice of an honor card as right component. For this sequence, there are four possible outcomes for the first activity; for each of these, there are four possibilities for the second activity; finally, after the first two activities in the sequence have been completed, the incomplete deck has only 19 honor cards remaining in it. The number of members in the truth set is $4 \cdot 4 \cdot 19 = 304$.

Since each of the 304 elements in the truth set has measure $1/132600$, the probability is $304/132600 = 38/16575$. A decimal approximation to the probability is 0.0023.

4 Basic principles in probability

We now develop some basic principles in probability.

We have been studying the probability of a given statement pertaining to the outcome of a given experiment. *The probability is a number.*

The probability is calculated by adding the measures assigned to the members of the truth set for the statement. Thus, if the measures have been already assigned, the probability of a statement actually depends only on the truth set for the statement. Any two statements having the same truth set have the same probability. Two statements with the same truth set are logically equivalent. Therefore, *any two logically equivalent statements have equal probabilities.*

The probability of a statement is the sum of the measures of the outcomes belonging to the truth set. In particular, if the truth set consists of all possible outcomes, then the total measure is prescribed to be 1. A statement whose truth set contains all logically possible results is a logically true statement. Therefore, *the probability of a logically true statement is 1.*

As another special situation, suppose the truth set of the given statement consists of only one outcome. Then the "sum" of the measures assigned to the members of the truth set is interpreted simply as the measure for the lone element in the subset.

One other particular case is worthy of mention. The truth set for a given statement may be the empty set. That is, the statement may be logically false; in other words, the statement is false for every logically possible outcome. In this situation, the "sum" of the measures assigned to the members of the truth set is interpreted to be the number 0. With this understanding, *the probability of a logically false statement is 0.*

In general, if p is any statement pertaining to the outcome of the given experiment, then the probability of p is a number. We may denote this number by $\Pr[p]$. Unless p is logically false, $\Pr[p]$ is the sum of finitely many positive numbers. Hence, in every case, $\Pr[p] \geq 0$. On the other hand, since the total measure assigned to all logical possibilities is 1, the sum of the measures assigned to those outcomes belonging to a subset cannot exceed 1. That is, $\Pr[p] \leq 1$.

5 The connective NOT

In Chapter 5 we studied several logical connectives. We stressed that the truth value for a compound statement is determined by the truth values for the simple statements from which the compound statement is built. We now inquire how the probability of a compound statement is related to the probabilities of its simple statements.

The easiest connective to analyze is NOT.

Theorem 1 For any statement p,
$$\Pr[\sim p] = 1 - \Pr[p].$$

In order to justify this assertion, let P be the truth set for p. Then the truth set for $\sim p$, the negation of p, is $\tilde{P}$, the complement of P. Every possible outcome belongs to exactly one of the two subsets, $\tilde{P}$ and P. The sum of the measures for elements in P is $\Pr[p]$; the sum of the measures for elements in $\tilde{P}$ is $\Pr[\sim p]$; the sum of all the measures is 1. Thus,
$$\Pr[p] + \Pr[\sim p] = 1.$$
This equation yields the desired result that
$$\Pr[\sim p] = 1 - \Pr[p].$$
We also observe that
$$\Pr[p] = 1 - \Pr[\sim p].$$
The latter formula is especially useful in certain computational problems.

Example 6 A true coin is tossed four times in succession. What is the probability that a tail appears at least once?

The experiment is a sequence of four simple activities, each activity being a single toss of the coin. Since each simple activity in the sequence has, after the tosses preceding it, two possible results, HEAD or TAIL, there are $2 \cdot 2 \cdot 2 \cdot 2 = 16$ possible outcomes for the experiment. Each of these 16 outcomes may be represented by an ordered quadruple, every component of which is either HEAD or TAIL. For example, one outcome is (HEAD, HEAD, TAIL, HEAD). We assign the number $\frac{1}{16}$ as a measure to each outcome.

The statement under consideration involves the phrase "at least once." The truth set for the statement contains every quadruple that has one component TAIL, or two components TAIL, or three components TAIL, or all four components TAIL. One method for finding how many elements belong to the truth set is to count (using the procedures of Chapter 6) the number of ordered quadruples of each of these classifications and to add; we obtain $4 + 6 + 4 + 1 = 15$. The desired probability is $\frac{15}{16}$.

A computationally simpler approach, however, is to note that the negation of the given statement is "No tail appears." This negation is logically equivalent to "Every toss yields a head." The truth set for the negation then consists of one ordered quadruple, namely, (HEAD, HEAD, HEAD, HEAD). The probability for the negation of the original statement is $\frac{1}{16}$. The probability of the given statement is $1 - \frac{1}{16} = \frac{15}{16}$.

Example 7 What is the probability that a poker hand, which has been fairly dealt from an ordinary deck of playing cards, contains at least one red face card?

The experiment consists in selecting a subset of five cards from the deck. The number of possible outcomes is given by a binomial coefficient, namely, $\binom{52}{5}$. We assign to each possible hand the same number, $\dfrac{1}{\binom{52}{5}}$, as its measure.

If p denotes the statement that the poker hand contains at least one red face card, then $\sim p$ is the statement that the hand has no red face card. There are in the deck six red face cards: the kings, queens, and jacks of hearts and of diamonds. There are, therefore, 46 other cards. The truth set for $\sim p$ consists of all hands that are contained in the subset composed of the 46 other cards. The number of possible hands contained in this subset is $\binom{46}{5}$. Consequently, the probability of the negation of p is $\binom{46}{5}/\binom{52}{5}$. Hence,

$$\Pr[p] = 1 - \frac{\binom{46}{5}}{\binom{52}{5}}.$$

In case we wish to express the probability as a simple fraction, we note that

$$\binom{46}{5} \Big/ \binom{52}{5} = \frac{46 \cdot 45 \cdot 44 \cdot 43 \cdot 42}{5!} \Big/ \frac{52 \cdot 51 \cdot 50 \cdot 49 \cdot 48}{5!}$$

$$= \frac{46 \cdot 45 \cdot 44 \cdot 43 \cdot 42}{52 \cdot 51 \cdot 50 \cdot 49 \cdot 48},$$

which (after suppression of common factors in the numerator and denominator) is 32637/61880. Thus,

$$\Pr[p] = 1 - \frac{32637}{61880} = \frac{29243}{61880}.$$

As a decimal, the probability is approximately 0.47. The chances are a little less than even that a poker hand, as dealt, has one or more red face cards.

An attempt to evaluate $\Pr[p]$ directly, without consideration of the statement $\sim p$, would be considerably more involved arithmetically in this example. The reader should outline the steps required for a direct computation and contrast the amount of labor with that of the above method.

Examples 6 and 7 illustrate that the probability of a statement may be determined indirectly—namely, by finding the probability of the negation of the statement and subtracting the result from 1. This indirect method is sometimes efficient. Of course, its use should be restricted to situations in which it saves effort. Often—but not always—such situations are suggested by a phrase like "at least" or "at most" in the wording of the problem.

In summary, for each statement p, $\Pr[p] + \Pr[\sim p] = 1$.

EXERCISES B

1. From an ordinary deck two cards are drawn in succession without replacement. Let each possible outcome be represented by an ordered pair, in which the left component is the first card drawn and the right component is the second card drawn. Find the probability of each of the following.
   (a) The first card is a diamond.
   (b) The second card is a diamond.
   (c) At least one of the cards is a diamond.
   (d) $\{(x, y) \mid x \text{ is a king}\}$.
   (e) The first card is a king, and the second card is a black queen.
   (f) The first card is a king, and the second card is a black king.
   (g) A black king appears first and is followed by a king.
   (h) Neither of the two cards is a red face card.
   (i) One of the cards is the heart ace.

2. In an ordinary roll of a pair of dice, let X be the total number of spots shown. Find the probability of each of the following statements.
   (a) $X = 3$.   (b) $(X - 8)(X - 11) = 0$.   (c) $X > 9$.   (d) $4 < X \leq 8$.

3. An urn contains ten balls, alike except for color; two of the balls are blue, three are yellow, and five are red. One ball is drawn and replaced; one ball is drawn and replaced; for the third time, one ball is drawn and replaced. Let each possible outcome be represented by an ordered triple, in which the left, middle, right components identify respectively the first, second, third balls drawn.
   (a) Show that the number of possible outcomes is 1000.
   (b) How many of these outcomes belong to the truth set of the statement that the second ball is yellow but neither of the other two is yellow?
   (c) What is the probability that all three balls drawn are yellow?
   (d) What is the probability that each of the yellow balls is drawn?
   (e) What is the probability of an ordered triple of the type (blue, red, yellow)?
   (f) What is the probability of an ordered triple of the type (red, yellow, blue)?
   (g) What is the probability that no two balls drawn are the same color?
   (h) What is the probability that at least one of the balls drawn is blue?
   (i) What is the probability that at most two of the balls drawn are red?
   (j) What is the probability that both of the blue balls are drawn?

4. A fair lottery has ten tickets, of which two are worth prizes. A person holds t tickets in the lottery. What is the probability that he wins at least one prize in each of the following situations?
   (a) $t = 1$.   (b) $t = 5$.   (c) $t = 2$.   (d) $t = 9$.   (e) $t = 8$.   (f) $t = 0$.

5. A factory packages 500 manufactured parts of the same type in a box. Suppose that 12 of the items are faulty. An inspector selects 20 items from the box as a sample for testing.
   (a) How many different samples are possible?
   (b) What probability measure may be reasonably assigned to each of the possible samples?
   (c) How many samples contain no faulty part?
   (d) How many samples contain all 12 of the faulty items?
   (e) What is the probability that the chosen sample has no faulty part in it?
   (f) What is the probability that the sample has (exactly) two faulty items?
   (g) What is the probability that there is at least one faulty item in the sample?
   (h) What is the probability that more than half of the items in the sample are faulty?

6. A factory packages 100 manufactured parts of the same type in a carton. Suppose that three of the items are faulty. A prospective buyer selects two items from the carton as a sample for testing.
   (a) How many different samples are possible?
   (b) What probability measure may be reasonably assigned to each of the possible samples?
   (c) How many samples contain no faulty part?
   (d) How many samples contain all three of the faulty items?
   (e) What is the probability that the chosen sample has no faulty part in it?
   (f) What is the probability that the sample has two faulty items?
   (g) What is the probability that there is at least one faulty item in the sample?

7. A company packages 30 cartons of material. In each of six cartons a special prize is enclosed. The firm ships ten cartons to each of three customers, Adams, Baker, and Clark.
   (a) Find the number of logically possible outcomes for the distribution of cartons.
   (b) What is the probability that Adams receives all of the prizes?
   (c) What is the probability that two customers do not receive any prize?
   (d) What is the probability that exactly one customer receives three cartons containing prizes?
   (e) What is the probability that each of the customers receives two prizes?

8. Let a, b, c, d be natural numbers such that $a < b < d$ and $a < c < d + a - b$. Suppose that d families live in a certain community and that c families in the community have more than one automobile. Let b families in the community be chosen by lot. What is the probability that a families among those chosen are multicar families?

6 Equiprobable measure

Many of our illustrations thus far have involved the assigning of an *equiprobable measure*. Let us examine in more detail this case which occurs so frequently. Suppose that the experiment under consideration has n logically possible outcomes. Suppose further that our experience and/or our sense of symmetry and/or our impression of lack of bias justify us in assigning the same measure to each outcome. If we denote this common measure by m, then the sum $m + m + \cdots + m = nm$ of the measures assigned to all the outcomes is 1. From the equation $nm = 1$, we find that each outcome is assigned the measure $1/n$. This result agrees with the measure $\frac{1}{2}$ for each of the two outcomes in the toss of a true coin, the measure $\frac{1}{36}$ for each of the 36 results in the roll of a pair of dice, the measure $\frac{1}{52}$ for each of the 52 possible cases in the draw of a card from an ordinary deck.

Now suppose we are interested in a certain statement p pertaining to the outcome of the above experiment. The probability of p depends upon its truth set. Since all the members of the truth set have the same measure $1/n$, the sum of the measures may be easily calculated as soon as we know how many outcomes belong to the truth set for p. If the number of outcomes that make p true is f, then the probability of p is $1/n + 1/n + \cdots + 1/n = f/n$.

For the case of equiprobable measure, the probability of a statement is the quotient obtained by dividing the number of elements in the truth set for the statement by the number of all possible outcomes for the experiment. The number f has sometimes been called the number of favorable cases, an outcome being "favorable" provided it belongs to the truth set of the statement.

Whenever we speak of a selection being made "at random," we will understand this phrase to mean that the equiprobable measure is assigned in the experiment.

In conclusion, note that the formula $\Pr[p] = f/n$ applies only in the important special case of equiprobable measure, but is not applicable in general.

7 Logically inconsistent pair of statements

We now begin our investigation of the relationship between probability and the ideas of the disjunction and the conjunction of statements.

A pair of statements pertaining to a certain activity is called a *logically inconsistent* pair provided the conjunction of the two statements is logically false.

In other words, a pair of statements is logically inconsistent if and only if there is no logically possible case in which both statements are true. In still other words, the criterion for a logically inconsistent pair of statements is that the truth sets for the respective statements be disjoint, that is, have an empty intersection.

> **Theorem 2** Let p and q be two statements pertaining to the outcome of a certain experiment. If the pair (p, q) is a logically inconsistent pair, then the probability of the disjunction $p \lor q$ is the sum of the probabilities of the two given statements.
>
> In symbols,
>
> $\Pr[p \lor q] = \Pr[p] + \Pr[q]$ if (p, q) is a logically inconsistent pair.

The verification of this important principle resembles the proof (see Section 5) of Theorem 1 concerning $\Pr[\sim p]$. Let T be the set of all possible outcomes for the given experiment; let the subsets P and Q be the respective truth sets for the statements p and q. By hypothesis, $P \cap Q$ is empty. (See Fig. 1.)

Figure 1

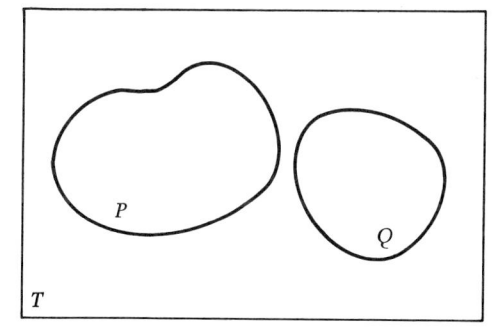

The truth set for the disjunction $p \lor q$ is the union $P \cup Q$. Every element in $P \cup Q$ belongs to exactly one of the two subsets P, Q because no element belongs to $P \cap Q$. The sum of the measures assigned to elements of $P \cup Q$ is therefore obtainable by adding the sum of the measures for members in P and the sum of the measures for outcomes in Q. In other words, $\Pr[p \lor q] = \Pr[p] + \Pr[q]$ in case $p \land q$ is logically false.

> **Example 8** Consider a single roll of a pair of dice. The probability that the total number of spots shown is 7 (see Example 4 in Section 3) is given by $\Pr[\text{SEVEN}] = \frac{6}{36}$. The corresponding probability for a total of 11 is $\Pr[\text{ELEVEN}] = \frac{2}{36}$. It is, of course, impossible for the total to be both 7 and 11 on a single roll; in other words, the above principle is applicable. We find that $\Pr[\text{SEVEN} \lor \text{ELEVEN}] = \frac{6}{36} + \frac{2}{36} = \frac{2}{9}$. The probability is $\frac{2}{9}$ that on a single roll, the total number of spots is either 7 or 11.

Example 9 In one urn there are two balls, alike except for color; one is black and the other is white. Another urn contains three balls, distinguishable only by color; one is black, one is red, and one is white. One of the two urns is selected at random. From the chosen urn one ball is drawn at random. What is the probability that the ball drawn is white?

The experiment consists of a sequence of two activities, the selection of the urn and the drawing of the ball. An outcome of the experiment may be conveniently identified by an ordered pair in which the left component names the urn selected and the right component tells the color of the ball drawn. If the urns mentioned in the problem are called first and second, respectively, then the five possible outcomes are the following ordered pairs: (FIRST, BLACK), (FIRST, WHITE), (SECOND, BLACK), (SECOND, RED), (SECOND, WHITE).

Since the urn is selected at random, each urn has probability $\frac{1}{2}$ of being selected. The truth set for the statement that the first urn is selected is {(FIRST, BLACK), (FIRST, WHITE)}. The sum of the measures assigned to the two outcomes in this truth set should be $\frac{1}{2}$. Since the ball is drawn at random from the chosen urn, the two outcomes in this truth set should have equal measures. In order for these two measures to be the same and to have sum $\frac{1}{2}$, each of them should be $\frac{1}{4}$.

The set {(SECOND, BLACK), (SECOND, RED), (SECOND, WHITE)} is the truth set for the statement that the selected urn is the second urn. The total measure assigned to the three outcomes in this truth set should be $\frac{1}{2}$. Furthermore, these three outcomes should have equal measures, since the ball is drawn from the chosen urn "at random." Thus, the measure assigned to each of these three outcomes should be $\frac{1}{6}$.

In summary, we list the possible outcomes and the measure assigned to each in Table 3. For the statement under consideration, namely,

Table 3

Outcome	(FIRST, BLACK)	(FIRST, WHITE)	(SECOND, BLACK)	(SECOND, RED)	(SECOND, WHITE)
Measure	$\frac{1}{4}$	$\frac{1}{4}$	$\frac{1}{6}$	$\frac{1}{6}$	$\frac{1}{6}$

"The ball drawn is white," the truth set is {(FIRST, WHITE), (SECOND, WHITE)}. The probability for the statement is the sum $\frac{1}{4} + \frac{1}{6} = \frac{5}{12}$.

This problem may be viewed as an application of our study of the probability of a disjunction. Suppose we let p be the statement that "The first urn is chosen and the color of the ball drawn is white." Suppose we let q be the statement that "The second urn is chosen and the color of the ball drawn is white." Then the statement whose probability is requested, namely, "The ball drawn is white," is logically equivalent to the disjunction $p \lor q$. Furthermore, the two statements p and q form a logically inconsistent pair. Since $\Pr[p] = \frac{1}{4}$ and $\Pr[q] = \frac{1}{6}$, we obtain $\Pr[p \vee q] = \frac{1}{4} + \frac{1}{6} = \frac{5}{12}$.

We observe, finally, that altogether there are five balls in the urns and two of them are white. Nevertheless, the probability of drawing a white ball is not $\frac{2}{5}$. The reason is that the description of the problem does not permit us to assign equal measures to the drawings of the five individual balls. Such an assignment would violate the requirement that the urn be chosen at random.

The preceding principle may be extended to a disjunction of more than two statements. If a finite collection of statements has the property that every two of them form a logically inconsistent pair, then the probability of the disjunction of all of them is the sum of the probabilities of the given statements. Thus, $\Pr[p \vee q \vee r] = \Pr[p] + \Pr[q] + \Pr[r]$, if each of the pairs (p, q), (p, r), (q, r) is logically inconsistent.

EXERCISES C

1. In a certain game played with an ordinary deck of playing cards, a hand consists of four different cards. Let X, Y, Z be the respective numbers of club cards, of red cards, and of aces in the hand. Which of the following are logically inconsistent pairs of statements?
   (a) $(X = 3, Y = 2)$.   (b) $(Y = 2, Z = 3)$.   (c) $(Y = 1, Z = 4)$.
   (d) $(X = 3, X > 2)$.   (e) $(X = 3, Z = 3)$.

2. If (p, q) is a logically inconsistent pair of statements pertaining to the outcome of an experiment, evaluate $\Pr[p \wedge q]$.

3. A tall urn contains nine balls, of which two are red, three are white, and four are blue. A short urn contains six balls, of which two are red, three are white, and one is blue. An urn is chosen at random. From the selected urn one ball is drawn at random. Find the probability of each of the following statements.
   (a) The tall urn is chosen.
   (b) The short urn is chosen and a white ball is drawn.
   (c) A red ball is drawn from the tall urn.
   (d) The color of the ball drawn is white.
   (e) The ball drawn is not red.
   (f) The ball drawn is not a blue ball from the short urn.

4. Three coin boxes are arranged in a stack. The top box contains a dime and a quarter; the middle box contains a nickel, a quarter, and a half dollar; the bottom box contains a nickel, a dime, a quarter, and a half dollar. One box is selected at random, and one coin is picked at random from the chosen box. Find the probability of each of the following statements.
   (a) The bottom box is selected.
   (b) A dime is picked from the top box.
   (c) A half dollar is picked.
   (d) A quarter is picked.
   (e) The coin picked is worth less than 15 cents.
   (f) The coin picked is not the nickel in the bottom box.

5. Let (p, q) be a pair of statements pertaining to the outcome of an experiment.
   (a) Show that $(q, \sim q)$ is a logically inconsistent pair.
   (b) Show that $(p \wedge q, p \wedge \sim q)$ is a logically inconsistent pair.
   (c) Under what circumstances is $(p \vee q, p \vee \sim q)$ a logically inconsistent pair?
   (d) If (p, q) is a logically inconsistent pair, under what circumstances is $(\sim p, q)$ also a logically inconsistent pair?

8 The connective OR

Let p and q be any two statements pertaining to the outcome of a certain experiment. As before, let T be the set of all possible outcomes for the experiment, and let P and Q be the truth sets for p and q, respectively. (See Fig. 2a.) The truth set for the disjunction $p \vee q$ is the union $P \cup Q$. The set $P \cup Q$ is expressible as the union of three subsets that are mutually disjoint.

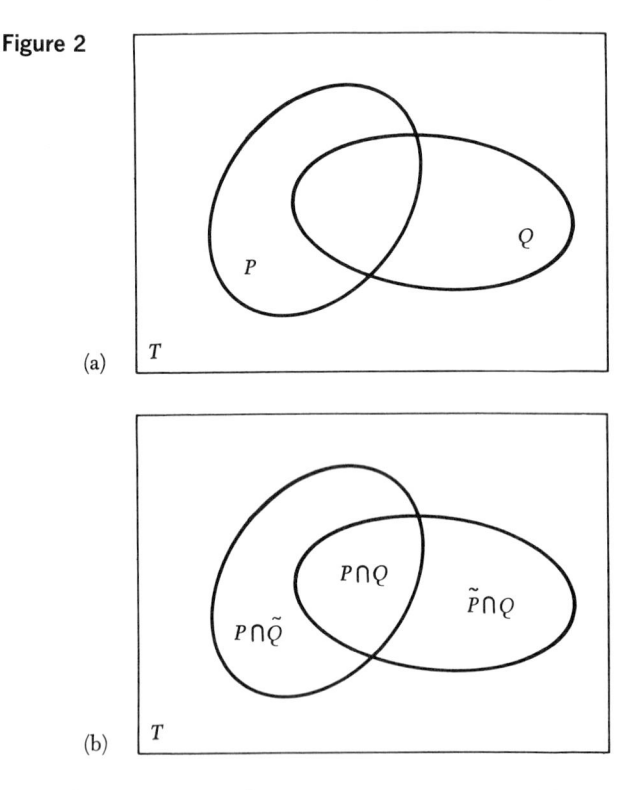

Figure 2

(a)

(b)

(See Fig. 2b.) They are $P \cap \tilde{Q}$, $P \cap Q$, and $\tilde{P} \cap Q$. In the language of logic, the compound statement $p \vee q$ is logically equivalent to the disjunction of the three statements $p \wedge \sim q$, $p \wedge q$, and $\sim p \wedge q$; each two of these three statements form a logically inconsistent pair. Hence,

$$\Pr[p \vee q] = \Pr[p \wedge \sim q] + \Pr[p \wedge q] + \Pr[\sim p \wedge q].$$

To each member of the equation we add the number $\Pr[p \wedge q]$. We obtain
$$\Pr[p \vee q] + \Pr[p \wedge q]$$
$$= \Pr[p \wedge \sim q] + \Pr[p \wedge q] + \Pr[\sim p \wedge q] + \Pr[p \wedge q].$$
The sum of the first two terms in the right member of this equality is $\Pr[p \wedge \sim q] + \Pr[p \wedge q] = \Pr[p]$, because the pair $(p \wedge \sim q, p \wedge q)$ is a logically inconsistent pair of statements whose disjunction is logically equivalent to the simple statement p. Likewise, the sum of the last two terms in the equation is $\Pr[q]$, because q is equivalent to the disjunction of the statements in the inconsistent pair $(\sim p \wedge q, p \wedge q)$. With these simplifications, we obtain
$$\Pr[p \vee q] + \Pr[p \wedge q] = \Pr[p] + \Pr[q].$$
This result may be expressed in the useful form
$$\Pr[p \vee q] = \Pr[p] + \Pr[q] - \Pr[p \wedge q].$$

Theorem 3 For any two statements pertaining to the outcome of an experiment, the sum of their probabilities is the sum of the probability of their disjunction and the probability of their conjunction.

Or, the probability of the disjunction of any two statements is obtainable by adding the probabilities of the individual statements and then subtracting the probability of their conjunction.

Earlier, we treated the special case in which the statements p, q form an inconsistent pair. In that situation the conjunction $p \wedge q$ is logically false and consequently has probability zero. Thus,
$$\Pr[p \vee q] = \Pr[p] + \Pr[q] - \Pr[p \wedge q]$$
$$= \Pr[p] + \Pr[q] - 0$$
$$= \Pr[p] + \Pr[q],$$
in agreement with the result obtained for the special case.

Example 10 One card is drawn from an ordinary deck of playing cards. What is the probability that the card drawn is a diamond or a face card?

Let the two statements that the card is a diamond and that it is a face card be denoted by d and f, respectively. We are asked to find $\Pr[d \vee f]$. For a fair drawing we choose the equiprobable measure, assigning the number $\frac{1}{52}$ to each outcome. Since the deck contains 13 diamond cards and 12 face cards, $\Pr[d] = \frac{13}{52}$ and $\Pr[f] = \frac{12}{52}$. The truth set for the conjunction $d \wedge f$ is {DIAMOND KING, DIAMOND QUEEN, DIAMOND JACK}. Hence, $\Pr[d \wedge f] = \frac{3}{52}$. Consequently,
$$\Pr[d \vee f] = \Pr[d] + \Pr[f] - \Pr[d \wedge f]$$
$$= \tfrac{13}{52} + \tfrac{12}{52} - \tfrac{3}{52}$$
$$= \tfrac{11}{26}.$$

Example 11 A lot of 12 items manufactured by a machine is packed in a box. Among the items in the lot, three are defective. A sample of two items is chosen at random from the lot. Let X be the number of defective items in the sample. List the logically possible values for X, and find for each possibility the probability that X has that value.

Since there are more defective items in the lot than there are items in the sample, the number of defective items in the sample may be any number up to the size of the sample. That is, the possible values of X are 0, 1, or 2.

Now, in the present context, the word "sample" means "subset." The experiment consists in selecting a subset with two members from the set (or "lot") with 12 members. The number of possible outcomes is the number of such subsets, namely, $\binom{12}{2}$. Since the sample is chosen at random, each of the $\binom{12}{2} = 66$ samples has the same probability of being picked, namely, $\frac{1}{66}$.

The statement $X = 0$ has a truth set consisting of those samples with no defective item. Such a sample is obtainable by choosing a subset with two members from the set formed by the nine nondefective items in the lot. The number of such samples is $\binom{9}{2} = 36$. Hence, the probability that $X = 0$ is $\frac{36}{66} = \frac{6}{11}$.

Consider now the truth set for the statement $X = 1$. A sample belongs to this truth set if it contains one defective item and (therefore) one good item. Such a sample arises if a subset of one element is chosen from the set of three defectives and if a subset of one member is picked from the set of nine nondefectives. The number of such samples is the product $\binom{3}{1}\binom{9}{1} = 3 \cdot 9 = 27$. Hence, the probability that $X = 1$ is $\frac{27}{66} = \frac{9}{22}$.

The truth set for the statement $X = 2$ consists of those samples which are contained in the set of defectives. In a set with three members the number of subsets with two elements is $\binom{3}{2} = 3$. Consequently, $\Pr[X = 2] = \frac{3}{66} = \frac{1}{22}$.

We may summarize our findings in a table listing the possible values k for X and their respective probabilities (Table 4). As a computational check, we note that $\frac{12}{22} + \frac{9}{22} + \frac{1}{22} = 1$. Since each of the 66 possible outcomes is accounted for exactly once in Table 4, the total of the probabilities listed in the table should be one.

Table 4

k	0	1	2
$\Pr[X = k]$	$\frac{12}{22}$	$\frac{9}{22}$	$\frac{1}{22}$

Probability

Example 12 Let X be the number of black aces in a poker hand as dealt. Make a table showing the possible values for X and the probability of each value.

This problem bears a strong resemblance to the preceding example, even though the language is different. We have initially a set: in one case, a lot of manufactured items; in the other case, a deck of cards. Some of the members of the set have a distinguishing characteristic: in one case, they are defective; in the other case, they are black aces. From the set we select at random a subset: in one case, a sample; in the other case, a poker hand. In each case, we let X be the number of distinguished elements contained in the chosen subset, and we ask for full information about the probabilities associated with X.

In the present example, the experiment consists in the deal of a poker hand, that is, a selection of five cards from the 52-card pack. The number of possible outcomes is $\binom{52}{5}$ and each is assigned the same measure $\dfrac{1}{\binom{52}{5}}$.

Since the number five of cards in a poker hand exceeds the number two of black aces in the deck, the number of distinguished cards in the hand cannot be more than the number of black aces in the entire deck. Thus, $X = 0$ or $X = 1$ or $X = 2$.

The truth set for $X = 0$ consists of those hands contained in the set of 50 cards other than black aces. There are $\binom{50}{5}$ such hands.

A hand satisfies the condition $X = 1$ if it is obtainable by choosing one black ace from a set of two members and by selecting four elements from the complementary set of 50 cards which are not distinguished. There are $\binom{2}{1}\binom{50}{4}$ such hands.

The truth set for $X = 2$ has $\binom{2}{2}\binom{50}{3}$ hands.

Table 5

k	0	1	2
$\Pr[X = k]$	$\dfrac{\binom{50}{5}}{\binom{52}{5}} = \dfrac{1081}{1326}$	$\dfrac{\binom{2}{1}\binom{50}{4}}{\binom{52}{5}} = \dfrac{235}{1326}$	$\dfrac{\binom{2}{2}\binom{50}{3}}{\binom{52}{5}} = \dfrac{10}{1326}$

In Table 5 are listed the three possible values k for X and the respective probabilities. As a partial check, we note that the sum

$$\Pr[X = 0] + \Pr[X = 1] + \Pr[X = 2] = \frac{1081}{1326} + \frac{235}{1326} + \frac{10}{1326} = 1.$$

Decimal approximations for the respective probabilities are 0.815, 0.177, 0.008. The "odds" against being dealt a black ace in a poker hand are roughly 4 to 1.

EXERCISES D

1. A bowl contains five tickets bearing the respective numbers 1, 2, 3, 4, 5. An experiment consists in selecting at random from the bowl three tickets together.
   (a) What measure is assigned to each of the possible outcomes in this experiment?
   (b) Let X be the sum of the numbers on the three selected tickets. Make a table, listing all the possible values for X and the probability for each value of X.
   (c) Find the probability that $X \in \{7, 9, 11\}$.
   (d) Find the probability that $X > 9$.
   (e) Find $\Pr[(X - 11)(X - 12) \neq 0]$.

2. For the experiment described in Exercise 1, let Y be the smallest of the numbers on the three selected tickets. Find the probability of each of the following.
   (a) $Y = 2$. (b) $Y \geq 3$. (c) $(X = 8) \wedge (Y = 1)$.
   (d) $(X = 8) \vee (Y = 1)$. (e) $(X, Y) = (9, 2)$. (f) $(X = 9) \vee (Y = 2)$.

3. For the experiment described in Exercise 1 and discussed in Exercise 2, let Z be the middle of the numbers on the three selected tickets. (By the "middle" we mean the number among the three which is neither the smallest nor the largest.) Find the probability of each of the following.
   (a) $Z = 3$. (b) $(Z = 3) \wedge (Y = 1)$.
   (c) $(Y = 1) \vee (Z = 3)$. (d) $(X = 10) \vee (Z = 4)$.
   (e) $(X = 10) \vee (Y = 2) \vee (Z = 3)$. (f) $(Y, Z) = (3, 2)$.

4. A red bowl contains three white balls and two red balls. A white bowl contains one white ball and two red balls. An experiment consists of two successive drawings. First, a ball is drawn at random from the red bowl and replaced. Second, a ball is drawn at random from the bowl which is colored the same as the first ball drawn. Represent each outcome of the experiment by an ordered pair in which the first component identifies the first ball drawn and the second component tells the second ball drawn.
   (a) Show that the number of possible outcomes is 19.
   (b) What total measure should be assigned to the outcomes belonging to the truth set of the statement that the first ball drawn is white?
   (c) How many outcomes belong to the truth set mentioned in part (b)?
   (d) What measure is assigned to an outcome of the type (white, red)?
   (e) What measure is assigned to an outcome of the type (red, white)?
   (f) How many outcomes belong to the truth set of the statement that the second ball drawn is white?
   (g) What is the probability that the second ball drawn is white?

9 Illustrations

Example 13 Suppose that a poker hand (as dealt) contains X diamond cards and Y heart cards. Evaluate the probability of the disjunction $(X = 1) \vee (Y = 1)$.

One approach to the problem is to find $\Pr[X = 1]$, find $\Pr[Y = 1]$, find $\Pr[(X = 1) \wedge (Y = 1)]$, and then use the formula $\Pr[X = 1] + \Pr[Y = 1] - \Pr[(X = 1) \wedge (Y = 1)]$ to calculate the desired probability. A poker hand with one diamond and four cards from the other suits is obtainable by selecting one element from a set with 13 members and by choosing four elements from a set with 39 members. The number of such poker hands is $\binom{13}{1}\binom{39}{4}$. Hence,

$$\Pr[X = 1] = 13 \cdot \binom{39}{4} \Big/ \binom{52}{5} = \frac{27417}{66640}.$$

Since the roles of the diamond suit and of the heart suit in a deck are symmetric, $\Pr[Y = 1]$ is the same as $\Pr[X = 1]$.

The truth set for the conjunction $(X, Y) = (1, 1)$ consists of hands each containing one diamond, one heart, and three black cards. The number of such hands is $\binom{13}{1}\binom{13}{1}\binom{26}{3}$ and, hence,

$$\Pr[(X, Y) = (1, 1)] = 13 \cdot 13 \cdot \binom{26}{3} \Big/ \binom{52}{5} = \frac{845}{4998}.$$

Finally,

$$\Pr[(X = 1) \vee (Y = 1)] = \frac{27417}{66640} + \frac{27417}{66640} - \frac{845}{4998} = \frac{65351}{99960}.$$

Example 14 A red die and a white die are rolled together. In the notation of Example 4 (in Section 3), find the probability of each of the following statements: (a) $r < w$; (b) $(r < w) \vee (r = 4)$.

The set of 36 logically possible outcomes has a partition consisting of three subsets, namely, the truth set for $r = w$, the truth set for $r < w$, and the truth set for $r > w$. There are six elements in the truth set for $r = w$, and the remaining 30 outcomes satisfy the condition $r \neq w$. By virtue of symmetry, the truth set for $r < w$ has as many members as the truth set for $r > w$. Each of these two cells in the partition contains $\frac{1}{2} \cdot 30 = 15$ outcomes. Hence, $\Pr[r < w] = \frac{15}{36} = \frac{5}{12}$.

If $r = 4$, there are six possible outcomes for the white die. Thus, the truth set for $r = 4$ has six elements, and $\Pr[r = 4] = \frac{6}{36} = \frac{1}{6}$.

The conjunction $(r < w) \wedge (r = 4)$ is true in case r is four and w is either five or six; that is, the truth set is $\{(4, 5), (4, 6)\}$. The probability of the conjunction is $\frac{1}{18}$.

Finally, $\Pr[(r < w) \vee (r = 4)] = \frac{5}{12} + \frac{1}{6} - \frac{1}{18} = \frac{19}{36}$.

Example 15 Suppose we roll a single die in a fair attempt to show an ace. In case we fail, we keep trying. We stop as soon as an ace appears. If we suffer four successive failures, we give up the attempt. Let Z be the number of rolls in the experiment. Make a table, listing the possible values of Z and the probability of each.

Each roll of the die is a trial, an attempt to achieve a certain goal, namely, obtaining an ace. Each trial results in either success or failure. One possible outcome for the experiment is (3, 6, 3, 1), a notation representing four attempts in which the numbers of spots appearing on successive rolls are three, six, three, one. This outcome is of the type (failure, failure, failure, success). Other possible outcomes are (2, 1), (3, 1), (4, 1), (5, 1), (6, 1); each of these is of the type (failure, success). For each of these five outcomes, the value of Z is two; furthermore, these are the only possible cases in which the number of rolls is two.

Every possible outcome is of one of the following types: (success), (failure, success), (failure, failure, success), (failure, failure, failure, success), or (failure, failure, failure, failure). Indeed, we recall that achieving a success stops the procedure and that we give up after four failures.

The only outcome of the type (success) is (1). The probability of an ace on the first roll is $\frac{1}{6}$. This is the only outcome for which $Z = 1$; hence, $\Pr[Z = 1] = \frac{1}{6}$.

We have previously listed the outcomes for which $Z = 2$. Since there are five ordered pairs and each has measure $\frac{1}{36}$, we conclude that $\Pr[Z = 2] = \frac{5}{36}$.

The truth set for the statement $Z = 3$ consists of all ordered triples of the type (failure, failure, success). Since five faces of the die yield failure while only one gives success, the number of ordered triples of type (failure, failure, success) is $5 \cdot 5 \cdot 1 = 25$. Since each ordered triple has measure $1/(6 \cdot 6 \cdot 6)$, we find that $\Pr[Z = 3] = 25/216$.

An outcome of either of the types (failure, failure, failure, success) or (failure, failure, failure, failure) satisfies the condition $Z = 4$. The number of ordered quadruples in which each of the first three components is not an ace is $5 \cdot 5 \cdot 5 \cdot 6$; note the factor 6 appears in the product because any number of spots may be a fourth component. Since there are 6^4 ordered quadruples, we conclude that $\Pr[Z = 4] = 5 \cdot 5 \cdot 5 \cdot 6/6^4 = 125/216$.

Table 6

h	1	2	3	4
$\Pr[Z = h]$	$\frac{1}{6}$	$\frac{5}{36}$	$\frac{25}{216}$	$\frac{125}{216}$

The desired result is Table 6. As a partial check, we note that

$$\sum_{h=1}^{4} \Pr[Z = h] = \Pr[Z = 1] + \Pr[Z = 2] + \Pr[Z = 3] + \Pr[Z = 4]$$
$$= \frac{1}{6} + \frac{5}{36} + \frac{25}{216} + \frac{125}{216} = 1.$$

Example 16 From an ordinary deck, remove and lay aside all cards except the kings, the queens, and the jacks. Consider an experiment involving the "face deck" consisting of the 12 face cards. Let each king and each queen be called a "royal card." From the face deck draw at random one card, in an attempt to secure a royal card. In case of failure, do not replace any previously drawn card and draw at random another card. Repeat as often as necessary, until a royal card is drawn. Let W be the number of draws required to obtain a royal card. Find the probabilities for the various possible values of W.

This problem has many resemblances to the preceding example, but there are also some marked contrasts, as we shall see. Each draw is a trial that may result in success or in failure; we may use the same style of analysis as before. If $W = 1$, a royal card is drawn on the first attempt; since there are eight royal cards in the 12-card deck, $\Pr[W = 1] = \frac{8}{12} = \frac{2}{3}$.

If $W = 2$, the first card drawn is a jack and the second card drawn is a king or queen. The number of possible outcomes of this type is $4 \cdot 8$. The number of ordered pairs of distinct cards is $12 \cdot 11$; the factor 11 arises because the first card drawn is not replaced before the second drawing. Hence, $\Pr[W = 2] = (4 \cdot 8)/(12 \cdot 11) = \frac{8}{33}$.

The truth set for the statement $W = 3$ consists of all ordered triples of the type (failure, failure, success). The number of such outcomes is $4 \cdot 3 \cdot 8$, because the left component is a jack, the middle component is a different jack, and the right component is a royal card. Since the number of ordered triples of cards, with no repetition permitted, is $12 \cdot 11 \cdot 10$, we find that $\Pr[W = 3] = (4 \cdot 3 \cdot 8)/(12 \cdot 11 \cdot 10) = \frac{4}{55}$.

In order to evaluate $\Pr[W = 4]$, we consider ordered quadruples. There are $4 \cdot 3 \cdot 2 \cdot 8$ outcomes of the type (failure, failure, failure, success) and there are $12 \cdot 11 \cdot 10 \cdot 9$ ordered quadruples of four distinct cards. Thus, $\Pr[W = 4] = (4 \cdot 3 \cdot 2 \cdot 8)/(12 \cdot 11 \cdot 10 \cdot 9) = \frac{8}{495}$.

In a similar manner,

$$\Pr[W = 5] = (4 \cdot 3 \cdot 2 \cdot 1 \cdot 8)/(12 \cdot 11 \cdot 10 \cdot 9 \cdot 8) = \frac{1}{495}.$$

Since four unsuccessful trials exhaust all the cards in the face deck other than the royal cards, it is not possible for success to be delayed beyond the fifth trial. In other words, no value of W exceeds five, and our analysis is complete. Summarizing our findings, we have Table 7.

Table 7

k	1	2	3	4	5
$\Pr[W = k]$	$\frac{2}{3}$	$\frac{8}{33}$	$\frac{4}{55}$	$\frac{8}{495}$	$\frac{1}{495}$

We apply the usual check to verify that $\sum_{k=1}^{5} \Pr[W = k] = 1$.

EXERCISES E

1. A lot of eight light bulbs has three bad bulbs. A sample of four bulbs is selected at random from the lot. Let W be the number of bad bulbs in the sample. Find each of the following numbers.
   (a) $\Pr[W = 0]$.
   (b) $\Pr[W = 1]$.
   (c) $\Pr[W = 3]$.
   (d) $\Pr[W = 4]$.

2. Six cards are selected from an ordinary deck: the kings, queens, and jacks of spades and of hearts. The remainder of the deck is discarded. The six-card deck is thoroughly shuffled. Then cards are withdrawn in succession, one at a time, without replacement. Let X be the number of the draw on which the spade queen appears.
   (a) Calculate $\Pr[X = 2]$.
   (b) Find $\Pr[X = 3]$.
   (c) Evaluate $\Pr[X = 6]$.
   (d) Give a careful argument, of the common-sense type, explaining why each value of X has the same probability.

3. In the experiment described in Exercise 2, let Y be the number of the draw on which a queen appears for the first time.
   (a) What are the logically possible values for Y?
   (b) For each possible value of Y, find its probability.
   (c) Perform a partial check on your calculations by computing the sum of the probabilities you found in part (b).

4. In the experiment described in Exercise 2, let Z be the number of the draw on which a spade appears for the first time.
   (a) Make a table, listing all the possible values for Z and the probability for each.
   (b) Perform a partial check on your computations in part (a) by calculating

   $$\sum_j \Pr[Z = j].$$

5. An urn contains two white balls and three green balls. A ball is drawn at random from the urn and replaced. Again a ball is drawn and replaced. The same activity is done for the third time, and then finally for the fourth time. Let W be the number of the draw on which a white ball appears for the first time; in case no white ball appears in any of the drawings, let $W = 0$.
   (a) Make a table, showing each possible value of W and its probability.
   (b) Partially check your calculations by computing $\sum_k \Pr[W = k]$.

6. A red bowl contains three white balls and two red balls. A white bowl contains one white ball and two red balls. An experiment consists of three successive drawings. Each time a ball is drawn at random from a bowl and replaced in the same bowl before the next drawing. The first drawing is made from the red bowl. The second drawing is made from the bowl of the same color as the first ball drawn. The third drawing is made from the bowl of the same color as the second ball drawn.
   (a) Considering each outcome to be given by an ordered triple of balls, find the number of possible outcomes for this experiment.
   (b) For each possible outcome, find the measure that should be assigned to it.
   (c) What is the probability that the first ball drawn is white?
   (d) What is the probability that the second ball drawn is white?
   (e) What is the probability that the third ball drawn is white?

10 Conditional probability

Thus far we have emphasized the relationships involving probability for the logical operations of negation and disjunction. We turn now in more detail to conjunction. We shall be led to the notion of conditional probability.

We begin with a simple example.

Example 17 An experiment consists in choosing at random one letter from the set of letters in the word REGULATIONS. Let p be the statement that the chosen letter is one of the letters in the word STRING. Let q be the statement that the chosen letter appears in the word GLARE.

Since REGULATIONS has 11 letters with no repetition, each outcome has measure $\frac{1}{11}$. Consequently, $\Pr[p] = \frac{6}{11}$ and $\Pr[q] = \frac{5}{11}$. Furthermore, since $p \wedge q$ is true if and only if the chosen letter is either R or G, $\Pr[p \wedge q] = \frac{2}{11}$.

The calculation of the three probabilities in the preceding paragraph depends upon the assigned measures, namely, a measure of $\frac{1}{11}$ for each of the 11 possible outcomes. Consider now the effect that additional information may have on the problem. Suppose we are given that the chosen letter is contained in the word STRING; in other words, suppose we are given that the statement p is true. With this supplementary knowledge at our disposal, it seems reasonable that the probabilities, expressing our degrees of belief, will be different from the probabilities determined without benefit of the extra information. How can we evaluate these changes?

One approach to the question is to start afresh: assign a new measure of $\frac{1}{6}$ to each of the six outcomes in the set {S, T, R, I, N, G} and a new measure of 0 to each of the five previously possible outcomes now excluded; observe that now q is true if and only if either G or R is chosen; hence, the new probability for q is $\frac{1}{6} + \frac{1}{6} = \frac{1}{3}$. Note that the new probability for p is 1, since every outcome, now accepted as possible, belongs to the truth set for p.

However, it is desirable for us to be able to calculate probabilities based on the supplementary information without the need for reassignment of measures. Given that p is true, the only outcomes worthy of consideration are those belonging to the truth set for p. Thus, the outcomes in the truth set for q that are worthy of consideration are actually the members of the truth set for the conjunction $p \wedge q$. The new probability for q, given that p is true, is obtainable by comparing $\Pr[p \wedge q]$ and $\Pr[p]$. Indeed, this new probability, which we shall write as $\Pr[q|p]$, is $\frac{1}{3}$, and we note that

$$\frac{\Pr[p \wedge q]}{\Pr[p]} = \frac{\frac{2}{11}}{\frac{6}{11}} = \frac{1}{3}.$$

Thus, the new probability may be expressed as a quotient of probabilities based on the original information.

Our next example offers further insight into these ideas by contrasting two illustrative situations.

Example 18 From an ordinary deck of playing cards, two cards are drawn in succession. Let p be the statement that the first card drawn is a club, and let q be the statement that the second card drawn is a club. Then $p \wedge q$ of course represents the statement that both cards are clubs. Now, $\Pr[p] = \frac{1}{4}$ and $\Pr[q] = \frac{1}{4}$. We are interested in evaluating $\Pr[p \wedge q]$.

Actually, we have not yet given a complete description of the experiment, for we have not specified whether the first card drawn is or is not replaced before the drawing of the second card. The number $\Pr[p \wedge q]$ depends upon this question of replacement. We wish to compare the two situations.

First, suppose that the initially drawn card is restored to the deck before the next drawing. Then the set of possible outcomes for the experiment is the set of all ordered pairs of cards, and there are $52 \cdot 52$ ordered pairs. The truth set for $p \wedge q$ is the set of ordered pairs of club cards, and there are $13 \cdot 13$ such pairs. Hence, in this case,

$$\Pr[p \wedge q] = \frac{13 \cdot 13}{52 \cdot 52} = \frac{1}{16}.$$

Next, suppose that the second drawing is made from the incomplete deck that remains after the first card is removed. Then the set of possible outcomes is the set of all ordered pairs of *distinct* cards, and there are $52 \cdot 51$ of these pairs. The truth set for $p \wedge q$ is the set of ordered pairs of *distinct* club cards, of which there are $13 \cdot 12$. Hence, in this case,

$$\Pr[p \wedge q] = \frac{13 \cdot 12}{52 \cdot 51} = \frac{1}{17}.$$

Now, the product of $\Pr[p] = \frac{1}{4}$ and $\Pr[q] = \frac{1}{4}$ is the number $\frac{1}{16}$. This product, we observe, is different from $\Pr[p \wedge q]$ in the case of successive drawings without replacement, but it is the same as $\Pr[p \wedge q]$ in the case of drawings with replacement. In the latter case, we say that (p, q) is a pair of independent statements in the sense of probability. The former case suggests conditional probability.

Let us look further at the experiment that does not involve replacement. The truth set P for the statement p consists of those ordered pairs of the type (club card, card different from left component). There are $13 \cdot 51$ members in this set P. The truth set S for

$p \wedge q$ is a subset of P consisting of those ordered pairs of the type (club card, club card different from left component). There are $13 \cdot 12$ members in S. Now, $\Pr[p] = (13 \cdot 51)/(52 \cdot 51)$ and

$$\Pr[p \wedge q] = \frac{13 \cdot 12}{52 \cdot 51} = \frac{13 \cdot 12}{13 \cdot 51} \cdot \frac{13 \cdot 51}{52 \cdot 51} = \frac{13 \cdot 12}{13 \cdot 51} \cdot \Pr[p].$$

What interpretation can we give to the number $(13 \cdot 12)/(13 \cdot 51)$? The denominator is the number of elements in P and the numerator is the number of elements in the subset S of the set P. If we had an experiment in which P is the set of all possible outcomes, and if we assigned equal measures to all the outcomes, then $(13 \cdot 12)/(13 \cdot 51)$ would be the probability of a statement whose truth set is S. Now, an experiment in which P is the set of all possible outcomes is an experiment in which p is accepted as true. If we accept p as true, then $p \wedge q$ becomes logically equivalent to q; hence, S becomes the truth set for q. Thus, $(13 \cdot 12)/(13 \cdot 51)$ is the probability for the statement q, subject to the restriction that p is accepted as true. We denote this number by the symbol $\Pr[q|p]$ and read it as "the conditional probability of q, given p." Substituting $\Pr[q|p]$ for $(13 \cdot 12)/(13 \cdot 51)$ in the equation above, we obtain $\Pr[p \wedge q] = \Pr[q|p] \cdot \Pr[p]$. Numerically, $\Pr[q|p] = \frac{4}{17}$, and the preceding equation becomes $\frac{1}{17} = \frac{4}{17} \cdot \frac{1}{4}$.

It is important to give an interpretation to this conditional probability. To accept the truth of p means that we are supposing the first card has already been drawn and has been observed to be a club. At the time of the second drawing, therefore, the incomplete deck of 51 cards contains only 12 club cards. On this basis, the probability that the second card drawn is a club is $\frac{12}{51} = \frac{4}{17}$. This probability for q is dependent on the assumption that p is true. This is the conditional probability for q, "given p," and we note that it is not the same as the number $\Pr[q] = \frac{1}{4}$.

Finally, let us return to the experiment in which the first card drawn is replaced before the second drawing. In this situation, suppose we accept p as true. The second card is drawn from a *complete* deck, and the probability that the card is a club is $\frac{13}{52} = \frac{1}{4}$. Here, $\Pr[q|p] = \frac{1}{4}$ and is the same as $\Pr[q]$. It appears that the acceptance of p as true has no influence on the probability of q. We have a pair (p, q) of independent statements in the sense of probability. Nevertheless, we may note that $\Pr[p \wedge q] = \Pr[q|p] \cdot \Pr[p]$ in this case also, for $\frac{1}{16} = \frac{1}{4} \cdot \frac{1}{4}$.

The examples we have been discussing suggest the following definition.

Definition Consider an experiment having a set of logically possible outcomes to each of which a probability measure has been assigned. Let p and q be statements pertaining to the outcome of the experiment such that p has positive probability. Then the *conditional probability for q, given p*, denoted by $\Pr[q|p]$, is the quotient

$$\frac{\Pr[p \wedge q]}{\Pr[p]}.$$

As a consequence of the preceding definition, we have the equation

$$\Pr[p \wedge q] = \Pr[p] \cdot \Pr[q|p] \quad \text{if } \Pr[p] \neq 0.$$

Since $p \wedge q$ and $q \wedge p$ are logically equivalent, we sometimes apply the definition to write

$$\Pr[p \wedge q] = \Pr[q] \cdot \Pr[p|q] \quad \text{if } \Pr[q] \neq 0.$$

In words, *the probability of the conjunction of two statements is the product of the probability of one of the statements (if its probability is positive) and the conditional probability of the other statement, given the former statement.*

A pictorial approach may help in understanding the notion of conditional probability. Let the set of all logically possible outcomes for an experiment be denoted by T. Let P and Q be the truth sets for the respective statements p and q pertaining to the outcome. Then P and Q are subsets of T, and we may use Venn diagrams to depict the situation (Fig. 3). The four sets $P \cap Q$, $P \cap \tilde{Q}$, $\tilde{P} \cap Q$, $\tilde{P} \cap \tilde{Q}$ have the properties (as illustrated in Fig. 3c) that their union is T itself and any two of them have an empty intersection. Thus, the four statements $p \wedge q$, $p \wedge {\sim}q$, ${\sim}p \wedge q$, ${\sim}p \wedge {\sim}q$ have the properties that their disjunction is logically true and any two of them form a logically inconsistent pair. The sum of the probabilities of these four statements is equal to 1.

Figure 3

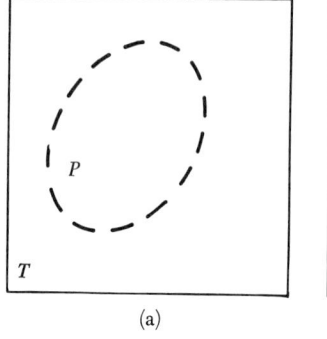

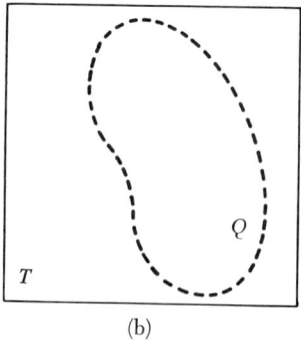

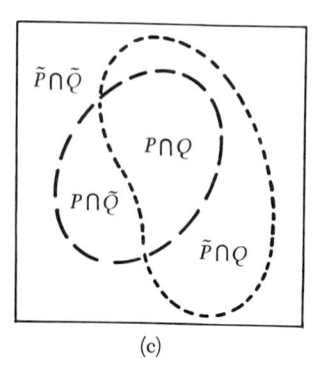

(a)   (b)   (c)

Now, acceptance of p as true may be interpreted as meaning that the set T of logically possible outcomes is reduced to the set P. In other words, outcomes belonging to $\tilde{P}$ are discarded from consideration. Modified diagrams are shown in Fig. 4, where the gray regions hint at the black-out of discarded possibilities. "Given p," the new probability for p is 1. Also, "given p," the outcomes for which q is true are those belonging to $P \cap Q$, that is, those for which $p \wedge q$

(a) (b) (c)

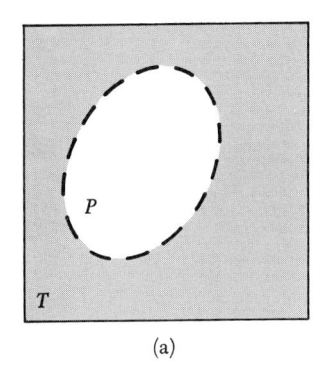

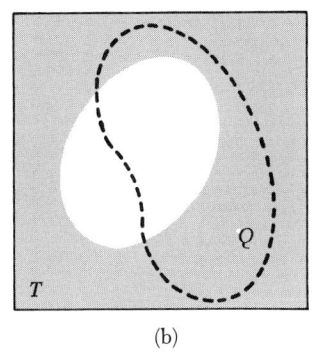

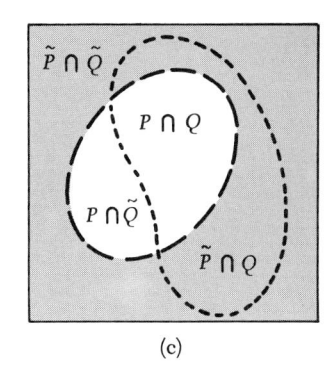

Figure 4

is true. Thus, "given p," the new probability for q may be obtained by comparing the (old) probability for $p \wedge q$ with the (old) probability for p. A reasonable method for comparing is by proportionality. The defining equation

$$\Pr[q\,|\,p] = \frac{\Pr[p \wedge q]}{\Pr[p]}$$

shows that, for pertinent statements such as q, the conditional probability of the statement, given p, is proportional to the probability of the conjunction of p and the statement. The constant of proportionality is $1/\Pr[p]$.

11 Independent pair of statements

We now formulate another idea suggested in Example 18.

> **Definition** Let p and q be statements pertaining to the outcome of an experiment for which a probability function has been assigned. The pair (p, q) is called an *independent pair*, relative to the assigned probability function, provided $\Pr[p \wedge q] = \Pr[p] \cdot \Pr[q]$.

Suppose that each of the statements p, q has positive probability. Then the equations

$$\Pr[p] \cdot \Pr[q\,|\,p] = \Pr[p \wedge q] = \Pr[q] \cdot \Pr[p\,|\,q]$$

show that the criterion for (p, q) to be an independent pair may be phrased either by the condition $\Pr[q\,|\,p] = \Pr[q]$ or by the equation $\Pr[p\,|\,q] = \Pr[p]$. In other words, *a pair of statements each of which has nonzero probability form an independent pair if and only if the probability of one of them is the same as its conditional probability, given the other.*

> **Example 19** Three balls in an urn are labeled 1, 2, 3, respectively. A ball is drawn at random from the urn, it is replaced, and

again a ball is drawn at random from the urn. Let X and Y be the numbers on the balls in the first and second drawings, respectively. Denote by p, q, r, respectively, the following three statements: $X + Y$ is an even number; $Y \neq 2$; $X + Y < 4$. Test each of the following pairs for independence in the sense of probability: (p, q), (q, r), (p, r), $(q, \sim r)$.

Table 8

Outcome	p	q	r	$p \wedge q$	$q \wedge r$	$p \wedge r$	$q \wedge \sim r$
(ONE, ONE)	T	T	T	T	T	T	F
(ONE, TWO)	F	F	T	F	F	F	F
(ONE, THREE)	T	T	F	T	F	F	T
(TWO, ONE)	F	T	T	F	T	F	F
(TWO, TWO)	T	F	F	F	F	F	F
(TWO, THREE)	F	T	F	F	F	F	T
(THREE, ONE)	T	T	F	T	F	F	T
(THREE, TWO)	F	F	F	F	F	F	F
(THREE, THREE)	T	T	F	T	F	F	T

We summarize the information by preparing a truth table (Table 8), showing the nine logically possible outcomes in the experiment and the corresponding truth value for pertinent statements. In Table 8, each entry in the column headed "p" has been determined by checking whether, for the outcome under consideration, the sum of the two components is even. Likewise, the entries in the column headed "q" are determined by whether the second component of the ordered pair representing the outcome is different from 2. In a similar manner, the entries T in the fourth column identify the members of the truth set for r. Entries in the last four columns are obtained by applying the methods of Chapter 5 to the entries in the columns headed "p," "q," "r."

Table 9

Statement	p	q	r	$p \wedge q$	$q \wedge r$	$p \wedge r$	$q \wedge \sim r$
Probability	$\frac{5}{9}$	$\frac{2}{3}$	$\frac{1}{3}$	$\frac{4}{9}$	$\frac{2}{9}$	$\frac{1}{9}$	$\frac{4}{9}$

Since each outcome is assigned the measure $\frac{1}{9}$, Table 8 enables us to count how many members belong to various truth sets and obtain the results in Table 9. Since $\Pr[p] \cdot \Pr[q] = \frac{5}{9} \cdot \frac{2}{3} = \frac{10}{27}$ is not the same as $\Pr[p \wedge q]$, the pair (p, q) is not an independent pair. Since $\Pr[q] \cdot \Pr[r] = \frac{2}{3} \cdot \frac{1}{3} = \frac{2}{9}$ is the same as $\Pr[q \wedge r]$, the pair (q, r) is an independent pair in the probability sense. Next, (p, r) fails to satisfy the condition of being an independent pair. Finally, since $\Pr[\sim r] = 1 - \Pr[r] = \frac{2}{3}$, we find that $(q, \sim r)$ is an independent pair.

It may seem reasonable that if q and r form an independent pair, then q and the negation of r should also form an independent pair.

We have found that, in this experiment, the statement $Y \neq 2$ and the statement $X + Y < 4$ form an independent pair in the sense of probability. To some readers this independence may seem reasonable. On the other hand, some readers may experience a common-sense feeling that the two statements are somehow interrelated. The phrase, "an independent pair," may suggest the common-sense idea of statements that "do not depend on each other" or that "have nothing to do with one another." Although this intuitive idea helps to motivate the notion of an independent pair, we warn the reader that the test for independence is prescribed in the definition. The criterion is not affected by the connotations an individual person ascribes to the word "independent." In brief, the decision concerning whether a pair of statements is independent in the probability sense is made by comparing probabilities: Does the probability of the conjunction of the two statements equal the product of the probabilities of the individual statements?

EXERCISES F

1. An urn contains five green balls, four brown balls, and three pink balls. One ball is drawn at random and replaced; afterward a ball is drawn at random.
   (a) What is the probability that the first ball drawn is green?
   (b) What is the probability that the second ball drawn is green?
   (c) What is the probability that both balls are green?
   (d) What is the conditional probability that the second ball is green, given that the first ball is green?
   (e) What is the probability that one of the balls drawn is brown and the other is pink?
   (f) What is the conditional probability that one of the balls drawn is brown, given that at least one of them is pink?
   (g) What is the conditional probability that both balls drawn are green, given that at least one of them is green?
   (h) What is the conditional probability that the second ball is pink, given that at least one of the balls is pink?

2. Repeat all of Exercise 1 with the following modification: In the description of the experiment, substitute the phrase "and not replaced" for the phrase "and replaced."

3. An experiment consists of three successive tosses of a single true coin. Let p, q, r be the statements that a tail appears on the first, second, third tosses, respectively. As usual, represent each outcome as an ordered triple whose kth component tells the face showing on the kth toss for $k \in \{1, 2, 3\}$.
   (a) What is the total number of possible outcomes?
   (b) What measure is assigned to each outcome?
   (c) How many outcomes belong to the truth set for q?
   (d) Evaluate $\Pr[q]$.
   (e) How many outcomes belong to the truth set for r?
   (f) Evaluate $\Pr[r]$.

(g) For how many outcomes is the conjunction $q \wedge r$ true?
(h) Evaluate $\Pr[q \wedge r]$.
(i) Decide whether (q, r) is an independent pair of statements in the probability sense. Justify your decision.
(j) Calculate $\Pr[p]$.
(k) Calculate $\Pr[p \wedge r]$.
(l) Decide whether (p, r) is an independent pair of statements in the probability sense. Justify your decision.
(m) Show that (p, q) is a probabilistically independent pair.
(n) Evaluate $\Pr[r|p]$ and $\Pr[r|q]$.
(o) Calculate $\Pr[r|p \wedge q]$.

4. A true coin is tossed ten times in succession.
   (a) What is the probability of a head appearing on the tenth toss?
   (b) If each of the first nine tosses shows a tail, what is the probability of a head appearing on the tenth toss?

5. Consider an ordinary deck of playing cards.
   (a) If an experiment consists in drawing one card at random from the deck, show that the drawing of a diamond and the drawing of a king form an independent pair of statements in the probability sense.
   (b) If an experiment consists in drawing two cards together at random from the deck, show that the drawing of two diamonds and the drawing of two kings do not form an independent pair of statements in the probability sense.
   (c) If an experiment consists in drawing two cards together at random from the deck, show that the drawing of at least one diamond and the drawing of at least one king do not form a probabilistically independent pair of statements.
   (d) If an experiment consists in drawing two cards together at random from the deck, decide whether the drawing of exactly one diamond and the drawing of exactly one king form a probabilistically independent pair. Justify your decision.

6. Consider a deck of ten cards consisting of the king, queen, jack, and ten of hearts and the jack, ten, nine, eight, seven, and six of clubs. An experiment consists of drawing two cards together at random from this deck. Find the probability of each of the following.
   (a) At least one club is drawn.
   (b) At least one heart is drawn, given that at least one club is drawn.
   (c) At least one club is drawn, given that at least one heart is drawn.
   (d) At least one jack is drawn.
   (e) Both jacks are drawn.
   (f) Both jacks are drawn, given that the heart jack is drawn.
   (g) Both jacks are drawn, given that the club jack is drawn.
   (h) Both jacks are drawn, given that at least one jack is drawn.
   (i) Exactly one club is drawn.
   (j) Exactly one club is drawn, given that the club jack is drawn.
   (k) Exactly one club is drawn, given that at least one jack is drawn.

7. Consider a 16-card deck consisting of the aces, kings, queens, and jacks from an ordinary deck of playing cards. An experiment consists in drawing cards, one at a time, in succession, without replacement, from the new deck until all 16 cards have been drawn. Let X be the number of the draw on which a king appears for the first time, let Y be the number of the draw on which an ace appears for the first time, and let Z be the number of the draw on which a king appears for the second time. Calculate each of the following probabilities.
   (a) $\Pr[X = 3]$.
   (b) $\Pr[(X = 3) \wedge (Y = 1)]$.
   (c) $\Pr[X = 3 | Y = 1]$.
   (d) $\Pr[Y = 1 | X = 3]$.
   (e) $\Pr[(X, Z) = (3, 4)]$.
   (f) $\Pr[(X, Z) = (13, 14)]$.

(g) $\Pr[X = Y]$.
(h) $\Pr[X > Y]$.
(i) $\Pr[Z > Y > 9]$.
(j) $\Pr[X = 3 | Z = 4]$.
(k) $\Pr[Z = 3 | X = 4]$.
(l) $\Pr[X < Z]$.
(m) $\Pr[X \leq 3]$.
(n) $\Pr[Z \geq 13]$.
(o) $\Pr[X + Y = 4]$.
(p) $\Pr[X + Z = 4]$.
(q) $\Pr[Y + Z = 4]$.

8. Let p and q be statements pertaining to the outcome of an experiment such that neither of the statements is logically false. Show that if (p, q) is a logically inconsistent pair, then (p, q) is not an independent pair in the probability sense.

9. Let p and q be statements pertaining to the outcome of an experiment. Show that if (p, q) is an independent pair in the probability sense, then $(\sim p, q)$ is also an independent pair.

12 A posteriori probability

Example 20 Three urns are numbered 1, 2, 3, respectively. Each urn contains several balls, some white, the others green: urn 1 contains one white ball and three green balls; urn 2 contains five white balls and one green ball; urn 3 contains four white balls and five green balls. The data are summarized in the following diagram.

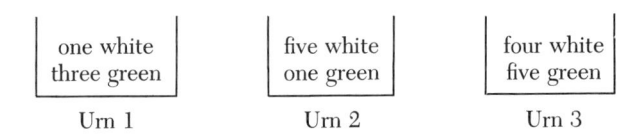

one white three green	five white one green	four white five green
Urn 1	Urn 2	Urn 3

An experiment consists in selecting one of the three urns at random and then drawing at random one ball from the chosen urn. Let U be the number of the urn chosen, and let w be the statement that the ball drawn is white. Calculate each of the following: $\Pr[w]$, $\Pr[U = 2 | w]$.

There are three different ways in which a white ball may be drawn: namely, from urn 1, from urn 2, or from urn 3. The drawing of a white ball from urn 2 is a member of the truth set of the conjunction $(U = 2) \wedge w$. Likewise, the conjunction $(U = 1) \wedge w$ is true in case urn 1 is selected and the ball drawn is white. Consider the three conjunctions $(U = 1) \wedge w$, $(U = 2) \wedge w$, $(U = 3) \wedge w$; each pair of them is logically inconsistent, and the disjunction of all three of them is logically equivalent to w. Consequently,

$$\Pr[w] = \Pr[(U = 1) \wedge w] + \Pr[(U = 2) \wedge w] + \Pr[(U = 3) \wedge w].$$

Now, the probability of a conjunction often may be evaluated by using a conditional probability. Thus, $\Pr[(U = 2) \wedge w] = \Pr[U = 2] \cdot \Pr[w | U = 2]$. Since the urn is selected at random, $\Pr[U = 2] = \frac{1}{3}$. If we accept the statement $U = 2$ as true, then the selection of a ball would be a random selection from five white balls and one green ball, and the probability of drawing a white ball would be $\frac{5}{6}$. The preceding sentence describes the conditional probability of the statement w, given that $U = 2$; consequently, $\Pr[w \mid U = 2] = \frac{5}{6}$. Hence, $\Pr[(U = 2) \wedge w] = \frac{1}{3} \cdot \frac{5}{6} = \frac{5}{18}$.

We apply the same type of reasoning to obtain

$$\Pr[(U = 1) \wedge w] = \Pr[U = 1] \cdot \Pr[w \mid U = 1] = \frac{1}{3} \cdot \frac{1}{4} = \frac{1}{12}$$

and

$$\Pr[(U = 3) \wedge w] = \Pr[U = 3] \cdot \Pr[w \mid U = 3] = \frac{1}{3} \cdot \frac{4}{9} = \frac{4}{27}.$$

In each case the conditional probability is calculated by supposing the corresponding urn actually has been chosen and considering the statement w in that restricted situation. Since $\Pr[w]$ is the sum of the probabilities of the three conjunctions, we obtain

$$\Pr[w] = \frac{1}{12} + \frac{5}{18} + \frac{4}{27} = \frac{55}{108}.$$

Finally,

$$\Pr[U = 2 \mid w] = \frac{\Pr[(U = 2) \wedge w]}{\Pr[w]} = \frac{5/18}{55/108} = \frac{6}{11}.$$

The conditional probability that $U = 2$, given w, may be interpreted in the following manner. Suppose that, after the experiment has been conducted, we are given partial information about the outcome, namely, the information that the ball drawn is actually a white ball. This supplementary fact presumably may alter the chances we believe should be ascribed to various other statements pertaining to the outcome. (As an extreme case, consider the statement that a green ball is drawn; its likelihood collapses to zero as soon as the announcement is made that w is observed to be true.) Since urn 2 has a much higher percentage of white balls than the other urns, it seems that there is more chance that the observed white ball came from urn 2 than from any one of the others. This greater chance is revealed in the number $\Pr[U = 2 \mid w] = \frac{6}{11}$. We recall that the statement $U = 2$ has, without the acceptance that w is true, a probability $\frac{1}{3}$; thus, the probability that $U = 2$ is significantly increased by additional information that w is true.

By contrast,

$$\Pr[U = 1 \mid w] = \frac{\Pr[(U = 1) \wedge w]}{\Pr[w]} = \frac{1/12}{55/108} = \frac{9}{55};$$

the likelihood that $U = 1$ is considerably decreased from its original, or "a priori," value $\frac{1}{3}$ to its "a posteriori" value $\frac{9}{55}$, which reflects the effect of accepting w as a true statement. In fact, $\frac{9}{55}$ is less than half of $\frac{1}{3}$.

13 Repeated trials

Very often in applications, an experiment consists of many repetitions of a certain activity. The repetitions may be made in chronological order; for example, we may toss a single penny 100 times in succession, or we may study the traffic flow at a busy intersection every day at noon for a year. On the other hand, the repetitions may be made simultaneously; as illustrations, we may toss 200 pennies together, or we may conduct a clinical study of the effects of some drug administered on the same day to rats from the same litter.

In this repetitious type of experiment, involving many trials of an activity, we are frequently interested in a statement that pertains to the individual activity. In such a case, we may say that an outcome for the activity that makes the statement true yields a *success*, while an outcome for the activity that does not belong to the truth set yields a *failure*. We may say that each occurrence of the activity is a *trial*. Customarily, in this experiment consisting of repeated trials, we are interested in how many successes are achieved. We may be also interested in any patterns for the successes; for example, the order in which they may appear chronologically or spatially.

14 Bernoulli experiments

In many instances, success on any one trial and success on any other trial in the experiment may form an independent pair of statements in the probability sense. If this is the situation, we speak of the experiment as having *independent trials*. If, say, the outcome HEAD yields success on a single toss of a true coin, then a sequence of 1000 fair tosses of this coin is an experiment consisting of independent trials. By contrast, if obtaining a queen yields success on the drawing of a single card, then a sequence of 20 successive draws, without replacement, from an originally complete deck of playing cards is not an experiment consisting of independent trials. Indeed, the drawing of a queen on the second trial and a success on the seventh trial (as an illustration) do not form an independent pair of statements.

Example 21 Suppose an experiment consists of four independent trials of a certain activity. Let p be the probability of a success on a single trial. Let X be the number of successes in the four trials. Possible values for X are 0, 1, 2, 3, and 4. Calculate the probability for each value of X.

For each index j, let the statement s_j correspond to success on the jth trial. By hypothesis, each pair of the four statements s_1, s_2, s_3, s_4 is an independent pair. The statement $X = 4$, meaning that the experiment produces four successes, is logically equivalent to the conjunction $s_1 \wedge s_2 \wedge s_3 \wedge s_4$. Now we utilize the definition of an independent pair to obtain

$$\Pr[X = 4] = \Pr[s_1 \wedge s_2 \wedge s_3 \wedge s_4]$$
$$= \Pr[s_1] \cdot \Pr[s_2] \cdot \Pr[s_3] \cdot \Pr[s_4] = p \cdot p \cdot p \cdot p = p^4.$$

In a similar manner, the statement $X = 0$ is logically equivalent to the conjunction of the four negations $\sim s_j$ for $j \in \{1, 2, 3, 4\}$. Since each pair of these negations is probabilistically independent,

$$\Pr[X = 0] = \Pr[\sim s_1] \cdot \Pr[\sim s_2] \cdot \Pr[\sim s_3] \cdot \Pr[\sim s_4]$$
$$= (1 - p)^4.$$

The statement $X = 3$ is true in case any three of the trials yield success and the remaining one yields failure. Since the number of subsets consisting of three trials from the entire set of four trials is the binomial coefficient $\binom{4}{3}$, the truth set for $X = 3$ has a partition consisting of $\binom{4}{3}$ cells: one cell contains outcomes that are ordered quadruples of the type (success, success, success, failure), another of the type (success, success, failure, success), another of the type (success, failure, success, success), and a fourth of the type (failure, success, success, success). The probability for each of these types is $p^3(1 - p)$. Thus,

$$\Pr[X = 3] = \binom{4}{3} p^3 (1 - p).$$

A corresponding analysis shows that

$$\Pr[X = 1] = \binom{4}{1} p (1 - p)^3 \quad \text{and} \quad \Pr[X = 2] = \binom{4}{2} p^2 (1 - p)^2.$$

Table 10

k	0	1	2	3	4
$\Pr[X = k]$	$(1-p)^4$	$4(1-p)^3 p$	$6(1-p)^2 p^2$	$4(1-p) p^3$	p^4

We summarize our findings in Table 10. With the aid of the exponent zero and the binomial coefficients, we may also write

$$\Pr[X = 0] = \binom{4}{0}(1-p)^4 p^0 \quad \text{and} \quad \Pr[X = 4] = \binom{4}{4}(1-p)^0 p^4.$$

Thus, all entries in Table 10 fit the same formula, namely,

$$\Pr[X = k] = \binom{4}{k}(1-p)^{4-k} p^k \quad \text{for } k \in \{0, 1, 2, 3, 4\}.$$

The theory of probability of course requires that $\sum_{k=0}^{4} \Pr[X = k] = 1$.

In our study of the binomial theorem in Chapter 7 we learned that

$$\sum_{k=0}^{4} \Pr[X = k] = \sum_{k=0}^{4} \binom{4}{k}(1 - p)^{4-k}p^k$$

is the expansion of $[(1 - p) + p]^4$. Since $[(1 - p) + p]^4 = 1^4 = 1$, we have a partial check on our work.

The discussion in Example 21 for an experiment with four trials may be adapted to an experiment in which the number of trials is any natural number.

Let n be a natural number, and let p be a real number such that $0 \leq p \leq 1$. Suppose an experiment consists of n independent trials of a certain activity, such that p is the probability for "success" on each trial. Let X be the number of successes in the n trials. Then, for each $k \in \{0, 1, 2, \ldots, n\}$, the probability of exactly k successes is given by

$$\Pr[X = k] = \binom{n}{k}(1 - p)^{n-k}p^k.$$

A situation of the type described in the preceding paragraph is called a *Bernoulli experiment* in honor of Jacob Bernoulli (1654–1705), whose pioneering work on probability was published posthumously in 1713. Some of our previous illustrations have involved Bernoulli experiments, and we now give other examples.

Example 22 A machine in a factory produces a certain metal part. The performance of the machine over a long period of time enables us to assign a probability of 0.02 that a part produced by the machine is defective. Experience also reveals that defects in two of the parts seem to be independent events in the sense of probability. What is the probability that, among 1000 parts produced one week, 20 are defective?

We may consider the production of each part as a trial in a Bernoulli experiment. Thinking of the production of a defective part as being a "success," the desired probability is

$$\binom{1000}{20}(1 - 0.02)^{980}(0.02)^{20} = \binom{1000}{20}(0.98)^{980}(0.02)^{20}.$$

The direct evaluation of this number is a formidable task. Fortunately, as we shall see in Section 10 of Chapter 12, another method is available. Approximately, the probability is 0.09.

Note that 20 is precisely the product of 0.02 and 1000. On this basis, some persons might guess that the probability of exactly 20 defectives should be quite high. Our calculations show how very wrong this common misconception can be. True, the number of defectives has a greater probability of being 20 than of being any other single number; but that does not mean that any specific number of defectives has very much chance of occurring.

Note that our general use of the word *success* in discussing a Bernoulli experiment refers to an outcome belonging to the truth set of the statement under consideration. As illustrated in Example 22, a "success" in this sense may, from the manufacturing viewpoint, mean that the machine produced unsuccessfully.

Finally, we emphasize that a Bernoulli experiment requires independent trials. Unless there is satisfactory evidence to substantiate the independence, we may not use the theory for a sequence of independent trials. In Example 22, this evidence is provided by observation of the past performance of the machine. How to acquire an adequate justification is not an easy statistical task and will not be treated here. In case the trials in an experiment are not independent, we may use more elaborate tools designed for handling such problems.

15 Assignment of measures

The worth of our probability theory depends significantly upon our ability to assign, in a reasonable manner, the measures to the outcomes of the experiment under analysis. We recall, from the beginning, that a probability discussion starts with an itemization of logically possible outcomes and an assignment of a measure to each. In our examples from games of chance, we choose measures based on common sense: the issue of symmetry is an important contributing factor. But in many applications we have only a nebulous basis for the assignment of measures. We often must rely on a well-balanced mixture of good judgment and past experience.

Example 23 Suppose we have an activity and a statement pertaining to its outcome. As usual, the possible outcomes may be classified according to whether they yield success or failure. Suppose we are justified in accepting that the probability for success is a definite number p, but suppose we do not know what p is. Suppose we believe that p is either $\frac{1}{3}$ or $\frac{2}{3}$; furthermore, suppose we have reason to think that $\frac{1}{3}$ has three times as much chance of being the correct value for p as $\frac{2}{3}$ has. We may conduct a Bernoulli experiment consisting of seven independent trials of this activity. Suppose that five successes appear in the experiment. How does this information affect our thinking about the value of p?

We have agreed that p is one of the two numbers in the set $\{\frac{1}{3}, \frac{2}{3}\}$. Originally, we assigned three times as much measure to the statement $p = \frac{1}{3}$ as to the statement $p = \frac{2}{3}$. Since the total measure is 1, the respective measures are $\frac{3}{4}$ and $\frac{1}{4}$. That is, the a priori probabilities are $\Pr[p = \frac{1}{3}] = \frac{3}{4}$ and $\Pr[p = \frac{2}{3}] = \frac{1}{4}$.

In any Bernoulli experiment consisting of seven independent trials of this activity, let X be the number of successes. We are interested particularly in the statement $X = 5$, because this outcome has actually been observed in the experiment as conducted. In general, the probability for $X = 5$ depends upon the correct value for p and is given by $\binom{7}{5}(1-p)^2 p^5 = 21(1-p)^2 p^5$.

In particular, if we accept $p = \frac{1}{3}$, then we have

$$\Pr\left[X = 5 \mid p = \frac{1}{3}\right] = 21\left(1 - \frac{1}{3}\right)^2\left(\frac{1}{3}\right)^5 = \frac{28}{729};$$

while, if we accept $p = \frac{2}{3}$, we obtain

$$\Pr\left[X = 5 \mid p = \frac{2}{3}\right] = 21\left(1 - \frac{2}{3}\right)^2\left(\frac{2}{3}\right)^5 = \frac{224}{729}.$$

Now the statement $X = 5$ is logically equivalent to the disjunction of the statement $(p = \frac{1}{3}) \wedge (X = 5)$ and the statement $(p = \frac{2}{3}) \wedge (X = 5)$. Since these latter two statements form a logically inconsistent pair, the sum of their probabilities is the probability of their disjunction. Thus,

$$\begin{aligned}\Pr[X = 5] &= \Pr[(p = \tfrac{1}{3}) \wedge (X = 5)] + \Pr[(p = \tfrac{2}{3}) \wedge (X = 5)] \\ &= \Pr[p = \tfrac{1}{3}] \cdot \Pr[X = 5 \mid p = \tfrac{1}{3}] \\ &\quad + \Pr[p = \tfrac{2}{3}] \cdot \Pr[X = 5 \mid p = \tfrac{2}{3}] \\ &= \frac{3}{4} \cdot \frac{28}{729} + \frac{1}{4} \cdot \frac{224}{729} = \frac{77}{729}.\end{aligned}$$

We are now ready to calculate the a posteriori probability for $p = \frac{1}{3}$ and for $p = \frac{2}{3}$, conditional probabilities that reflect the observed outcome of five successes in the experiment. We have

$$\Pr\left[p = \frac{1}{3} \mid X = 5\right] = \frac{\Pr[(p = \frac{1}{3}) \wedge (X = 5)]}{\Pr[X = 5]} = \frac{21/729}{77/729} = \frac{3}{11}$$

and

$$\Pr\left[p = \frac{2}{3} \mid X = 5\right] = \frac{\Pr[(p, X) = (\frac{2}{3}, 5)]}{\Pr[X = 5]} = \frac{56/729}{77/729} = \frac{8}{11}.$$

The statement $p = \frac{1}{3}$, which originally was given the probability $\frac{3}{4}$, has a conditional probability, accepting the experimental result, of only $\frac{3}{11}$. Rephrased, the statement $p = \frac{1}{3}$ had, before the experiment was conducted, odds of 3 to 1 in favor, but after the outcome of the experiment becomes known, the odds are 8 to 3 against. Thus, the particular experiment, as conducted, offers information that leads us to change significantly our belief concerning the correct value of p.

Note, however, that we still do not have any right to state which number the probability p actually is.

EXERCISES G

1. In a Bernoulli experiment consisting of eight independent trials of an activity such that the probability for success on each individual trial is $\frac{1}{3}$, calculate the probability of exactly five successes.

2. In a Bernoulli experiment consisting of $n = 5$ independent trials of an activity such that the probability of success on each individual trial is $p = \frac{1}{4}$, let X be the total number of successes. Make a table, showing all the possible values for X and the probability of each of them.

3. Repeat Exercise 2, with the modification that $(n, p) = (4, \frac{1}{2})$.

4. Describe an experiment and give a statement pertaining to the outcome of the experiment such that the probability of the statement is
$$\binom{400}{350}\left(\frac{1}{6}\right)^{50}\left(\frac{5}{6}\right)^{350}.$$

5. Describe an experiment and give a statement pertaining to the outcome of the experiment such that the probability of the statement is
$$\sum_{k=80}^{100}\binom{100}{k}\left(\frac{1}{6}\right)^{100-k}\left(\frac{5}{6}\right)^{k}.$$

6. Describe an experiment and give a statement pertaining to the outcome of the experiment such that the probability of the statement is
$$\sum_{k=290}^{310}\binom{750}{k}(0.6)^{750-k}(0.4)^{k}.$$

7. Write, but do not compute, the probability that in a sequence of 10,000 tosses of a penny the number of heads that appear is between 4950 and 5100, inclusive.

8. Suppose that "success" on a single roll of a pair of gambling dice means the outcome belongs to the set {SEVEN, ELEVEN}. Suppose the dice are rolled five times.
   (a) What is the probability of at least two successes?
   (b) What is the conditional probability of at least four successes, given that at least two successes occur?
   (c) What is the conditional probability of at least two successes, given that the number of successes is at least four?

9. For the experiment described in Exercise C-3, find each of the following probabilities.
   (a) The conditional probability that the ball drawn is blue, given that the short urn is selected.
   (b) The conditional probability that the ball is white, given that it is not red.
   (c) The conditional probability that the short urn is selected, given that the ball is blue.
   (d) The conditional probability that the tall urn is selected, given that the ball is red.
   (e) The conditional probability that a red ball is drawn from the tall urn, given that the ball drawn is not a blue ball.

(f) The conditional probability that a white ball is drawn from the short urn, given that the ball drawn is not a red ball from the short urn.

10. For a certain activity and a certain statement pertaining to its outcome, let the number p be the unknown probability that an outcome belongs to the truth set of the statement. Suppose that p is one of the three numbers $\frac{5}{8}, \frac{6}{8}, \frac{7}{8}$, and suppose that, on the basis of available evidence, the respective probabilities of the statements $p = \frac{5}{8}$, $p = \frac{6}{8}, p = \frac{7}{8}$ are $\frac{1}{2}, \frac{1}{3}, \frac{1}{6}$. Let r be the statement that, in a Bernoulli experiment consisting of four independent trials of the activity, three successes occur.
   (a) Calculate $\Pr[r|p = \frac{5}{8}]$.   (b) Calculate $\Pr[r|p = \frac{6}{8}]$.
   (c) Calculate $\Pr[r|p = \frac{7}{8}]$.   (d) Find $\Pr[r]$.
   (e) Calculate $\Pr[p = \frac{5}{8}|r]$.   (f) Calculate $\Pr[p = \frac{6}{8}|r]$.
   (g) Calculate $\Pr[p = \frac{7}{8}|r]$.
   (h) Compare the a posteriori probabilities in parts (e), (f), (g) with the a priori probabilities that $p = \frac{5}{8}, p = \frac{6}{8}, p = \frac{7}{8}$, respectively.

11. For a certain activity and a certain statement pertaining to its outcome, let the number p be the unknown probability of success. Suppose that p is known to be one of the three numbers $\frac{3}{6}, \frac{4}{6}, \frac{5}{6}$ and that the probabilities of these three values are respectively $\frac{1}{8}, \frac{2}{8}, \frac{5}{8}$. Let s be the statement that, in a Bernoulli experiment consisting of four independent trials of the activity, success occurs four times.
   (a) Calculate $\Pr[s]$.   (b) Calculate $\Pr[p = \frac{3}{6}|s]$.
   (c) Calculate $\Pr[p = \frac{4}{6}|s]$.   (d) Calculate $\Pr[p = \frac{5}{6}|s]$.
   (e) Compare the a posteriori probabilities in parts (b), (c), (d) with the corresponding a priori probabilities that $p = \frac{3}{6}, p = \frac{4}{6}, p = \frac{5}{6}$, respectively.

16 Summary of probability

Given an experiment with a finite nonempty set of logically possible outcomes, we assign to each outcome a positive number as its measure, subject to the condition that the sum of all the assigned measures is 1.

Although the assignment of the same measure to each possible outcome is a very important special case, this definitely is not required—other assignments are useful in many situations. Theoretically, any method of assignment that meets the specified conditions is acceptable.

The probability of a given statement pertaining to the outcome of the experiment is the sum of the measures of the outcomes belonging to the truth set for the statement. As consequences of this definition, we have the following relationships, in which p and q denote pertinent statements.

1. $0 \leq \Pr[p] \leq 1$ for every p.
2. If p and q are logically equivalent statements, then $\Pr[p] = \Pr[q]$.
3. If p is logically true, then $\Pr[p] = 1$.
4. If p is logically false, then $\Pr[p] = 0$.
5. $\Pr[p] + \Pr[\sim p] = 1$ for every p.
6. $\Pr[p \vee q] + \Pr[p \wedge q] = \Pr[p] + \Pr[q]$ for every pair (p, q).
7. If (p, q) is a logically inconsistent pair, then $\Pr[p \vee q] = \Pr[p] + \Pr[q]$.

The next two relationships define conditional probability and an independent pair, respectively.

8. $\Pr[p \wedge q] = \Pr[p] \cdot \Pr[q|p]$ for every pair (p, q) such that $\Pr[p] \neq 0$.
9. $\Pr[p \wedge q] = \Pr[p] \cdot \Pr[q]$ if and only if (p, q) is an independent pair in the sense of probability.

EXERCISES H

1. A gambling die is rolled successively until a six-spot appears for the first time. However, if no six-spot appears in the first four rolls, the process is stopped. Let X be the number of rolls in this experiment. Make a table, listing the logically possible values for X and the probability for each value.

2. From an ordinary deck of playing cards, three cards are drawn in succession, without replacement.
    (a) What is the probability that each of the first two cards is a spade and the third card is a diamond?
    (b) What is the probability that each of the first two cards is a spade and the third card is a king?
    (c) What is the probability that each of the first two cards is black and the third card is a king?
    (d) What is the conditional probability that the third card is a diamond, given that each of the first two cards is a spade?
    (e) What is the conditional probability that the third card is a king, given that each of the first two cards is a spade?
    (f) What is the conditional probability that the third card is a heart, given that each of the first two cards is red?
    (g) What is the conditional probability that each of the first two cards is a spade, given that the third card is a diamond?
    (h) What is the conditional probability that each of the first two cards is black, given that the third card is a king?
    (i) What is the conditional probability that the second card is the king of spades, given that the first card is a spade and the third card is a king?

3. Consider a Bernoulli experiment consisting of 100 independent trials of a certain activity in which the probability of success on a single trial is a number p such that $0 < p < 1$. Let Y be the number of the trial on which the first success occurs (or, in case no success ever occurs, let Y be the number 100).
    (a) Show that $\Pr[Y = 1] = p$.
    (b) Show that $\Pr[Y = 2] = p(1 - p)$.
    (c) Find a formula giving $\Pr[Y = k]$ for $k \in \{1, 2, \ldots, 99\}$.
    (d) Find $\Pr[Y = 100]$.
    (e) Find $\Pr[1 \leq Y \leq 99]$ by simplifying the sum $\sum_{k=1}^{99} \Pr[Y = k]$.
    (f) Combine the results of parts (d) and (e) to show that $\sum_{k=1}^{100} \Pr[Y = k] = 1$.

4. A bowl contains six balls, of which five are red and one is white. An experiment consists of five successive drawings at random, one ball at a time, from the bowl. After

each drawing, the ball drawn is replaced before the next drawing. Furthermore, following the first and the third drawings one red ball is removed from the bowl, while immediately after the second and the fourth drawings one extra red ball is placed in the bowl. Let Y be the number of white balls drawn during the experiment. Make a table, showing the possible values for Y and the probability of each value.

5. A bowl contains six balls, of which five are green and one is white. An experiment consists of three successive drawings at random, one ball at a time, from the bowl. After each drawing, the ball drawn is replaced before the next drawing, and furthermore at the same time there is placed in the bowl an extra ball of the same color as the ball drawn. Let Z be the number of white balls drawn during the experiment. Make a table, showing the possible values for Z and the probability of each value.

6. In a certain game the movement of a pawn is governed by the toss of a coin. If the coin shows a head, the pawn moves one space to the right; if the outcome is the appearance of a tail, the pawn moves one space to the left. The game consists of four tosses of the coin and the corresponding pawn movements.
    (a) What is the conditional probability that the pawn does not return to its starting position at any stage during the game, given that the first movement of the pawn is to the right?
    (b) What is the probability that the pawn does not return to its starting position at any stage during the game?
    (c) What is the probability that the pawn, at the end of the game, is in the same position as at the beginning?
    (d) What is the probability that the pawn returns to its starting position at least twice during the game?
    (e) What is the probability that the pawn returns to its starting position exactly once during the game?
    (f) What is the probability that, at different stages during the game, the pawn is on the right side and on the left side of the starting position?

7. Repeat Exercise 6, with the following modification: the number of coin tosses and corresponding pawn movements is six.

8. Repeat Exercise 6, with the following modification: the number of coin tosses and corresponding pawn movements is five.

9. Let p and q be statements pertaining to the outcome of a certain experiment.
    (a) Show that $\Pr[\sim p | q] = 1 - \Pr[p | q]$.
    (b) Find, if possible, a relationship between the two numbers $\Pr[p | \sim q]$, $\Pr[p | q]$.
    (c) Let P and Q be the truth sets for the respective statements p and q. If $Q \subset P$, show that $\Pr[p | q] = 1$.
    (d) Evaluate $\Pr[p | q]$ in case the conditional statement $q \to p$ is logically true, and explain your conclusion.

10. A lot of 12 manufactured items contains two defectives. A sample of two items is selected at random from the lot. Let X be the number of defectives in the sample. This sample is not replaced. Afterward another sample of three items is chosen from the remaining portion of the lot. Let Y be the number of defectives in the second sample. Calculate $\Pr[Y > X]$.

11. Repeat Exercise 10 with the following modification: the original lot contains three defectives.

12. Repeat Exercise 6, with the following modification: the movement of the pawn is governed by the roll of a single gambling die. If the die shows an ace or a deuce, the pawn moves one space to the left; if the die shows more than two spots, the

pawn moves one space to the right. The game consists of four rolls of the die and the corresponding pawn movements.

13. For each of the following, describe an experiment and find a pair (p, q) of statements pertaining to the outcome of the experiment and satisfying the prescribed condition.
    (a) $\Pr[p \wedge q] = \Pr[p] \cdot \Pr[q] > 0$.
    (b) $\Pr[p \wedge q] > \Pr[p] \cdot \Pr[q] > 0$.
    (c) $\Pr[p] \cdot \Pr[q] > \Pr[p \wedge q] > 0$.

14. Let p, q, r be statements pertaining to the outcome of a certain experiment, and suppose that no pair of these statements is logically inconsistent.
    (a) Show that $\Pr[p \wedge q \wedge r] = \Pr[q \wedge r] \cdot \Pr[p | q \wedge r]$.
    (b) Show that $\Pr[p \wedge q \wedge r] = \Pr[r] \cdot \Pr[q | r] \cdot \Pr[p | q \wedge r]$.
    (c) Show that $\Pr[p \wedge q \wedge r] = \Pr[p] \cdot \Pr[q | p] \cdot \Pr[r | p \wedge q]$.

17 Random variable

We have already begun our study of the probability distribution for a random variable. In order to pursue this investigation, we need to enlarge our vocabulary.

> **Definition** Suppose that an experiment has finitely many possible outcomes and that measures in the probability sense have been assigned to the outcomes. A correspondence that associates a real number to each outcome is called a *random variable*.

We customarily denote a random variable by a capital letter such as X. Many of our previous examples and exercises have involved one or more random variables. In fact, any occurrences in this chapter of the letters W, X, Y, or Z have been illustrations of the idea of a random variable. Very often the values of a random variable are integers, such as when we have said, "Let X be the number of . . . ."

> **Example 24** If X is the total number of spots showing on a single roll of a pair of dice, then X is a random variable whose values belong to the set $\{2, 3, 4, \ldots, 11, 12\}$. If Y is the number of heads appearing in a sequence of four successive tosses of a coin, then Y is a random variable whose values belong to $\{0, 1, 2, 3, 4\}$. If a petty bettor bets 50 cents on each of five football bowl games, then the number Z of dollars he reaps describes a random variable whose values are the members of $\{-\frac{5}{2}, -\frac{3}{2}, -\frac{1}{2}, \frac{1}{2}, \frac{3}{2}, \frac{5}{2}\}$.

Occasionally, a random variable for an experiment may have a different value for each possible outcome, but commonly several different outcomes belong to the truth set for a statement of the type $X = h$, where h is one particular value of the random variable X.

Definition If we are given a random variable for a certain experiment, the specification of the set of all possible values for the random variable and the probability for each value describes the *probability distribution* of the random variable.

The two most important types of probability distributions we have studied so far are the *uniform* distribution and the *binomial* distribution. A random variable has a uniform distribution provided each of its possible values has the same probability. Thus, if n different numbers are possible values for the random variable Y with a uniform distribution and if h is any one of those numbers, then $\Pr[Y = h] = 1/n$.

The random variable X that gives the number of successes in a Bernoulli experiment has a binomial distribution. If there are n trials in the experiment, if p is the probability of success on each single trial, and if $k \in \{0, 1, \ldots, n\}$, then we have learned that $\Pr[X = k] = \binom{n}{k}(1-p)^{n-k}p^k$. We have already noted a connection between the preceding formula and the binomial theorem, a relationship which explains the name of "binomial" probability distribution.

Other types of probability distributions also are of great importance in theory and in applications. Several of these we encounter occasionally in our illustrative problems, but not frequently enough to study in detail their special features. In Chapter 12, we shall be able, with an extended notion of a random variable, to investigate briefly the standard normal distribution, which has primary significance in many phases of statistical work.

18 Expected value of a random variable

If we bet a dollar on a fair toss of a true coin, we may "expect" to come out even. Although the actual outcome yields either a dollar won or a dollar lost, the "50-50 chance" enables us subconsciously to "expect" that we will neither gain nor lose. We may rationalize that "in the long run" our average winnings will be approximately zero; this, of course, supposes that we bet many times in succession on repeated tosses.

From the somewhat vague notion conveyed by the word "expect" in the preceding paragraph we may extract a very precise concept of *expected value*. The precision is necessary in order that the concept be mathematically useful. Although we retain the name expected value, the idea has been broadened and no longer is confined to express what we may "expect."

Definition Suppose that a random variable is given for an experiment with finitely many possible outcomes. The *expected value* of the random variable is the sum of all the products obtainable by multiplying each value of the random variable by its probability.

We denote the expected value of the random variable X by $E[X]$. We emphasize that $E[X]$ is a number. In symbols, this number is expressible by the formula

$$E[X] = \sum_k k \cdot \Pr[X = k],$$

where the summation applies to every value k which X assumes.

Example 25 Tom bets a dollar on a fair toss of a true coin. Let X be the number of dollars he wins on his bet. Then $\Pr[X = 1] = \frac{1}{2}$ and $\Pr[X = -1] = \frac{1}{2}$. Consequently, $E[X] = 1 \cdot \frac{1}{2} + (-1)\frac{1}{2} = 0$. Thus, the value computed according to the definition agrees with the intuitive value discussed in the opening paragraph of this section.

Example 26 In a Bernoulli experiment involving three trials, the probability for success on each trial is $\frac{3}{4}$. Find the expected value of the number of successes.

Table 11

k	0	1	2	3
$\Pr[X = k]$	$\frac{1}{64}$	$\frac{9}{64}$	$\frac{27}{64}$	$\frac{27}{64}$

The probability distribution for the number X of successes is given in Table 11. Indeed,

$$\Pr[X = k] = \binom{3}{k}\left(\frac{1}{4}\right)^{3-k}\left(\frac{3}{4}\right)^k \quad \text{for each } k \in \{0, 1, 2, 3\}.$$

We obtain $E[X]$ by multiplying the two numbers in each column of the table and adding the four products; the sum is $0 \cdot \frac{1}{64} + 1 \cdot \frac{9}{64} + 2 \cdot \frac{27}{64} + 3 \cdot \frac{27}{64} = \frac{9}{4}$.

Example 27 Let m be a natural number, and let $\{1, 2, \ldots, m\}$ be the set of values for a random variable X with a uniform probability distribution. Show that $E[X] = (m + 1)/2$.

For each number $h \in \{1, 2, \ldots, m\}$, the condition that X has a uniform distribution means that $\Pr[X = h] = 1/m$. Consequently,

$$E[X] = \sum_{h=1}^{m} h \cdot \Pr[X = h] = \sum_{h=1}^{m} h \cdot \frac{1}{m}.$$

Since the sequence of numbers $1(1/m), 2(1/m), \ldots, m(1/m)$ is an arithmetic sequence, we may use the results of Section 11 in Chapter 7 to find the sum. We obtain

$$E[X] = \sum_{h=1}^{m} h \frac{1}{m} = \frac{1}{m} \sum_{h=1}^{m} h = \frac{1}{m} \frac{m(m+1)}{2} = \frac{m+1}{2}.$$

As an application of Example 27, if Y is the number of spots showing on a roll of a single die, then Y has a uniform distribution and
$$E[Y] = \frac{6+1}{2} = \frac{7}{2}.$$

Example 28 Let Z be the total number of spots appearing on a roll of a pair of dice. The probability distribution for Z is given in Table 12 [see Example 4 (in Section 3) and Exercise B-2]. If we multiply each number in the top row by the corresponding probability beneath it and if we add the 11 products, the sum $E[Z]$ is 7.

Table 12

k	2	3	4	5	6	7	8	9	10	11	12
$\Pr[Z=k]$	$\frac{1}{36}$	$\frac{2}{36}$	$\frac{3}{36}$	$\frac{4}{36}$	$\frac{5}{36}$	$\frac{6}{36}$	$\frac{5}{36}$	$\frac{4}{36}$	$\frac{3}{36}$	$\frac{2}{36}$	$\frac{1}{36}$

We note that the expected value of the total number of spots appearing on a roll of two dice is twice the expected value of the number of spots appearing on a roll of one die.

Example 29 During a coffee break three clerks, two salesmen, and one junior executive arrive together at the coffee table and form a queue in a random fashion. Let X be the number of the position in the queue occupied by the "leading" clerk, that is, the clerk who is nearest the head of the lineup. Find the probability distribution and the expected value for this random variable X.

The statement $X = 1$ means that a clerk leads the queue; since three of the six people are clerks, the probability that $X = 1$ is $\frac{1}{2}$. The statement $X = 2$ means that the leading clerk is second in the queue; in other words, the first position is not occupied by a clerk and the second position is. The probability of this conjunction (note the key word "and") may be evaluated with the aid of a conditional probability. The statement that a clerk does not lead the queue has probability $\frac{1}{2}$. The conditional probability of the statement that a clerk is in the second position, given that a clerk is not in the first position, is $\frac{3}{5}$; for, when we accept as true the statement that the leading position is filled by a person not a clerk, then three of the remaining five persons are clerks. Thus, $\Pr[X = 2] = \frac{1}{2} \cdot \frac{3}{5} = \frac{3}{10}$.

A member of the truth set for the statement $X = 3$ is an ordered sextuple of the six employees such that the first two components are not clerks and the third component is a clerk. If p, q, r are the statements that a clerk occupies the first, the second, the third position, respectively, then $X = 3$ is logically equivalent to $\sim p \wedge \sim q \wedge r$. Thus,
$$\begin{aligned}\Pr[X=3] &= \Pr[\sim p \wedge \sim q \wedge r] = \Pr[\sim p \wedge \sim q] \cdot \Pr[r \mid \sim p \wedge \sim q] \\ &= \Pr[\sim p] \cdot \Pr[\sim q \mid \sim p] \cdot \Pr[r \mid \sim p \wedge \sim q] \\ &= \tfrac{1}{2} \cdot \tfrac{2}{5} \cdot \tfrac{3}{4} = \tfrac{3}{20}.\end{aligned}$$

Finally, the statement $X = 4$ means that all three of the people who are not clerks precede the leading clerk in the queue. In other words, $X = 4$ is logically equivalent to the statement $\sim p \wedge \sim q \wedge \sim r$. Hence,

$$\Pr[X = 4] = \Pr[\sim p] \cdot \Pr[\sim q | \sim p] \cdot \Pr[\sim r | \sim p \wedge \sim q]$$
$$= \tfrac{1}{2} \cdot \tfrac{2}{5} \cdot \tfrac{1}{4} = \tfrac{1}{20}.$$

Table 13

h	1	2	3	4
$\Pr[X = h]$	$\tfrac{1}{2}$	$\tfrac{3}{10}$	$\tfrac{3}{20}$	$\tfrac{1}{20}$

We summarize our results in Table 13. Finally, we calculate

$$E[X] = 1 \cdot \tfrac{1}{2} + 2 \cdot \tfrac{3}{10} + 3 \cdot \tfrac{3}{20} + 4 \cdot \tfrac{1}{20} = \tfrac{7}{4}.$$

EXERCISES I

1. Let the probability distribution for a random variable X be given by the accompanying table. Find $E[X]$.

k	0	1	2	3
$\Pr[X = k]$	0.4	0.3	0.2	0.1

2. Suppose that a random variable Y satisfies the conditions that $\Pr[Y = -1] = \tfrac{1}{3}$, $\Pr[Y = 0] = \tfrac{1}{6}$, $\Pr[Y = 1] = \tfrac{1}{2}$. Calculate $E[Y]$.

3. If Z is the number of successes in a Bernoulli experiment consisting of three trials such that the probability of success on each trial is $\tfrac{1}{2}$, calculate $E[Z]$ by applying the definition.

4. Find (by applying the definition) the expected value of the number of successes in a Bernoulli experiment consisting of $n = 2$ trials such that the probability of success on each trial is the number p between 0 and 1.

5. Repeat Exercise 4, with the following modification: $n = 3$.

6. Repeat Exercise 4, with the following modification: $n = 4$.

7. Find the expected value of each of the following random variables.
   (a) The random variable W in Exercise E-1.
   (b) The random variable W in Exercise E-5.
   (c) X in Exercise E-2.
   (d) Y in Exercise E-3.
   (e) Z in Exercise E-4.
   (f) X in Exercise D-1.
   (g) Y in Exercise D-2.
   (h) Z in Exercise D-3.
   (i) X in Exercise H-1.
   (j) The number of dollars that the coin picked in Exercise C-4 is worth.

8. An experiment consists in tossing two coins simultaneously. Let the random variables X, Y, Z for this experiment be as follows: X is the number of heads; Y is the number

of tails; Z is the number of matches (that is, $Z = 0$ in case one coin shows a head and the other shows a tail, while $Z = 1$ in case the two coins "match"). By applying the definition in each case, calculate the expected value of each of the following random variables.

(a) X.
(b) $W = X - E[X]$.
(c) $(X - E[X])^2$.
(d) X^2.
(e) $X + Y$.
(f) $X \cdot Y$.
(g) $X + Z$.
(h) $(X + Z - E[X + Z])^2$.
(i) $X \cdot Z$.

9. Let X, Y, Z, W be the random variables described in Exercise 8. Decide the answer to each of the following yes-no questions, and justify each answer.
   (a) Is $E[X + Y]$ the same as $E[X] + E[Y]$?
   (b) Is $E[X \cdot Y]$ the same as $E[X] \cdot E[Y]$?
   (c) Is $E[X \cdot Z]$ the same as $E[X] \cdot E[Z]$?
   (d) Is $E[W^2]$ the same as $(E[W])^2$?
   (e) Is $E[W^2]$ the same as $E[X^2] - (E[X])^2$?

19 Properties of expected value

There may be several random variables associated with a given experiment. Suppose that X and Y are random variables for a certain experiment. Then Z is another random variable, where Z is given by the formula $Z = (X - 3)^2$. Likewise, W is a random variable, if $W = X + Y$. Again, $3X - 4$ describes a random variable. To each random variable we have assigned a number, namely, its expected value. We wish to study the elementary properties of expected value and to learn how the expected value of a combination of random variables is related to their individual expected values.

First, consider an extreme case in which a random variable Z assumes only one value, say the number c. Then $\Pr[Z = c] = 1$, since the truth set for $Z = c$ contains all possible outcomes. Consequently, $E[Z] = c \cdot 1 = c$. We may sometimes write $E[c] = c$, meaning that c is the expected value of a random variable whose sole value is c.

Second, let c be a number, and let X be a random variable. Then $E[cX] = c \cdot E[X]$. Indeed, for each value k that X has, the statements $X = k$ and $cX = ck$ are logically equivalent and have the same probability. Thus

$$E[cX] = \sum_k (ck) \cdot \Pr[cX = ck] = \sum_k c \cdot k \cdot \Pr[X = k]$$

$$= c \sum_k k \cdot \Pr[X = k] = c \cdot E[X].$$

(The preceding reasoning is applicable unless $c = 0$; if $c = 0$, then $E[cX]$ and $c \cdot E[X]$ are each equal to zero and hence are equal to one another.)

Example 30 If the expected value of the number D of dollars won in a certain game is 7, then the expected value of the number Q of quarter dollars won is 28; for $Q = 4D$, and $E[Q] = E[4D] = 4E[D] = 4 \cdot 7 = 28$.

Example 31 A coin is tossed three times in succession. Let X be the number of heads that appear. Let Y be the number of heads that appear in succession at the beginning of the sequence. Each outcome is an ordered triple with measure $\frac{1}{8}$. In Table 14 are listed each of the eight possible outcomes and the corresponding values of certain random variables. Since the sum of the entries in the column

Table 14

Outcome	X	Y	$X + Y$	$X \cdot Y$
(HEAD, HEAD, HEAD)	3	3	6	9
(HEAD, HEAD, TAIL)	2	2	4	4
(HEAD, TAIL, HEAD)	2	1	3	2
(HEAD, TAIL, TAIL)	1	1	2	1
(TAIL, HEAD, HEAD)	2	0	2	0
(TAIL, HEAD, TAIL)	1	0	1	0
(TAIL, TAIL, HEAD)	1	0	1	0
(TAIL, TAIL, TAIL)	0	0	0	0

headed X is 12, $E[X] = \frac{12}{8} = \frac{3}{2}$; similarly, the numbers in the column headed Y have total 7 and hence $E[Y] = \frac{7}{8}$. From the column headed $X + Y$, we find that $E[X + Y] = \frac{19}{8}$. Thus, $E[X + Y] = E[X] + E[Y]$. By contrast, the sum of the entries in the column headed $X \cdot Y$ is 16; hence, $E[XY] = \frac{16}{8} = 2$ and $E[X \cdot Y]$ is not the same as $E[X] \cdot E[Y]$.

Example 31 illustrates an alternative method for computing the expected value of a random variable. By definition, $E[X] = \sum_k k \cdot \Pr[X = k]$. Now the factor $\Pr[X = k]$ is itself a sum, namely, the sum of the measures of all outcomes belonging to the truth set for $X = k$. In order to calculate $E[X]$, we may multiply the measure of each possible outcome by the value of X associated with the outcome, and then add all the resulting products. This is the procedure we adopted in Example 31, as we multiplied each entry in the column headed X by the measure $\frac{1}{8}$ and added the products.

In view of this alternative method, we conclude that $E[X + Y]$ should be the same as $E[X] + E[Y]$ for any two random variables for an experiment. Indeed, $E[X] + E[Y]$ is obtainable by multiplying the measures of the outcomes by the corresponding values for X and adding, then multiplying the measures by the respective values of Y and adding, and finally adding the two sums. The resulting number is the same as that found by multiplying each measure by the respective value of the random variable $X + Y$ and adding the various products, a sum that is $E[X + Y]$.

In summary, the expected value of the sum of two random variables for an experiment is the sum of the expected values of the individual random variables. In symbols,

$$E[X + Y] = E[X] + E[Y] \quad \text{for every pair } (X, Y).$$

This result may be extended to a sum of several random variables. Moreover, we may combine this observation with the previous property of expected value and conclude that the expected value of any linear combination of random variables is the same linear combination of their expected values. For example, if X, Y, Z are random variables for an experiment and if a, b, c are numbers, then $aX + bY + cZ$ is a random variable whose expected value is $aE[X] + bE[Y] + cE[Z]$.

20 Mean, variance, and standard deviation

The expected value of a random variable is sometimes called its *mean*. We often denote the mean of a random variable by μ (the lower-case Greek letter "mu").

Example 32 Let X be a random variable, and let $\mu = E[X]$. Show that $E[X - \mu] = 0$.

We keep in mind that μ is a number. However, we may also think of a random variable whose only value is the number μ. From the latter viewpoint we obtain $E[X - \mu] = E[X] - E[\mu]$; now $E[\mu] = \mu$ as an application of our first property of the symbol E; on the other hand, $E[X] = \mu$ from the given hypotheses. Thus,

$$E[X - \mu] = E[X] - E[\mu] = \mu - \mu = 0.$$

Stated in words, Example 32 says that zero is the expected value of the *deviation* of any random variable from its own mean. The mean of a random variable is a sort of average or typical value. It is often important to have some idea of how closely the values of the random variable cluster about the mean or of how widely spread out they are. In other words, are the values concentrated near the mean or are they dispersed? To what extent?

The difference between the random variable and its mean, namely, $X - \mu$, measures the deviation away from the mean. Since no information can be gleaned from the expected value of the first power of this difference $X - \mu$ (as Example 32 has shown), we may turn to the expected value of the second power of $X - \mu$. This expected value is an important way of measuring the extent to which values of X are dispersed (or spread out) about the mean of X.

Definition Let X be a random variable for an experiment. The expected value of the second power of the difference between X and its mean is called the *variance* of the random variable X.

If we adopt the notation Var[X], then the definition, in symbols, is

$$\text{Var}[X] = E[(X - E[X])^2].$$

The formula is simpler if we use the abbreviation $\mu = E[X]$; with this understanding we may write

$$\text{Var}[X] = E[(X - \mu)^2].$$

Since every value of the random variable $(X - \mu)^2$ is nonnegative, its expected value is nonnegative. Indeed Var[X] is positive except in the very special case where X has only one value. Thus, Var[X] has a square root that is nonnegative.

Definition The *standard deviation* of a random variable is the square root of the variance of the random variable.

We customarily denote the standard deviation of a random variable by σ (the lower-case Greek letter "sigma"). Notice that σ is a number and that $\sigma \geq 0$. Since $\sigma = (\text{Var}[X])^{1/2}$, we conclude that $\text{Var}[X] = \sigma^2$. The number σ, which is the square root of the mean of the square of the deviation of X from its mean, is occasionally known as the "root-mean-square" deviation, abbreviated as "rms" deviation.

EXERCISES J

1. For each of the following pairs of random variables, calculate the expected value of the sum of the two random variables (by applying the definition), and compare the result with the sum of the expected values of the individual random variables.
   (a) The random variables in Exercises I-7g and I-7h.
   (b) The random variables in Exercises I-7f and I-7h.

2. For each of the following pairs of random variables, calculate the expected value of the product of the two random variables, and compare the result with the product of the expected values of the random variables.
   (a) The random variables in Exercises I-7g and I-7h.
   (b) The random variables in Exercises I-7f and I-7g.

3. Let r be a positive number. Let W be a random variable with a uniform distribution such that W has only two possible values, namely, r and $-r$. Calculate Var[W].

4. Let r be a positive number. Let Z be a random variable with a uniform distribution such that Z has only three possible values, namely, $-r, 0, r$. Calculate Var[Z].

5. Let Y be the number of heads in five successive tosses of a penny. Calculate Var[Y], using the definition.

6. Using the definition, calculate the variance of the random variable Z in Example 28.

7. A bag contains ten tickets bearing the respective numbers $0, 1, 2, \ldots, 9$. A contestant

in a game of chance pays his fee and then draws at random one ticket from the bag. He receives an amount of money depending on the number of the ticket he draws, as given in the accompanying table. Let the fee, f dollars, for playing the game be chosen so that the game is "fair." This means that the expected value of the net winnings of the customer is zero.

Ticket number	0	1	2	3	4	5	6	7	8	9
Payoff in dollars	6	6	0	0	2	0	12	0	2	2

(a) Show that the number f for a fair game should satisfy the equation $\frac{4}{10}(-f) + \frac{3}{10}(2-f) + \frac{2}{10}(6-f) + \frac{1}{10}(12-f) = 0$.
(b) Find the fee for a fair game.
(c) Calculate the expected value of the payoff to the contestant.
(d) Explain why, for a fair game, the expected value of the payoff, as computed in part (c), should be numerically equal to the fee for playing the game.

8. A bag contains 11 numbered tickets. A contestant in a game of chance pays to the concessionaire a fee of g dollars and then draws at random one ticket from the bag. He receives a payoff providing he has drawn a ticket with a lucky number. The accompanying table specifies each possible payoff and the corresponding probability. Let the price of g dollars for playing the game be chosen so that the expected value of the concessionaire's take on a game is $\frac{1}{2}$ dollar. Find g.

Payoff in dollars	0	3	5
Probability	$\frac{8}{11}$	$\frac{2}{11}$	$\frac{1}{11}$

21 Direct calculation of variance

Let μ be the expected value of a random variable X for a certain experiment. Then, $W = (X - \mu)^2$ is a random variable for the same experiment. By definition, the variance of X is the expected value of W, that is, $\text{Var}[X] = E[W]$. To compute $E[W]$, we may evaluate the sum $\sum_k k \cdot \Pr[W = k]$. An alternative method is to multiply the measure of each possible outcome by the corresponding value of W and to add the products. Since $W = (X - \mu)^2$, this procedure is the same as multiplying each measure by the corresponding value of $(X - \mu)^2$ and adding. Another way to express the latter sum is

$$\sum_h (h - \mu)^2 \Pr[X = h].$$

Consequently,

$$\text{Var}[X] = \sum_h (h - \mu)^2 \Pr[X = h].$$

This equation enables us to calculate σ by the following sequence of steps: give the probability distribution for X; find $\mu = E[X]$; for each possible value

h of X, subtract μ, then find the second power of the difference, afterward multiply by the corresponding probability; next add the various products; finally extract the square root.

Example 33 An experiment consists of four successive tosses of a pair of coins. Let X be the number of double heads. Find the mean and the standard deviation of X.

In this Bernoulli experiment there are four trials and for each trial the probability of success [that is, of the outcome (HEAD, HEAD)] is $\frac{1}{4}$. In Table 15 are summarized the binomial distribution for X and the calculations for determining μ and σ. Thus, $\mu = 1$ and $\sigma = \sqrt{3}/2$.

Table 15

h	$\Pr[X = h]$	$h \cdot \Pr[X = h]$	$h - \mu$	$(h - \mu)^2$	$(h - \mu)^2 \Pr[X = h]$
0	$\binom{4}{0}\left(\frac{3}{4}\right)^4\left(\frac{1}{4}\right)^0 = \frac{81}{256}$	0	-1	1	$\frac{81}{256}$
1	$\binom{4}{1}\left(\frac{3}{4}\right)^3\left(\frac{1}{4}\right)^1 = \frac{108}{256}$	$\frac{27}{64}$	0	0	0
2	$\binom{4}{2}\left(\frac{3}{4}\right)^2\left(\frac{1}{4}\right)^2 = \frac{54}{256}$	$\frac{27}{64}$	1	1	$\frac{54}{256}$
3	$\binom{4}{3}\left(\frac{3}{4}\right)^1\left(\frac{1}{4}\right)^3 = \frac{12}{256}$	$\frac{9}{64}$	2	4	$\frac{48}{256}$
4	$\binom{4}{4}\left(\frac{3}{4}\right)^0\left(\frac{1}{4}\right)^4 = \frac{1}{256}$	$\frac{1}{64}$	3	9	$\frac{9}{256}$
Sum		$\mu = \frac{64}{64} = 1$			$\sigma^2 = \frac{192}{256} = \frac{3}{4}$

EXERCISES K

1. Using the method illustrated in Example 33, calculate the standard deviation of each of the following random variables.
   (a) X, from Example 26. (b) X, from Example 29. (c) Y, from Example 31.

2. Let m be a natural number, and let $\{1, 2, \ldots, m\}$ be the set of values for a random variable X with a uniform probability distribution. Use the method of Example 33 to calculate $\text{Var}[X]$ in each of the following special cases.
   (a) $m = 3$. (b) $m = 5$. (c) $m = 2$. (d) $m = 8$. (e) $m = 1$.

22 Alternate calculation of variance

Calculation of the variance of a random variable X by a direct procedure involving values of the variable $(X - \mu)^2$ often requires more arithmetical labor than an indirect method that we now proceed to develop.

Let X be a random variable for an experiment. Then the variance of X is expressible in terms of the expected value of the random variable X^2 as follows:
$$\mathrm{Var}[X] = E[X^2] - (E[X])^2.$$

To establish this equation, we apply several properties of expected value. Let $\mu = E[X]$, and keep in mind that μ is a number. Then

$$\begin{aligned}
\mathrm{Var}[X] &= E[(X-\mu)^2] && \text{(by definition of variance)} \\
&= E[X^2 - 2\mu X + \mu^2] && \text{(by the binomial theorem)} \\
&= E[X^2] - 2\mu E[X] + E[\mu^2] && \text{(by the linear properties of expected value)} \\
&= E[X^2] - 2\mu \cdot \mu + \mu^2 && \text{(since } \mu = E[X] \text{ and since } \mu^2 \text{ is a number)} \\
&= E[X^2] - \mu^2 \\
&= E[X^2] - (E[X])^2 && \text{(since } \mu = E[X]\text{).}
\end{aligned}$$

Example 33 (again) In the Bernoulli experiment the respective values of X^2 are 0, 1, 4, 9, 16; hence,

$$E[X^2] = 0\left(\frac{81}{256}\right) + 1\left(\frac{108}{256}\right) + 4\left(\frac{54}{256}\right) + 9\left(\frac{12}{256}\right) + 16\left(\frac{1}{256}\right)$$
$$= \frac{448}{256} = \frac{7}{4}.$$

Consequently, $\mathrm{Var}[X] = \frac{7}{4} - (1)^2 = \frac{3}{4}$.

Example 34 Let Z be the number of spots appearing on a roll of a pair of dice. In Example 28 we listed the probability distribution for Z and found that $E[Z] = 7$. Now, $E[Z^2] = \sum_k k^2 \Pr[Z = k]$ is the sum of the products obtainable from the table in Example 28 by multiplying each k^2 by the appropriate probability;

$$E[Z^2] = 4\left(\frac{1}{36}\right) + 9\left(\frac{2}{36}\right) + \cdots + 121\left(\frac{2}{36}\right) + 144\left(\frac{1}{36}\right) = \frac{329}{6}.$$

Hence, $\mathrm{Var}[Z] = \frac{329}{6} - (7)^2 = \frac{35}{6}$.

Example 35 Let p be a real number such that $0 \leq p \leq 1$. Let W be a random variable that has only two possible values: the value 1 with probability p, or the value 0 with probability $1 - p$. The probability distribution for W is given in Table 16.

Table 16

k	1	0
$\Pr[W = k]$	p	$1 - p$

Then $E[W] = 1 \cdot p + 0 \cdot (1 - p) = p$. If we apply the method of Example 33 to calculate the variance of W, we obtain Table 17.

Table 17

k	$\Pr[W = k]$	$k - \mu$	$(k - \mu)^2$	$(k - \mu)^2 \Pr[W = k]$
1	p	$1 - p$	$(1 - p)^2$	$(1 - p)^2 p$
0	$1 - p$	$-p$	$(-p)^2$	$p^2(1 - p)$

Thus, $\text{Var}[W] = (1 - p)^2 p + p^2(1 - p)$, which simplifies to $p(1 - p)$. Since the preceding discussion applies to any number p satisfying $0 \leq p \leq 1$, we shall have several opportunities to utilize the results $E[W] = p$ and $\text{Var}[W] = p - p^2$ in later developments.

EXERCISES L

1. Use the relationship expressed by the formula $\text{Var}[X] = E[X^2] - (E[X])^2$ to calculate the standard deviation of each of the following random variables.
   (a) W, from Exercise J-3.
   (b) Z, from Exercise J-4.
   (c) Y, from Exercise J-5.
   (d) X, from Example 26.
   (e) X, from Example 29.
   (f) X, from Example 31.
   (g) Y, from Example 31.
   (h) $X + Y$, from Example 31.
   (i) $X \cdot Y$, from Example 31.

2. After referring to Exercise I-7, calculate the variance of each of the following random variables.
   (a) W in Exercise E-1.
   (b) W in Exercise E-5.
   (c) X in Exercise E-2.
   (d) Y in Exercise E-3.
   (e) Z in Exercise E-4.
   (f) X in Exercise D-1.
   (g) Y in Exercise D-2.
   (h) Z in Exercise D-3.
   (i) X in Exercise H-1.
   (j) The random variable in Exercise I-7j.

3. The following equation is true for every natural number m:
$$\sum_{h=1}^{m} h^2 = \frac{m(m + 1)(2m + 1)}{6}.$$
   Use this formula to calculate the standard deviation of the random variable X given in Example 27.

4. Consider the random variable W in Example 35.
   (a) Calculate $E[W^2]$.
   (b) Then calculate $\text{Var}[W]$ by using the equation $\text{Var}[W] = E[W^2] - (E[W])^2$.
   (c) Show that the variance, as found in part (b), is equal to $p(1 - p)$.

23 Properties of variance

The variance of a random variable is a measure of the "spread" or "variability" of the random variable about its mean. If the probability distribution for X gives relatively high probabilities to values of X near μ, then $\text{Var}[X]$, the spread of the values of X, is relatively small. On the other hand, if there are moderately high probabilities for values of X far from μ, then there is greater dispersion and Var[X] is relatively large.

Example 36 In order to gain a better appreciation of the notion of variance, we consider four different random variables, all of which have the same mean, but which do not have the same variance. Let the probability distributions for the random variables U, V, W, Z be given by Table 18. Graphs of these probability distributions are shown in Fig. 5.

Either computation or symmetry reveals that $E[U] = E[V] = E[W] = E[Z] = 3$; each of these random variables has mean 3. Also, notice that all the random variables have the same least value 1 and the same greatest value 5.

Table 18

(a)

k	1	2	3	4	5
Pr[U = k]	$\frac{1}{5}$	$\frac{1}{5}$	$\frac{1}{5}$	$\frac{1}{5}$	$\frac{1}{5}$

(b)

k	1	2	3	4	5
Pr[V = k]	$\frac{1}{10}$	$\frac{1}{10}$	$\frac{6}{10}$	$\frac{1}{10}$	$\frac{1}{10}$

(c)

k	1	2	3	4	5
Pr[W = k]	$\frac{2}{6}$	$\frac{1}{6}$	0	$\frac{1}{6}$	$\frac{2}{6}$

(d)

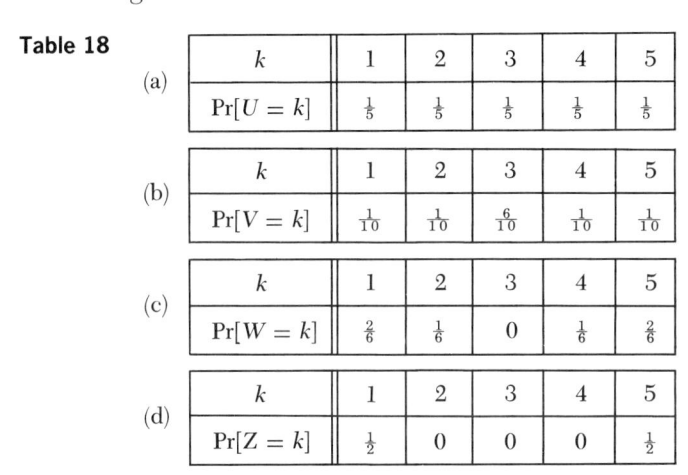

k	1	2	3	4	5
Pr[Z = k]	$\frac{1}{2}$	0	0	0	$\frac{1}{2}$

Figure 5

(a) Distribution for U

(b) Distribution for V

(c) Distribution for W

(d) Distribution for Z

By contrast, their variances are quite different. In fact, we find by straightforward calculation that Var[U] = 2, Var[V] = 1, Var[W] = 3, and Var[Z] = 4. Observe in Fig. 5 the relative "concentration" of probability near the mean in the distribution for V, the moderate "variability" in the equiprobable distribution for U, the more pronounced "spread" in the distribution for W, and the extreme case of "dispersion" in the distribution for Z.

We now turn our attention to studying the properties of variance. We shall see that they are significantly different from the properties of expected value.

If Y is a random variable that has only one value, say c, then Var[Y] = 0. Indeed, $E[Y] = c$ and the only value of $(Y - E[Y])^2$ is 0. Hence, Var[Y] = $E[(Y - E[Y])^2]$ = 0. We may abbreviate our result by saying Var[c] = 0.

If X and Y are two random variables such that $Y = X + c$, where c is a number, then Var[Y] = Var[X]. Since $E[Y] = E[X] + c$, we conclude that

$$Y - E[Y] = (X + c) - (E[X] + c) = X - E[X].$$

Hence, the expected value of the second power of $Y - E[Y]$ is the same as the expected value of the second power of $X - E[X]$; in other words, Var[Y] = Var[X]. In summary, the variance of a random variable is unaffected if all its values are increased (or decreased) by adding the same number.

If X and Z are two random variables such that $Z = cX$, where c is a number, then Var[Z] = c^2 Var[X]. We prove this result as follows. If $E[X]$ is denoted by μ, then $E[Z] = c\mu$. Hence,

$$\text{Var}[Z] = E[(Z - c\mu)^2] = E[(cX - c\mu)^2]$$
$$= E[c^2(X - \mu)^2] = c^2 E[(X - \mu)^2] = c^2 \text{Var}[X].$$

A warning is now appropriate. If X and Y are two random variables for a certain experiment, the following equation is often false: Var[$X + Y$] = Var[X] + Var[Y]. Although the equation is true for some pairs (X, Y), there are many instances in which the equation is not satisfied.

Example 37 A single card is drawn from an ordinary deck of playing cards. Let R be the number of red cards drawn, and let S be the number of spade cards drawn. Then with the aid of Table 19 we may show that Var[$R + S$] = $\frac{3}{16}$, while Var[R] + Var[S] = $\frac{1}{4} + \frac{3}{16} = \frac{7}{16}$.

Table 19

Suit of card drawn	Probability	R	S	R + S
Spade	$\frac{1}{4}$	0	1	1
Heart	$\frac{1}{4}$	1	0	1
Diamond	$\frac{1}{4}$	1	0	1
Club	$\frac{1}{4}$	0	0	0

We make three applications of the result in Example 35 concerning variance. The random variable R has only two values, 1 and 0, each with probability $\frac{1}{2}$; hence, $\mathrm{Var}[R] = (\frac{1}{2}) - (\frac{1}{2})^2 = \frac{1}{4}$. Since only two values, the value 1 with probability $\frac{1}{4}$ and the value 0, are possible for S, its variance is $(\frac{1}{4}) - (\frac{1}{4})^2 = \frac{3}{16}$. Finally, the sum $R + S$ is either 1 or 0, with probability $\frac{3}{4}$ for the value 1; consequently, $\mathrm{Var}[R + S] = (\frac{3}{4}) - (\frac{3}{4})^2 = \frac{3}{16}$.

Example 38 A card is drawn from an ordinary deck and replaced. A second card is drawn from the deck. Let S be the number of spade cards drawn on the first drawing, and let T be the number of red cards drawn on the second drawing. As in Example 37, we find that $\mathrm{Var}[S] = \frac{3}{16}$ and $\mathrm{Var}[T] = \frac{1}{4}$; hence, $\mathrm{Var}[S] + \mathrm{Var}[T] = \frac{7}{16}$. Now, $\Pr[S + T = 2] = \frac{1}{8}, \Pr[S + T = 1] = \frac{1}{2}, \Pr[S + T = 0] = \frac{3}{8}$. Hence,

$$E[S + T] = \tfrac{3}{4}, \qquad E[(S + T)^2] = 1,$$

and

$$\mathrm{Var}[S + T] = E[(S + T)^2] - (E[S + T])^2 = 1 - (\tfrac{3}{4})^2 = \tfrac{7}{16}.$$

In this case, $\mathrm{Var}[S + T]$ is the same number as $\mathrm{Var}[S] + \mathrm{Var}[T]$.

As Examples 37 and 38 show, the variance of the sum of two random variables may or may not be the same as the sum of their variances. A partial explanation of this phenomenon is that the pair of random variables in Example 38 is an independent pair, while the pair in Example 37 is not. In order to gain a better understanding, we introduce now the idea of an independent pair of random variables. The notion is closely related to that of an independent pair of statements.

Definition Let X and Y be random variables for a certain experiment. The pair (X, Y) is called an *independent pair of random variables* provided, for every ordered pair (h, k) of numbers, the statements $X = h$ and $Y = k$ form an independent pair of statements in the probability sense.

In a Bernoulli experiment, the number of successes on any one trial and the number of successes on any other trial form an independent pair of random variables. The trials are "independent" trials.

Two important special properties apply to an independent pair of random variables. These properties are summarized by the equations

$$E[X \cdot Y] = E[X] \cdot E[Y] \quad \text{and} \quad \mathrm{Var}[X + Y] = \mathrm{Var}[X] + \mathrm{Var}[Y].$$

Although it is rather rare in practice, these properties may apply even though the pair (X, Y) is not independent.

We briefly suggest why the first-mentioned property holds for an independent pair. The definition of an independent pair guarantees that
$$\Pr[(X = h) \wedge (Y = k)] = \Pr[X = h] \cdot \Pr[Y = k].$$
Hence,
$$\begin{aligned}
E[X \cdot Y] &= \sum_m m \cdot \Pr[XY = m] \\
&= \sum_{(h,k)} hk \cdot \Pr[(X, Y) = (h, k)] \\
&= \sum_h \sum_k h \cdot k \cdot \Pr[X = h] \cdot \Pr[Y = k] \\
&= \left(\sum_h h \Pr[X = h]\right)\left(\sum_k k \Pr[Y = k]\right) \\
&= E[X] \cdot E[Y].
\end{aligned}$$

We use this property to show that an independent pair of random variables also has the second-mentioned property.

$\text{Var}[X + Y]$
$$\begin{aligned}
&= E[(X + Y)^2] - (E[X + Y])^2 \\
&= E[X^2 + 2XY + Y^2] - (E[X] + E[Y])^2 \\
&= \{E[X^2] + 2E[XY] + E[Y^2]\} - \{(E[X])^2 + 2E[X]E[Y] + (E[Y])^2\} \\
&= \{E[X^2] - (E[X])^2\} + \{E[Y^2] - (E[Y])^2\} + 2\{E[XY] - E[X]E[Y]\} \\
&= \text{Var}[X] + \text{Var}[Y] + 2 \cdot 0 \\
&= \text{Var}[X] + \text{Var}[Y].
\end{aligned}$$

We may observe that in the preceding development the hypothesis concerning the independence of the pair (X, Y) is not utilized until the next to last step, where $E[XY] - E[X]E[Y]$ is replaced by 0. The previous part of the development applies for any pair (X, Y).

EXERCISES M

1. For each of the following pairs, decide whether it is an independent pair of random variables and, in each case, justify your decision.
   (a) The random variables X and Y in Exercises I-8 and I-9.
   (b) The random variables X and Z in Exercises I-8 and I-9.
   (c) The random variables in Exercise J-2a.
   (d) The random variables in Exercise J-2b.
   (e) The random variables X and Y in Example 31.

2. Let Y be a random variable with probability distribution given by the accompanying table.

k	1	2	3	4	5
$\Pr[Y = k]$	0.02	0.24	0.48	0.24	0.02

(a) Draw a bar graph of the probability distribution as in Example 36.
(b) Evaluate $E[Y]$.
(c) Evaluate $\text{Var}[Y]$.
(d) If μ and σ are the mean and standard deviation for Y, compute $\mu - 2\sigma$ and $\mu + 2\sigma$.
(e) Show that there is a very high probability that the random variable lies in the interval $\{t \,|\, \mu - 2\sigma \leq t \leq \mu + 2\sigma\}$.
(f) Show that $\{t \,|\, \mu - 2\sigma \leq t \leq \mu + 2\sigma\} = \{t \,|\, -2\sigma \leq t - \mu \leq 2\sigma\}$.
(g) Consider the random variables U, V, W, Z in Example 36. For each of them, show that the probability is 1 that the random variable differs numerically from its mean by no more than twice its standard deviation.

3. For each of the following random variables, let μ denote its mean and σ denote its standard deviation. Find the probability that the random variable is between the numbers $\mu - \sigma$ and $\mu + \sigma$, inclusive.
(a) The random variable U in Example 36.
(b) The random variable V in Example 36.
(c) W in Example 36.   (d) Z in Example 36.   (e) Y in Exercise 2.

24 Summary of mean and variance

We collect, as a summary, some fundamental properties of expected value and of variance. We suppose that c is a number and that X and Y are random variables for an experiment.

1. $E[X] = \sum_{k} k \cdot \Pr[X = k]$.
2. $E[c] = c$.
3. $E[cX] = cE[X]$.
4. $E[X + Y] = E[X] + E[Y]$.
5. $E[X \cdot Y] = E[X] \cdot E[Y]$ may be a false statement, but it is true for every independent pair (X, Y) of random variables.
6. $\text{Var}[X] = E[(X - E[X])^2]$.
7. $\text{Var}[X] = E[X^2] - (E[X])^2$.
8. $\text{Var}[c] = 0$.
9. $\text{Var}[cX] = c^2 \text{Var}[X]$.
10. $\text{Var}[X + c] = \text{Var}[X]$.
11. $\text{Var}[X + Y] = \text{Var}[X] + \text{Var}[Y]$ may be a false statement, but it is true for every independent pair (X, Y) of random variables.

We consider a simple application of these results. Let X be a random variable. Let μ and σ be the mean and the standard deviation for X, respectively. Suppose $\sigma > 0$. Then $T = (X - \mu)/\sigma$ is a random variable for the same experiment. We calculate the mean and the standard deviation for T.

First, $E[T] = E[\sigma^{-1}(X - \mu)] = \sigma^{-1}E[X - \mu] = 0$, in accordance with Example 32. Next, $\text{Var}[T] = \text{Var}[\sigma^{-1}(X - \mu)] = (\sigma^{-1})^2 \cdot \text{Var}[X - \mu] = \sigma^{-2} \cdot \text{Var}[X] = \sigma^{-2} \cdot \sigma^2 = 1$. Thus, T has mean 0 and standard deviation 1. We may call T the *standard* random variable associated with the given X.

25 Binomial distribution

Consider a Bernoulli experiment consisting of n trials on each of which the probability for success is the number p. Let S be the total number of successes in the experiment. Then S is a random variable whose mean is np and whose variance is $np(1-p)$. In symbols,

$$E[S] = np \quad \text{and} \quad \text{Var}[S] = np(1-p).$$

We wish to establish these two important properties of a binomial distribution.
    One approach is by means of the definitions. The mean is given by

$$E[S] = \sum_{k=0}^{n} k \cdot \Pr[S = k] = \sum_{k=0}^{n} k \cdot \binom{n}{k}(1-p)^{n-k}p^{k}.$$

We may find the variance of S by computing

$$E[S^2] = \sum_{k=0}^{n} k^2 \binom{n}{k}(1-p)^{n-k}p^{k}.$$

Although evaluation of the sums for $E[S]$ and $E[S^2]$ can be done without great effort (in fact, we have seen examples for small n), there is another method which involves much less calculation. Instead, it relies heavily upon the theory that we have developed.
    For each $j \in \{1, 2, \ldots, n\}$, introduce the random variable X_j which is the number of successes on the jth trial in the experiment. Then X_j has only two possible values, 1 with probability p, and 0 with probability $1-p$. Consequently, by Example 35,

$$E[X_j] = p.$$

Now S, the total number of successes in all the trials, is simply $X_1 + X_2 + \cdots + X_n$. Thus,

$$E[S] = E[X_1 + \cdots + X_n] = E[X_1] + \cdots + E[X_n] = p + \cdots + p = np.$$

Also, by Example 35, we find for each $j \in \{1, 2, \ldots, n\}$ that

$$\text{Var}[X_j] = p - p^2 = p(1-p).$$

Now, every pair of the n random variables $X_1, \ldots, X_n$ is an independent pair. Hence,

$$\begin{aligned}\text{Var}[S] &= \text{Var}[X_1 + \cdots + X_n] = \text{Var}[X_1] + \cdots + \text{Var}[X_n] \\ &= p(1-p) + \cdots + p(1-p) = np(1-p).\end{aligned}$$

For a binomial distribution, $\mu = np$ and $\sigma = [np(1-p)]^{1/2}$. If we adopt the frequently used abbreviation q for $1-p$, the probability of failure on each individual trial in a Bernoulli experiment, then $\sigma = (npq)^{1/2}$.

Example 39 Suppose that a Bernoulli experiment consists of 100 trials and that the probability of success on each single trial is $\frac{1}{10}$. Then the expected value of the number of successes is $100(\frac{1}{10}) = 10$; the variance is $100(\frac{1}{10})(\frac{9}{10}) = 9$, and the standard deviation of the number of successes is 3. If S is the number of successes, then $(S - 10)/3$ is the corresponding standard random variable.

EXERCISES N

1. Calculate the mean and the standard deviation for the number of successes in a Bernoulli experiment involving n trials with probability p of success on each trial, in each of the following special cases.
   (a) $n = 50$ and $p = \frac{1}{5}$. (b) $(n, p) = (1000, \frac{3}{4})$.
   (c) $n = 2$ and $p = 0.9$. (d) $(n, p) = (200, \frac{1}{3})$.

2. Let W be a random variable given by the probability distribution $\Pr[W = 1] = \frac{4}{5}$, $\Pr[W = 0] = \frac{1}{5}$.
   (a) Find the expected value of the random variable $(5W - 4)/2$.
   (b) Find the variance of the random variable $(5W - 4)/2$.
   (c) Show that $(5W - 4)/2$ is the standard random variable associated with W.

3. Find the standard random variable associated with each of the following random variables.
   (a) W in Exercise J-3.   (b) Z in Exercise J-4.
   (c) Y in Exercise J-5.   (d) X in Example 27 (compare Exercise L-3).

4. Let X, Y, Z be random variables pertaining to a certain experiment. Suppose that $E[X] = 3$ and $\text{Var}[X] = 2$, that $E[Y] = -1$ and $E[Y^2] = 3$, that $E[Z] = 0$ and $E[Z^2] = 16$. Suppose that (X, Z) is an independent pair of random variables. For each of the following, decide whether the given information is adequate to determine the number; if so, calculate the number; if not so, explain why there is insufficient information.
   (a) $E[5X]$.   (b) $E[X + 2]$.   (c) $E[Y + 1]$.   (d) The standard deviation for Z.
   (e) $\text{Var}[Y + 1]$.   (f) $E[7X^2]$.   (g) $E[Y + Z]$.   (h) $\text{Var}[X + Y]$.
   (i) $\text{Var}[2X + 5Z]$.   (j) $E[Y \cdot Z]$.   (k) $\text{Var}[X - Z]$.   (l) $E[(X + 2Z)^2]$.

5. Three lottery tickets in a hat are labeled 1, 2, 3, respectively. Consider two different experiments. In one experiment, a single ticket is drawn at random from the hat; let X be the number on the ticket drawn. In another experiment (which begins with all three tickets in the hat), two tickets are drawn at random, in succession, without replacement of the first before the drawing of the second; let X_1 and X_2 be the respective numbers on the first and second tickets drawn in this experiment; let $\bar{X} = (X_1 + X_2)/2$, and let $W = [(X_1 - \bar{X})^2 + (X_2 - \bar{X})^2]/2$.
   (a) Make a table listing all the possible outcomes for the second experiment and the corresponding values of the random variable $\bar{X}$.
   (b) Use the table in part (a) to calculate $E[\bar{X}]$.
   (c) Find $E[X_1]$ and $E[X_2]$.
   (d) Use the results of part (c) to calculate $E[\bar{X}]$.
   (e) Calculate $E[X]$, and compare $E[X]$ with $E[\bar{X}]$.
   (f) Calculate $\text{Var}[X]$.   (g) Calculate $\text{Var}[\bar{X}]$.   (h) Calculate $E[W]$.
   (i) For the three numbers $E[W]$, $\text{Var}[X]$, $\text{Var}[\bar{X}]$, show that one of them is the sum of the other two.

6. From an ordinary deck of playing cards, four cards are drawn at random, in succession. Each card drawn in turn is replaced before the next drawing is made. For each $j \in \{1, 2, 3, 4\}$, let the random variable X_j be the number of cards in the spade suit on the jth drawing. Let $\bar{X} = (X_1 + X_2 + X_3 + X_4)/4$, and let $W = \frac{1}{4} \sum_{j=1}^{4} (X_j - \bar{X})^2$.

   (a) Make a table, listing all the possible values for $\bar{X}$ and the probability of each.
   (b) Use the information in part (a) to calculate $E[\bar{X}]$.
   (c) Calculate $\text{Var}[\bar{X}]$.
   (d) Calculate $E[W]$.

7. A bowl contains one white ball and one red ball. Consider two different experiments. In one experiment, a single ball is drawn at random from the bowl and replaced; let X be the number of white balls drawn. In another experiment, a ball is drawn at random and replaced, again a ball is drawn at random and replaced, and then a third trial is made. For each $j \in \{1, 2, 3\}$, let the random variable X_j be the number of white balls drawn on the jth trial. Let $\bar{X} = \frac{1}{3} \sum_{j=1}^{3} X_j$, and let $W = \frac{1}{3} \sum_{j=1}^{3} (X_j - \bar{X})^2$.

   (a) Make a table, listing all the possible outcomes for the second experiment and the corresponding values of the random variable $\bar{X}$.
   (b) Use the table in part (a) to calculate $E[\bar{X}]$.
   (c) Find $E[X]$, and compare $E[X]$ with $E[\bar{X}]$.
   (d) Find $E[X_j]$ for each j.
   (e) Use the results of part (d) to find $E[\bar{X}]$.
   (f) Calculate $\text{Var}[X]$.
   (g) Calculate $\text{Var}[\bar{X}]$.
   (h) Find a simple numerical relationship connecting the three numbers: $\text{Var}[X]$, $\text{Var}[\bar{X}]$, the number of trials in the experiment.
   (i) The relationship discovered in part (h) is applicable because each pair of the random variables X_1, X_2, X_3 is an independent pair. Determine whether the analogous relationship applies to the situation in Exercise 5.
   (j) Calculate $E[W]$.
   (k) In Exercise 5i, a relationship connecting three numbers for that experiment was discovered. Decide whether the corresponding relationship applies to the numbers $E[W]$, $\text{Var}[X]$, $\text{Var}[\bar{X}]$ in the present situation.

8. Use the results of Section 25 in order to show that if n is a natural number and p is a real number satisfying $0 < p < 1$, then $\sum_{k=0}^{n} k^2 \binom{n}{k}(1-p)^{n-k} p^k = np(np - p + 1)$.

26 Samples

Consider an experiment that consists of a sequence of n activities. For each $j \in \{1, 2, \ldots, n\}$, let X_j be a random variable for the jth activity. Suppose that the experiment is performed and that v_j is the observed value of X_j for each j. Then the vector $(v_1, v_2, \ldots, v_n)$ represents the observed outcome for the entire experiment. We may call this vector a *sample point*. The average of the numbers $v_1, \ldots, v_n$, namely, $(v_1 + \cdots + v_n)/n$, may be called the *sample average*. The sample average, which we may denote by $\bar{v}$, is the observed value of the random variable $\bar{X} = (X_1 + \cdots + X_n)/n$.

The difference $v_j - \bar{v}$ measures the deviation of the jth observation from the sample average. The average of the numbers $(v_1 - \bar{v})^2, (v_2 - \bar{v})^2, \ldots, (v_n - \bar{v})^2$ is the number $[(v_1 - \bar{v})^2 + \cdots + (v_n - \bar{v})^2]/n$; we may call this number the *sample variance*. The square root of the sample variance is called the *sample standard deviation*. If we denote the sample standard deviation by s, then the sample variance is s^2.

The preceding definitions may be summarized as follows:

$$\bar{v} = \frac{1}{n} \sum_{j=1}^{n} v_j,$$

$$s^2 = \frac{1}{n} \sum_{j=1}^{n} (v_j - \bar{v})^2,$$

$$s = \left[\frac{1}{n} \sum_{j=1}^{n} (v_j - \bar{v})^2 \right]^{1/2}$$

In some statistical investigations the sample variance is defined slightly differently, namely, as $(n-1)^{-1} \sum_{j=1}^{n} (v_j - \bar{v})^2$. There are strong arguments to support each of the two viewpoints. Whichever definition of sample variance is adopted, its square root is the sample standard deviation.

Example 40 A single die is rolled four times in succession. The observed result of this experiment is given by the ordered quadruple $(5, 2, 4, 5)$. The sample average is $\bar{v} = (5 + 2 + 4 + 5)/4 = 4$. The sample variance (as we have defined it) is the average of $(5-4)^2$, $(2-4)^2$, $(4-4)^2$, $(5-4)^2$; thus, $s^2 = (1 + 4 + 0 + 1)/4 = 3/2$; hence, the sample standard deviation is $\sqrt{6}/2$.

Example 41 An activity consists of ten successive tosses of a coin. An experiment consists of ten successive trials of the activity.

In Table 20 are presented the results for a certain actual performance of this experiment on a nickel. We abbreviate the outcomes

Table 20

Trial number	Outcome	Number of heads
1	(H, T, T, H, T, H, H, T, H, T)	$v_1 = 5$
2	(H, H, T, T, T, H, H, H, T, H)	$v_2 = 6$
3	(H, T, H, T, H, T, H, H, H, T)	$v_3 = 6$
4	(T, T, T, T, H, T, H, T, T, H)	$v_4 = 3$
5	(T, H, T, T, H, H, H, T, H, H)	$v_5 = 6$
6	(T, T, T, H, T, T, H, T, H, T)	$v_6 = 3$
7	(T, T, H, T, T, H, H, T, T, H)	$v_7 = 4$
8	(H, H, H, T, H, T, H, T, H, H)	$v_8 = 7$
9	(T, H, H, T, H, T, H, T, T, H)	$v_9 = 5$
10	(H, T, T, H, T, T, T, T, T, H)	$v_{10} = 3$

HEAD and TAIL by H and T, respectively, and we let v_j be the number of heads that appear in the jth trial of the activity, for

$$j \in \{1, 2, \ldots, 10\}.$$

The observed sample point is

$$(v_1, \ldots, v_{10}) = (5, 6, 6, 3, 6, 3, 4, 7, 5, 3).$$

The sample average is

$$\bar{v} = \tfrac{1}{10}(v_1 + \cdots + v_{10}) = \tfrac{1}{10}(48) = 4.8.$$

The sample variance is

$$s^2 = \tfrac{1}{10}[(v_1 - 4.8)^2 + \cdots + (v_{10} - 4.8)^2]$$
$$= \tfrac{1}{10}[(0.2)^2 + (1.2)^2 + \cdots + (-1.8)^2]$$
$$= \tfrac{1}{10}(19.60) = 1.96.$$

The sample standard deviation is

$$s = \sqrt{1.96} = 1.4.$$

EXERCISES O

1. A pair of gambling dice are rolled five times. The observed results are given by the vector (7, 11, 7, 9, 5). Calculate the sample average, the sample variance, and the sample standard deviation.

2. The pair of dice in Exercise 1 are rolled five more times, yielding the sample point (12, 10, 10, 4, 9). Calculate the sample average, the sample variance, and the sample standard deviation.

3. Suppose that a survey is conducted to determine the number of books a student reads in a year. Ten students are selected; their replies are given by the vector (5, 6, 4, 12, 6, 4, 28, 60, 1, 4). Calculate the sample average, the sample variance, and the sample standard deviation.

4. Consider the ten successive rolls of the pair of dice for which the results are given in Exercises 1 and 2. For this combined experiment, calculate the sample average, the sample variance, and the sample standard deviation.

5. Discover, if possible, a simple relationship connecting the sample averages in Exercises 1, 2, 4.

6. Discover, if possible, a simple relationship connecting the sample standard deviations in Exercises 1, 2, 4.

7. Using the defining formulas for the sample average and the sample variance, namely,

$$\bar{v} = \frac{1}{n}\sum_{j=1}^{n} v_j \quad \text{and} \quad s^2 = \frac{1}{n}\sum_{j=1}^{n}(v_j - \bar{v})^2,$$

respectively, derive the (computationally useful) alternate formula

$$s^2 = \left(\frac{1}{n}\sum_{j=1}^{n} v_j^2\right) - \bar{v}^2.$$

27 Median and mode

Consider an experiment that consists of a sequence of n activities. For each $j \in \{1, \ldots, n\}$, let X_j be a random variable for the jth activity. An actual performance of the experiment determines an observed sample point $(v_1, \ldots, v_n)$. We may think of having made a sequence of n observations or measurements. Suppose now that n is not small. We may arrange the components of the vector in ascending order of magnitude, beginning with the smallest and ending with the largest. When the observed data are thus arranged, the "one in the middle" is called the median. Specifically, if n is an odd number, then the *median* of the sample is the number occupying the $[(n+1)/2]$th position in the list; if n is an even number, then the *median* of the sample is the average of the numbers appearing in the $[n/2]$th position and the $[n/2 + 1]$th position in the list.

Example 42 Let $(17, 5, 8, 13, 15, 9, 9, 8, 16, 20, 18)$ be an observed sample point. Arranging the components in ascending order, we obtain 5, 8, 8, 9, 9, 13, 15, 16, 17, 18, 20. Among the 11 entries, the middle position is the sixth; note that $6 = (11+1)/2$. The median of the sample is the number 13.

Example 43 Suppose that $(0, 2, 1, 1, 3, 1, 1, 4, 2)$ is an observed sample point. Arranged in ascending order, the components are 0, 1, 1, 1, 1, 2, 2, 3, 4. The median of the sample is the number appearing in the fifth position, namely, the number 1.

Example 44 Let $(82, 91, 90, 86, 42, 86)$ be a sample point. Among six entries in a list, the "middle" position is midway between the third and the fourth. (Note that $3 = \frac{6}{2}$ and $4 = \frac{6}{2} + 1$.) The median of the sample is 86.

Example 45 The median of the sample point $(0.3, -0.2, 0.7, -0.8, -0.4, 0.1, 0.0, -0.6)$ is the average of -0.2 and 0.0, namely, -0.1.

Again consider an experiment in which we make n observations or measurements. Again suppose that the number n is not small. Let a sample point be observed. Suppose that there are several numbers, each of which is a component of the vector several times. The number (if any) that is a component of the vector most often may be called the *mode* of the sample.

Example 46 Suppose that $n = 160$ and that the frequency of occurrence of each number b as a component of the observed sample

Table 21

b	230	240	250	260	270	280	290	300
Frequency (number of times b is a component of the vector)	16	21	35	12	9	22	12	33

point is given in Table 21. The mode of the sample is 250, because 35 is the greatest frequency. The median of the sample is 260, because the 80th and 81st positions, arranged from bottom to top, are each occupied by 260. The sample average is 266, because

$$(230 \cdot 16 + 240 \cdot 21 + 250 \cdot 35 + 260 \cdot 12 + 270 \cdot 9 \\ + 280 \cdot 22 + 290 \cdot 12 + 300 \cdot 33)/160 = 266.$$

Example 47 Suppose that 100 observations yield the data in Table 22.

Table 22

Observed number	1	2	3	4	5	6	7
Frequency	11	19	23	12	28	5	2

The mode of the sample is 5, which has the maximum frequency. Since $11 + 19 < 50 < 11 + 19 + 23$, the median is 3. The sample average is $\bar{v} = \frac{1}{100} \sum v_j = 3.5$. The sample variance is $s^2 = \frac{1}{100} \sum v_j^2 - (\bar{v})^2 = 14.64 - (3.5)^2 = 2.39$. Thus, the sample standard deviation s is approximately 1.55.

We may note that $\bar{v} - 2s = 3.5 - 3.1 = 0.4$ and $\bar{v} + 2s = 3.5 + 3.1 = 6.6$. Almost all of the observed data belong to the set $\{x \mid 0.4 < x < 6.6\} = \{x \mid -2s < x - \bar{v} < 2s\}$. In other words, almost all of the data are closer to the sample average than two sample standard deviations away. Indeed, only two of the 100 observations lie farther from the sample average.

The property mentioned in the preceding paragraph is typical of a great many practical situations. Customarily, the set $\{x \mid \bar{v} - 2s < x < \bar{v} + 2s\}$ contains almost all of the observed data when a reasonably large number of observations have been made.

EXERCISES P

1. An experiment consisting of 25 rolls of a pair of dice yielded the vector (5, 8, 6, 7, 6, 10, 11, 8, 10, 12, 2, 6, 8, 2, 5, 10, 5, 9, 7, 9, 5, 5, 3, 11, 9). Calculate the median, the mode, and the sample average.

2. Consider the experiment performed and recorded in Example 41. For the observed sample point, find the percentage of trials that belong to each of the following sets.
    (a) $\{x | \bar{v} - s < x < \bar{v} + s\}$.   (b) $\{x | \bar{v} - 2s < x < \bar{v} + 2s\}$.

3. Take a coin. Perform for yourself the experiment described in Example 41, keeping a record of the outcomes of the 100 tosses. For the data you observe, determine each of the following.
    (a) The observed sample point $(v_1, \ldots, v_{10})$.
    (b) The sample average $\bar{v}$.
    (c) The sample variance s^2.
    (d) The sample standard deviation s.
    (e) The percentage of trials belonging to the set $\{x | \bar{v} - s < x < \bar{v} + s\}$.
    (f) The percentage of trials belonging to the set $\{x | \bar{v} - 2s < x < \bar{v} + 2s\}$.

4. Repeat Exercise N-6, with the following modification: in the description of the experiment, substitute the phrase "is not replaced" for the phrase "is replaced."

5. Repeat Exercise 4 with the following modification: in the description of the random variables, substitute the phrase "red kings" for the phrase "cards in the spade suit." (Retain the feature of successive drawings without replacement.)

PART III

Elementary Functions

Functions and Their Graphs

Distance

Inverse and Rational Functions

Transcendental Functions

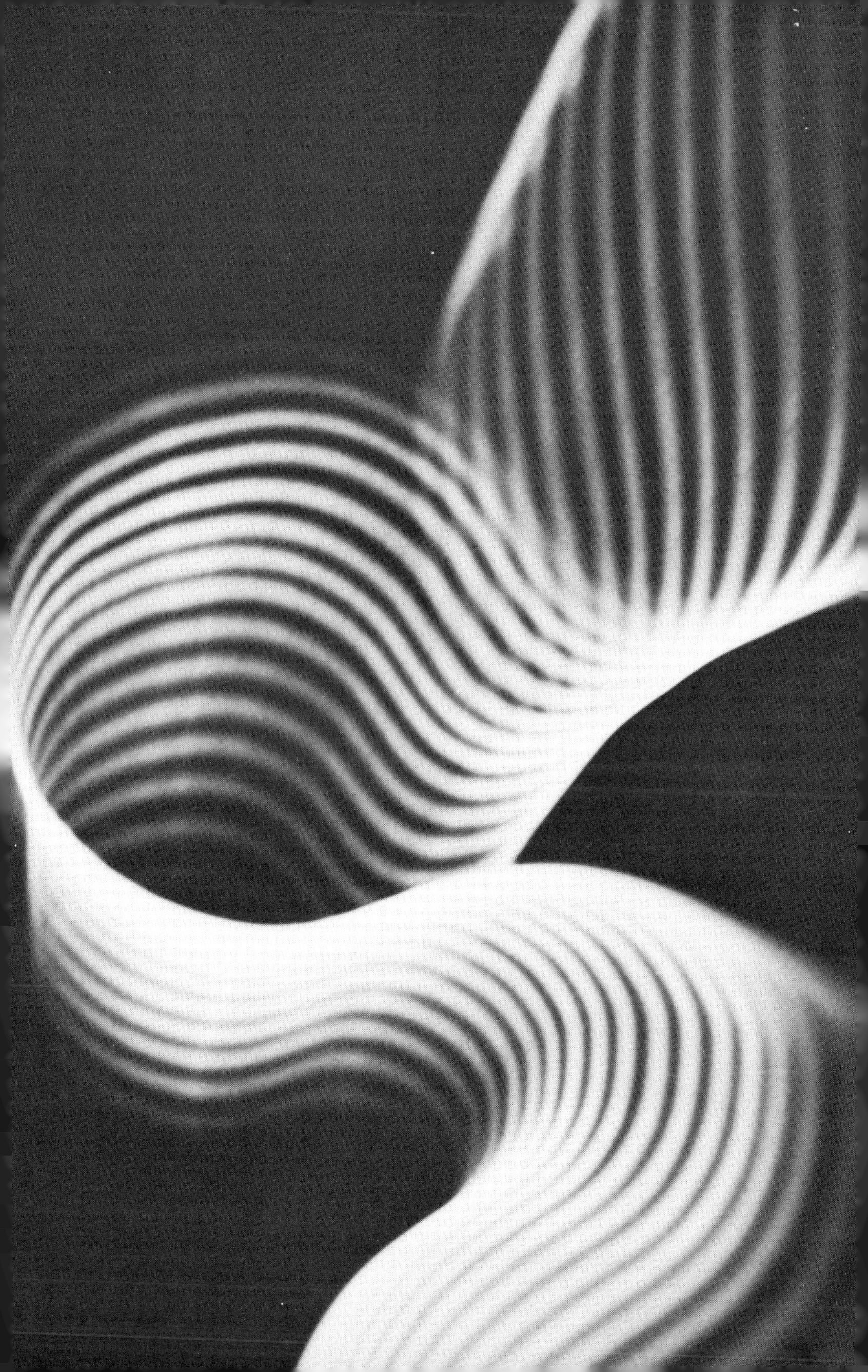

Functions and Their Graphs

1 Introduction

In Chapter 1 we discussed linear functions. A linear function L is often given by a formula of the type $L(x) = mx + b$, where m and b are real numbers such that $m \neq 0$. A major aid in our appreciation of a linear function is provided by its graph; on a two-dimensional coordinate system the graph of a linear function is a line.

Example 1 The formula $2x + 1$ describes a linear function. Some of the ordered pairs that belong to this linear function are $(-2, -3)$, $(-1, -1)$, $(0, 1)$, $(\frac{3}{2}, 4)$, and $(68, 137)$. The graph of the linear function is also called the graph of the equation $y = 2x + 1$. The graph is a line with slope 2. (See Fig. 1.) The line consists of all points

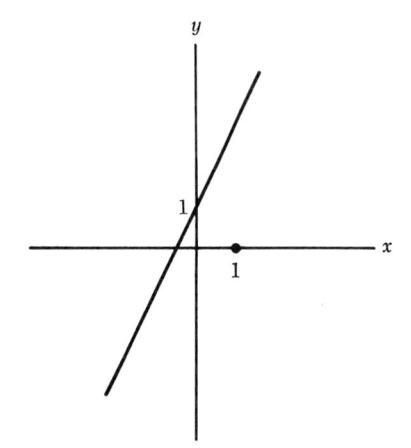

Figure 1

whose coordinates (x, y) satisfy the equation $y = 2x + 1$. In other words, the linear function is the set of all ordered pairs (x, y) of real numbers such that $y = 2x + 1$. In the language of set theory, we may express the linear function as $\{(x, y) \mid y = 2x + 1\}$.

In the present chapter we begin a detailed study of other functions. In each case a picture representing the function offers a vivid presentation of the characteristic features of the function. Such a picture is a graph of the function on a two-dimensional coordinate system.

2 Congruent parabolas

We turn our attention to examining and comparing the graphs of several functions that can be expressed by rather simple algebraic formulas. Next after the linear relationships discussed in Chapter 1, perhaps the simplest algebraic expression is x^2.

Example 2 Sketch the graph of the equation $y = x^2$.

An elementary method of attacking the problem is to make a table of corresponding values for x and y. If we select a real number for x, the formula x^2 tells us that the corresponding value for y is calculated by multiplying the chosen number by itself. Thus if $x = 3$, then $y = 3 \cdot 3 = 9$; if $x = -2$, then $y = (-2)^2 = 4$. In this manner we obtain Table 1.

Table 1

x	-4	-3	-2	-1	$-\frac{1}{2}$	0	$\frac{1}{3}$	$\frac{2}{3}$	1	$\frac{3}{2}$	2	4	7
y	16	9	4	1	$\frac{1}{4}$	0	$\frac{1}{9}$	$\frac{4}{9}$	1	$\frac{9}{4}$	4	16	49

We now plot on a coordinate system each ordered pair of corresponding numbers. These include $(-2, 4)$, $(-1, 1)$, $(-\frac{1}{2}, \frac{1}{4})$, $(0, 0)$, $(\frac{2}{3}, \frac{4}{9})$, $(\frac{3}{2}, \frac{9}{4})$, and as many more as we like. Each of the points plotted in Fig. 2 has coordinates of the type (a, a^2), that is, the right coordinate is the second power of the left coordinate. The desired graph

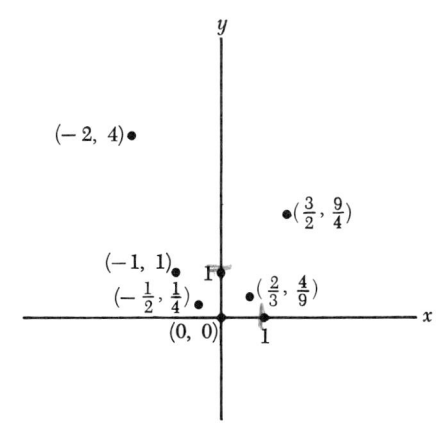

Figure 2

Figure 3

Figure 4

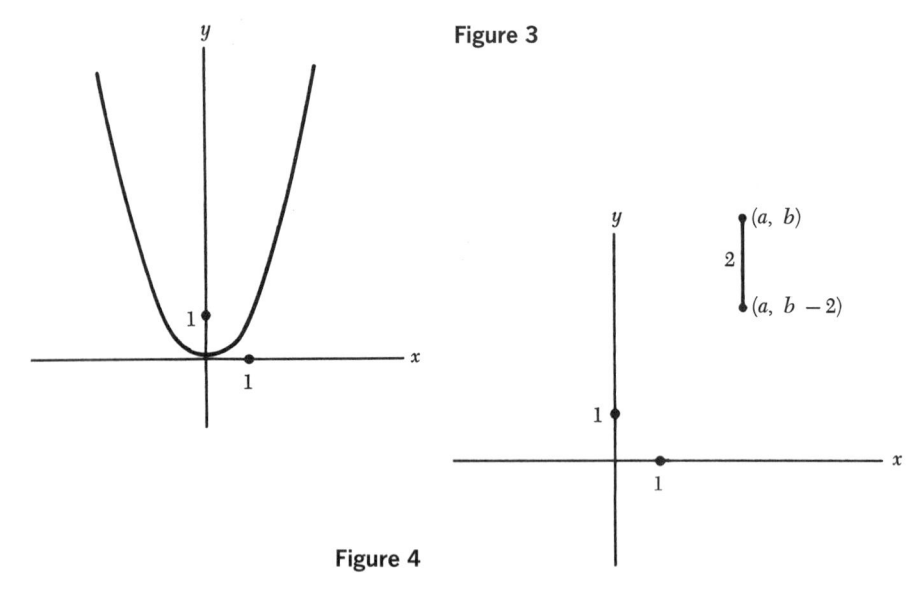

consists of *all* points of the form (a, a^2). There is one of these points for each real number a. If we imagine the result obtained by plotting all these points, it seems reasonable that they may form a curve. This, indeed, is the case, and Fig. 3 shows a part of the curve. The entire curve is called a *parabola*. The parabola is the graph of a certain set of ordered pairs of real numbers, namely, the set $\{(x, y) \mid y = x^2\}$.

Example 3 Sketch the graph of the equation $y = x^2 - 2$.

Although we can solve this problem in the same manner as Example 2, it is much more instructive to proceed as follows. Suppose we are given a real number as a value for x and wish to calculate the corresponding value of y. The formula $x^2 - 2$ instructs us to multiply the given number by itself and afterwards subtract 2 from the product. The first of these calculations is the same as we did in Example 2. In the present example, then, we may take each result obtained in the preceding example and subtract 2. In other words, for each number x, the associated y in the present example is two less than the corresponding y in the former situation.

What is the geometrical meaning of the preceding statement? It states that if a point (a, b) belongs to the graph in Example 2, then the point $(a, b - 2)$ belongs to the graph in Example 3. Now $(a, b - 2)$ is two units below (a, b) on the coordinate system (see Fig. 4). Thus, if we take each point belonging to the graph in Example 2 and lower it two units on the coordinate system, we will obtain a point on the desired graph. Collectively, then, the set of all points on the desired graph looks like the graph in Example 2 insofar as size and shape are concerned, but is located two units lower. The desired graph is suggested by Fig. 5.

Figure 5

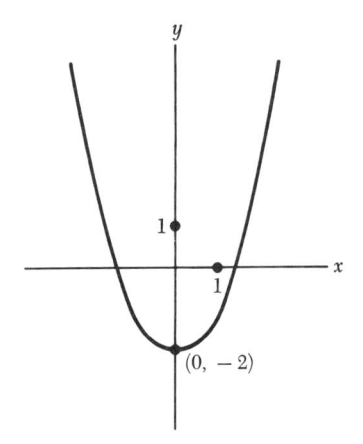

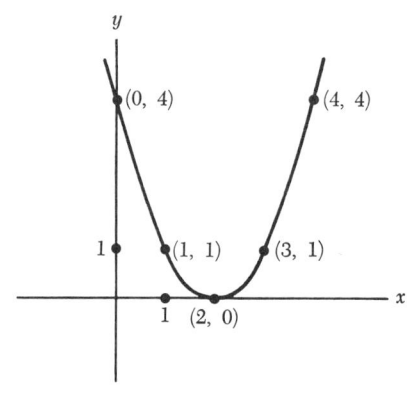

Figure 6

The parabola that is the graph of the set $\{(x, y) \mid y = x^2 - 2\}$ is congruent to the graph of the equation $y = x^2$ and is located two units below it.

The lowest point on the graph of the equation $y = x^2$ is the origin. Except for this point, the graph lies entirely in the upper coordinate halfplane $\{(x, y) \mid y > 0\}$. The lowest point on the parabola represented in Fig. 5 is $(0, -2)$. The parabola is contained in the closed halfplane $\{(x, y) \mid y \geq -2\}$.

Example 4 Sketch the graph of the equation $y = (x - 2)^2$.

We prepare Table 2 showing various members of the set $\{(x, y) \mid y = (x - 2)^2\}$. We plot these points and join them by the smooth curve that seems to be formed by them and other points whose coordinates satisfy the equation $y = (x - 2)^2$ (Fig. 6).

Table 2

x	-2	-1	0	1	$\frac{3}{2}$	2	$\frac{5}{2}$	3	4	5
y	16	9	4	1	$\frac{1}{4}$	0	$\frac{1}{4}$	1	4	9

A comparison of Fig. 6 with Fig. 3 reveals a strong resemblance. The two curves appear to be congruent to one another. It seems that the curve in Fig. 6 is a parabola that can be obtained from the parabola in Fig. 3 by sliding, or shifting, the curve two units toward the right.

We wish to present an analytical argument to demonstrate that the observations mentioned in the preceding paragraph are justifiable. Let Q denote the set $\{(x, y) \mid y = x^2\}$ whose graph is the parabola shown in Fig. 3, and let S be the set $\{(x, y) \mid y = (x - 2)^2\}$ under consideration in the present example. Suppose that the point (a, b) belongs to the graph of Q; this means that $b = a^2$. We contend that then the point $(a + 2, b)$ belongs to the graph of S; to substantiate

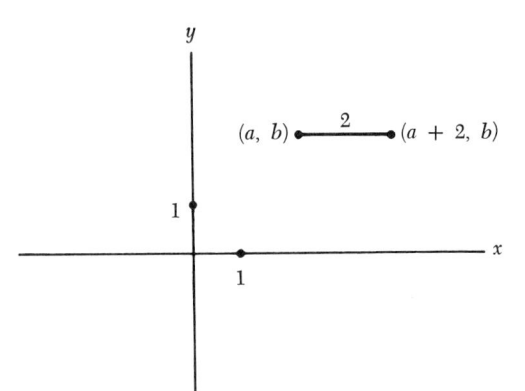

Figure 7

our claim, we check that the coordinates of $(a + 2, b)$ satisfy the equation $y = (x - 2)^2$; indeed, substituting $a + 2$ for x in the right member of the equation, we obtain $[(a + 2) - 2]^2 = a^2$, which agrees with the result obtained if we replace y in the left member by b.

In summary, for any point (a, b) in the graph of Q, the point $(a + 2, b)$ belongs to the graph of S. On a coordinate system, the point $(a + 2, b)$ is two units to the right of the point (a, b) (see Fig. 7). This means that each point on the graph of Q, if shifted two units to the right, becomes a point on the graph of S. Collectively then the graph of S is congruent to, and is located on the coordinate system two units to the right of, the graph of Q.

EXERCISES A

1. Sketch the graph of each of the following equations.
   (a) $y = -x$. (b) $y = x^2$. (c) $y = -x^2$.

2. Sketch the graph of each of the following sets.
   (a) $\{(x, y) \mid y = -x + 3\}$. (b) $\{(x, y) \mid y = x^2 + 3\}$. (c) $\{(x, y) \mid y = -x^2 + 3\}$.

3. For each of the following pairs of curves, explain the relationship between the two curves, telling in what ways they are alike and in what ways they are different.
   (a) The curves in Exercises 1a and 2a.   (b) The curves in Exercises 1b and 2b.
   (c) The curves in Exercises 1c and 2c.

4. Sketch the graph of each of the following.
   (a) $y = (x + 3)^2$. (b) $y = -(x + 3)^2$.

5. Repeat Exercise 3 for each of the following pairs of curves.
   (a) The curves in Exercises 1b and 4a.   (b) The curves in Exercises 1c and 4b.

6. Compare and contrast the three curves that are the respective graphs of the equations $y = x^2$, $y = (x + 3)^2$, $y = (x - 3)^2$.

7. Compare and contrast the graphs of $y = x^2$, $y = x^2 + 4$, $y = x^2 - 4$.

8. Sketch the graph of each of the following equations.
   (a) $y = x^3$. (b) $y = -x^3$. (c) $y = x^4$.

9. Sketch the graph of the set $\{(x, y) \mid y = x^2 + 2x + 1\}$. [HINT: Use the binomial theorem.]
10. Find an equation for each curve described below.
    (a) The curve is congruent to the graph of $y = x^2$ and is located five units below the graph of $y = x^2$.
    (b) The curve is congruent to, and is located ten units to the left of, the graph of the set $\{(x, y) \mid y = x^2\}$.
    (c) The curve is congruent to, and is located on the coordinate system one unit to the right of, the graph of the equation $y = x^3$.
    (d) The curve is a reflection of the graph of $y = x^4$, namely, the reflection in the x-coordinate axis.

3 Other examples

Example 5 Sketch the graph of the equation $y = 1/x$.

Several ordered pairs belonging to the set $\{(x, y) \mid y = 1/x\}$ are presented in Table 3. The equation $y = 1/x$ is equivalent to the

Table 3

x	1	2	3	$\frac{1}{2}$	$\frac{1}{4}$	$-\frac{1}{4}$	$-\frac{2}{3}$	-1	$-\frac{3}{2}$	-3	-5
y	1	$\frac{1}{2}$	$\frac{1}{3}$	2	4	-4	$-\frac{3}{2}$	-1	$-\frac{2}{3}$	$-\frac{1}{3}$	$-\frac{1}{5}$

equation $xy = 1$. A noteworthy feature about this equation is that neither x nor y can assume the value 0. Indeed, no product is 1 if any of the factors is 0. Thus, the graph of the equation $y = 1/x$ contains no point with a zero coordinate; that is, no point on either coordinate axis belongs to the graph. (See Fig. 8.)

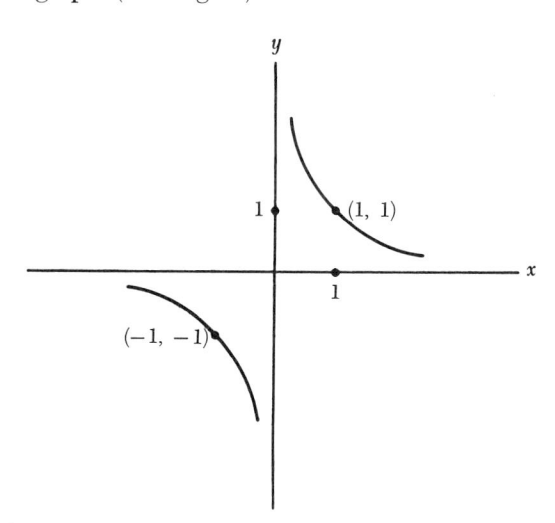

Figure 8

We note that the curve in Fig. 8 consists of two separate branches. One branch lies in the positive quadrant and the other in the negative quadrant. The curve is called a *hyperbola* (more specifically, an equilateral hyperbola or a rectangular hyperbola).

Example 6 Compare and contrast the graphs of the three equations $y = x^2$, $y = 3x^2$, $y = \frac{1}{3}x^2$.

For each real number a, we have a corresponding point on each of the graphs. For the equation $y = x^2$, the point is (a, a^2); for the equation $y = 3x^2$, the point is $(a, 3a^2)$; for the equation $y = \frac{1}{3}x^2$, the point is $(a, a^2/3)$. As shown in Fig. 9, the point $(a, 3a^2)$ is three times as far from the x-coordinate axis as the point (a, a^2), while the point $(a, a^2/3)$ is one-third as far from the horizontal coordinate axis as (a, a^2).

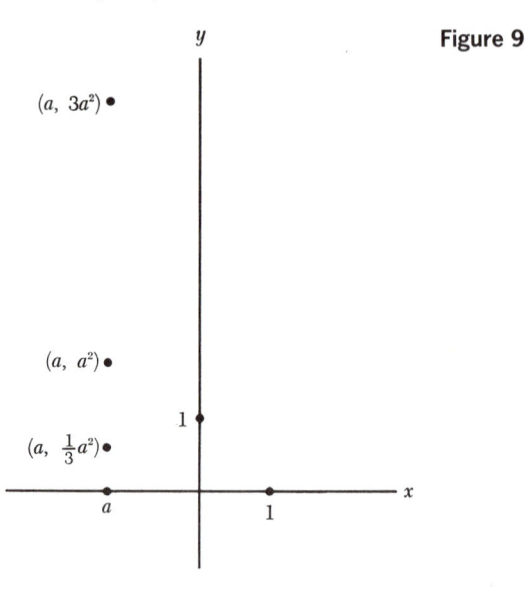

Figure 9

The graph of the set $Q = \{(x, y) | y = x^2\}$ is the parabola pictured in Fig. 3. The graph of the set $\{(x, y) | y = 3x^2\}$ resembles this parabola in general shape, but all distances from the x-coordinate axis are three times as large. The graph of the equation $y = \frac{1}{3}x^2$ also has the general appearance of the graph of Q, but in this case distances from the x-coordinate axis of corresponding points are only one-third as great. Each of these three curves, as shown in Fig. 10, is a parabola.

EXERCISES B

1. Sketch the graph of each of the following equations.
    (a) $y = x^{-1}$.   (b) $y = x^{-1} + 4$.   (c) $y = x^{-1} - 4$.

Figure 10

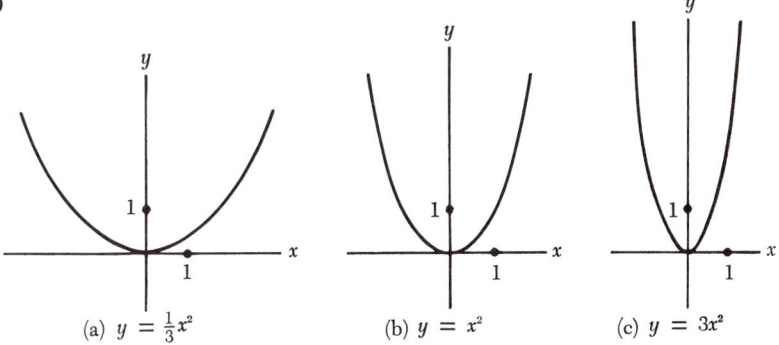

(a) $y = \frac{1}{3}x^2$ (b) $y = x^2$ (c) $y = 3x^2$

2. Find an equation for the hyperbola that is congruent to the graph of the set $\{(x, y) \mid xy = 1\}$ and that is located as described below.
   (a) Two units below the graph of $y = x^{-1}$.
   (b) Three units above the graph of $y = x^{-1}$.
   (c) Four units to the right of the graph of $y = x^{-1}$.
   (d) Five units to the left of the graph of $y = x^{-1}$.

3. Sketch the graph of each of the following sets.
   (a) $\{(x, y) \mid y = (x - 1)^{-1}\}$.   (b) $\{(x, y) \mid y(x + 2) = 1\}$.

4. Sketch the graph of the equation $y = x^{-2}$.

5. Sketch the graph of the set $\{(x, y) \mid xy = -1\}$.

6. Sketch the graph of each of the following equations.
   (a) $y = \frac{4}{x}$.   (b) $y = \frac{1}{4x}$.   (c) $y = -\frac{4}{x}$.

7. For each of the following pairs of curves, compare and contrast the two curves, explaining in what ways they are alike and in what ways they are different.
   (a) The graph of $y = x^{-1}$ and the graph of $y = 4x^{-1}$.
   (b) The graph of $y = 3x^{-1}$ and the graph of $y = -3x^{-1}$.
   (c) The graphs of $y = 5x^2$ and $y = 5x^2 + 8$.
   (d) The graphs of $y = \frac{4}{3}x^2$ and $y = \frac{3}{4}x^2$.
   (e) The graphs of $y = 4(x - 3)^2$ and $y = 4x^2 - 24x + 36$.

8. Sketch the graph of each of the following equations.
   (a) $y = x^3 - x$.   (b) $y = x - x^3$.   (c) $y = x^3 - 1$.
   (d) $y = x + x^{-1}$.   (e) $y = x - x^{-1}$.

4   Function

The preceding examples in this chapter have been illustrations of the graphs of functions. We now introduce some convenient terminology for our future discussion of functions.

First of all, what do we mean by a "function"? In each of our examples, we have had a formula that expresses y in terms of x. Essentially, the formula gives us a rule for computation. If a real number c is selected as a value for x, the rule (or formula) usually tells us how to calculate the associated value for y. If the number we calculate in this manner is d, then we may consider the ordered pair (c, d). The first component c is chosen, and then the second component d in the ordered pair is determined. The rule specifies what number d should be paired, or matched, with c.

In our graphing examples, we have imagined taking all the real numbers c to which the rule is applicable and, for each of them, forming the ordered pair (c, d). If we recall the equation $y = 1/x$ and remember that this formula is not meaningful when $x = 0$, we realize why the preceding sentence has the qualifying clause "to which the rule is applicable."

In our examples, we then imagined plotting the points corresponding to all the ordered pairs (c, d) we had formed. These points formed a subset of the plane (usually a curve), which we spoke of as the graph of the equation involving x and y. This graph is a picture of the set of ordered pairs. We may think of such a set of ordered pairs as being a function. Then the subset of the coordinate plane becomes the graph of a function.

The set of ordered pairs in each of our illustrations has the property that, if the first component c of a pair is specified, then the second component d is determined—in fact, determined by the rule. Another way to express this property is to say that any number that is the first component of some pair can be the first component of only one pair in the set. Rephrased, no number is the first component of two different pairs in the set. Stated still another way, any two ordered pairs in the set have different first components.

On the basis of the above remarks, we can formulate a description of what we mean by a function. There are (at least) two highly useful viewpoints that we can adopt. We shall state both in the form of definitions.

> **Definition** A *function* is a nonempty set of ordered pairs of real numbers such that any two of the pairs have different first components.

Often it is more convenient not to concentrate on the set of ordered pairs themselves, but rather to emphasize the relationship between the components of the various pairs. This association is prescribed by the rule that determines the second component of a pair when the first component is given. If we consider the set of all real numbers that are first components, the rule assigns to each of these numbers a number. The given number and the assigned (or corresponding) number are the respective components of an ordered pair in the function.

> **Definition** Given a nonempty set of real numbers, a *function* is prescribed by a rule that assigns to each member of the set one real number.

The number that a function assigns to a given number may be called the *image* of the given number under the function. Thus, for each ordered pair

Figure 11

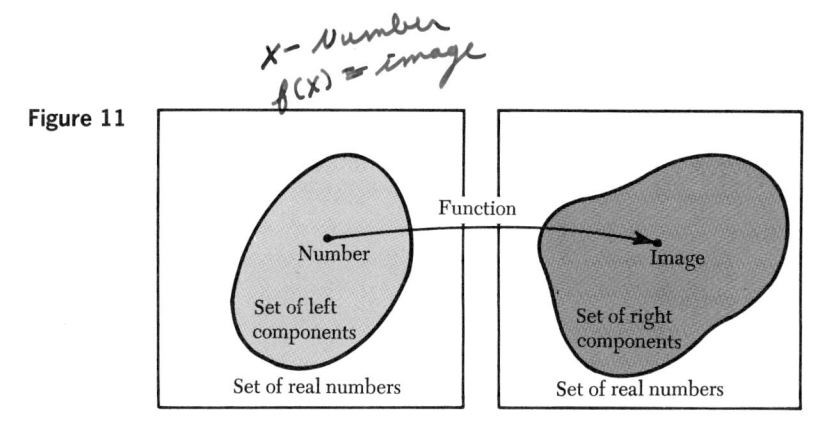

[handwritten annotations: x – Number, f(x) = image]

Number — Function → Image
Set of left components / Set of right components
Set of real numbers / Set of real numbers

belonging to a function, the second component is the image of the first component in the pair.

Although we seldom deem it desirable to offer two definitions for a concept, we have made an exception for the broadly encompassing and highly versatile notion of a function. The two viewpoints are intimately related to one another, and each of them enlightens our development. We shall use the two descriptions interchangeably. For example, in the preceding paragraph, the first sentence exploited the idea of a function as a rule or correspondence, while the second sentence interpreted a function as a certain set of ordered pairs.

Figure 11 suggests the viewpoint that a function maps each number in a certain set onto its image. We should imagine (in general) many arrows in the diagram; each element in the set of left components is the source of a tie that terminates at its image in the set of right components. The pictorial arrow serves to link together the left component and its image to form an ordered pair. The function is the entire matching or correspondence between the two sets.

Example 7 The set $\{(1, 3), (2, 4), (4, 0), (5, 4)\}$ is a function. The image of 1 is 3; the image of 2 is 4; the number 4 is also the image of 5. The number 0 does not have an image under this function. The graph of the function consists of just four points and is shown in Fig. 12.

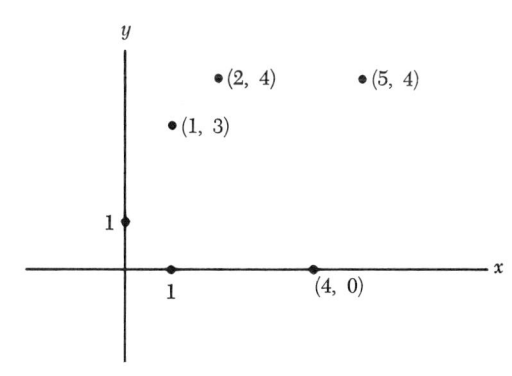

Figure 12

4 / Function

Consider the set {1, 2, 4, 5} of real numbers. Each member of this set is assigned an image under the rule described by the set {(1, 3), (2, 4), (4, 0), (5, 4)}. In this case the rule is not given conveniently by a formula.

In general, the idea of a function does not require that the relationship be expressed by a formula. Although a formula is a very common method for identifying a function, the description of a function by means of a formula frequently is either not convenient or not known. The essential feature about a function is that the rule should assign an image to every member of the given set of numbers.

Sometimes a function is specified in the following manner. We have a partition of the given set of numbers into two or more cells; the rule is given by several formulas, with one formula applying to the members of one cell, but a different formula applying to the numbers in another cell.

Example 8 Consider the function given by the following information: if $x < 1$, then the image of x is $2x$; if $x \geq 1$, then the image of x is $3 - x$.

We may denote this function by the letter g. The image of a real number b under this function may be written $g(b)$. Since -2 is a root of the inequality $x < 1$, the image of -2 is $2 \cdot (-2)$; we summarize this observation by writing $g(-2) = -4$. Since 5 satisfies the condition $x \geq 1$, the image of 5 may be calculated by the formula $3 - x$; thus, $g(5) = -2$. Some other images are: $g(-1) = -2$; $g(\frac{1}{2}) = 1$; $g(4) = -1$. Since -1 is the image of 4, the ordered pair $(4, -1)$ belongs to the function g. Using the language of set theory $(4, -1) \in g$. Some other members of the set g are $(0, 0)$, $(1, 2)$, $(3, 0)$.

We plot the points whose coordinates have been identified (Fig. 13a). These points suggest the shape of the graph of the function g (see Fig. 13b).

Figure 13

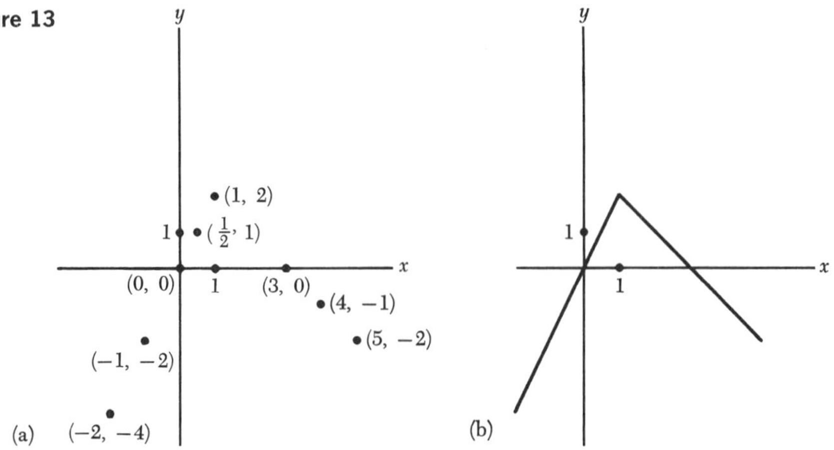

340   Functions and Their Graphs

Figure 14

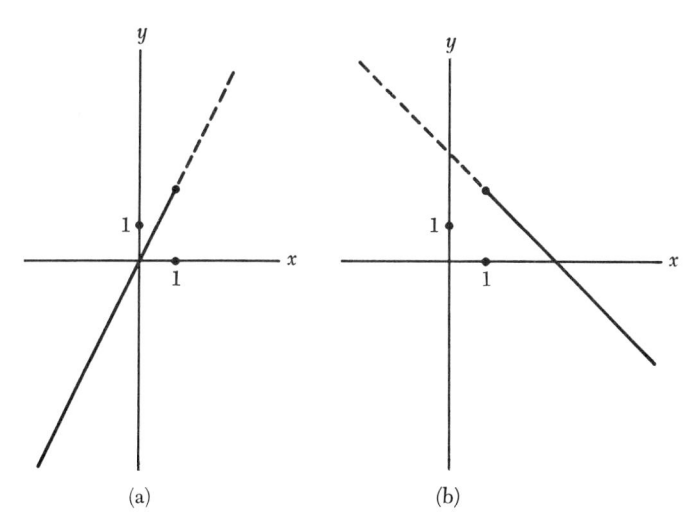

(a)   (b)

We confirm the appearance of the graph of g as follows. We are given the condition that if $x < 1$, then $g(x) = 2x$. Thus, the portion of the graph of g for x-coordinates less than 1 should be the same as the portion of the graph of the equation $y = 2x$ for those x-coordinates. Now $2x$ is the formula for a linear function, a function whose graph is a line with slope 2. The portion of the line to the left of the point (1, 2) is an open halfline. (See Fig. 14a.)

If $x \geq 1$, then $g(x) = 3 - x$. Thus, the other portion of the graph of g should be the same as the portion of the graph of the equation $y = 3 - x$ for $x \geq 1$. This is a closed halfline contained in the line with slope -1 which is the graph of the linear function given by the formula $3 - x$. (See Fig. 14b.)

In summary, we have a partition of the set of real numbers, consisting of the cell $\{x \mid x < 1\}$ and the cell $\{x \mid x \geq 1\}$. In each of the cells, the function g assigns images that are, respectively, the same as those assigned by a linear function. Thus, the graph of g is the union of two subsets each of which is a halfline. The graph is a "curve" that we recognize as an angle with the point (1, 2) as vertex. These observations, together with the remark that the sides of the angle have respective slopes 2 and -1, permit a rapid sketching of the graph of our function g.

EXERCISES C

1. For each of the following rules, write a formula for the function described.
   (a) Every real number has an image which is three more than twice the second power of the number.
   (b) The function consists of all ordered pairs such that the sum of the components in each ordered pair is four.
   (c) The image of each real number under the function is $\frac{3}{2}$.
   (d) Every number is the image of itself under the function.

2. Tell which of the following sets are functions, and explain each decision.
   (a) $\{(1, 2), (2, 1), (3, 5), (7, 6), (8, 1)\}$.   (b) $\{(x, y) \mid y = x^{-2}\}$.
   (c) $\{(x, y) \mid x = y^{-2}\}$.   (d) $\{(1, 2), (2, 1), (3, 1), (1, 3)\}$.
   (e) $\{(x, y) \mid y = 0\}$.   (f) $\{(0, 0)\}$.
   (g) $\{(x, y) \mid x = 0\}$.   (h) $\{(x, y) \mid x + 2y = 3\}$.
   (i) $\{(x, y) \mid x + 2y < 3\}$.   (j) $\{(x, y) \mid (x > 0) \wedge (y = x^4)\}$.

3. Sketch the graph of each of the following functions.
   (a) The function g given as follows: if $x \geq 0$, then $g(x) = 1$; if $x < 0$, then $g(x) = 0$.
   (b) $h = \{(x, y) \mid [(x < 0) \wedge (y = -1)] \vee [(x, y) = (0, 0)] \vee [(x > 0) \wedge (y = 1)]\}$.
   (c) The function G, where $G(x) = x$ if $x \geq 0$, but $G(x) = 0$ if $x < 0$.
   (d) $\{(x, y) \mid x$ is a positive integer and $y = 1/x\}$.
   (e) $\{(x, y) \mid (0 \leq x \leq 1) \wedge (y = 1 - x)\}$.
   (f) The function H given as follows: if $x \geq 0$, then $H(x) = x$;
       if $x < 0$, then $H(x) = -x$.

4. Sketch the graph of the function p that gives the postage (in cents) for first class mail in terms of the weight w (in ounces):
   if $0 < w \leq 1$, then $p(w) = 5$; if $1 < w \leq 2$, then $p(w) = 10$;
   if $2 < w \leq 3$, then $p(w) = 15$; if $3 < w \leq 4$, then $p(w) = 20$; etc.

5. Explain why a "linear function," as described in Chapter 1, is also a "function," as described in the present chapter.

6. Let X be a random variable for a certain experiment. Explain why a listing of the possible values for X and the probability for each (such as we often presented in tabular form in Chapter 8) describes a function.
   (REMARK: This function is naturally called the probability function for the random variable X; the image of each number k under the function is $\Pr[X = k]$.)

7. Sketch the graph of each of the following functions.
   (a) The set of all ordered pairs (k, a) such that $k \in \{1, 2, \ldots, 10\}$ and $a = \frac{1}{2}k + 5$.
   (b) The set of all ordered pairs (n, t) such that $n \in \{0, 1, 2, 3, 4\}$ and $t = 6(\frac{1}{2})^n$.
   (c) The set of all ordered pairs (n, w) such that $n \in \{0, 1, \ldots, 6\}$ and $w = -\frac{1}{6}n + 2$.
   (d) The set of all ordered pairs (k, g) such that $k \in \{1, 2, 3\}$ and $g = \frac{1}{25}(10)^k$.
   (e) The set of all ordered pairs (p, q) such that $p \in \{-2, -1, \ldots, 2\}$ and $q = p^2$.

8. Sketch the graph of each of the following functions.
   (a) $\{(x, y) \mid y = x^2\}$.   (b) $\{(x, y) \mid (x \geq 0) \wedge (y = x^2)\}$.
   (c) $\{(x, y) \mid (x < 0) \wedge (y = x^2)\}$.

5 Domain, range, image

A function is frequently denoted by a single letter, such as L or f or F or g. As we shall learn later, several common functions have special notations.

> **Definition** The *domain* of a function is the set of all real numbers each of which is the first component of an ordered pair belonging to the function.

The domain of the function f may be denoted by Dom f.

Definition The *range* of a function is the set of all real numbers each of which is the second component of an ordered pair belonging to the function.

The range of the function f may be denoted by the symbol Rg f.

A function f assigns to each real number in its domain an image that is in its range. If c is an element in the domain of f, then the image of c under f is customarily written $f(c)$. We usually read the symbol $f(c)$ as "f of c."

Let us recall the second description of a function: "Given a nonempty set of real numbers, a function is prescribed by a rule that assigns to each member of the set one real number." The given nonempty set is the domain of the function. The set of all images under the function is the range of the function. The pictorial viewpoint of Fig. 11 (see Section 4) is repeated in Fig. 15, in which the two sets, the domain and the range of the function, are labeled.

Figure 15

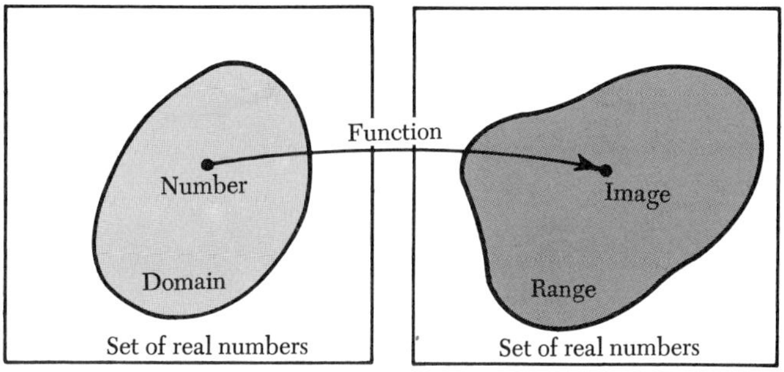

Example 9 If Q is the second-power function $\{(x, y) \mid y = x^2\}$ discussed in Example 2, then the domain of Q is the set of all real numbers because every real number has a second power. However, Rg $Q = \{t \mid t \geq 0\}$, since the second power of a real number is nonnegative and every nonnegative number has a square root. The fact that Rg Q contains no negative numbers is portrayed in Fig. 3 by the feature that the parabola contains no point below the x-coordinate axis. We note that $Q(3) = 9$, that $Q(-3) = 9$, that $Q(\frac{1}{3}) = \frac{1}{9}$, and that $Q(0) = 0$.

The domain of the reciprocal function $\{(x, y) \mid y = 1/x\}$ is the set of all numbers different from zero, namely, $\{x \mid x \neq 0\}$. We remarked in Example 5 that the formula $1/x$ is not applicable in case $x = 0$. The rule does however assign an image to every number except zero.

In Example 7, the domain of the function is $\{1, 2, 4, 5\}$ and the range of the function is $\{0, 3, 4\}$. The members of the domain are the x-coordinates of points on the graph; the range is the set of all numbers that are y-coordinates of points on the graph.

5 / Domain, range, image

If g is the function discussed in Example 8, then Rg $g = \{y\,|\,y \leq 2\}$. Geometrically, this means that no point on the graph is at a height greater than two and that any number not exceeding two is the y-coordinate of at least one point belonging to the angle.

Example 10 Discuss the function $\{(x, y)\,|\,y = 3\}$.

The rule specifying this function is extremely simple. It says that the image of a number is 3. The image is 3, regardless of what the given number may be. Thus, 3 is the image of 5, the image of 8, the image of -2, as well as the image of every other number. The graph of the function consists of all points at a height of three units. In other words, the graph is a horizontal line three units above the x-coordinate axis. (See Fig. 16.)

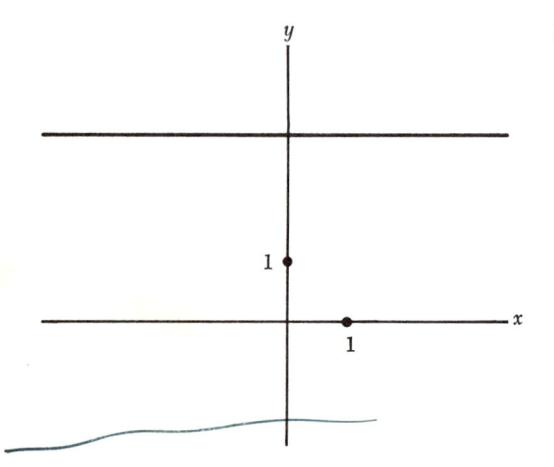

Figure 16

Example 11 Suppose an urn contains five balls that are alike except for color; three of the balls are white and two are green. A simple activity consists in drawing a ball at random from the urn, noting its color, and replacing it. Consider an experiment consisting of three successive trials of this activity. Let the random variable X denote the number of white balls drawn in the experiment. The logically possible outcomes for the experiment are: $X = 0$, $X = 1$, $X = 2$, $X = 3$. For each number $c \in \{0, 1, 2, 3\}$, the probability of the statement $X = c$ is the image of c under the binomial distribution function b given by the formula

$$b(x) = \binom{3}{x}\left(\frac{3}{5}\right)^x\left(\frac{2}{5}\right)^{3-x}.$$

The domain of b is the set $\{0, 1, 2, 3\}$. The function b is a set consisting of four ordered pairs, namely, the set $\{(0, 0.064), (1, 0.288), (2, 0.432), (3, 0.216)\}$. The graph of b has just four points. (See Fig. 17.)

344   Functions and Their Graphs

Figure 17

(0, 0.064)
(1, 0.288)
(2, 0.432)
(3, 0.216)

Figure 18

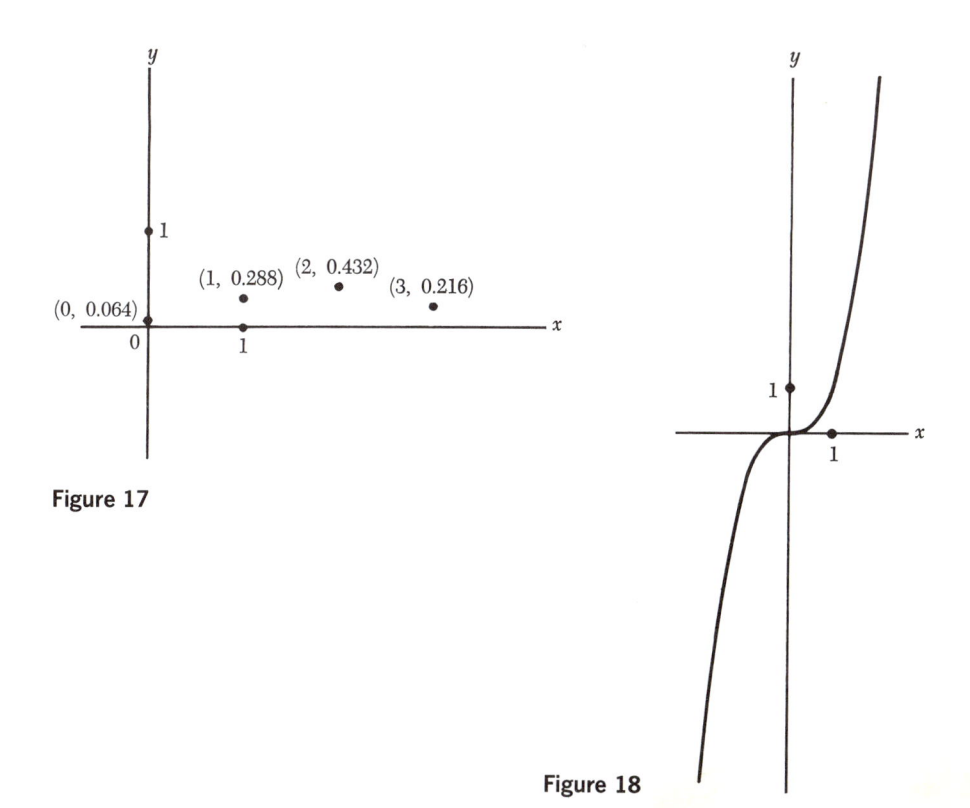

An important class of functions is the collection of power functions. For each natural number n, the nth power function is the function given by the formula x^n. We have already noted the second-power function $Q = \{(x, y) \mid y = x^2\}$. It consists of all ordered pairs in which the second component is the square of the first component. A simpler illustration is the identity function $\{(x, y) \mid y = x\}$, which is the case $n = 1$. Its graph is the line consisting of all points whose two coordinates are equal to one another. Another example is the third-power function, whose formula is x^3. The graph of the equation $y = x^3$ appears in Fig. 18.

EXERCISES D

1. Several illustrations of the notion of a function have been discussed in Chapter 7.
   (a) Explain why an arithmetic sequence, as described in Chapter 7, is a function, as defined in the present discussion.
   (b) Explain why a geometric sequence is a function.
   (c) Explain why every sequence is a function.
   (d) Explain why $\{(r, p) \mid r$ is a rational number and $p = 2^r\}$ is a function.
   (e) Explain why a function is specified by the rule that r is a rational number and $p = (\frac{1}{3})^r$.
   (f) Let b be a positive number. Show that the rule that assigns to each rational number r the image b^r is a function.

2. Identify the domain of each of the following functions.
   (a) $\{(x, y) | y = x^3 - 1\}$.
   (b) $\{(x, y) | y = 6\}$.
   (c) $\{(x, y) | y = x^{1/2}\}$.
   (d) $\{(x, y) | y = \dfrac{5}{x^2 - 4}\}$.
   (e) $\{(0, 1), (2, 5), (3, 5), (6, 1)\}$.
   (f) The function described in Exercise 1e.

3. Identify the range of each of the following functions.
   (a) $\{(0, 1), (2, 5), (3, 5), (6, 1)\}$.
   (b) $\{(x, y) | y = -4\}$.
   (c) $\{(x, y) | [(x < 0) \wedge (y = 1)] \vee [(x > 0) \wedge (y = -1)]\}$.
   (d) $\{(x, y) | y = 2x - 1\}$.
   (e) $\{(x, y) | xy = 1\}$.
   (f) $\{(x, y) | [(x < 2) \wedge (y = -x)] \vee [(x \geq 2) \wedge (y = 3x - 8)]\}$.
   (g) $\{(x, y) | [(x < 2) \wedge (y = -x)] \vee [(x \geq 2) \wedge (y = 3x - 7)]\}$.

4. A certain random variable Z has a binomial distribution with $n = 5$ and $p = \tfrac{1}{3}$. The probability function for Z is $\{(k, t) | t = \Pr[Z = k]\}$.
   (a) How many ordered pairs belong to this probability function?
   (b) Give the domain of the function.
   (c) Give the range of the function.
   (d) Give another formula for the function.
   (e) Draw carefully the graph of the function.

5. Repeat Exercise 4 for a random variable having a binomial distribution with $(n, p) = (6, \tfrac{1}{2})$.

6. Repeat Exercise 4 for a random variable having a binomial distribution with $(n, p) = (7, 0.1)$.

7. Repeat Exercise 4 for a random variable having a binomial distribution with $(n, p) = (5, \tfrac{1}{2})$.

8. Sketch the graph of each of the following equations.
   (a) $y = \dfrac{x^3}{4}$.
   (b) $y = -x^4$.
   (c) $y = 1 - x^3$.

6 Graph

We now summarize our discussion of graphs.

> **Definition** Let f be a function. On a two-dimensional coordinate system, the set of all points each of which has coordinates forming an ordered pair belonging to f is called the *graph of the function*. The graph of the function f is also called the graph of the equation $y = f(x)$.

We recall that a function does not contain two different ordered pairs with the same first component. Hence, the graph of a function does not contain two different points with the same x-coordinate. In other words, the graph of a function does not contain two different points on any line parallel to the y-coordinate axis. We may rephrase these observations as follows.

Theorem 1 A nonempty subset of the coordinate plane is the graph of a function if and only if every vertical line in the plane intersects the subset in not more than one point. *(only one y for each x)*

Example 12 Let Q be the second-power function, given by the *and* formula $Q(x) = x^2$. Let Q^* be the set $\{(x, y) \mid (x \geq 0) \wedge (y = x^2)\}$. Thus, Q^* contains such ordered pairs as $(5, 25)$, $(3, 9)$, $(\frac{1}{2}, \frac{1}{4})$, $(0, 0)$. On the other hand, even though $16 = (-4)^2$, the ordered pair $(-4, 16)$ does not belong to Q^* because the left component -4 does not satisfy the requirement of being nonnegative. To each nonnegative number, Q^* assigns an image, namely, the second power of the given number. Thus, Q^* is a function. Each ordered pair (a, a^2) that belongs to Q^* also is a member of Q. Hence Q^*, considered as a set of ordered pairs, is a subset of Q. However, Q^* is not the same as Q, because all real numbers (including the negative numbers) have images under Q while Dom Q^* consists of nonnegative numbers only.

Figure 19 shows the contrasting graphs of Q and Q^*. The graph of Q has the property that every vertical line in the plane intersects the curve in one point (and, therefore, in "not more than one" point). The graph of Q^* has the feature that each vertical line in the plane

Figure 19

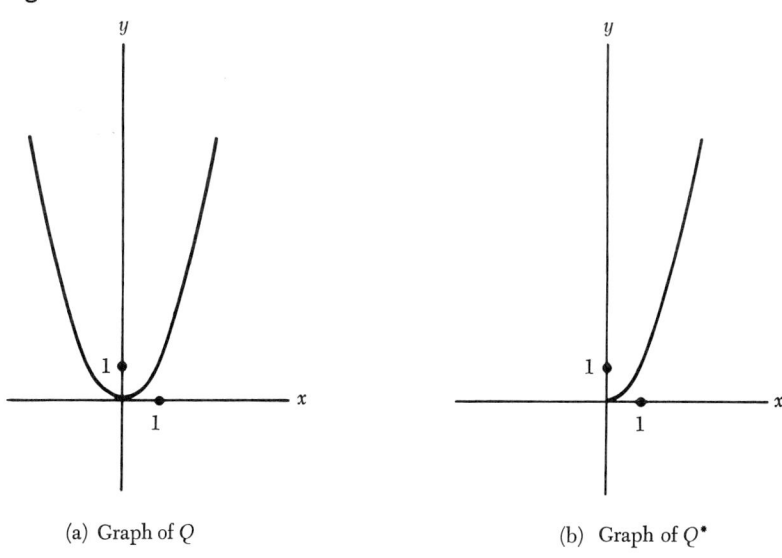

(a) Graph of Q

(b) Graph of Q^*

intersects the curve in not more than one point; indeed, although some vertical lines do not intersect the graph at all, none has more than one point in common with the curve.

Theorem 2 Every nonempty subset of a function is itself a function.

PROOF Since the function cannot contain two different ordered pairs with the same left component, neither can the subset.

Any ordered pair belonging to a nonempty subset of a function also belongs to the function. Hence, a number that belongs to the domains of both functions has the same image under both of them. In general, however, there are numbers that serve as left components of ordered pairs belonging to the given function but that are not members of the domain of the subset. Geometrically, these numbers appear as x-coordinates of points on the graph of the original function—points that are discarded in constructing the graph of the subset.

7 The absolute-value function

A particularly useful function is the absolute-value function. This rule focuses on the numerical size of a number and ignores its sign. Under the absolute-value function, the image of a nonzero number, either positive or negative, is the associated positive number obtained by suppressing the sign, that is, disregarding whether the given number is positive or negative. Thus, the image of a positive number p is just p itself. The image of 0 is agreed to be 0.

Now consider a negative number n; this number may be expressed in the form $-p$, where p is positive. Furthermore, p represents the numerical size of the number $n = -p$, while the symbol "$-$" conveys the information that n is negative. The positive number associated with n by disregarding the negative character of n is p. We take p to be the image of n under the absolute-value function. We note that p is expressible in terms of n; indeed, since $n = -p$, we obtain the result that $p = -n$. The image of a negative number n under the absolute-value function is $-n$.

Special symbolism is ordinarily used for the image of a number x under the absolute-value function: we write $|x|$, which we read as "the absolute value of x." On the basis of the two preceding paragraphs, we summarize the absolute-value function as follows:

$$\text{if } x \geq 0, \text{ then } |x| = x;$$
$$\text{if } x < 0, \text{ then } |x| = -x.$$

Note that the rule is given by two separate cases. One formula is applicable

348   Functions and Their Graphs

for determining the image of a nonnegative number; a different formula specifies the image of a negative number.

Example 13 Since 6 is positive, its absolute value is itself; that is, $|6| = 6$. Since -4 is negative, its absolute value is obtained from the formula $-x$; the result is $-(-4) = 4$, and we observe that 4 is the same as the result we would achieve by dropping the sign "$-$" from the number -4. Thus, $|-4| = 4$. Also, $|0| = 0$.

Example 14 The concept of the absolute-value function is vital in arithmetic. Consider the product of two numbers, such as -36 and 23. The usual method for multiplying is to begin by ignoring the signs of the factors, then to calculate $36 \cdot 23 = 828$, and finally to attach the sign $-$ to the number 828. In general, the product of two nonzero real numbers is determined by multiplying their absolute values and by deciding whether the product is positive or negative according to whether the factors have like or unlike signs.

If we confine our attention to nonnegative numbers in the domain, then the graph of the absolute-value function agrees with the graph of the linear function given by the formula x. Thus, part of the graph of the absolute-value function is the curve shown in Fig. 20a. Now consider negative numbers in the domain. For each of these the image under the absolute-value function is the same as the image under the linear function given by the formula $-x$. Thus, in the left coordinate halfplane, the graph of the absolute-value function agrees with the graph of the equation $y = -x$. This is the portion of the line with slope -1 to the upper left from the origin (see Fig. 20b). The entire graph of the absolute-value function is the union of these two pieces.

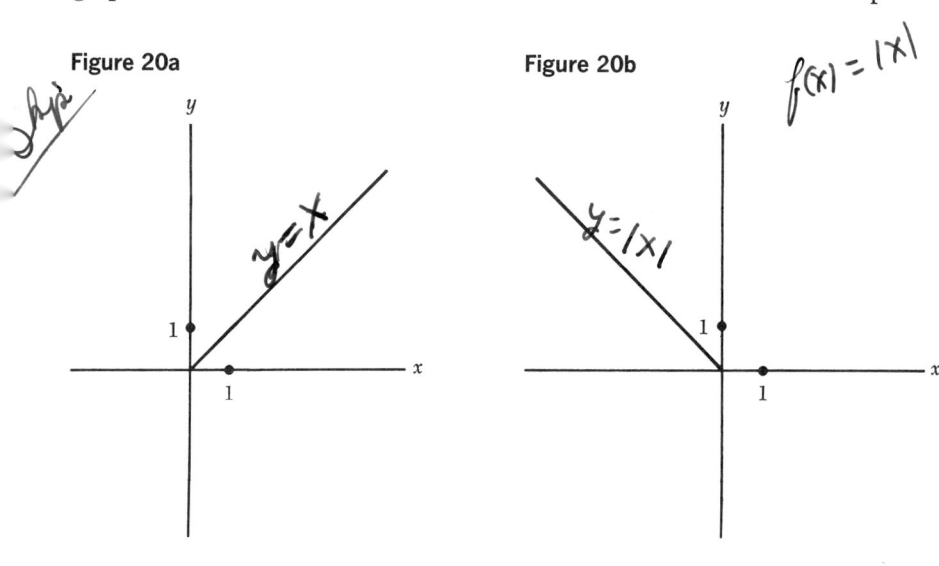

Figure 20a

Figure 20b

7 / The absolute-value function

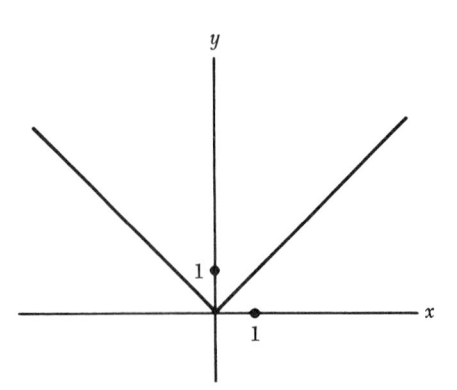

Figure 20c

Figure 20c shows the union of the two halflines appearing in Figs. 20a and 20b. Indeed, the absolute-value function is the union of the function $\{(x, y) \mid (x \geq 0) \wedge (y = x)\}$ and the function $\{(x, y) \mid (x < 0) \wedge (y = -x)\}$.

The graph of the absolute-value function is an angle with vertex at the origin. In fact, since the sides of the angle bisect the respective quadrants of the coordinate system, the angle is a right angle. The right angle is entirely contained in the upper closed halfplane of the coordinate system. The appearance of the graph strikingly emphasizes that the absolute value of a number cannot be negative.

Example 15 Solve the inequality $|x| < 1$.

Let r be a real number. We consider two cases separately. First, suppose that $r \geq 0$. Then $|r| = r$. Hence, in this case, r is a root of the inequality if and only if $r < 1$. Thus, each number satisfying

$$0 \leq r < 1$$

is a root. Second, suppose that $r < 0$. Then $|r| = -r$. Therefore, in this case, r is a root of the inequality if and only if $-r < 1$, or, equivalently, $r > -1$. Thus, each number satisfying

$$-1 < r < 0$$

is a root. Combining the results obtained in the separate cases, we find that the roots of the inequality $|x| < 1$ are the members of the set

$$\{x \mid -1 < x < 1\},$$

that is, the set of all numbers between -1 and 1. (See Fig. 21.)

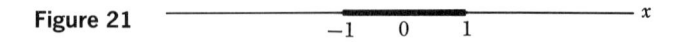

Figure 21

EXERCISES E

1. (a) Give the absolute value of 8.   (b) Find the absolute value of -8.

Functions and Their Graphs

2. Show that $|-3| = |3|$.

3. Calculate each of the following numbers.
   (a) $|3.54| + |-3.54|$.
   (b) $|(3.54) + (-3.54)|$.
   (c) $-|3.54| + |-3.54|$.

4. Sketch the graph of the function $\{(x, y) | y = -|x|\}$.

5. Show that:
   (a) The absolute value of every positive number is positive.
   (b) The absolute value of every negative number is positive.
   (c) $\text{Rg}\{(x, y) | y = |x|\} \subset \{t | t \geq 0\}$.

6. Sketch the graph of the function given by the formula $x - |x|$.   *et $y = x - |x|$*

7. Solve each of the following equations.
   (a) $x = |x|$.
   (b) $x = -|x|$.
   (c) $x = 2|x|$.

8. For each of the following inequalities, find the set of all roots.
   (a) $|x| > 0$.
   (b) $|x| \geq 2$.
   (c) $|x| < 3$.
   (d) $|x| \geq x$.
   (e) $|x| \geq -x$.

9. Let a and b be two different real numbers. Let the difference obtained by subtracting the smaller of these two numbers from the larger be called d.
   (a) In case $a = 4$ and $b = 3$, calculate d.
   (b) In case $a > b$, find d.
   (c) In case $(a, b) = (2, 6)$, calculate d.
   (d) In case $a < b$, find d.
   (e) In case $a > b$, show that d is the same as $|b - a|$.
   (f) In case $a < b$, show that $d = |b - a|$.
   (g) Show that $d = |b - a|$, no matter which of the given numbers a, b is the larger.

10. Sketch the graph of each of the following functions.
    (a) $\{(x, y) | (-1 \leq x \leq 1) \wedge (y = 1 - |x|)\}$.
    (b) $\{(x, y) | (x > 0) \wedge (xy = 1)\}$.
    (c) $\{(x, y) | (|x| \leq 2) \wedge (y = x^2 - 4)\}$.
    (d) $\{(x, y) | (|x| \leq 2) \wedge (y = 4 - x^2)\}$.

11. Sketch the graph of each of the following functions.
    (a) $\{(x, y) | (x < 0) \wedge (y = -x^2)\} \cup \{(x, y) | (x \geq 0) \wedge (y = x^2)\}$.
    (b) $\{(x, y) | y = x \cdot |x|\}$.
    (c) The function w, given as follows: if $x \leq -2$, then $w(x) = -6x - 14$; if $-2 < x < 0$, then $w(x) = 2x + 2$; if $0 \leq x < 2$, then $w(x) = -2x + 2$; if $x \geq 2$, then $w(x) = 6x - 14$.

12. Sketch the graph of the function h given as follows: if $x < -1$, then $h(x) = -1$; if $|x| \leq 1$, then $h(x) = x$; if $x > 1$, then $h(x) = 1$.

8 Increasing function

A property of the graph of the third-power function $\{(x, y) | y = x^3\}$ that is strikingly suggested by inspection of Fig. 18 is the following: the farther to the right we move on the curve, the higher up we go. This "rising" characteristic of the curve may be described by saying that the third-power function is an increasing function.

Figure 19b shows that the graph of the function $Q^* = \{(x, y) | (x \geq 0) \wedge (y = x^2)\}$ has this same rising feature. For any two points on the curve, the one with the larger x-coordinate has the greater y-coordinate.

Figure 1 displays the graph of the linear function L given by the formula

8 / Increasing function   351

$L(x) = 2x + 1$. This line is a rising "curve." If (a, b) and (c, d) are any two points belonging to the graph of L and if $c > a$, then $d > b$. The point with the greater left coordinate also has the larger right coordinate.

Definition Let F be a function, and let U be a nonempty subset of the domain of F. We say that F is an *increasing function on U* provided, for any two numbers in U, the greater of them has the greater image under F. In particular, if U is equal to Dom F, then we simply say that F is an *increasing function*.

In symbols, the condition that F be an increasing function on U may be expressed as follows: for any $a \in U$ and any $c \in U$, if $c > a$, then $F(c) > F(a)$.

A natural contrast to the notion of an increasing function is the concept of a *decreasing* function. If H is a function and if $U \subset$ Dom H, we say that H is a decreasing function on U provided that, for any two numbers a and c in U, if $c > a$, then $H(c) < H(a)$. In other words, the condition is that the greater of any two members of U has the smaller image under H. The graph of a decreasing function is a "falling" curve: the farther to the right we move on the curve, the lower down we go.

Example 16 Let Q be the second-power function $\{(x, y) \mid y = x^2\}$. The function Q is increasing on the set of nonnegative numbers. It is a decreasing function on the set $\{x \mid x \leq 0\}$. Notice, however, that Q is neither an "increasing function" nor a "decreasing function," because it is neither increasing nor decreasing on its entire domain.

Example 17 The graph of a certain function E is depicted in Fig. 22. List several properties of the function that are suggested by the picture.

Any drawing of the graph of a function ought to reveal, in a pictorial manner, the significant properties of the function that can be readily portrayed. It is frequently impossible to draw an entire graph, especially if the graph extends on and on without bound. Consequently, when we choose the portion of a graph that we will draw,

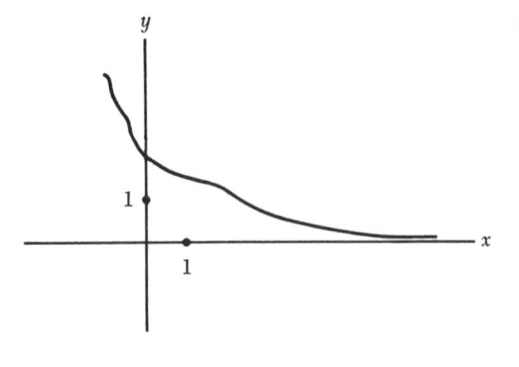

Figure 22

we must be careful not to draw a part that will misrepresent the true nature of the function.

We presume then that Fig. 22 not only is a reasonably accurate drawing of the part of the graph that fits in the allotted space, but also conveys the proper impression about any part of the graph beyond what is shown. On this basis, we are led to ascribe the following properties to E.

1. The function E is a decreasing function. Note the falling character of the curve.

2. The domain of E is contained in the halfline $\{x \mid x \geq -1\}$. Notice that the curve appears to have an endpoint at the point with x-coordinate -1. There is no evidence of any portion of the curve farther to the left.

3. The range of E consists of positive numbers only. Notice that the curve apparently lies entirely above the x-coordinate axis.

4. The range of E contains no number greater than 4. Observe that the above-mentioned apparent endpoint of the curve has y-coordinate 4.

5. Rg $E = \{t \mid 0 < t \leq 4\}$. Note that the curve falls gradually from a height of 4 toward the x-coordinate axis. Every positive number not exceeding 4 appears to be an image under the function E.

6. $(0, 2) \in E$.

EXERCISES F

1. A certain function H has all of the following properties. Draw a curve that displays all these features and therefore might be a graph of H.
   (i) Dom $H = \{x \mid -1 < x < 2\}$.
   (ii) H is an increasing function.
   (iii) $H(0) = 3$.

2. Show each of the following.
   (a) The reciprocal function $\{(x, y) \mid xy = 1\}$ is a decreasing function on the set of positive numbers.
   (b) The reciprocal function is a decreasing function on the set of negative numbers.
   (c) The reciprocal function is not a decreasing function.
   (d) The absolute-value function $\{(x, y) \mid y = |x|\}$ is a decreasing function on the set of negative numbers.
   (e) The absolute-value function is an increasing function on $\{x \mid x > 0\}$.
   (f) The fourth-power function $\{(x, y) \mid y = x^4\}$ is an increasing function on $\{x \mid x > 0\}$.
   (g) The fourth-power function is a decreasing function on $\{x \mid x < 0\}$.
   (h) The fifth-power function is an increasing function.

3. A certain function F has all of the following properties. Draw a curve that displays all these features and therefore might be the graph of F.
   (i) The domain of F is the set of all nonnegative numbers.
   (ii) Rg $F = \{y \mid y > 0\}$.
   (iii) F is decreasing on the set $\{x \mid 0 < x < 4\}$.
   (iv) F is increasing on the set $\{x \mid x > 4\}$.
   (v) $(0, 2) \in F$.

4. Each of the following diagrams shows the graph of a certain function. List several properties of the function which are suggested by the picture.

(a) Graph of function f.

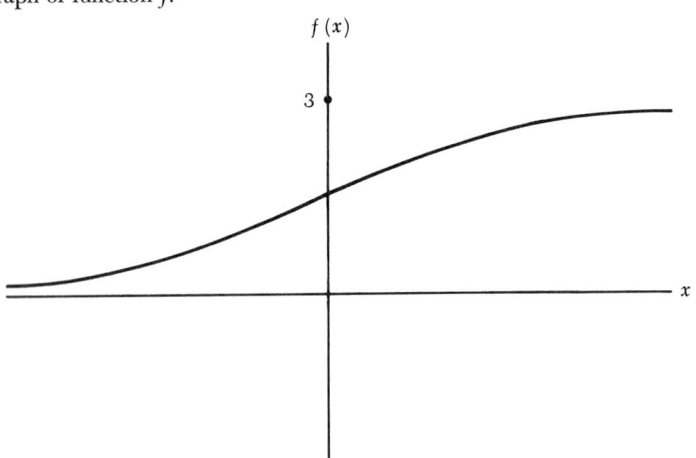

(b) Graph of function g.

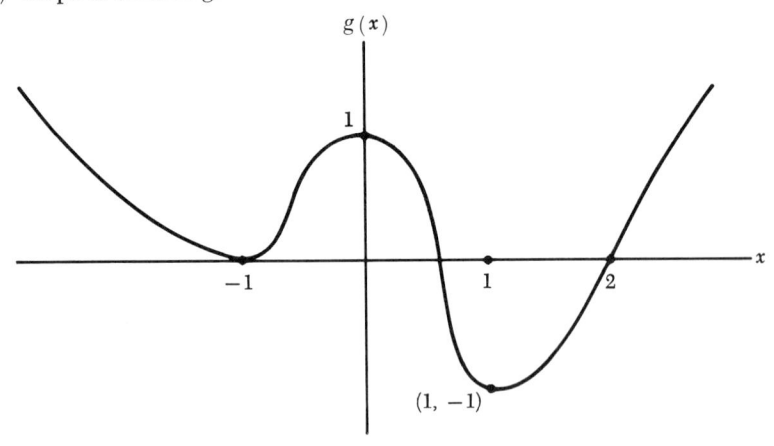

(c) Graph of function h.

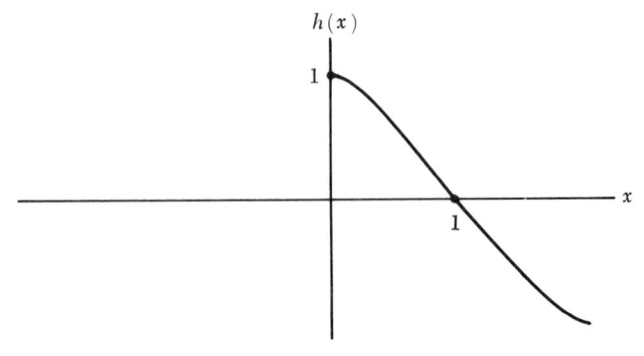

9 Summary

A function has a domain and a range. Each member of the domain has an image under the function. The range is the set of all these images. A member of the domain and its image form an ordered pair belonging to the function. The function consists of all these ordered pairs.

The function is prescribed by a rule that identifies the image of each element in the domain. Often this rule is expressed by a formula. Thus, as an example, we have the doubling function given by the formula $2x$. Sometimes we shorten our terminology and speak simply of "the function $2x$"; this abbreviation means "the function given by the formula $2x$," or more specifically, "the function $\{(x, y) \mid y = 2x\}$." As another illustration, the second-power function sometimes may be called "the function x^2" provided we understand clearly that x^2 is not the function but rather that $\{(x, y) \mid y = x^2\}$ is the function, and that x^2 is a formula for the function.

Finally, we observe that in this chapter we have described a function as an ordered pair of real numbers. The domain and the range have been subsets of the set of all real numbers. It is not necessary that we make this restriction. In a more general situation, the domain of a function may be any nonempty set and the range likewise may be any nonempty set. The members of the domain or of the range may be numbers or may be elements of another kind—points, people, vectors, etc. A random variable in probability is a function whose domain consists of all the possible outcomes for a certain experiment. In Chapter 12 we shall extensively discuss a function whose range consists of points.

The essential feature of a function is that it has a domain and that it assigns to each member of its domain a unique image.

Distance 10

1 Introduction

The notion of distance is fundamental in Euclidean geometry.

Mathematically, the distance between any two distinct points is a positive real number. Physically, we customarily express distance in terms of a suitable unit of length; we may speak of 5 meters, or 50 feet, or $\frac{3}{4}$ mile. However, in our work, we shall suppose that some appropriate unit has been fixed, and all our distances will be simply numbers—such as 67, $\frac{5}{2}$, or $\sqrt{3}$.

Although basically distance applies to pairs of points, we may combine it with perpendicularity in order to discuss the distance between a point and a line, or the distance between a point and a plane, or the distance between two parallel lines, or other extensions of the notion. As applications of the distance concept we shall describe in this chapter the parabola, the perpendicular-bisector, the circle, the sphere, and related topics.

2 Distance and absolute value

Subtract left from Right or bottom from top.

An important application for the absolute-value function is its usefulness in representing distances.

Example 1 Consider a one-dimensional coordinate system on a line. The distance between two points is obtainable from the coordinates of the points by subtracting the smaller coordinate from the larger. In Fig. 1, the distance between the points C and E is the difference $7 - 2 = 5$. The distance between B and D may be calcu-

Figure 1

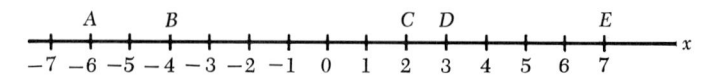

lated by subtracting the smaller coordinate -4 from the larger coordinate 3; the difference $3 - (-4)$ is 7. The difference between the greater number -4 and the smaller number -6, namely, $(-4) - (-6) = 2$, is the distance between A and B.

Example 2 Let r be a real number not 2. What is the distance between the points whose respective coordinates are 2 and r?

Although we are given the information that $r \neq 2$, we do not know which of the numbers r, 2 is the greater. First, suppose $r > 2$. Then the graph has the appearance of Fig. 2a, in which the point r is to the right of the point 2. In this case the distance between the two points is $r - 2$. On the other hand, suppose $r < 2$. Then on the coordinate system the point r is, as shown in Fig. 2b, to the left of the point 2. In this situation the distance between the two points is the difference obtained by subtracting the smaller number r from the larger number 2, namely, $2 - r$. In summary, the distance is either $r - 2$ or $2 - r$, according as $r > 2$ or $r < 2$.

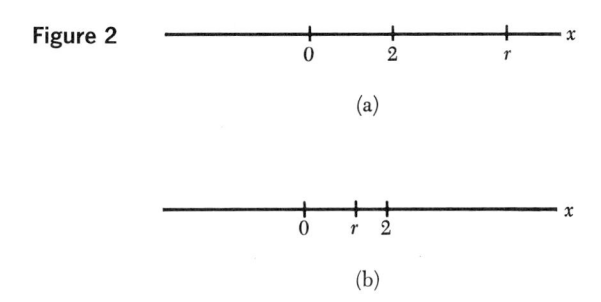

Figure 2

(a)

(b)

The two different formulas for the distance, each applicable in certain situations, may be fused into a single formula with the aid of the absolute-value concept. In case $r < 2$, the distance is the positive number $2 - r$. Now $2 - r = -(r - 2)$. The positive number $2 - r$ is the absolute value of the negative number $r - 2$; that is, $2 - r = |r - 2|$. Hence, $|r - 2|$ is the distance if $r < 2$. On the other hand, in case $r > 2$, the distance is the positive number $r - 2$, which is equal to its own absolute value. That is, the distance is $|r - 2|$ if $r > 2$. In other words, in every case, the distance is given by the single formula $|r - 2|$. Finally, we may observe that $|r - 2|$ gives the correct distance also in the degenerate situation in which $r = 2$.

The discussion in the preceding example may be adapted to the general case. Let a and b be real numbers. The difference obtained by subtracting the smaller of these numbers from the larger is either $a - b$ or $b - a$, depending on whether $a > b$ or $a < b$. More simply, the difference is $|a - b|$. On a one-dimensional coordinate system, the distance between the points

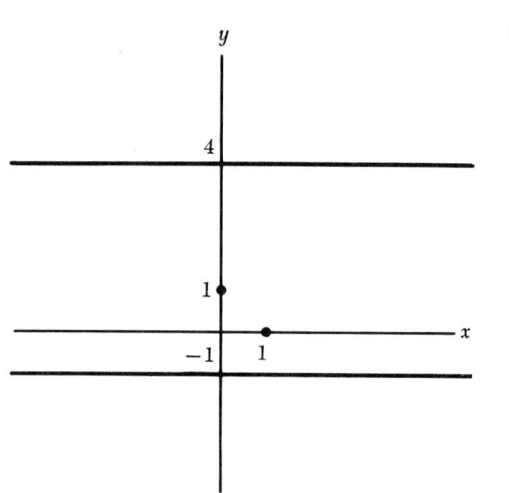

Figure 3

with respective coordinates a and b is $|a - b|$. This applies no matter whether $a > b$ or $a < b$ or even perhaps $a = b$.

Example 3 On a two-dimensional coordinate system, consider the respective graphs of the two constant functions $\{(x, y) \mid y = 4\}$ and $\{(x, y) \mid y = -1\}$. Each of the graphs (see Fig. 3) is a line, and the two lines are parallel to one another. Therefore, we may speak of the distance between the two lines. The distance between the lines is the same as the distance between the two points on the y-axis with respective y-coordinates 4 and -1. The distance is $4 - (-1)$, namely, 5.

Example 4 What is the distance between the lines that are respectively the graphs of $\{(x, y) \mid y = c\}$ and $\{(x, y) \mid y = d\}$, where c and d are constants?

As in Example 3, the desired distance is equal to the distance between the two points on the y-coordinate axis with respective y-coordinates c and d. If we knew that $c < d$, we would give $d - c$ as the answer; if we knew that $d < c$, we would express the distance as $c - d$. Since we have no information permitting us to compare c and d, the distance may be conveniently given by $|c - d|$.

Example 5 On a two-dimensional coordinate system, find the distance between the line whose equation is $x = 3$ and the point whose coordinates are $(7, 2)$.

The distance between a line and a point not on the line is the length of the segment that joins the point and the line and that is perpendicular to the line. The length of the horizontal segment in Fig. 4 is the same as the distance between the points on the hori-

Figure 4

Figure 5

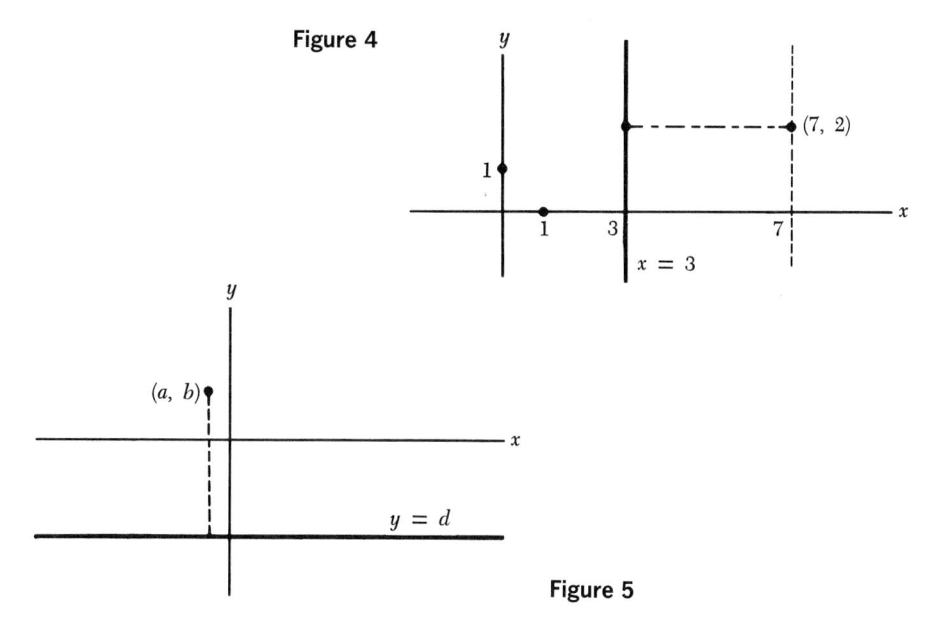

zontal axis with x-coordinates 3 and 7, respectively. Thus, the line $x = 3$ and the point $(7, 2)$ are 4 units apart.

Example 6 On a two-dimensional coordinate system, the distance between the line $y = d$ and the point (a, b) is the absolute value of the difference between the second coordinates b and d, namely, $|b - d|$.

In Fig. 5, it appears that $d < b$. If this information is known, then the distance between the line and the point may be more simply expressed as $b - d$. In general, however, if we lack knowledge about the comparison of b and d, the absolute-value notation in the expression $|b - d|$ is highly desirable.

EXERCISES A

1. On a two-dimensional coordinate system, the graphs of the equations $x = -2$ and $x = 6$ are vertical lines. What is the distance between these two parallel lines?
2. On a two-dimensional coordinate system, how far apart are the parallel lines that are the respective graphs of the equations $x = a$ and $x = b$?
3. What is the distance between the two lines in each of the following pairs of parallel lines?
    (a) The graphs of $\{(x, y) | y = 8\}$ and $\{(x, y) | y = -5\}$.
    (b) The graphs of $y = -3$ and $y = -8$.
    (c) The x-axis and the line whose equation is $y = -16$.
    (d) The lines given by $y = c$ and $y = d$, respectively, in case $d > c$.
    (e) The graphs of $\{(x, y) | y = c\}$ and $\{(x, y) | y = d\}$.

2 / Distance and absolute value

4. What is the distance between the line whose equation is given and the point whose coordinates are given?
   (a) $y = 2$; $(1, 6)$.
   (b) $x = -4$; $(1, 7)$.
   (c) $y = p$; (a, b).
   (d) $x = q$; (a, b).
   (e) $x = r$; the origin $(0, 0)$.

5. Sketch the graph of each of the following equations.
   (a) $y = x^2$.
   (b) $y = |x|^2$.
   (c) $y = |x|^2 - x^2$.

6. Establish each of the following properties.
   (a) If p is a positive number, then $|p| \cdot |p| = p \cdot p$.
   (b) If n is a negative number, then $|n| \cdot |n| = n \cdot n$.
   (c) For every number a, $|a|^2 = a^2$.

7. Let b be a fixed real number. Explain why each of the following is true.
   (a) $|x - 2|^2 = (x - 2)^2$ for every real number x.
   (b) For every real number y, $|y + 3|^2 = (y + 3)^2$.
   (c) For every real number t, $|t - b|^2 = (t - b)^2$.

3 Distance between points

In Section 10 of Chapter 1 we applied the theorem of Pythagoras in order to find the distance between the origin and a point on a two-dimensional coordinate system. A similar discussion will enable us to determine the distance between any two points in the coordinate plane. By a repeated application of the theorem of Pythagoras, we may also express the distance between any two points on a three-dimensional coordinate system.

Example 7 What is the distance between the points $(-2, 4)$ and $(7, 1)$?

The extra appreciation gleaned from a graph often is worth the effort of drawing the picture. Let P and Q denote the points $(-2, 4)$ and $(7, 1)$, respectively. (See Fig. 6a.) For each of the two given points, consider the lines through the point and parallel to the respective coordinate axes, as indicated in Fig. 6b by the dashed lines. The distance between the vertical lines through P and Q, respectively, is

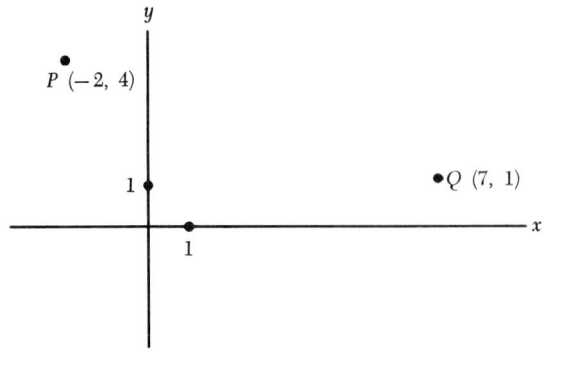

Figure 6a

Figure 6b

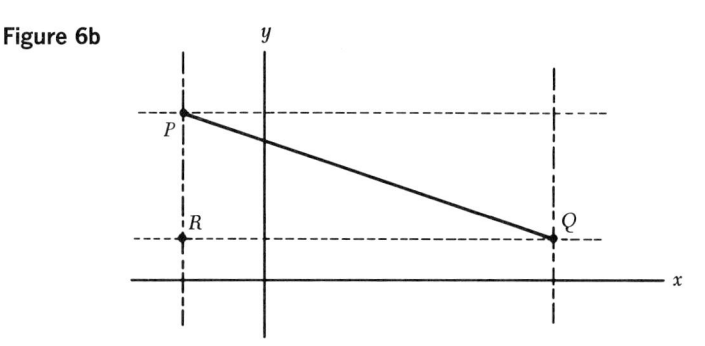

the difference between the x-coordinate 7 at Q and the x-coordinate -2 at P; thus, the two vertical lines are 9 units apart. The distance between the respective horizontal lines is the absolute value of the difference between y-coordinates at Q and at P, namely, $4 - 1 = 3$. If we let h be the distance between P and Q, the theorem of Pythagoras applied to the right triangle with vertices P, R, Q tells us that $h^2 = 9^2 + 3^2$. Since $h^2 = 90$, the distance between P and Q is $90^{1/2} = 3\sqrt{10}$.

We follow the pattern in Example 7 to derive a formula for the distance between any two points on a two-dimensional coordinate system. Let the given points be P with coordinates (a, b) and Q with coordinates (d, e).

We first consider the case in which $a \neq d$ and $b \neq e$. Then P and Q determine a line that is not parallel to either of the coordinate axes. The horizontal line containing Q and the vertical line containing P intersect at a point we may call R (see Fig. 7a). The distance between Q and R is the distance between the vertical line on which P lies and the point Q; this distance is $|a - d|$, the absolute value of the difference between the x-coordinates of P and of Q. The distance between P and R is the distance between the horizontal line on which Q lies and the point P; this distance is the absolute value of the difference between the y-coordinates of P and of Q, namely, $|b - e|$.

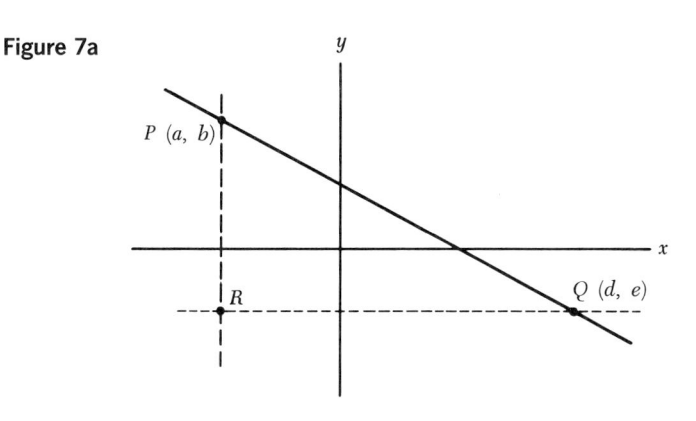

Figure 7a

3 / Distance between points

Figure 7b

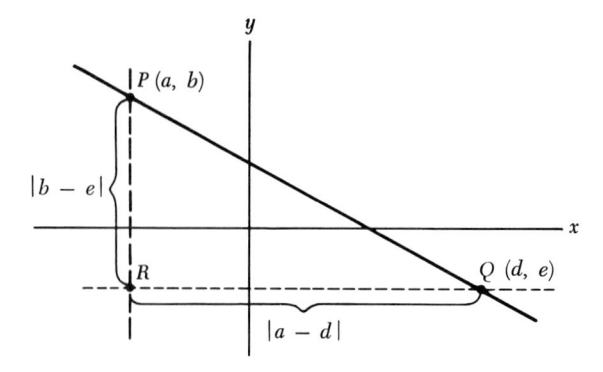

(See Fig. 7b.) The segment joining P and Q is the hypotenuse of the right triangle with vertices P, R, Q; let h be the distance between the given points P and Q. Since the lengths of the legs of the right triangle are $|a - d|$ and $|b - e|$, respectively, the theorem of Pythagoras yields $h^2 = |a - d|^2 + |b - e|^2$. Since the second power of a number is the same as the second power of its absolute value, we may simplify the preceding equation as follows:

$$h^2 = (a - d)^2 + (b - e)^2.$$

Now we examine the special cases that were excluded in the previous discussion. Suppose first that $b = e$. Then (see Fig. 8a) the points P and Q with respective coordinates (a, b) and (d, b) lie on a horizontal line, and the distance between them is $h = |a - d|$. Since $b - e = 0$, we have

$$h^2 = |a - d|^2 = |a - d|^2 + 0 = (a - d)^2 + (b - e)^2.$$

This expression for h^2 agrees with the formula obtained in the previous case.

Finally, suppose that $a = d$. Then (see Fig. 8b) an analogous argument shows that in this case also

$$h^2 = (a - d)^2 + (b - e)^2.$$

In summary, the second power of the distance between (a, b) and (d, e) is, in every case, given by $(a - d)^2 + (b - e)^2$. This, we recognize, is the sum of two squares, the square of the difference between the x-coordinates of the points and the square of the difference between their y-coordinates. The distance between the two points is

$$[(a - d)^2 + (b - e)^2]^{1/2}.$$

We state the conclusion as a theorem.

> **Theorem 1** On a two-dimensional coordinate system, the distance between any two points is the square root of the sum of two squares—the square of the difference between the left coordinates of the points, and the square of the difference between the right coordinates of the points.

Figure 8

(a)

(b)

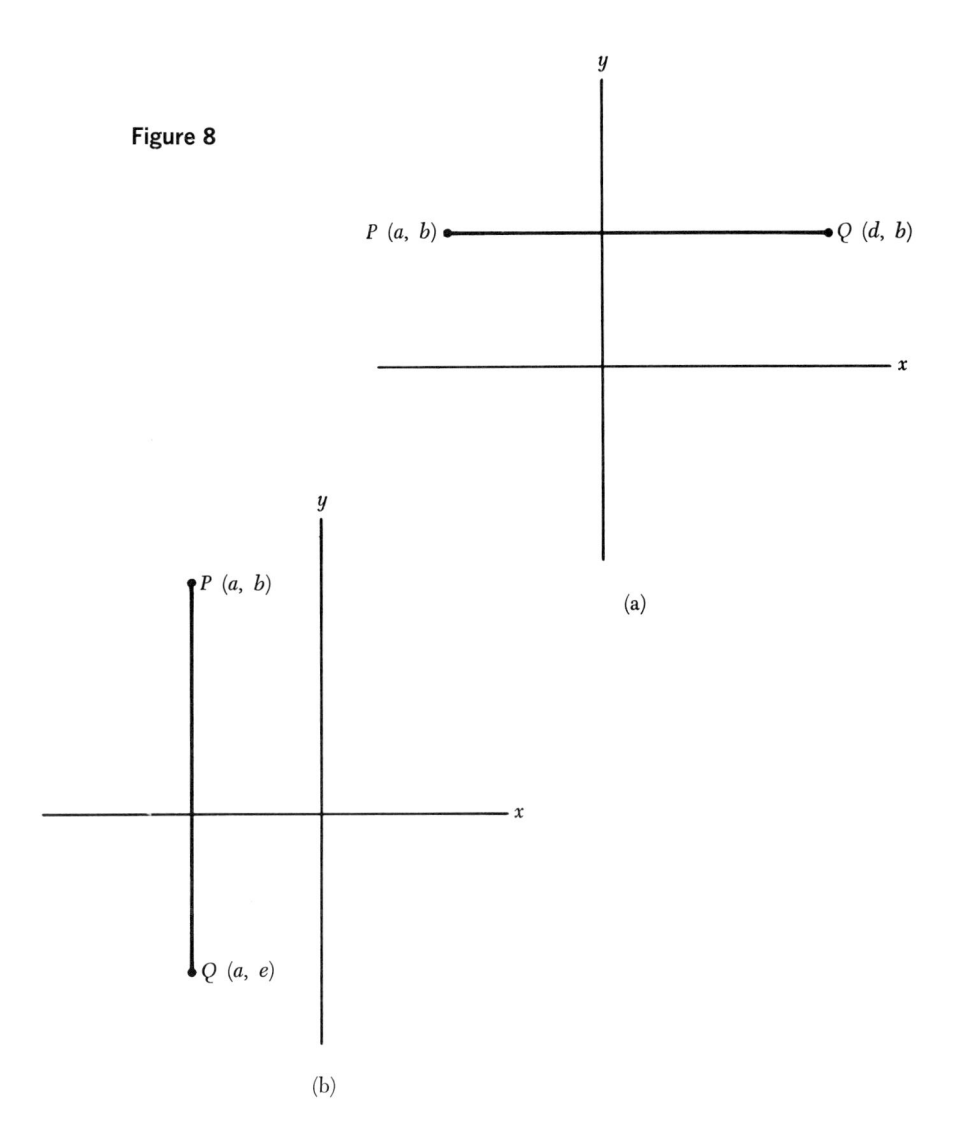

Example 8 Using the formula, find the distance between the points $(-1, 7)$ and $(4, -5)$.

The difference between the x-coordinates of the points is $(-1) - (4) = -5$; the difference between their y-coordinates is $(7) - (-5) = 12$. If h is the desired distance, then $h^2 = (-5)^2 + (12)^2 = 25 + 144 = 169$. Hence, the desired distance is $\sqrt{169} = 13$.

Example 9 Using the formula, find the distance between the point (c, d) and the origin.

If h is the number requested, then $h^2 = (c - 0)^2 + (d - 0)^2 = c^2 + d^2$. That is, $h = (c^2 + d^2)^{1/2}$. We observe that this agrees with the result obtained in Section 10 of Chapter 1.

EXERCISES B

1. What is the distance between the points in each of the following pairs?
   (a) $(1, 0)$ and $(-5, 0)$. (b) $(1, 0)$ and $(1, 7)$. (c) $(1, 0)$ and $(0, 1)$.
   (d) $(-12, 5)$ and $(0, 0)$. (e) $(a, 0)$ and $(0, b)$. (f) $(-2, 1)$ and $(1, -3)$.
   (g) (a, b) and (a, c). (h) $(-\frac{2}{3}, \frac{7}{3})$ and $(1, -\frac{5}{3})$.

2. In each part, the three points are the vertices of a triangle. Find the perimeter of the triangle.
   (a) $(2, 9), (4, 1), (5, 5)$. (b) $(-2, 0), (0, 3), (1, -4)$.

3. The final case in the proof of Theorem 1 treats the case where $a = d$ and is illustrated in Fig. 8b. Supply the details of the argument for this case by adapting the discussion for the previous special case.

4. Let P be the point $(-2, 1)$, let s be the line whose equation is $3x - 4y = 15$, and let r be the line that contains P and is perpendicular to s.
   (a) Find the slope of the line s.   (b) Find the slope of the line r.
   (c) Find an equation for the line r. (d) Find the point Q of intersection of s and r.
   (e) Find the distance between P and Q.
   (f) How far apart are the point P and the line s?

5. Repeat Exercise 4 in case P is the point $(4, -3)$ and s is the graph of $\{(x, y) \mid y = 2x - 1\}$.

6. Let A be the point with coordinates (x, y).
   (a) Write an equation that states that the distance between A and $(1, 4)$ is 9.
   (b) Write an equation that states that the point A and the line $y = 3$ are five units apart.
   (c) Write an equation that states that the distance between A and $(-2, 1)$ is the same as the distance between A and $(1, -3)$.

7. On a three-dimensional coordinate system, let A be the point $(3, 2, 0)$ and let B be the point $(2, 6, 8)$.
   (a) Make a large diagram showing the two points.
   (b) The point A lies in the plane $z = 0$. On the diagram, draw the shortest segment that joins B and the plane.
   (c) What is the length of the segment identified in part (b)?
   (d) Give the coordinates of the other endpoint D of the segment described in part (b).
   (e) Since both A and D lie in the plane $z = 0$, apply Theorem 1 (for left and middle coordinates) to find the distance between A and D.
   (f) On the diagram, draw the segment joining A and D.
   (g) Explain why the triangle with vertices A, D, B is a right triangle.
   (h) Use the theorem of Pythagoras to find the distance between the points A and B.

8. Let a, b, c be nonzero numbers. On a three-dimensional coordinate system, let Q be the point (a, b, c) and consider the origin with coordinates $(0, 0, 0)$.
   (a) Make a large diagram showing the point Q and the origin.
   (b) Draw the shortest segment joining Q and a point in the horizontal coordinate plane $z = 0$.
   (c) What is the length of the segment identified in part (b)?
   (d) Give the coordinates of the other endpoint S of the segment described in part (b).
   (e) Apply Theorem 1 (for left and middle coordinates) to find the distance between S and the origin.
   (f) Draw the segment joining S and the origin.
   (g) Explain why the origin and the points Q and S are vertices of a right triangle.
   (h) Use the theorem of Pythagoras to find the distance between the point Q and the origin.

4 Distance in three-dimensional space

It is instructive to examine Fig. 7a from a different point of view. For that purpose we again draw a picture showing the points P and Q and the segment joining them whose length is the number h we wish to evaluate (Fig. 9).

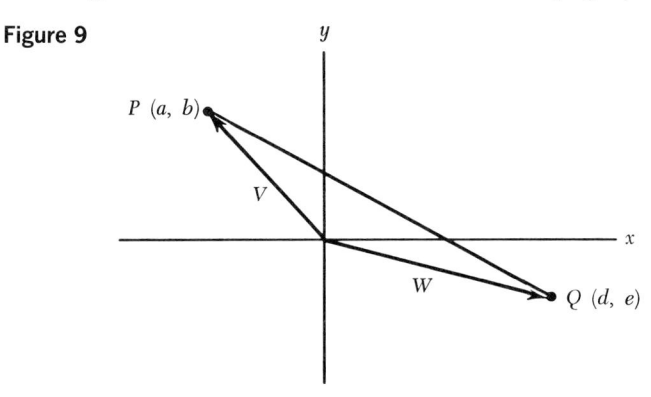

Figure 9

[Figure: Coordinate system with point $P(a, b)$ in the second quadrant, vector V from origin to P, point $Q(d, e)$ in the fourth quadrant, vector W from origin to Q, and a segment connecting P and Q.]

This time we choose to emphasize other supplementary features in the picture. Our formula expresses h in terms of the two differences $a - d$ and $b - e$. These two numbers are the components of the two-dimensional vector $(a - d, b - e)$. Now $(a - d, b - e) = (a, b) - (d, e)$. The vector (a, b) is the vector labeled V in Fig. 9 and conveniently represents a displacement from the origin to the point P. Likewise, the vector (d, e) may represent a displacement from the origin to the point Q; this vector is abbreviated W in the diagram. The difference $V - W = (a - d, b - e)$ is pictured geometrically as the displacement from Q to P, namely, as a segment directed from Q to P. The length of this segment is the number h, specifying the distance between the given points. Thus, the distance between two points is the square root of the sum of the squares of the components of the vector from one of the points to the other. This interpretation of the formula $h = [(a - d)^2 + (b - e)^2]^{1/2}$ we keep in mind during the next development.

Now we turn our attention to the concept of distance in three-dimensional space. On a three-dimensional coordinate system, let P be a point with coordinates (a, b, c) and let Q be a point with coordinates (d, e, f). The formula expressing the distance h between the two given points P and Q is a natural extension of the corresponding formula in the two-dimensional case. The interpretation cited at the end of the previous paragraph leads us to develop the three-dimensional analog as follows.

The vector associated with the point P is the vector with components (a, b, c). The vector associated with Q is (d, e, f). The vector from Q to P is the difference $(a, b, c) - (d, e, f) = (a - d, b - e, c - f)$. The distance between P and Q is the square root of the sum of the second powers of the components of this vector; in symbols, $h = [(a - d)^2 + (b - e)^2 + (c - f)^2]^{1/2}$.

Note that the important formula
$$h^2 = (a - d)^2 + (b - e)^2 + (c - f)^2$$
is the natural extension to the three-dimensional situation for the relationship $h^2 = (a - d)^2 + (b - e)^2$ in the case of two dimensions.

Our reasoning by analogy has suggested what formula in three dimensions we should expect. To establish that the result actually is correct, we apply the theorem of Pythagoras. We examine in detail the general case in which $a \neq d$ and $b \neq e$ and $c \neq f$. The special cases in which one or more of these inequalities is not satisfied may be handled by a suitable modification of the following procedure.

Let R be the point (a, e, f), let N be the point (a, b, f), and let n be the distance between Q and N (see Fig. 10). The distance between R and Q is $|a - d|$; the distance between N and R is $|b - e|$; since Q, R, N are the vertices of a right triangle (in the plane $z = f$), the theorem of Pythagoras yields

$$n^2 = |a - d|^2 + |b - e|^2.$$

The distance between P and N is $|c - f|$; since Q, N, P are the vertices of a right triangle (with right angle at N), the theorem of Pythagoras yields

$$h^2 = n^2 + |c - f|^2.$$

Combining the two equations, we obtain

$$h^2 = n^2 + |c - f|^2 = |a - d|^2 + |b - e|^2 + |c - f|^2$$
$$= (a - d)^2 + (b - e)^2 + (c - f)^2.$$

Thus, h is the square root of the sum of the squares of the components of the vector $(a - d, b - e, c - f)$; and this vector is the difference (a, b, c)

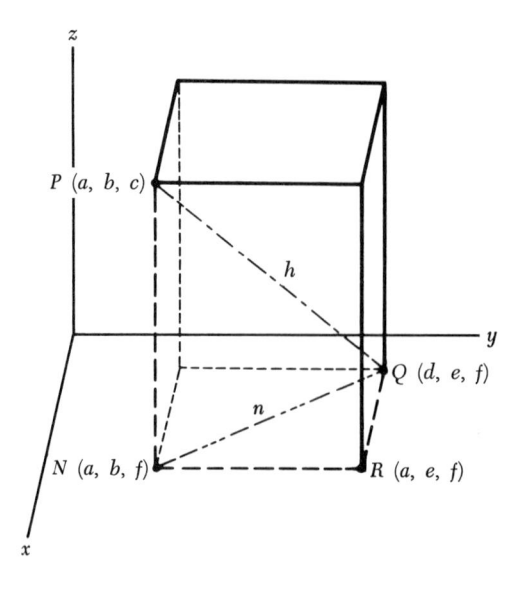

Figure 10

— (d, e, f), the difference between the vector determined by P and the vector associated with Q.

Our conclusion is stated in the next theorem. Note the analogy between it and the two-dimensional case stated in Theorem 1.

Theorem 2 On a three-dimensional coordinate system, the distance between any two points is the square root of the sum of three squares—the square of the difference between the left coordinates of the points, the square of the difference between the middle coordinates of the points, and the square of the difference between the right coordinates of the points.

Example 10 The distance between the points $(5, 8, -2)$ and $(-1, 6, 1)$ is

$$[([5] - [-1])^2 + (8 - 6)^2 + ([-2] - [1])^2]^{1/2}$$
$$= (36 + 4 + 9)^{1/2} = 7.$$

EXERCISES C

1. Find the distance between each of the following pairs of points.
   (a) $(1, 2, -2)$ and $(0, 0, 0)$. (b) $(5, 2, -7)$ and $(9, -1, 5)$.
   (c) $(1, 1, 1)$ and $(-1, -1, -1)$. (d) $(2, -1, -1)$ and $(-4, 1, 2)$.

2. Find the distance between each of the following pairs of points.
   (a) (a, b, c) and $(0, 0, 0)$. (b) (a, b, c) and (a, b, f).
   (c) $(0, b, c)$ and $(a, b, 0)$. (d) (a, b, c) and $(-a, -b, -c)$.

3. Let P be the point $(2, 5, 6)$, and let T be the plane whose equation is $z = 1$.
   (a) On a three-dimensional coordinate system, sketch the plane and the point.
   (b) Find the point on the plane T that is nearest to P.
   (c) How far apart are the plane T and the point P?

4. For each of the following, find the distance between the plane and the point.
   (a) Plane $z = -1$, point $(-3, 2, 1)$. (b) Plane $y = 4$, point $(1, 7, -2)$.
   (c) Plane $x = r$, point (a, b, c). (d) Plane $x + y = 3$, point $(4, -1, 5)$.

5. The vertices of a triangle are $(-0.2, 1.3, 0.6)$, $(-0.3, 1.7, 1.4)$, $(0, 1.7, 1)$. Find the perimeter of the triangle.

5 Definition of a parabola

In Sections 2 and 3 of Chapter 9, we encountered several members of the class of parabolas. Each parabola we discussed appeared as the graph of a suitable algebraic equation. The notion of distance enables us to give a geometrical description for a parabola.

In a plane, we choose any point and any line not containing the point. We consider the chosen line and the chosen point as fixed. Call the point F.

Figure 11

(a) (b)

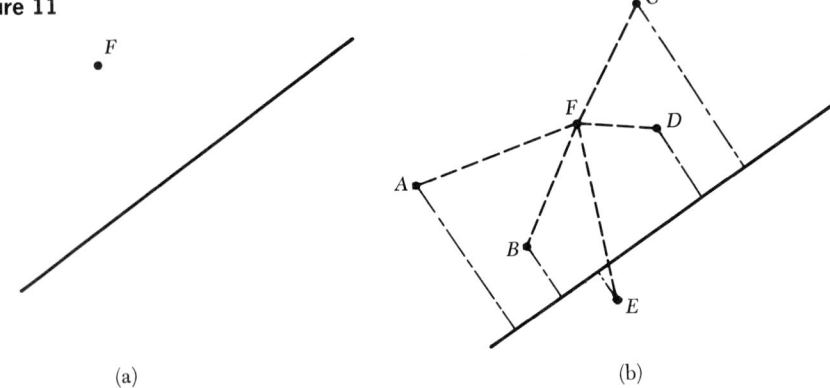

(See Fig. 11a.) For each point in the plane, we may compare its distance from the fixed point F with its distance from the fixed line. In Fig. 11b, five typical points are tested. We observe that B is nearer to the line than to F. Certainly E also is not so far from the line as from F. On the other hand, C is farther from the line than it is from F. By contrast, each of the points A and D has the property that its distance from F is equal to its distance from the line. Points A and D belong to the parabola determined by the fixed line and the fixed point. The other three points tested do not belong to the parabola. Indeed, the parabola contains a point of the plane if and only if the point is equally far from the line and from F. We formalize this discussion in the following definition.

> **Definition** Let a line and a point not on the line be chosen. In the plane that contains the given line and the given point, the set of all points that are equally distant from the given line and from the given point is a curve called a *parabola*.

A parabola, according to the definition, is a certain subset of a plane. Our next goal is to discover how a parabola may be described in terms of a suitable coordinate system in the plane.

We begin by elaborating on the definition. Any line and any point such that the line does not contain the point determine a parabola. It is useful to assign names to the given line and the given point. They are called the *directrix* of the parabola and the *focus* of the parabola, respectively. The parabola lies in the plane determined by its directrix and its focus. A point in

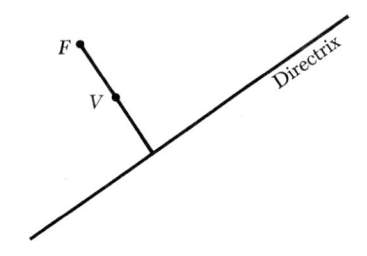

Figure 12

368   Distance

this plane belongs to the parabola if and only if it is equally distant from the chosen line and chosen point. Certainly, then, one point that belongs to the parabola is the point midway between the directrix and the focus. In Fig. 12, the point V is the midpoint of the segment that joins the focus to the directrix and that is perpendicular to the directrix. The point V is known as the *vertex* of the parabola.

EXERCISES D

1. The directrix of a parabola separates the plane into two halfplanes. Explain why the parabola is entirely contained in one of these halfplanes, namely, the halfplane that contains the focus.

2. Let the distance between the focus and the directrix of a parabola be the positive number g. If d is any number such that $d > g/2$, show that there are two points on the parabola that are d units distant from the directrix.

3. On the two-dimensional coordinate system, let $(1, 0)$ be the focus of a parabola and let the line $x = -1$ be the directrix of the parabola.
   (a) Show that the vertex of the parabola is $(0, 0)$.
   (b) Show that if the point (x, y) belongs to the parabola, then $[(x - 1)^2 + y^2]^{1/2}$ is the same number as $x + 1$.
   (c) Show that if the coordinates (x, y) of a point Q have the property that $[(x - 1)^2 + y^2]^{1/2} = x + 1$, then Q belongs to the parabola.

6 Equation for a parabola

In the plane containing a parabola, we introduce a coordinate system in the following convenient manner. Choose the origin to be the point midway between the directrix and the focus; in other words, the vertex of the parabola is selected as the origin. Choose the y-coordinate axis to be the line joining the vertex and the focus of the parabola. Figure 13 illustrates a possible orientation of the coordinate system in the plane of the parabola. Since the y-coordinate axis is perpendicular to the directrix, the x-coordinate axis is parallel to the directrix. The vertex of the parabola, having been selected as the origin, of course lies on the x-coordinate axis.

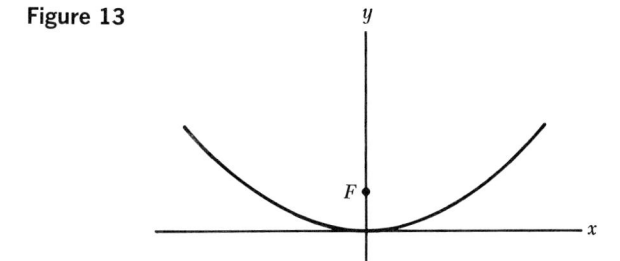

Figure 13

Since the focus lies on the y-coordinate axis, its x-coordinate is 0; if we denote its y-coordinate by p, then F is the point $(0, p)$. Since the vertex is the point $(0, 0)$ and since the vertex is not the focus, we observe that $p \neq 0$. Since the vertex is midway between the directrix and the focus, the directrix intersects the y-coordinate axis at the point $(0, -p)$. Since the directrix is parallel to the x-coordinate axis, an equation for the directrix is $y = -p$.

Now consider any point, say A, in the plane. Let the coordinates of A be (x, y). We seek an algebraic condition on the numbers x and y that states that A belongs to the parabola. To say that A is on the parabola means, in accordance with the definition, that the distance between the directrix and A is the same as the distance between the focus and A. (See Fig. 14.) The distance

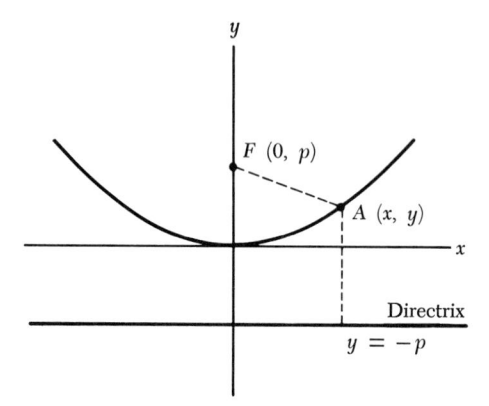

Figure 14

between the directrix and the point A is the absolute value of the difference between y-coordinates, namely, $|y - (-p)|$, or $|y + p|$. The distance between the two points A and F with respective coordinates (x, y) and $(0, p)$ is $[(x - 0)^2 + (y - p)^2]^{1/2}$. The statement that A belongs to the parabola is equivalent to the algebraic equation

$$|y + p| = [x^2 + (y - p)^2]^{1/2}.$$

From this equation we derive, by the following sequence of steps, an equivalent but much simpler equation.

$$|y + p|^2 = x^2 + (y - p)^2,$$
$$y^2 + 2py + p^2 = x^2 + y^2 - 2py + p^2,$$
$$4py = x^2,$$
$$y = \frac{1}{4p}x^2.$$

Thus, the parabola is the graph of the function $\{(x, y) \mid y = (1/4p)x^2\}$.

Figure 14 illustrates the situation in which the number p is positive. In this case, the focus belongs to the positive half of the y-coordinate axis, while the directrix is "below" the x-coordinate axis. The equation $y = (1/4p)x^2$ shows that y is the product of the positive number $(4p)^{-1}$ and the second

Distance

power of x. Since $x^2 \geq 0$ for all x, we deduce that $y \geq 0$ for all points on the parabola. In other words, the parabola lies in the closed upper coordinate halfplane $\{(x, y) \mid y \geq 0\}$.

The derivation of the equation $y = (1/4p)x^2$, as given above, is applicable for any nonzero number p; that is, it does not depend upon whether p is positive or negative. In case $p < 0$, then the focus lies on the negative half of the y-coordinate axis and the directrix is above the x-coordinate axis. Also, since $x^2 \geq 0$ for all x and since $(4p)^{-1} < 0$ in this case, we find that $y \leq 0$ for each point on the parabola. Thus, each point of the curve lies on or below the horizontal coordinate axis. Figure 15 shows a parabola which "opens down," in contrast to the parabola in Fig. 14, which "opens up."

Figure 15

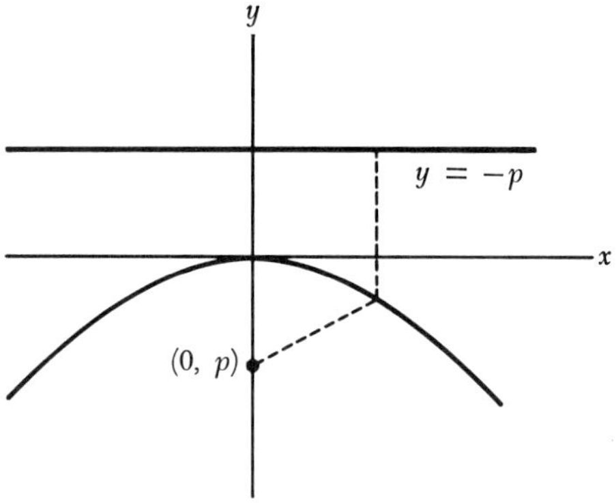

$y = -p$

$(0, p)$

In summary, every parabola whose vertex is the origin and whose directrix is horizontal has an equation of the form $y = ax^2$, where a is a nonzero number. Since $a = 1/(4p)$, we find that the distance between the vertex and the focus is $|p| = |1/4a|$. Thus, as portrayed in Fig. 10 of Chapter 9, the smaller $|a|$ is, the larger $|p|$ is, and hence the less sharply the parabola appears to bend near the origin.

On the other hand, if any nonzero number a is given, then the graph of the function given by the formula ax^2 is a parabola. Indeed, it is a parabola with vertex at the origin and with directrix parallel to the x-coordinate axis. By setting $a = 1/(4p)$ and solving for p, we find specifically that the graph is the parabola determined by the focus $\left(0, \dfrac{1}{4a}\right)$ and the directrix $y = -\dfrac{1}{4a}$.

EXERCISES E

1. Let F be the point $(0, 3)$, and let d be the line whose equation is $y = -3$.
   (a) Find the distance between a point A with coordinates (x, y) and the point F.
   (b) Find the distance between the same point A and the line d.
   (c) Write an equation that states that the distance between A and d is the same as the distance between A and F.

(d) Simplify the equation written in part (c), and show that a simplified form is $12y = x^2$.
(e) The set of all points whose coordinates satisfy the equation is a parabola. What is the vertex of the parabola?
(f) Let Q be the point on the parabola with x-coordinate 8. Calculate the distance between Q and d and the distance between Q and F, and verify that the two distances are the same.

2. Let G be the point $(0, -2)$, and let d be the graph of the equation $y = 2$.
   (a) If A is a point with coordinates (x, y), write an equation that states that the distance between A and d is the same as the distance between A and G.
   (b) Simplify the equation in part (a), obtaining $-8y = x^2$.
   (c) Let the point B on the parabola with focus G and directrix d have x-coordinate 4. Calculate the distance between B and G and the distance between B and d, and verify that the distances are equal.

3. The graph of the equation $y = \frac{1}{20}x^2$ is a parabola.
   (a) What is the vertex of the parabola?
   (b) What is the focus of the parabola?
   (c) Give an equation for the directrix of the parabola.

4. Repeat Exercise 3 for the equation $y = -\frac{1}{36}x^2$.

5. Repeat Exercise 3 for the equation $y = 3x^2$.

6. Recall that the parabolas given by the equations $y = \frac{1}{8}x^2$ and $y = \frac{1}{8}x^2 + 5$ are congruent. Give the vertex, the focus, and the directrix for the parabola whose equation is $y = \frac{1}{8}x^2 + 5$.

7. Give the vertex, the focus, and the directrix for the parabola whose equation is $y = 4x^2 - 2$. [HINT: The given curve is congruent to the graph of the equation $y = 4x^2$.]

8. Give the vertex, the focus, and the directrix for the parabola whose equation is $y = 3 - x^2$.

9. Let F be the point $(4, 1)$, and let d be the line $y = -1$.
   (a) If the point A has coordinates (x, y), write an equation stating that the distance between A and F is the same as the distance between A and d.
   (b) Simplify the equation written in part (a).
   (c) Let the parabola with focus F and directrix d intersect the y-coordinate axis at the point R. Calculate the distance between R and F and the distance between R and d, and verify that the distances are the same.

10. A parabola has the point $(0, p)$ as focus and the line $y = -p$ as directrix. Let M and N be the two distinct points on the parabola each of which has y-coordinate p. Find the distance between M and N.

7 Translation

Let Q be the second-power function $\{(x, y) \mid y = x^2\}$. In Section 2 of Chapter 9, we compared the parabola whose equation is $y = (x - 2)^2$ with the graph of Q. We noted that these two curves are congruent and that the graph of the equation $y = (x - 2)^2$ may be obtained from the graph of Q by translating (that is, shifting or sliding) two units to the right. These observations illustrate the following statement.

Let a and h be numbers such that $a \neq 0$. The graph of the function given by the formula $a(x - h)^2$ is congruent to the parabola whose equation is $y = ax^2$; furthermore it is obtainable from the parabola $y = ax^2$ by translating h units in the x direction.

Example 11 Figure 16a shows the graph of the function $\{(x, y) \mid y = \frac{1}{2}x^2\}$. This curve, translated $\frac{3}{2}$ units to the right, becomes the graph of the function given by the formula $\frac{1}{2}(x - \frac{3}{2})^2$. For example, the point $(1, \frac{1}{2})$ satisfies the equation $y = \frac{1}{2}x^2$. The corresponding point $\frac{3}{2}$ units to the right is $(1 + \frac{3}{2}, \frac{1}{2}) = (\frac{5}{2}, \frac{1}{2})$; as a check on our contention, we may verify that $(\frac{5}{2}, \frac{1}{2}) \in \{(x, y) \mid y = \frac{1}{2}(x - \frac{3}{2})^2\}$ by calculating $\frac{1}{2}(\frac{5}{2} - \frac{3}{2})^2 = \frac{1}{2}(1)^2 = \frac{1}{2}$. The two curves shown in Figs. 16a and 16b are a pair of congruent parabolas.

The vertex of the parabola $y = \frac{1}{2}x^2$ is the origin. To the right of the origin $\frac{3}{2}$ units, namely, at the point $(\frac{3}{2}, 0)$, is the vertex of the parabola $y = \frac{1}{2}(x - \frac{3}{2})^2$. Since $\frac{1}{2} = 1/(4p)$, we find that $p = \frac{1}{2}$. The focus of the parabola in Fig. 16a is the point $(0, \frac{1}{2})$. Consequently, the focus of the parabola in Fig. 16b is $\frac{3}{2}$ unit to the right of $(0, \frac{1}{2})$, namely, the point $(\frac{3}{2}, \frac{1}{2})$. The directrix of the parabola $y = \frac{1}{2}x^2$ is the horizontal line whose equation is $y = -\frac{1}{2}$. A horizontal translation sends a horizontal line into itself; thus, the two parabolas have the same directrix.

Figure 16

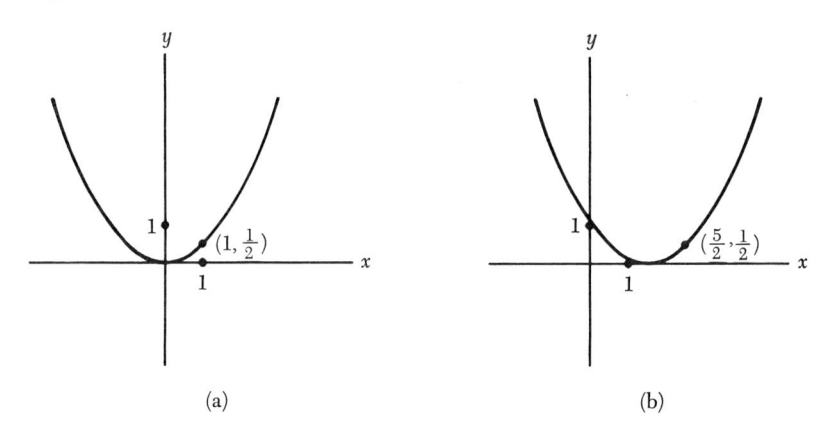

(a)   (b)

Example 12 Suppose the curve in Fig. 16a is translated two units up on the coordinate system. Then each point (c, d) on the curve is shifted to the point with its y-coordinate two greater, namely, the point $(c, d + 2)$. The statement that (c, d) satisfies the equation $y = \frac{1}{2}x^2$ is equivalent to the statement that $(c, d + 2)$ satisfies the equation $y = \frac{1}{2}x^2 + 2$. Thus, the parabola, in its new location on the

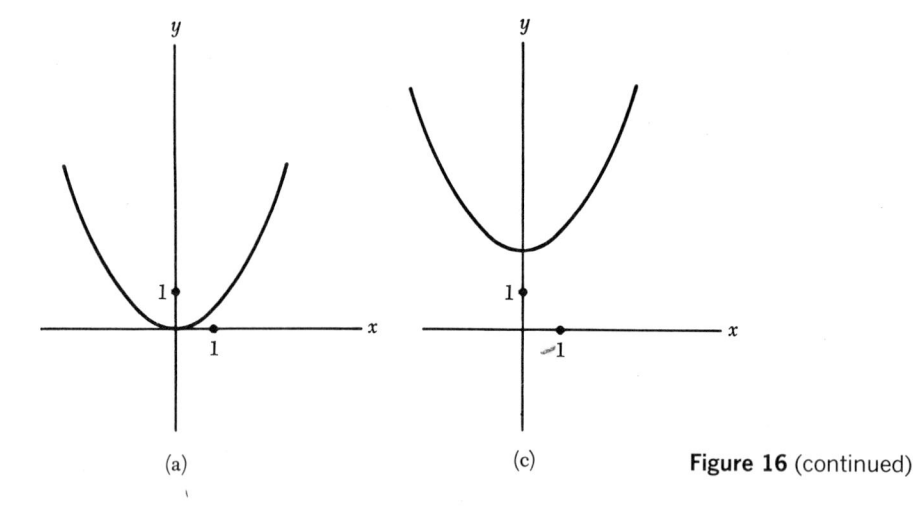

(a)   (c)   **Figure 16** (continued)

coordinate system, as shown in Fig. 16c, is the graph of the function $\{(x, y) \mid y = \frac{1}{2}x^2 + 2\}$.

Since the curve in Fig. 16c is two units above the curve in Fig. 16a, the vertex, the focus, and the directrix of the parabola $y = \frac{1}{2}x^2 + 2$ are two units above the corresponding feature of the parabola $y = \frac{1}{2}x^2$. Thus, the graph of the function given by the formula $\frac{1}{2}x^2 + 2$ is the parabola whose vertex is $(0, 2)$, whose focus is $(0, \frac{5}{2})$, and whose directrix has the equation $y = \frac{3}{2}$.

Example 12 illustrates the following remark. Let a and k be numbers such that $a \neq 0$. The graph of the function given by the formula $ax^2 + k$ is congruent to the parabola whose equation is $y = ax^2$; furthermore, it is obtainable from the parabola $y = ax^2$ by translating k units in the y direction.

A parabola with vertex at the origin, say the graph of $y = ax^2$, where $a \neq 0$, may be subjected to a translation that is neither in the x nor in the y direction, but rather that has components in both directions. Suppose the curve is translated h units in the x direction and k units in the y direction. Then the equation for the parabola in its new location is $y = a(x - h)^2 + k$. Notice that we have replaced x in the equation by $x - h$ and that we have added k to each image under the function. The substitution of $x - h$ for x represents the horizontal shift of h units, and the addition of k to each functional value is required by the vertical component k of the displacement vector (h, k).

In general, consider the graph of any function. If the curve represents the function f, that is, if the curve is the graph of the equation $y = f(x)$, then the displacement vector (h, k) translates each point of the curve to a new location h units to the right and k units up. The curve in its new position is the graph of the equation $y - k = f(x - h)$, that is, represents the function given by the formula $f(x - h) + k$ (see Fig. 17). We summarize these remarks in the following theorem.

Figure 17

Figure 18

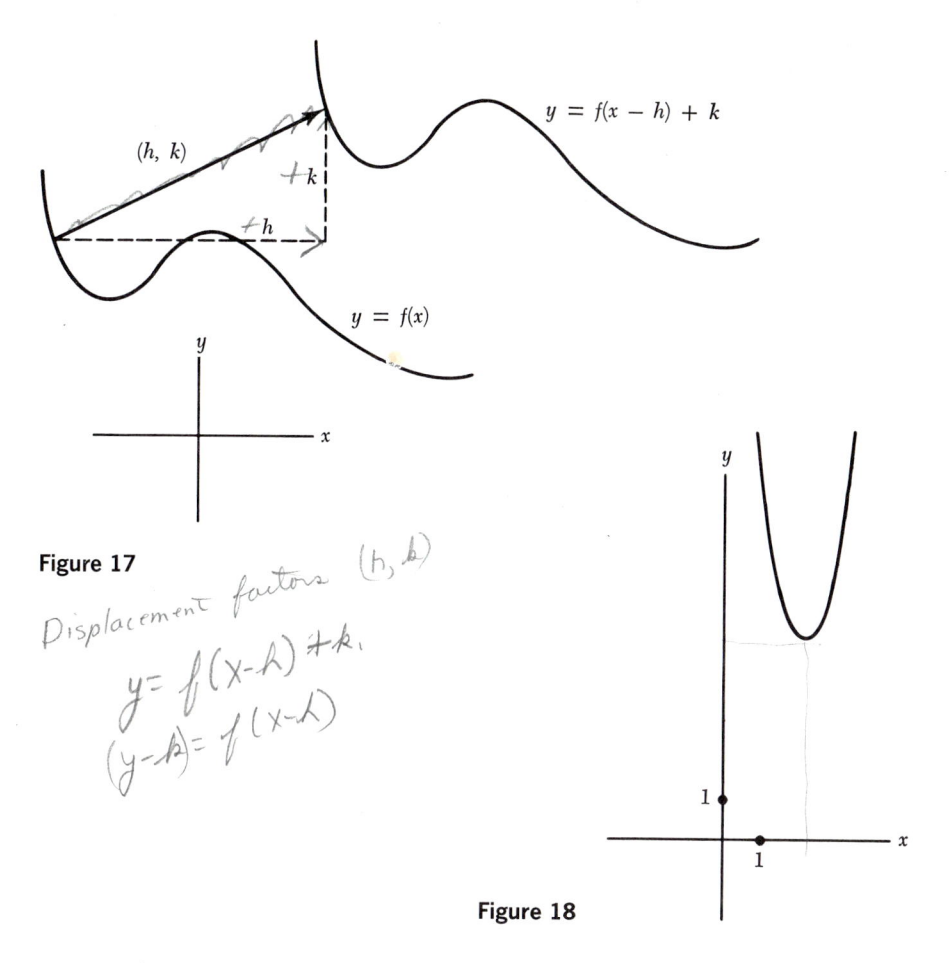

Theorem 3 Let h, k be real numbers, and let f be a function. Suppose the graph of the equation $y = f(x)$ is subjected to the displacement vector (h, k). Each point on the curve is translated h units in the x direction and k units in the y direction; and the curve in its new position is the graph of the equation $y - k = f(x - h)$.

Example 13 The graph of the equation $y = 3(x - 2)^2 + 5$ is a parabola. It is congruent to the parabola $y = 3x^2$ and is obtainable from the latter by applying the displacement vector $(2, 5)$. The parabola $y = 3x^2$ has vertex $(0, 0)$, focus $(0, \frac{1}{12})$, and directrix $y = -\frac{1}{12}$. These become, under application of the displacement vector $(2, 5)$, respectively the vertex, the focus, and the directrix of the parabola $y = 3(x - 2)^2 + 5$. Thus, the vertex is the point $(2, 5)$, the focus is the point $(2, \frac{61}{12})$, and the directrix is the line $y = \frac{59}{12}$. (See Fig. 18.)

The formula $3(x - 2)^2 + 5$ is equivalent to $3(x^2 - 4x + 4) + 5$ and also to $3x^2 - 12x + 17$. Thus, the parabola in Fig. 18 is the graph of the function $3x^2 - 12x + 17$.

7 / Translation   375

$y = \frac{1}{4p} x^2$

EXERCISES F

1. For each of the following equations, give the vertex, the focus, and the directrix of the parabola.
   (a) $8y = (x - 1)^2$.
   (b) $-16y = (x + 2)^2$.
   (c) $y = \frac{1}{6}x^2 - 3$.
   (d) $y = 10x^2 + 40$.
   (e) $y = 3(x - 1)^2 + 5$.
   (f) $3y = (x - 1)^2 + 5$.

2. Find an equation of the parabola whose focus is (9, 2) and whose vertex is (9, 0).

3. The focus of a certain parabola is the point (8, 1) and the directrix is the line whose equation is $y + 5 = 0$.
   (a) Find the vertex of the parabola.
   (b) Find an equation for the parabola.

4. Find an equation of the parabola whose directrix is the line $y = 8$ and whose focus is the point (3, 2).

5. If the directrix of a parabola is horizontal and if the vertex is the point $(-2, 1)$, then the parabola has an equation of the type $y = a(x + 2)^2 + 1$. Find the number a in case (6, 2) belongs to the parabola.

6. A parabola has a horizontal directrix and its vertex is $(-1, -3)$. Find an equation for the parabola if the curve contains the point $(0, -5)$.

7. Find an equation of the parabola whose vertex is (0, 3), whose directrix is horizontal, and which contains the point $(-2, -4)$.

8. Find an equation of the parabola that contains the two points $(-3, 5)$ and (5, 5) and whose directrix is the line $y = 1$.

8 Quadratic function

Suppose a, h, k are numbers such that $a \neq 0$. The equation

$$y = a(x - h)^2 + k$$

is equivalent to

$$y = a(x^2 - 2hx + h^2) + k$$

or to

$$y = ax^2 - 2ahx + (ah^2 + k).$$

Any parabola in the coordinate plane having a horizontal directrix is obtainable by translation of a parabola with horizontal directrix and with vertex at the origin. Thus, any parabola with horizontal directrix is the graph of an equation of the type $y = a(x - h)^2 + k$. In other words, any parabola with horizontal directrix is the graph of an equation of the form

$$y = ax^2 + bx + c,$$

where a, b, c are numbers such that $a \neq 0$.

Definition Let a, b, c be real numbers such that $a \neq 0$. Then the function $\{(x, y) \mid y = ax^2 + bx + c\}$ is called a *quadratic function*.

376 Distance

Example 14 Find a quadratic function f such that $(-1, 6) \in f$, $(1, 0) \in f$, and $(4, 21) \in f$.

There are real numbers a, b, c such that f is given by the formula

$$f(x) = ax^2 + bx + c.$$

We wish to determine these numbers a, b, c. Since $(-1, 6)$ is an element of the function f, we obtain $6 = f(-1) = a(-1)^2 + b(-1) + c$. Since $(1, 0)$ belongs to f, we obtain $0 = f(1) = a \cdot 1^2 + b \cdot 1 + c$. The third requirement is that $(4, 21) \in f$; hence, $21 = f(4) = a \cdot 4^2 + b \cdot 4 + c$. We now have a system of three equations involving the three numbers a, b, c that we are seeking:

$$a - b + c = 6,$$
$$a + b + c = 0,$$
$$16a + 4b + c = 21.$$

Applying the method of Chapter 2, we find that the root of the system is $(a, b, c) = (2, -3, 1)$. Hence, a formula for the quadratic function is

$$f(x) = 2x^2 - 3x + 1.$$

EXERCISES G

1. For each of the following pairs of formulas, show that the two formulas represent the same function.
   (a) $x^2 + 6x + 11$ and $(x + 3)^2 + 2$.
   (b) $4x^2 + 8x$ and $4(x + 1)^2 - 4$.
   (c) $-x^2 + 5x$ and $-(x - \frac{5}{2})^2 + \frac{25}{4}$.
   (d) $6x^2 + 18x + 16$ and $6(x + \frac{3}{2})^2 + \frac{5}{2}$.

2. Each of the following equations represents a parabola. Referring to Exercise 1, identify the vertex and the focus of each parabola.
   (a) $y = x^2 + 6x + 11$.
   (b) $y = 4x^2 + 8x$.
   (c) $x^2 - 5x + y = 0$.
   (d) $y = 6x^2 + 18x + 16$.

3. Find the distance between the two points that belong to the intersection of the parabola $\{(x, y) \mid y = a(x - h)^2 + k\}$ and the horizontal line containing the focus of the parabola.

4. Find a quadratic function that contains each of the ordered pairs $(1, -2)$, $(-1, 0)$, $(3, 20)$.

5. Find a quadratic function that contains the subset $\{(3, 23), (-1, 7), (0, 5)\}$.

6. Let q and n be nonzero real numbers.
   (a) Expand $(x + q)^2$ by the binomial theorem.
   (b) Observe that the last term q^2 in the expansion is the second power of half of the coefficient of x in the middle term.
   (c) Show that the sum of $(x^2 + nx)$ and $(n/2)^2$ is a perfect square.
   (d) Find a number such that the sum of it and $x^2 + 6x$ is a perfect square.
   (e) What number has the property that $x^2 + 3x$ is converted into a perfect square by adding the number?
   (f) Given $y^2 - 11y$, find the number that may be added to the given formula in order to complete the square.
   (g) Apply the procedure of completing the square to $t^2 - \frac{3}{5}t$.

9 Graph of a quadratic function

We have observed above that every parabola with a horizontal directrix is the graph of a quadratic function. We now wish to discuss the converse, namely, that the graph of every quadratic function is a parabola with a horizontal directrix.

Example 15 Consider the quadratic function given by the formula $x^2 + 8x + 9$. Now $x^2 + 8x + 9 = (x^2 + 8x + 16) - 7 = (x + 4)^2 - 7$. Thus, the graph of the function is a parabola with vertex $(-4, -7)$. The coefficient of x^2 is 1; since $1/(4p) = 1$, we find that $p = \frac{1}{4}$. The focus is $\frac{1}{4}$ unit above the vertex, namely, the point $(-4, -\frac{27}{4})$. The directrix is the horizontal line $\frac{1}{4}$ unit below the vertex; its equation is $y = -\frac{29}{4}$. (See Fig. 19.)

Our decision to express $x^2 + 8x + 9$ in the form $(x^2 + 8x + 16) - 7$ has been justified by the information concerning the parabola that can be gleaned from the form $(x + 4)^2 - 7$. We chose the number 16 in the parentheses in order that $x^2 + 8x + 16$ should be a second power. We obtained the number 16 by "completing the square": we noted the coefficient 8 of x, halved it to obtain 4, and squared the latter number. The number -7 outside the parentheses was selected because the sum of it and 16 is 9, the term in the given formula.

Example 16 Consider the quadratic function given by the formula $-x^2 + 3x + 2$. We seek to express this same function by a formula of the type $a(x - h)^2 + k$, in which form we will be able

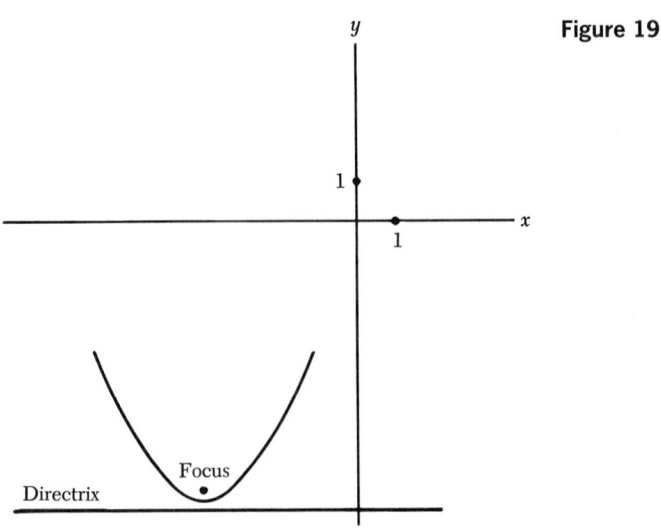

Figure 19

378 Distance

to recognize various features of the parabola that is the graph of the function. We must, however, in our algebraic manipulations, be very careful not to alter the given quadratic function.

The steps are as follows:

$$-x^2 + 3x + 2 = -(x^2 - 3x) + 2$$
$$= -(x^2 - 3x + \tfrac{9}{4}) + [2 + \tfrac{9}{4}]$$
$$= -(x - \tfrac{3}{2})^2 + \tfrac{17}{4}.$$

In the first step, we recognize that the multiplier of x^2 is -1 and we factor $-x^2 + 3x$ as the product of the coefficient -1 and $x^2 - 3x$. The coefficient -3 of x determines the number $\tfrac{9}{4}$ as follows: the second power of one-half of -3 is $[\tfrac{1}{2}(-3)]^2 = \tfrac{9}{4}$. In the second step, we add $\tfrac{9}{4}$ inside the parentheses to "complete the square" and we add $\tfrac{9}{4}$ inside the brackets to preserve the same function; notice that the insertion of $\tfrac{9}{4}$ inside parentheses preceded by a minus sign effectively means the addition of $-\tfrac{9}{4}$, and hence the counterbalancing act is to add $+\tfrac{9}{4}$ outside the parentheses. In the third step, we express $x^2 - 3x + \tfrac{9}{4}$ as a power of $x - \tfrac{3}{2}$.

Since the given quadratic function has the formula $-(x - \tfrac{3}{2})^2 + \tfrac{17}{4}$, the graph is a parabola with vertex $(\tfrac{3}{2}, \tfrac{17}{4})$. Furthermore $p = -\tfrac{1}{4}$, since $1/(4p) = -1$. The focus is $\tfrac{1}{4}$ unit below the vertex, and is therefore the point $(\tfrac{3}{2}, 4)$. The directrix is $\tfrac{1}{4}$ unit above the vertex, and its equation is $y = \tfrac{9}{2}$. The parabola (see Fig. 20) opens down; the vertex is the highest point on the curve.

Example 17 Show that the graph of the function $\{(x, y) \mid y = 3x^2 + 30x + 83\}$ is a parabola, and find its vertex, its focus, and its directrix.

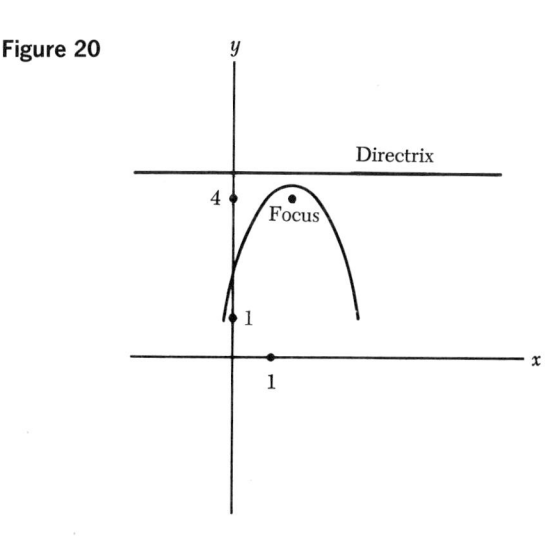

Figure 20

We wish to express the function by a formula of the type $a(x - h)^2 + k$:

$$3x^2 + 30x + 83 = 3(x^2 + 10x) + 83$$
$$= 3(x^2 + 10x + 25) + [83 - 75]$$
$$= 3(x + 5)^2 + 8.$$

In the first step, we factor $3x^2 + 30x$ as the product of 3 and $x^2 + 10x$; inside the parentheses the coefficient of x^2 is 1. The second power of one-half of 10 is $(\frac{1}{2} \cdot 10)^2 = 25$. In the second step, we add 25 inside the parentheses; since 25 is affected by the multiplier 3, we subtract 75 in order to preserve the function. In the third step, we express $x^2 + 10x + 25$ as a power of $x + 5$.

The graph of the equation $y = 3(x + 5)^2 + 8$ is a parabola with vertex $(-5, 8)$. Since $1/(4p) = 3$, the focus and directrix are $\frac{1}{12}$ unit above and below, respectively, the vertex. The focus has coordinates $(-5, \frac{97}{12})$, and the directrix has equation $y = \frac{95}{12}$.

An important elementary application of the quadratic function occurs in the study of moving bodies in physics. The motion of a body, if produced by the action of a constant force on the body, can be conveniently described in terms of quadratic relationships. The planar path of a projectile, such as a bullet, a baseball, or a missile, is approximately an arc of a parabola.

Searchlights and telescopes often have a reflecting surface with a parabolic cross section. In the telescope, the purpose is to concentrate the incoming light at the focus. In the searchlight with the light source located at the focus, the dispersion of the light is uniformly best. These applications are based on the principle that, at every point Q on a parabola, the line joining Q and the focus and the line that contains Q and is perpendicular to the directrix make congruent angles with the line that is tangent to the parabola at Q. In Fig. 21 the line joining A and B is tangent to the parabola at Q. The line joining Q and L is perpendicular to the directrix. Angle AQF and angle BQL are congruent to each other.

Another application of the quadratic function that is particularly useful in social sciences and economics is in approximating a more complicated function. Often in the study of natural phenomena, functional relationships may

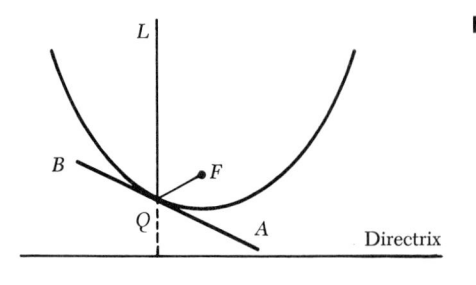

Figure 21

be either unwieldy or perhaps only partially known. It may be desirable to approximate such relationships by simple functions. The linear function is one of the best functions for this purpose. However, the linear function may not afford a good enough approximation, and the quadratic function, although less simple, may be needed for satisfactory results.

As a brief review of Sections 6 through 9, we state the following important conclusion.

Theorem 4 The graph of a function is a parabola with horizontal directrix if and only if the function is a quadratic function.

EXERCISES H

1. For each of the following functions, find three numbers a, h, k such that the given function is the same as the function given by the formula $a(x - h)^2 + k$.
   (a) $x^2 + 10x + 16$.
   (b) $x^2 - 3x + 4$.
   (c) $2x^2 + 20x + 57$.
   (d) $2x^2 - 4x - 3$.
   (e) $3x^2 + 2x + 4$.
   (f) $4x^2 - 20x + 34$.
   (g) $-x^2 + 4x$.
   (h) $-x^2 - x$.
   (i) $-3x^2 - 12x + 1$.
   (j) $-\frac{1}{4}x^2 - 2x + 1$.
   (k) $-\frac{1}{12}x^2 + \frac{1}{2}x - 1$.
   (l) $\frac{1}{3}x^2 + \frac{1}{6}x$.
   (m) $-6x^2 + 6x - 1$.
   (n) $1 - x^2$.

2. Use the information obtained in each part of Exercise 1 to find the vertex, the focus, and the directrix of the parabola that is the graph of the quadratic function.

3. Use the information obtained in Exercises 1 and 2 to sketch each of the parabolas.

4. For each of the functions in Exercise 1, decide which, if any, of the following properties the function has.
   (i) The function is increasing on the set of all positive numbers.
   (ii) The function is decreasing on the set $\{x \mid x > 3\}$.
   (iii) The function is increasing on the set $\{x \mid x < 1\}$.

5. Find the distance between the focus and the directrix of the parabola that is the graph of the quadratic function containing $(0, -\frac{1}{4})$, $(3, \frac{1}{8})$, and $(6, -\frac{7}{4})$.

6. Find an equation for the parabola whose vertex is $(-2, 3)$ and that contains the points $(0, 8)$ and $(-4, 8)$.

7. Let s be a positive number. A certain parabola has a horizontal directrix and contains the points $(-1, 0)$, $(0, 1)$, and $(s, 0)$.
   (a) Find an equation for the parabola.
   (b) Give the coordinates of the vertex of the parabola.
   (c) Give the coordinates of the focus of the parabola.
   (d) Give an equation for the directrix of the parabola.

8. Let h, k be nonzero real numbers. Find an equation for the parabola whose vertex is the point (h, k) and whose directrix is the horizontal coordinate axis.

9. Find an equation for a parabola whose focus lies on the vertical coordinate axis and whose directrix is the horizontal coordinate axis.

10. Find an equation for a parabola whose directrix is horizontal and whose focus is at the origin of the coordinate system.

11. Consider a quadratic function G given by the formula $ax^2 + bx + c$. Let the x-coordinate of the vertex of the parabola that is the graph of G be called v.
   (a) Express v in terms of a and/or b and/or c.
   (b) Show that, if $a > 0$, then G is an increasing function on the set $\{x \mid x > v\}$ and is a decreasing function on the set $\{x \mid x < v\}$.
   (c) Show that, if $a < 0$, then G is an increasing function on $\{x \mid x < v\}$ and is a decreasing function on $\{x \mid x > v\}$.

12. Let p be a positive number, and consider the parabola R with focus $(0, p)$ and vertex $(0, 0)$. Let S be the convex polyhedral set given by the inequality
$$\begin{bmatrix} -1 & 1 \\ 1 & 1 \end{bmatrix} \begin{bmatrix} x \\ y \end{bmatrix} \geq \begin{bmatrix} -p \\ -p \end{bmatrix}.$$
   (a) Show that the extreme point of S lies on the directrix of R.
   (b) Show that $R \subset S$.

10 Midpoint

Let n be a natural number. Let V and W be two (distinct) vectors in n-dimensional space. Each vector, being an ordered n-tuple of numbers, identifies a point in the n-dimensional space. If V' and W' are the points associated respectively with the vectors V and W, then the midpoint of the segment joining V' and W' is given by the average of the vectors V and W; that is, the midpoint is given by the vector $\frac{1}{2}(V + W)$.

Example 18 In one-dimensional space, the point midway between the points 2 and 9 has coordinate $\frac{1}{2}(2 + 9) = \frac{11}{2}$. (See Fig. 22.)

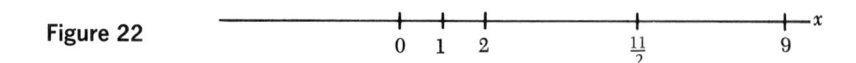

Figure 22

The midpoint of the segment joining -3 and 4 has coordinate $\frac{1}{2}(-3 + 4) = \frac{1}{2}$.

Example 19 In two-dimensional space, the point midway between the points $(3, -1)$ and $(-2, 5)$ is $\frac{1}{2}[(3, -1) + (-2, 5)] = \frac{1}{2}(1, 4) = (\frac{1}{2}, 2)$. (See Fig. 23.)

Example 20 In three-dimensional space, the two points $(2, 3, 5)$ and $(6, 10, -1)$ are the endpoints of a segment. The midpoint of the segment has three coordinates, each of which may be calculated by averaging the corresponding coordinates of the endpoints. The average of 2 and 6 is 4, the average of 3 and 10 is $\frac{13}{2}$, the average of 5 and -1 is 2. The midpoint is $(4, \frac{13}{2}, 2)$. (See Fig. 24.)

(−2, 5)

($\frac{1}{2}$, 2)

(3, −1)

Figure 23

(2, 3, 5)

(4, $\frac{13}{2}$, 2)

(6, 10, −1)

Figure 24

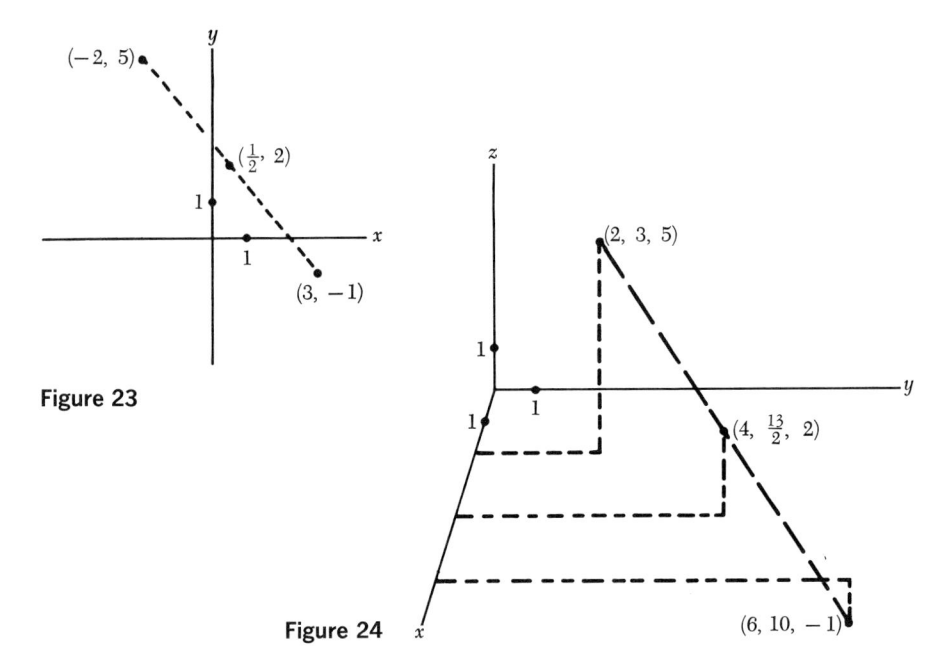

Having inspected several illustrations, we wish to justify the method for determining the midpoint of a segment whose endpoints V' and W' are given by the respective vectors V and W. Let the vector associated with the midpoint be M. The vector from V' to W' is $W - V$. The vector from V' to the midpoint is $M - V$. Since the desired point is halfway from V' to W', the vector $M - V$ is one-half of $W - V$. The equation $M - V = \frac{1}{2}(W - V)$ gives

$$\begin{aligned} M &= V + \tfrac{1}{2}(W - V) \\ &= \tfrac{1}{2}(2V + W - V) \\ &= \tfrac{1}{2}(V + W). \end{aligned}$$

Thus, M is the average of V and W, and we have established the following result.

Theorem 5 Each coordinate of the midpoint of a segment is the average of the corresponding coordinates of the endpoints of the segment.

11 Perpendicular-bisector

Example 21 On a two-dimensional coordinate system, let A and B be the respective points $(7, -1)$ and $(1, 5)$. Find the set of all points in the plane that are equally distant from A and from B.

Let P be a point in the plane, and let its coordinates be (x, y). The distance between A and P is $[(x - 7)^2 + (y + 1)^2]^{1/2}$. The distance between B and P is $[(x - 1)^2 + (y - 5)^2]^{1/2}$. (See Fig. 25.) The verbal

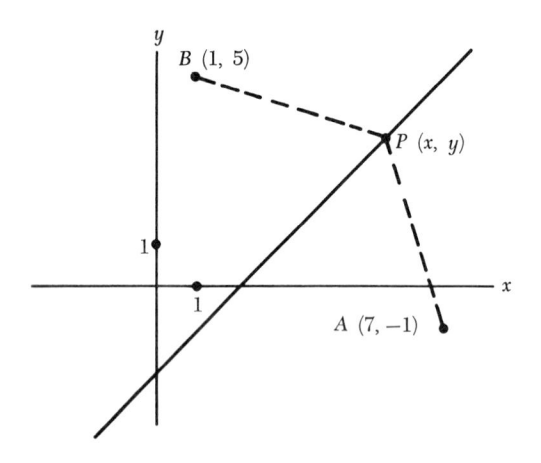

Figure 25

statement that P is equally distant from A and from B becomes the equation

$$[(x - 7)^2 + (y + 1)^2]^{1/2} = [(x - 1)^2 + (y - 5)^2]^{1/2}.$$

We obtain an equivalent but simpler equation by the following steps:

$$(x - 7)^2 + (y + 1)^2 = (x - 1)^2 + (y - 5)^2,$$
$$x^2 - 14x + 49 + y^2 + 2y + 1 = x^2 - 2x + 1 + y^2 - 10y + 25,$$
$$-12x + 12y + 24 = 0,$$
$$-x + y + 2 = 0.$$

The simplified equation $-x + y + 2 = 0$ we recognize as the equation of a line. Thus, the set l of all points in the plane equidistant from A and from B is a line. The line is shown in Fig. 25.

The midpoint of the segment joining A and B has coordinates $\frac{1}{2}[(7, -1) + (1, 5)] = (4, 2)$. These coordinates satisfy the equation $-x + y + 2 = 0$, as they should do since certainly the midpoint is equally distant from A and from B. The line joining A and B has equation $x + y = 6$ and slope -1. Since the line l has slope 1, it is perpendicular to the line joining A and B. Thus, l is the so-called *perpendicular-bisector* of the segment joining A and B.

The preceding example has illustrated two basic geometric results. Given two points in a plane, the set of all points that are equally distant from the given points is a line. Furthermore, it is the same as the perpendicular-bisector of the segment joining the given points.

The pattern applied to the particular numerical case in Example 21 may be used for any two points. Let V' be a point identified by the vector $V = (a, b)$, and let W' be another point identified by the vector $W = (d, e)$.

Figure 26

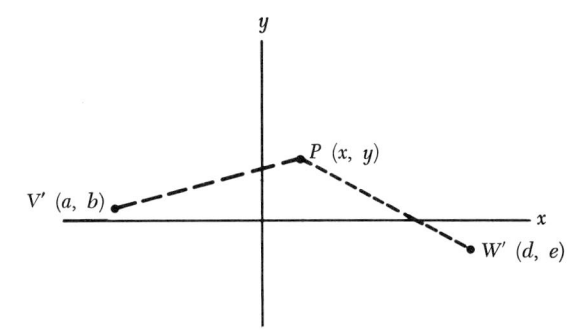

We wish to express algebraically the statement that the vector $S = (x, y)$ identifies a point P that is equally distant from V' and from W'. (See Fig. 26.)

The distance between V' and P is the length of the vector $S - V$; since $S - V = (x - a, y - b)$, its length is $[(x - a)^2 + (y - b)^2]^{1/2}$. The distance between W' and P is the length of the vector $S - W = (x, y) - (d, e) = (x - d, y - e)$, namely, $[(x - d)^2 + (y - e)^2]^{1/2}$. The condition that P is equally distant from V' and from W' is expressed by the equation

$$[(x - a)^2 + (y - b)^2]^{1/2} = [(x - d)^2 + (y - e)^2]^{1/2}.$$

We simplify:

$$(x - a)^2 + (y - b)^2 = (x - d)^2 + (y - e)^2,$$
$$x^2 - 2ax + a^2 + y^2 - 2by + b^2 = x^2 - 2dx + d^2 + y^2 - 2ey + e^2,$$
$$2(d - a)x + 2(e - b)y = d^2 - a^2 + e^2 - b^2.$$

Since V' and W' are distinct points, the vector $W - V = (d - a, e - b)$ is not the zero vector. Consequently, the graph of the equation $2(d - a)x + 2(e - b)y = d^2 - a^2 + e^2 - b^2$ is a line l.

The midpoint of the segment joining V' and W' has coordinates

$$\frac{1}{2}[(a, b) + (d, e)] = \left(\frac{d + a}{2}, \frac{e + b}{2}\right).$$

By straightforward substitution, we may verify that these coordinates satisfy the equation of the line l.

In case neither $d - a$ nor $e - b$ is zero, then the line l has slope $-(d - a)/(e - b)$ and the line l' joining V' and W' has slope $(e - b)/(d - a)$. Thus, the two lines are perpendicular to each other. In case one of the two numbers $d - a$ or $e - b$ is zero, then, of the two lines l and l', one is horizontal and the other vertical; consequently, they are perpendicular to each other.

In every case, the set of all points in the plane that are equally distant from the given points V' and W' is the line l which is the perpendicular-bisector of the segment joining V' and W'.

The preceding discussion may be readily adapted to the three-dimensional case. Suppose V' and W' are two given points identified respectively by the vectors $V = (a, b, c)$ and $W = (d, e, f)$. If $S = (x, y, z)$ is the vector associated with the point P in three-dimensional space, then the statement that P is equally distant from V' and from W' means that the vector $S - V$ and the vector $S - W$ have the same length. This equality, expressed in terms of the components of the vectors, is $[(x - a)^2 + (y - b)^2 + (z - c)^2]^{1/2} = [(x - d)^2 + (y - e)^2 + (z - f)^2]^{1/2}$. After simplification, we obtain

$$2(d - a)x + 2(e - b)y + 2(f - c)z = d^2 - a^2 + e^2 - b^2 + f^2 - c^2.$$

The set of all points in space that are equally distant from two given points is recognized as a plane. We may verify that it contains the midpoint of the segment joining the given points, for the equation is satisfied by the coordinates of the point

$$\frac{1}{2}(V + W) = \left(\frac{d + a}{2}, \frac{e + b}{2}, \frac{f + c}{2}\right).$$

Furthermore the plane is perpendicular to the line joining the given points. In brief, it is the perpendicular-bisecting plane of the segment.

EXERCISES I

1. Find the point midway between the two points in each of the following pairs.
   (a) 15.2, 8.6.
   (b) $a, -c$.
   (c) $(0, 2), (0, -\frac{5}{4})$.
   (d) $(3, 1), (-5, 5)$.
   (e) $(2a, 0), (0, 2b)$.
   (f) $(2, 1, -3), (0, 6, 1)$.
   (g) $(a, b, c), \left(5a, -b, \frac{1}{c}\right)$.
   (h) $(0, 0, 0), (p, q, r)$.

2. Let P be the point $(2, 5)$ and Q the point $(8, 1)$.
   (a) Show that the line joining P and Q is the graph of the set $\{(x, y) \mid t$ is a real number and $(x, y) = (2, 5) + t(6, -4)\}$.
   (b) What value of t in part (a) gives the point P?
   (c) What value of t gives Q?
   (d) What value of t gives the midpoint of the segment with endpoints P, Q?
   (e) What is the midpoint?
   (f) Describe geometrically the location of the point given by $t = \frac{1}{3}$.
   (g) Describe geometrically the location of the point identified by $t = \frac{2}{3}$.

3. Let P and Q be the points $(-1, 5, 2)$ and $(3, 7, -6)$, respectively.
   (a) Show that the graph of
   $$\{(x, y, z) \mid t \text{ is a real number and } (x, y, z) = (-1, 5, 2) + t(4, 2, -8)\}$$
   is the line joining P and Q.
   (b) What value of t corresponds to the point Q?
   (c) What geometrical set corresponds to the set of numbers $\{t \mid 0 \leq t \leq 1\}$?
   (d) What value of t identifies the midpoint of the segment joining P and Q?
   (e) Describe geometrically the three points whose respective t-coordinates are $\frac{1}{4}, \frac{1}{2}, \frac{3}{4}$.

4. Let A and B be the respective points $(3, 4)$ and $(-1, -6)$.
   (a) Find the slope of the line r joining A and B.
   (b) Find an equation for r.
   (c) Find the slope of a line perpendicular to r.
   (d) Find the midpoint of the segment with endpoints A and B.
   (e) Find an equation for the line which is the perpendicular-bisector of the segment joining A and B.

5. Let A and B be the same points as in Exercise 4; let P be a point with coordinates (x, y).
   (a) What is the distance between P and A?
   (b) What is the distance between P and B?
   (c) Write an equation stating that P is equally distant from A and from B.
   (d) Simplify the equation in part (c).
   (e) Show that the equations in Exercises 4e and 5d are equations for the same line.

6. Let C and D be the respective points $(2, -1, 1)$ and $(-4, 0, 3)$; let Q be the point (x, y, z).
   (a) Find the distance between Q and C.
   (b) Find the distance between Q and D.
   (c) Write, and simplify, an equation expressing the condition that Q is equally distant from C and from D.
   (d) Show that the graph of the equation in part (c) is a plane.
   (e) Find the midpoint M of the segment with endpoints C, D.
   (f) Show that the plane in part (d) contains M.
   (g) Fill the blank: The plane in part (d) is the _____-bisector of the segment joining C and D.

7. The union of the graphs of the four functions $x, -2x + 3, x + 3, -2x - 9$ contains a parallelogram. By finding the coordinates of the midpoint of each diagonal, show that the diagonals of this parallelogram bisect each other.

12 Circle

The notion of distance is fundamental to the discussion of circles and spheres.

Example 22 Find an equation for the circle with center at the point $(2, -5)$ and radius 3. (See Fig. 27.)

Figure 27

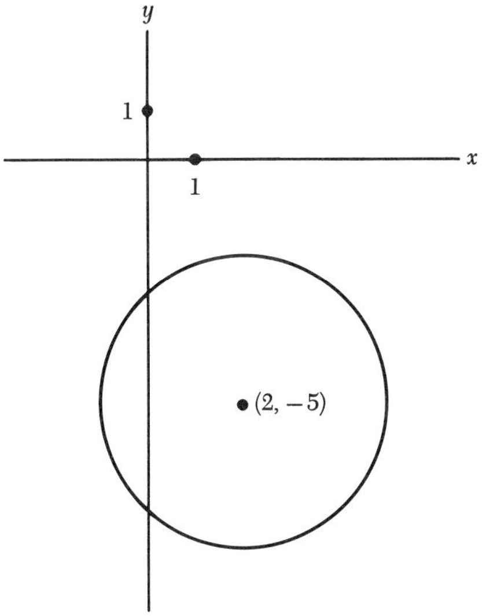

12 / Circle 387

Let P be a point in the plane of the circle, and let its coordinates be (x, y). The criterion for P to belong to the circle is the condition that the distance between P and the center should be 3. The distance between (x, y) and $(2, -5)$ is $[(x - 2)^2 + (y + 5)^2]^{1/2}$. Thus, an equation that expresses the statement that P belongs to the circle is

$$[(x - 2)^2 + (y + 5)^2]^{1/2} = 3.$$

A simpler, but equivalent, equation is

$$(x - 2)^2 + (y + 5)^2 = 9.$$

Definition Let a point and a positive number be given. In a plane containing the point, the set of all points such that the distance between each of them and the given point is the given number is a curve called a *circle*.

The given point is the *center* of the circle, and the given positive number is called the *radius* of the circle.

Suppose that a circle on a two-dimensional coordinate system has radius r and center (a, b). If the point P in the plane has coordinates (x, y), then the distance between the center and the point P is $[(x - a)^2 + (y - b)^2]^{1/2}$. The condition that P belongs to the circle is expressed algebraically by the equation $[(x - a)^2 + (y - b)^2]^{1/2} = r$. An equivalent and simpler equation is

$$(x - a)^2 + (y - b)^2 = r^2.$$

Example 23 By comparison with the preceding equation, we recognize that the graph of the set $\{(x, y) \mid (x - 8)^2 + (y + 2)^2 = 50\}$ is a circle whose center is $(8, -2)$ and whose radius is $\sqrt{50} = 5\sqrt{2}$.

Example 24 The two points $(2, 1)$ and $(-4, 3)$ are the endpoints of a diameter of a circle. Find an equation for the circle.

The center of the circle is the midpoint of every diameter of the circle; thus, the center is given by the vector $\frac{1}{2}[(2, 1) + (-4, 3)]$ $= \frac{1}{2}(-2, 4) = (-1, 2)$. The distance between the center and any point on the circle is the radius r; hence, $r^2 = ([2] - [-1])^2 + ([1] - [2])^2 = 3^2 + 1^2 = 10$. Since the center and the second power of the radius are now known, we may find an equation for the circle:

$$(x + 1)^2 + (y - 2)^2 = 10.$$

An alternative form for the equation is obtainable in two steps with the aid of the binomial theorem:

$$x^2 + 2x + 1 + y^2 - 4y + 4 = 10,$$
$$x^2 + y^2 + 2x - 4y - 5 = 0.$$

As a partial check on the correctness of the latter equation, we verify that the coordinates $(-4, 3)$ satisfy the equation:

$$(-4)^2 + (3)^2 + 2(-4) - 4(3) - 5 = 16 + 9 - 8 - 12 - 5 = 0.$$

EXERCISES J

1. Let C be the point $(1, -2)$, and let P be the point with coordinates (x, y). Write an equation stating that the distance between P and C is 4. Simplify the equation.

2. Write an equation for the circle whose center is $(-5, 7)$ and whose radius is 3.

3. Write an equation for the circle whose center is $(0, 3)$ and whose diameter is 6.

4. Identify the graph of each of the following sets.
   (a) $\{(x, y) \mid x^2 + (y - 3)^2 = 16\}$.
   (b) $\{(x, y) \mid (x - 5)^2 + (y + 3)^2 = 1\}$.
   (c) $\{(x, y) \mid (x + 6)^2 + y = 25\}$.
   (d) $\{(x, y) \mid x^2 + y^2 = -1\}$.
   (e) $\{(x, y) \mid 4x^2 + 4y^2 = 9\}$.

5. Write an equation for the circle described by each of the following.
   (a) Point $(6, 0)$ is the center; point $(1, 4)$ is on the circle.
   (b) Center is $(-1, -3)$; the diameter is 7.
   (c) Points $(5, -3)$ and $(3, -5)$ are endpoints of a diameter.
   (d) Center is $(2, -4)$; point $(3, 2)$ belongs to the circle.
   (e) Points $(2, -9)$ and $(-2, 9)$ are endpoints of a diameter.
   (f) Circle contains the points $(2, 8)$ and $(6, 0)$; center lies on the line whose equation is $x + 2 = 0$.

6. Sketch the graph of each of the following sets.
   (a) $\{(x, y) \mid x^2 + y^2 = 1\}$.
   (b) $\{(x, y) \mid (x - 2)^2 + (y + 1)^2 = \frac{16}{9}\}$.
   (c) $\{(x, y) \mid (y \geq 0) \wedge (x^2 + y^2 = 9)\}$.
   (d) $\{(x, y) \mid y \leq 1\} \cap \{(x, y) \mid (x + 2)^2 + (y - 1)^2 = \frac{25}{4}\}$.

7. Let C be the point $(1, -3, 4)$, and let P be the point with coordinates (x, y, z). Write, and simplify, an equation stating that P is six units away from C.

8. Sketch the graph of each of the following functions.
   (a) $\{(x, y) \mid [y > 0] \wedge [x^2 + y^2 = 1]\}$.
   (b) $\{(x, y) \mid [y > 0] \wedge [y^2 = 2 - x^2]\}$.
   (c) $\{(x, y) \mid y = (8 - x^2)^{1/2}\}$.
   (d) $\{(x, y) \mid y = \sqrt{18 - x^2}\}$.
   (e) $\{(x, y) \mid [(x, y) \geq (0, 0)] \wedge [x^2 + y^2 = 25]\}$.
   (f) $\{(x, y) \mid [(-x, y) > (0, 0)] \wedge [x^2 + y^2 = 16]\}$.

13 Equation for a circle

If a circle has center (a, b) and radius r, then the circle is the graph of the equation

$$(x - a)^2 + (y - b)^2 = r^2.$$

This equation is useful for displaying the center and the radius. With the aid of the binomial theorem, we may obtain a different equation for the circle, namely, $x^2 + y^2 - 2ax - 2by + a^2 + b^2 - r^2 = 0$. Since $-2a$ and $-2b$ and $a^2 + b^2 - r^2$ are constants, this equation has the form

$$x^2 + y^2 + dx + ey + g = 0,$$

where d, e, g are numbers. Thus, every circle in the coordinate plane is the graph of a set $\{(x, y) \mid x^2 + y^2 + dx + ey + g = 0\}$ for suitable numbers d, e, g.

Conversely, every set of the type $\{(x, y) \mid x^2 + y^2 + dx + ey + g = 0\}$ that contains more than one ordered pair has a graph that is a circle. We illustrate this assertion by the following two examples.

Example 25 Identify the graph of the equation $x^2 + y^2 - 3x + 5y + 7 = 0$.

We "complete the square," beginning with $x^2 - 3x$ and also beginning with $y^2 + 5y$. The successive transformations of the given equation are:

$$(x^2 - 3x \quad) + (y^2 + 5y \quad) = -7,$$
$$(x^2 - 3x + \tfrac{9}{4}) + (y^2 + 5y + \tfrac{25}{4}) = -7 + \tfrac{9}{4} + \tfrac{25}{4},$$
$$(x - \tfrac{3}{2})^2 + (y + \tfrac{5}{2})^2 = \tfrac{3}{2}.$$

The graph is a circle; its center is $(\tfrac{3}{2}, -\tfrac{5}{2})$ and its radius is $(\tfrac{3}{2})^{1/2}$.

Example 26 Identify the set $\{(x, y) \mid x^2 + y^2 - 2x + 4y + 7 = 0\}$.

We use the same method as in Example 25, and transform the equation by successive steps as follows.

$$(x^2 - 2x \quad) + (y^2 + 4y \quad) = -7,$$
$$(x^2 - 2x + 1) + (y^2 + 4y + 4) = -7 + 1 + 4,$$
$$(x - 1)^2 + (y + 2)^2 = -2.$$

However, $(x - 1)^2 \geq 0$ for all x, and $(y + 2)^2 \geq 0$ for all y; since the sum of two nonnegative numbers cannot be -2, there is no ordered pair (x, y) which satisfies the equation. In other words, the given set is the empty set.

Suppose that three points are given that are not collinear. These points determine a circle, that is, there is one and only one circle which contains all of them.

One method for finding the circle determined by three noncollinear points is the following. Select any two of the points, and find the perpendicular-bisector of the segment joining them. Select another pair of points, and find the perpendicular-bisector of the segment joining them. Find the point of intersection of these two lines; this point is the center of the circle and its distance from one of the given points is the radius. This method will be illustrated in Example 28.

Another method for finding the circle that contains three given points exploits the fact that the circle has an equation of the type $x^2 + y^2 + dx + ey + g = 0$.

Example 27 A circle contains the three points $(9, 0)$, $(-3, -6)$, $(1, 6)$. Find an equation for the circle, find its center, and find its radius. (See Fig. 28.)

Figure 28

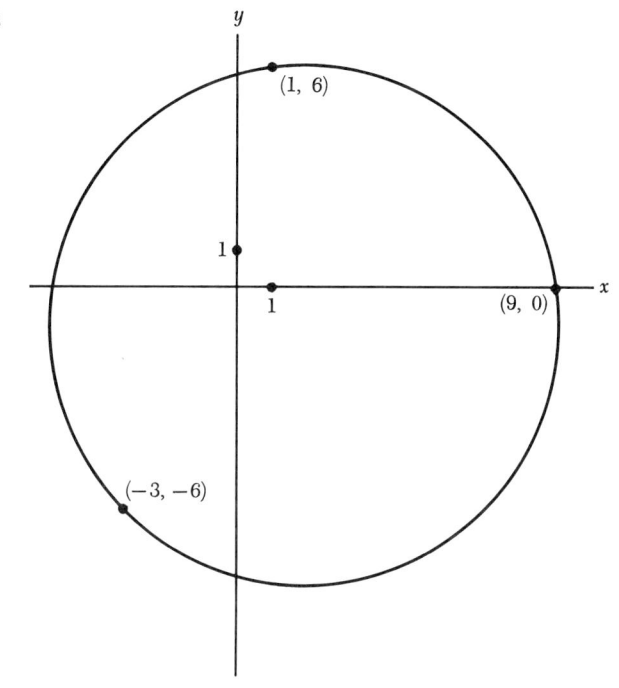

There are three numbers d, e, g such that the circle is the graph of the equation

$$x^2 + y^2 + dx + ey + g = 0.$$

Since the point $(9, 0)$ belongs to the circle, its coordinates must satisfy the equation; hence, $9^2 + 0^2 + d \cdot 9 + e \cdot 0 + g = 0$. Thus,

$$9d + g = -81.$$

Likewise, the coordinates of the point $(-3, -6)$ satisfy the equation, that is, $(-3)^2 + (-6)^2 + d(-3) + e(-6) + g = 0$. Thus,

$$-3d - 6e + g = -45.$$

13 / Equation for a circle   391

From the requirement that the circle contain the point (1, 6), we obtain similarly the equation

$$d + 6e + g = -37.$$

We now have three simultaneous conditions on the unknown numbers d, e, g. Since each condition is a linear equation, we may apply our technique for solving a system of linear equations. We note that the coefficient of g is 1 in each equation, and we take advantage of this by using the Gauss elimination method to solve for the column vector $\begin{bmatrix} g \\ e \\ d \end{bmatrix}$. That is, we express the system in the matrix form

$$\begin{bmatrix} 1 & 0 & 9 \\ 1 & -6 & -3 \\ 1 & 6 & 1 \end{bmatrix} \begin{bmatrix} g \\ e \\ d \end{bmatrix} = \begin{bmatrix} -81 \\ -45 \\ -37 \end{bmatrix}.$$

Subsequent tableaus in the condensed notation are as follows.

$$\left[\begin{array}{ccc|c} 1 & 0 & 9 & -81 \\ 0 & -6 & -12 & 36 \\ 0 & 6 & -8 & 44 \end{array}\right],$$

$$\left[\begin{array}{ccc|c} 1 & 0 & 9 & -81 \\ 0 & 1 & 2 & -6 \\ 0 & 0 & -20 & 80 \end{array}\right],$$

$$\left[\begin{array}{ccc|c} 1 & 0 & 0 & -45 \\ 0 & 1 & 0 & 2 \\ 0 & 0 & 1 & -4 \end{array}\right].$$

Since $\begin{bmatrix} g \\ e \\ d \end{bmatrix} = \begin{bmatrix} -45 \\ 2 \\ -4 \end{bmatrix}$, an equation for the circle is

$$x^2 + y^2 - 4x + 2y - 45 = 0.$$

In order to find the center and the radius, we proceed, as in Example 25, to complete the squares:

$$(x^2 - 4x \quad) + (y^2 + 2y \quad) = 45,$$
$$(x^2 - 4x + 4) + (y^2 + 2y + 1) = 50.$$

Since another equation for the circle is
$$(x - 2)^2 + (y + 1)^2 = 50,$$
the center is the point $(2, -1)$ and the radius is the number
$$\sqrt{50} = 5\sqrt{2}.$$

Example 28 Solve the same problem as in Example 27 by another method.

Let r be the line consisting of all points in the plane that are equally distant from $(9, 0)$ and $(1, 6)$. Let s be the perpendicular-

Figure 29

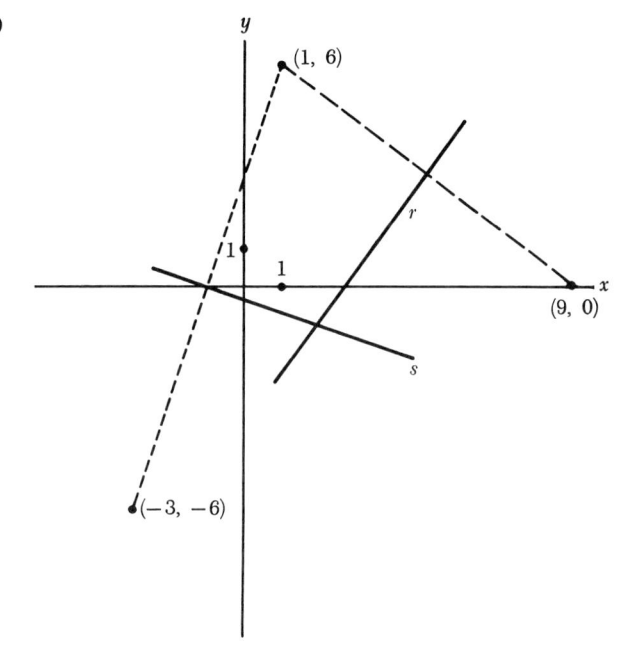

bisector of the segment joining $(1, 6)$ and $(-3, -6)$. (See Fig. 29.) An equation for the line r is
$$(x - 9)^2 + (y - 0)^2 = (x - 1)^2 + (y - 6)^2,$$
which may be simplified to
$$4x - 3y = 11.$$
Similarly, an equation for the line s is $(x - 1)^2 + (y - 6)^2 = (x + 3)^2 + (y + 6)^2$; a simpler form is
$$x + 3y = -1.$$

13 / Equation for a circle 393

To find $r \cap s$, apply the method of Chapter 2.

$$\begin{bmatrix} 1 & 3 & \vdots & -1 \\ 4 & -3 & \vdots & 11 \end{bmatrix},$$

$$\begin{bmatrix} 1 & 3 & \vdots & -1 \\ 0 & -15 & \vdots & 15 \end{bmatrix},$$

$$\begin{bmatrix} 1 & 0 & \vdots & 2 \\ 0 & 1 & \vdots & -1 \end{bmatrix}.$$

The point $(2, -1)$ is equally distant from all three of the given points. Thus, the center of the circle is $(2, -1)$. The radius of the circle is $[(9-2)^2 + (0-[-1])^2]^{1/2} = (7^2 + 1^2)^{1/2} = 50^{1/2}$. An equation for the circle is $(x-2)^2 + (y+1)^2 = 50$.

Notice that the method in Example 27 focuses attention on the equation. After obtaining an equation, we use it to find the center and the radius. By contrast, in Example 28, the method is to locate the center of the circle first. With this information, we may then easily determine the radius and an equation for the circle.

EXERCISES K

1. Each of the following is the equation of a circle. For each circle, identify the center, find the radius, and give the coordinates of one point on the circle.
   (a) $x^2 + y^2 - 100 = 0$.
   (b) $9(x+3)^2 + 9y^2 = 4$.
   (c) $x^2 - 2x + 1 + y^2 = 2$.
   (d) $(x - \frac{5}{2})^2 + (y + \frac{7}{2})^2 = \frac{1}{4}$.
   (e) $(2x-5)^2 + (2y+7)^2 = 1$.
   (f) $4x^2 + 20x + 4y^2 - 28y + 73 = 0$.
   (g) $x^2 + y^2 + 10x - 4y = 9$.
   (h) $9x^2 + 9y^2 - 6x + 24y = 10$.
   (i) $16(x^2 + y^2) + 16x - 24y + 9 = 0$.
   (j) $x(x+60) + y(y-80) = 0$.

2. Let Q, R, S be the respective points $(2, 9)$, $(7, -6)$, $(-4, 5)$. Use the method of Example 27 to find an equation for the circle containing the three given points.

3. Let Q, R, S be the points $(2, 9)$, $(7, -6)$, $(-4, 5)$, respectively.
   (a) Find an equation for the line s that is the perpendicular-bisector of the segment with endpoints Q and R.
   (b) Find an equation for the perpendicular-bisector r of the segment joining Q and S.
   (c) Find an equation for the perpendicular-bisector q of the segment with R and S as endpoints.
   (d) Find the point C belonging to the intersection $r \cap s$.
   (e) Show that $C \in q$.
   (f) Explain geometrically why the three lines q, r, s are concurrent, that is, why their intersection is a point.
   (g) Explain why C is the center of the circle containing the three given points Q, R, S.

(h) Finish applying the method of Example 28 to find an equation for the circle.
(i) Verify that the equations obtained in Exercises 2 and 3h are equivalent.

4. Use the method of Example 27 to find an equation for the circle containing the three points $(2, 1)$, $(-4, 2)$, $(-1, -5)$.

5. Use the method of Example 28 to find an equation for the circle to which $(-2, -1)$, $(4, 3)$, and $(2, -5)$ belong.

6. The three points $(0, 4)$, $(2, 3)$, $(8, 0)$ are collinear.
    (a) Use the method of Example 27 in an attempt to find a circle containing the given three points. Explain how the method breaks down.
    (b) Use the method of Example 28 in an attempt to find a circle containing the given points. Explain how this method fails.

14 Sphere

In a two-dimensional space, the subset consisting of all points at a prescribed distance from a fixed point is a circle. In a three-dimensional space, the analogous subset is a sphere.

> **Definition** Let a point and a positive number be given. In space the set of all points such that the distance between each of them and the given point is the given number is a surface called a *sphere*.
>
> **Example 29** Consider the sphere whose center is $(5, 3, -1)$ and whose radius is 4. (See Fig. 30.)
> Let P be a point in space, and let the coordinates of P be (x, y, z).

Figure 30

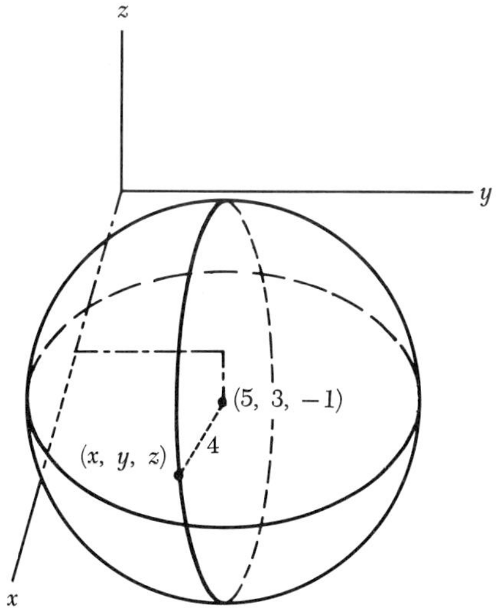

The distance between P and the center is $[(x - 5)^2 + (y - 3)^2 + (z + 1)^2]^{1/2}$. The criterion for P to belong to the sphere is that this distance be 4. Thus, an equation for the sphere is

$$[(x - 5)^2 + (y - 3)^2 + (z + 1)^2]^{1/2} = 4.$$

A simpler equation is

$$(x - 5)^2 + (y - 3)^2 + (z + 1)^2 = 16.$$

Suppose that a sphere on a three-dimensional coordinate system has center (a, b, c) and radius r. The distance between the center and a point P with coordinates (x, y, z) is $[(x - a)^2 + (y - b)^2 + (z - c)^2]^{1/2}$. The condition that P belongs to the sphere is expressed algebraically by the equation

$$[(x - a)^2 + (y - b)^2 + (z - c)^2]^{1/2} = r.$$

Thus, the sphere is the graph of the equation

$$(x - a)^2 + (y - b)^2 + (z - c)^2 = r^2.$$

With three applications of the binomial theorem, the preceding equation for a sphere can be converted into a different form. We obtain

$$x^2 - 2ax + a^2 + y^2 - 2by + b^2 + z^2 - 2cz + c^2 = r^2.$$

If we let the constants $-2a$, $-2b$, $-2c$, $a^2 + b^2 + c^2 - r^2$ be denoted respectively by d, e, f, g, then we observe that the sphere is the graph of the set

$$\{(x, y, z) \mid x^2 + y^2 + z^2 + dx + ey + fz + g = 0\}.$$

Conversely, every set of the type $\{(x, y, z) \mid x^2 + y^2 + z^2 + dx + ey + fz + g = 0\}$ has a graph that either is a sphere or else has no more than one point.

Example 30 Identify the graph of the equation

$$4x^2 + 4y^2 + 4z^2 + 12x - 2y + 8z + 7 = 0.$$

After dividing each member of the equation by 4, we complete the square three times.

$(x^2 + 3x\quad) + (y^2 - \tfrac{1}{2}y\quad) + (z^2 + 2z\quad) = -\tfrac{7}{4},$
$(x^2 + 3x + \tfrac{9}{4}) + (y^2 - \tfrac{1}{2}y + \tfrac{1}{16}) + (z^2 + 2z + 1)$
$\qquad\qquad = -\tfrac{7}{4} + \tfrac{9}{4} + \tfrac{1}{16} + 1,$
$(x + \tfrac{3}{2})^2 + (y - \tfrac{1}{4})^2 + (z + 1)^2 = \tfrac{25}{16}.$

The graph is a sphere; its center is $(-\tfrac{3}{2}, \tfrac{1}{4}, -1)$, and its radius is $\sqrt{\tfrac{25}{16}} = \tfrac{5}{4}$.

Example 31 Identify, on a two-dimensional coordinate system, the graph of the inequality $(x - 3)^2 + (y - 1)^2 < 16$.

The given inequality is equivalent to the condition

$$[(x-3)^2 + (y-1)^2]^{1/2} < 4.$$

Now $[(x-3)^2 + (y-1)^2]^{1/2}$ is the distance between the point (x, y) and the point $(3, 1)$. The inequality therefore is satisfied by each point whose distance from $(3, 1)$ is less than four units. The points that are less than four units away from a fixed point are the points inside a circle, namely, the circle with the given point as center and with radius 4. Thus, as shown in Fig. 31, the graph is the interior of a certain circle.

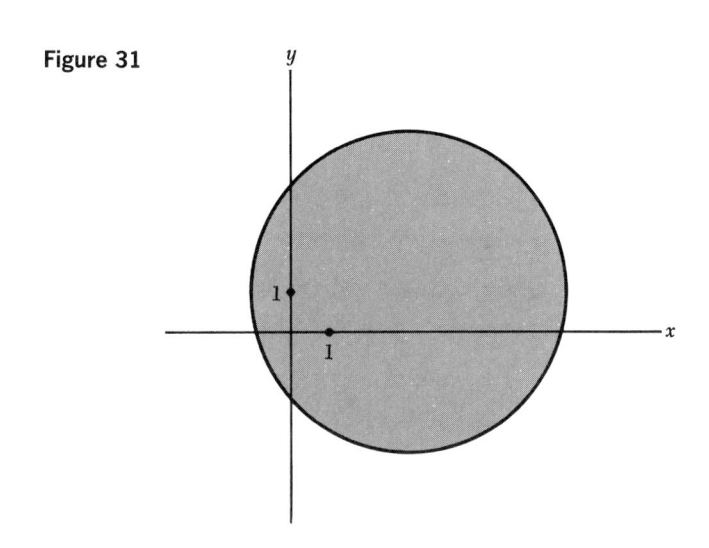

Figure 31

Example 32 Identify the graph of the set

$$\{(x, y, z) \mid x^2 + y^2 + z^2 + 2x - 6z \leqq 26\}.$$

We use the technique of completing the square in order to express the given set in a manner that will enable us to recognize its graph more readily.

$$\begin{aligned}
\{(x, y, z) \mid x^2 &+ y^2 + z^2 + 2x - 6z \leqq 26\} \\
&= \{(x, y, z) \mid (x^2 + 2x + 1) + y^2 + (z^2 - 6z + 9) \leqq 36\} \\
&= \{(x, y, z) \mid (x + 1)^2 + y^2 + (z - 3)^2 \leqq 36\} \\
&= \{(x, y, z) \mid [(x + 1)^2 + y^2 + (z - 3)^2]^{1/2} \leqq 6\}.
\end{aligned}$$

Since $[(x+1)^2 + y^2 + (z-3)^2]^{1/2}$ is the distance between (x, y, z) and $(-1, 0, 3)$, the inequality is the condition that a point be no more than six units from $(-1, 0, 3)$. Consequently, the graph is the union of a certain sphere and its interior, namely, the sphere with center $(-1, 0, 3)$ and radius 6.

Table 1

	Vector
Point	V
Linear equation	$\{X \mid V \neq 0 \wedge V \cdot X = d\}$
Line	$\{X \mid p \text{ is real and } W \neq 0 \text{ and } X = V + pW\}$
Distance	Length of $V - W$
Midpoint of segment	$\frac{1}{2}(V + W)$
Sphere	
Linear inequality	$\{X \mid V \neq 0 \wedge V \cdot X \leq d\}$
Interior of sphere	

15 Summary

We have often relied heavily on the analogy between the two-dimensional and the three-dimensional cases whenever we have been studying the algebraic description of geometric ideas. This strong resemblance leads to the unified approach by the use of vector notions and also suggests that we may extend the principles to geometry in higher dimensions. The brief summary of some of our work in Table 1 may be helpful in appreciating the marked similarities.

EXERCISES L

1. Write an equation for the sphere described by each of the following.
   (a) Point $(3, -4, 1)$ is the center; 6 is the radius.
   (b) Point $(-2, 0, 1)$ is the center; 8 is the diameter.
   (c) Point $(0, -5, 1)$ is the center; point $(2, -4, 3)$ is on the sphere.
   (d) Points $(6, -1, 3)$ and $(-2, 5, 7)$ are endpoints of a diameter.

2. The graph of each of the following equations is a sphere. In each case, find the center, find the radius, and give one point that belongs to the sphere.
   (a) $(x - 4)^2 + (y + 2)^2 + z^2 = 9$.
   (b) $(x + \frac{1}{2})^2 + (y - \frac{3}{2})^2 + (z + 2)^2 = \frac{25}{4}$.
   (c) $4x^2 + 4y^2 + 4z^2 + 12y + 20z = 15$.
   (d) $x^2 + y^2 + z^2 = 625$.

3. The four points $(-1, 1, 5)$, $(0, 5, 4)$, $(6, -4, 1)$, $(7, -3, -3)$ lie on a sphere. Find an equation for the sphere by using the method that is the analog of the method in Example 27; that is, consider an equation of the type
$$x^2 + y^2 + z^2 + dx + ey + fz + g = 0,$$
where the numbers d, e, f, g are to be found.

	Two-dimensional case	Three-dimensional case
	(a, b)	(a, b, c)
	$\{(x, y) \mid (a, b) \neq (0, 0) \wedge ax + by = d\}$ LINE	$\{(x, y, z) \mid (a, b, c) \neq (0, 0, 0) \wedge ax + by + cz = d\}$ PLANE
	$\{(x, y) \mid p$ is real and $(d, e) \neq (0, 0)$ and $(x, y) = (a, b) + p(d, e)\}$	$\{(x, y, z) \mid p$ is real and $(d, e, f) \neq (0, 0, 0)$ and $(x, y, z) = (a, b, c) + p(d, e, f)\}$
	$[(a - d)^2 + (b - e)^2]^{1/2}$	$[(a - d)^2 + (b - e)^2 + (c - f)^2]^{1/2}$
	$\left(\dfrac{a + d}{2}, \dfrac{b + e}{2}\right)$	$\left(\dfrac{a + d}{2}, \dfrac{b + e}{2}, \dfrac{c + f}{2}\right)$
	$(x - a)^2 + (y - b)^2 = r^2$ CIRCLE	$(x - a)^2 + (y - b)^2 + (z - c)^2 = r^2$ SPHERE
	$\{(x, y) \mid (a, b) \neq (0, 0) \wedge ax + by \leq d\}$ CLOSED HALFPLANE	$\{(x, y, z) \mid (a, b, c) \neq (0, 0, 0) \wedge ax + by + cz \leq d\}$ CLOSED HALFSPACE
	$(x - a)^2 + (y - b)^2 < r^2$ INTERIOR OF CIRCLE	$(x - a)^2 + (y - b)^2 + (z - c)^2 < r^2$ INTERIOR OF SPHERE

4. The points $(5, 1, 8)$, $(7, -3, -6)$, $(-2, 6, 6)$, $(9, -4, 1)$ lie on a sphere. Find an equation for the sphere by using the method that is an adaptation of the method in Example 28. Recall that in three-dimensional space the perpendicular-bisector of a segment is a plane.

5. Repeat Exercise 3 (using the method prescribed) for the sphere containing $(3, -6, 1)$, $(0, -2, 8)$, $(5, 5, -6)$, $(-7, -4, 3)$.

6. Repeat Exercise 4 (using the method mentioned) for the points $(-1, -2, 3)$, $(6, 1, -5)$, $(0, 0, 2)$, $(-2, 2, -6)$.

7. Let P and Q be the points $(1, -1, 3)$ and $(5, 2, -2)$, respectively. Consider the set of all points of the form $(5, 2, -2) + t(-4, -3, 5)$, where t is a number.
   (a) What value of t corresponds to Q?
   (b) What value of t gives the point P?
   (c) What value of t gives the point $(-7, -7, 13)$?
   (d) Fill the blanks: The set of all points under consideration is the _____ joining _____ and _____.
   (e) What value of t gives the point N such that P is midway between N and Q?
   (f) What value of t corresponds to the midpoint M of the segment with endpoints P and Q?
   (g) What are the coordinates of M?
   (h) If a segment has one endpoint P and midpoint Q, what is the other endpoint?
   (i) Decide whether $(\frac{11}{3}, 1, -\frac{4}{3})$ belongs to the line joining P and Q. Explain your decision.
   (j) Repeat part (i) for the point $(21, 14, -21)$.

8. Write an inequality that states that the point (x, y) is less than eight units away from $(2, -5)$.

9. Describe in geometric language the graph of each of the following sets.
   (a) $\{(x, y) \mid x^2 + y^2 < 10\}$.
   (b) $\{(x, y) \mid x^2 + y^2 \leq 10\}$.
   (c) $\{(x, y) \mid (x + \frac{3}{2})^2 + (y - \frac{4}{5})^2 \leq 6\}$.
   (d) $\{(x, y, z) \mid x^2 + (y + 1)^2 + (z - 3)^2 < 4\}$.

(e) $\{(x, y, z) | 4x^2 + 4(y - \frac{1}{2})^2 + 4(z + 5)^2 \leq 3\}$.
(f) $\{(x, y, z) | 16x^2 + (4y + 7)^2 + 4(2z - 9)^2 < 25\}$.
(g) $\{(x, y) | x^2 + y^2 + 2x > 5\}$.
(h) $\{(x, y) | x^2 - y + 2x < 5\}$.

10. The equation of a certain sphere is $x^2 + y^2 + z^2 - 6x + 4y - 2z = 10$.
   (a) Find the center of the sphere.
   (b) Give the radius of the sphere.
   (c) Name one point on the sphere.
   (d) Find a point that is inside the sphere but that is not the center.
   (e) Give the coordinates of a point that is outside the sphere.
   (f) Write an inequality such that the graph of the set of roots of the inequality is the interior of the sphere.
   (g) Write an inequality whose graph is the union of the sphere and the set of all points outside the sphere.

11. Sketch the graph of each of the following.
   (a) $\{(x, y) | x^2 + y^2 < 1\}$.
   (b) $\{(x, y) | (x - 2)^2 + y^2 \leq 16\}$.
   (c) $\{(x, y) | 4x^2 + (2y + 1)^2 \leq 9\}$.
   (d) $\{(x, y) | x^2 < 1\}$.
   (e) $\{(x, y) | x^2 + (y - 3)^2 > 25\}$.
   (f) $\{(x, y) | y < x^2 + 2\}$.
   (g) $\{x | x^2 < 1\}$.

12. Sketch the graph of each of the following.
   (a) $\{(x, y) | y^2 = x^2\}$.
   (b) $\{(x, y) | y^2 < x^2\}$.
   (c) $\{(x, y) | y < |x|\}$.
   (d) $\{(x, y) | |y| < |x|\}$.

Inverse and Rational Functions

1 Introduction

A situation of extreme importance in mathematics is the equation with exactly one root. The process of solving such an equation often illustrates or introduces a new operation or a new function, a sort of reversal procedure.

In arithmetic, the question concerning addition that is posed by the illustrative sentence $3 + x = 8$ leads to the notion of subtraction. The root of the equation $a + x = b$ is obtained by subtracting a from b. A new operation, namely, subtraction, is introduced to solve addition problems in which the sum and one summand are known. Subtraction is a sort of reversal of addition.

In a similar fashion, division arises when we wish to answer inquiries about multiplication. The equation $3x = 8$ states that 8 is a product of two factors, one of which is 3; we identify the other factor by dividing 8 by 3. Division is a sort of reversal of multiplication.

In Chapter 3, we studied linear transformations on a vector space. Suppose that, associated with a certain linear transformation $\mathfrak{T}$ on a two-dimensional vector space is the matrix $\begin{bmatrix} 1 & 2 \\ 2 & 5 \end{bmatrix}$. Suppose that $\mathfrak{T}$ assigns to a vector U the image W, that is,

$$W = \begin{bmatrix} 1 & 2 \\ 2 & 5 \end{bmatrix} U;$$

this equation may be solved for U as follows:

$$U = \begin{bmatrix} 5 & -2 \\ -2 & 1 \end{bmatrix} W,$$

for $\begin{bmatrix} 5 & -2 \\ -2 & 1 \end{bmatrix}$ is the matrix associated with the inverse of $\mathfrak{T}$.

More generally, consider any linear transformation $\mathfrak{T}$ that has an inverse, say $\mathfrak{R}$. If $\mathfrak{T}$ assigns to a vector U the image W, that is, if

$$\mathfrak{T}(U) = W,$$

401

then we may express U by the relationship
$$U = \mathcal{R}(W).$$
To find the image of a known vector U, we apply the linear transformation $\mathcal{T}$ to it; to find a vector whose image W is known, we apply the inverse $\mathcal{R}$. The linear transformations $\mathcal{T}$ and $\mathcal{R}$ are inverses of each other. Each of them effectively reverses the transformation that the other accomplishes. If we apply to any vector U the linear transformation $\mathcal{T}$ and then apply to the resulting image $\mathcal{T}(U)$ the linear transformation $\mathcal{R}$, we recover the original vector U, because
$$\mathcal{R}(\mathcal{T}(U)) = U \quad \text{for every vector } U \text{ in the space.}$$

Moreover, the matrices that are associated with $\mathcal{T}$ and $\mathcal{R}$, respectively, are inverses of each other. Their product is the identity matrix that is associated with the identity transformation that maps every vector onto itself.

The first part of this chapter is devoted to studying the reversal of the mapping accomplished by a function. The second part discusses rational functions of a more advanced type than those in previous chapters.

2 Inverse functions

Many functions have inverses; many others do not.

Example 1 Consider the linear function D given by the formula $D(x) = 2x$. The rule for this function is to find the image of any number by doubling it. The image of 5 is twice 5, or 10. The image of $-\frac{3}{4}$, obtained by doubling, is $-\frac{3}{2}$. In general, the image of a known number c is $2c$.

Now suppose the image of a number is known to be d, and we seek to find the number. Calling the number x, we obtain the condition that $2x = d$. Thus, the desired number is $d/2$. To find the root, we have performed a process that may be considered as the opposite of multiplying by 2, namely, the process of dividing by 2. Thus, a reversal of doubling is halving.

The rule of halving a number describes a function H; a formula for H is $H(x) = \frac{1}{2}x$. Now $H(10) = 5$; we have already observed that $D(5) = 10$. Also $H(-\frac{3}{2}) = -\frac{3}{4}$; we previously noted that $D(-\frac{3}{4}) = -\frac{3}{2}$.

In general, suppose that c and d are two numbers. The statements $2c = d$ and $\frac{1}{2}d = c$ are equivalent to one another. Thus, the statement $D(c) = d$ and the statement $H(d) = c$ are equivalent to each other. Rephrased, $(c, d) \in D$ if and only if $(d, c) \in H$. See Fig. 1.

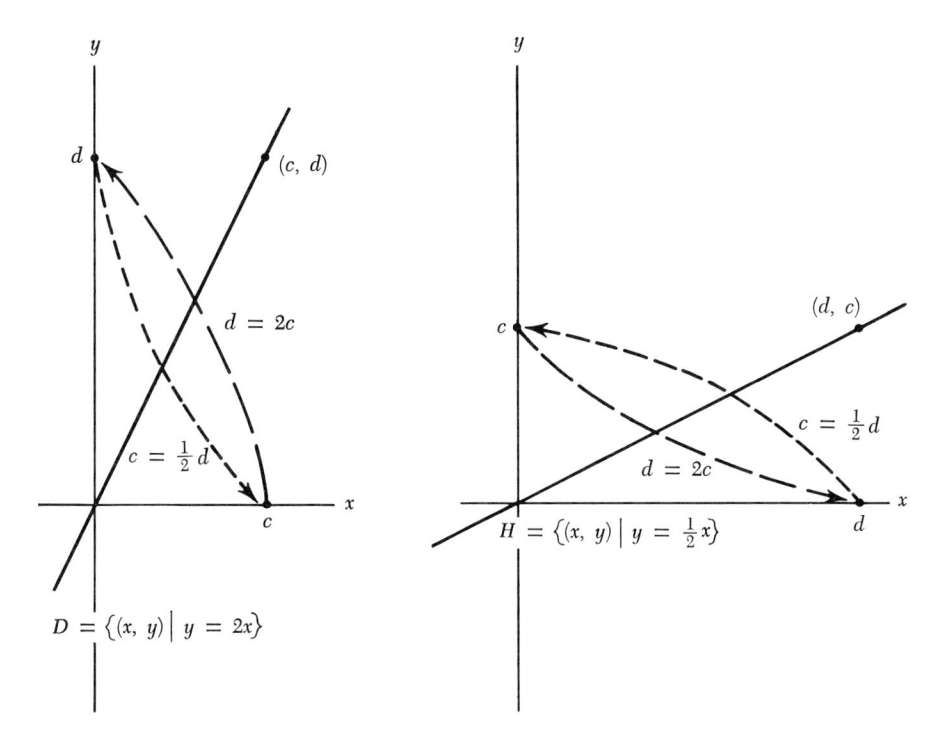

Figure 1

Left graph: $D = \{(x, y) \mid y = 2x\}$, with points (c, d), $d = 2c$, $c = \frac{1}{2}d$.

Right graph: $H = \{(x, y) \mid y = \frac{1}{2}x\}$, with points (d, c), $d = 2c$, $c = \frac{1}{2}d$.

Example 2 Consider the linear function S given by the formula $S(x) = x - 2$. The rule for this function is that the image of a number may be calculated by subtracting 2. For example, $S(9) = 7$ and $S(\frac{1}{2}) = -\frac{3}{2}$. Any number c belongs to the domain of S and its image is $c - 2$.

If a number d is given, does d belong to the range of S? If d is an image, is it the image of more than one number? Find each number that has d as its image. These problems ask us to solve the equation $S(x) = d$. Now, the equation $x - 2 = d$ has exactly one root, namely, $d + 2$. To compute this root, we employ a process that may be regarded as the opposite of subtracting 2, namely, the process of adding 2. A reversal of subtracting a number is adding the same number.

The rule of adding 2 to a number describes a function, which we may call A; a formula for A is $A(x) = x + 2$. Compare the statement $A(7) = 9$ with the above-mentioned statement $S(9) = 7$. Likewise, observe that $A(-\frac{3}{2}) = \frac{1}{2}$, and recall the previous remark that $S(\frac{1}{2}) = -\frac{3}{2}$.

In general, suppose that c and d are two numbers. The statements $c - 2 = d$ and $d + 2 = c$ have the same truth set. That is,

2 / Inverse functions

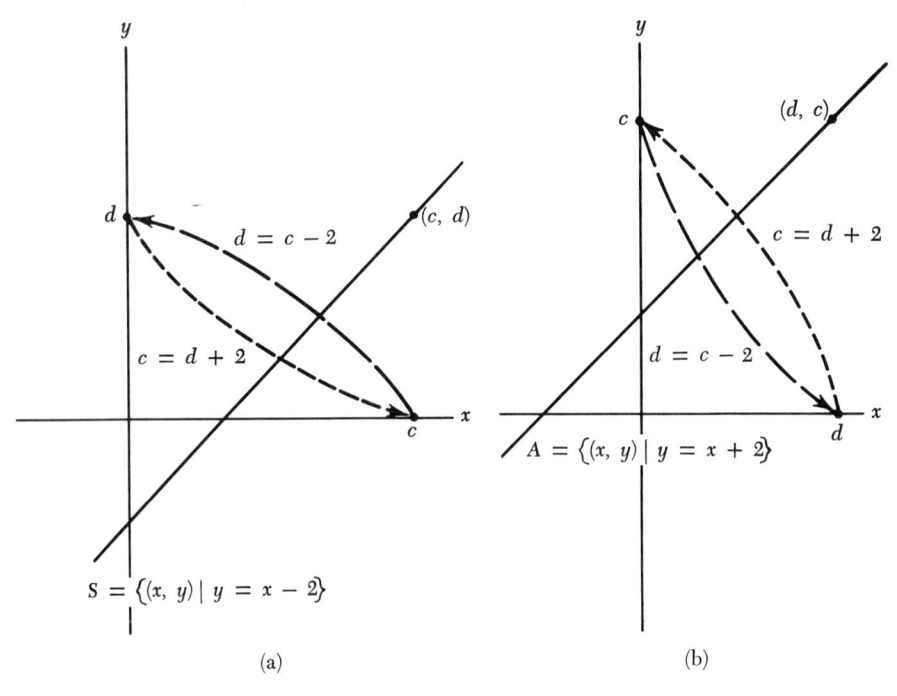

Figure 2

$S(c) = d$ if and only if $A(d) = c$. In other words, $(c, d) \in S$ and $(d, c) \in A$ are logically equivalent statements. See Fig. 2.

Example 3 Let L be the linear function $\{(x, y) \,|\, y = 3x + 2\}$. The domain of L consists of all real numbers. If c is a known number, its image is $3c + 2$. If d is the known image of an unknown number c, then we may determine c by solving an equation as follows:

$$3c + 2 = d,$$
$$3c = d - 2,$$
$$c = \tfrac{1}{3}(d - 2).$$

The image of a number c under L, namely, $3c + 2$, is obtained by first multiplying by 3 and then adding 2 to the product. If an image d under L is given and we wish to find the original number, we first subtract 2 and then divide the difference by 3. This latter rule describes a function; if we call this function G, then a formula for G is given by $G(x) = \tfrac{1}{3}(x - 2)$.

Note that if $L(c) = d$, then $G(d) = c$. Observe also that if $G(d) = c$, then $L(c) = d$. Thus, $(c, d) \in L$ if and only if $(d, c) \in G$. In other words, G contains an ordered pair of numbers if and only if the pair obtained by interchanging components belongs to L.

Inverse and Rational Functions

As a numerical illustration, $(2, 8) \in L$, since $L(2) = 3 \cdot 2 + 2 = 8$. Interchanging the components of $(2, 8)$ yields the ordered pair $(8, 2)$; since $G(8) = \frac{1}{3}(8 - 2) = 2$, we find that $(8, 2) \in G$.

The preceding three examples are simple illustrations of the notion of an inverse for a function. The two functions H and D in Example 1 are inverses of each other. In Example 2, the function A is the inverse of the function S. Likewise, in Example 3, each of the functions is the inverse of the other. We shall soon see that every linear function has an inverse, and that a function that is not a linear function may have an inverse. First, however, we present a definition. In order to appreciate the following definition more fully, you may wish to review the discussion of functions in Chapter 9.

Definition Suppose F is a function such that any two ordered pairs belonging to F have different second components. Then the set of all ordered pairs (d, c) for which $(c, d) \in F$ is a function called the *inverse* of F.

In symbolism, the inverse of F (in case it exists) is the set $\{(x, y) \mid (y, x) \in F\}$.

Notice, in accordance with the definition, that the criterion for any ordered pair (d, c) to belong to the inverse of F is that the pair (c, d), obtained by interchanging left and right components, should belong to F itself.

Not every function has an inverse. A function that fails to satisfy the supposition prescribed in the definition has no inverse.

Example 4 A familiar function that has no inverse is the second-power function Q given by the formula $Q(x) = x^2$. Among the ordered pairs that belong to Q are $(2, 4)$ and $(-2, 4)$. If we interchange components in each of these pairs, we obtain $(4, 2)$ and $(4, -2)$. Since these two different pairs have the same left component, they cannot both be members of a function. That is, the set of all ordered pairs (d, c) for which $(c, d) \in Q$ fails to be a function. Because of this failure, we say that Q has no inverse.

The assumption stated in the definition is precisely the requirement that guarantees that $\{(x, y) \mid (y, x) \in F\}$ actually is a function. We emphasize that the inverse of a function is itself a function. Usually a function and its inverse are different from one another. However, as the next example illustrates, it is possible that a function is the same as its own inverse.

Example 5 Let R be the reciprocal function given by the formula x^{-1}. The domain of R consists of all nonzero numbers, and Rg R = Dom R. Suppose then that c and d are numbers different from 0. The statement $R(c) = d$ means the same as $c^{-1} = d$, hence the same as $cd = 1$, therefore the same as $d^{-1} = c$, and consequently

Figure 3

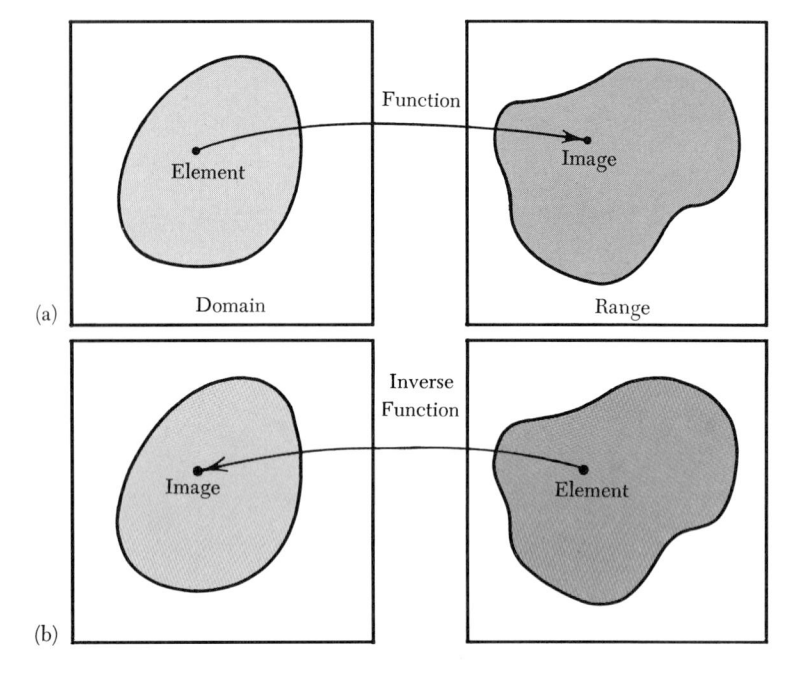

(a)

(b)

the same as $R(d) = c$. In other words, $(c, d) \in R$ if and only if $(d, c) \in R$. Thus, the set of all pairs (d, c) for which $(c, d) \in R$ is the same as the set R. The reciprocal function R is the inverse of itself.

Suppose that F is a function that has an inverse. Then the statement that d is the image of c under F is logically equivalent to the statement that c is the image of d under the inverse of F. Under the inverse the roles of an element and its image are interchanged from their roles under the original function. Figure 3a is an adaptation of Figs. 11 and 15 in Chapter 9. In case each image in the range of the function is the tip of only one arrow, then we may reverse the direction of the arrow. Contrast Fig. 3b with Fig. 3a to appreciate the reversal and interchange of roles.

EXERCISES A

1. Let L be the linear function given by $L(x) = 4x$.
   (a) Show that L contains each of the ordered pairs $(2, 8)$ and $(-\frac{1}{2}, -2)$.
   (b) Show that each of the ordered pairs $(0, 0)$ and $(\frac{2}{3}, \frac{8}{3})$ belongs to L.
   (c) Since L contains $(2, 8)$, the inverse of L contains $(8, 2)$. Explain why the inverse of L contains the ordered pair $(-2, -\frac{1}{2})$.
   (d) Explain why each of the couples $(0, 0)$ and $(\frac{8}{3}, \frac{2}{3})$ belongs to the inverse of L.
   (e) Fill the blank with a number: Each member of L is an ordered pair whose second component is _____ multiplied by its first component.
   (f) Fill the blank with a number: Each member of the inverse of L is an ordered pair whose second component is _____ multiplied by its first component.
   (g) The formula for the inverse of L is $\frac{1}{4}x$. How does the slope of the graph of the inverse of L compare with the slope of the graph of L?

2. Let L be the linear function given by $L(x) = x + 3$.
   (a) Find an ordered pair belonging to L.
   (b) Find an ordered pair that is a member of the inverse of L.
   (c) Fill the blank: The rule for finding the image of a number under L is to add 3 to the number; the rule for finding the image of a number under the inverse of L is to add _____ to the number.
   (d) Give a formula for the inverse of L.

3. Let G be the linear function $\{(x, y) | y = 3x - 1\}$.
   (a) $(c, d) \in G$ if and only if $d = 3c - 1$. Solve the equation for c in terms of d.
   (b) $(c, d) \in G$ if and only if (d, c) belongs to the inverse of G. Show that (d, c) belongs to the inverse of G if and only if $c = \frac{1}{3}d + \frac{1}{3}$.
   (c) If the function F is the inverse of G, show that $F(x) = \frac{1}{3}x + \frac{1}{3}$ for every number x.
   (d) What function is the inverse of F?

4. Let G be the linear function $\{(x, y) | y = -2x + 5\}$.
   (a) $(c, d) \in G$ if and only if $d = -2c + 5$. Solve the equation for c in terms of d.
   (b) $(c, d) \in G$ if and only if (d, c) belongs to the inverse of G. Show that (d, c) belongs to the inverse of G if and only if $c = -\frac{1}{2}d + \frac{5}{2}$.
   (c) Fill the blank with a formula: If the function F is the inverse of G, then $F(x) =$ _____ for every number x.
   (d) What function is the inverse of F?

5. Each of the following is a formula for a function. In each case find a formula for the inverse of the given function.
   (a) $\frac{1}{2}x$.   (b) $x - 7$.   (c) $9x + 1$.   (d) $3/x$.   (e) $-x + 12$.

6. Find the inverse of each of the following functions.
   (a) $\{(-1, 2), (1, 1), (3, 4), (5, 0)\}$.   (b) $\{(2, 7)\}$.
   (c) $\left\{(x, y) \mid [(x < 1) \wedge (y = -x)] \vee \left[(x \geq 1) \wedge \left(y = -\frac{x + 3}{4}\right)\right]\right\}$.

7. In each part of this exercise, sketch the graphs of the two given functions on a single coordinate system. Note that the functions in each pair are inverses of one another.
   (a) $\{(x, y) | y = -2x\}$ and $\{(x, y) | y = -x/2\}$.
   (b) $\{(x, y) | y = x + 4\}$ and $\{(x, y) | y = x - 4\}$.
   (c) $\{(x, y) | y = 2x + 6\}$ and $\{(x, y) | y = \frac{1}{2}x - 3\}$.

8. Suppose that the function G has an inverse and that the inverse is H.
   (a) Directly from the definition of an inverse, show that Dom $H =$ Rg G.
   (b) Likewise, show that Rg $H =$ Dom G.
   (c) If Figs. 3a and 3b represent the respective functions G and H, where in the second diagram should the label "Domain" be placed?
   (d) As in part (c), where should the label "Range" be inserted in Fig. 3b?

3 Geometric interpretation

Suppose a given function F has an inverse, say G. Each ordered pair (d, c) in G is obtainable from an ordered pair (c, d) in F by interchanging components. If now we interchange the components of each member (d, c) in G, the resulting set of ordered pairs will be the original function F. Thus, G also has an inverse; in fact, the inverse of G is F.

In other words, the relationship of being an inverse is a mutual one; func-

Figure 4

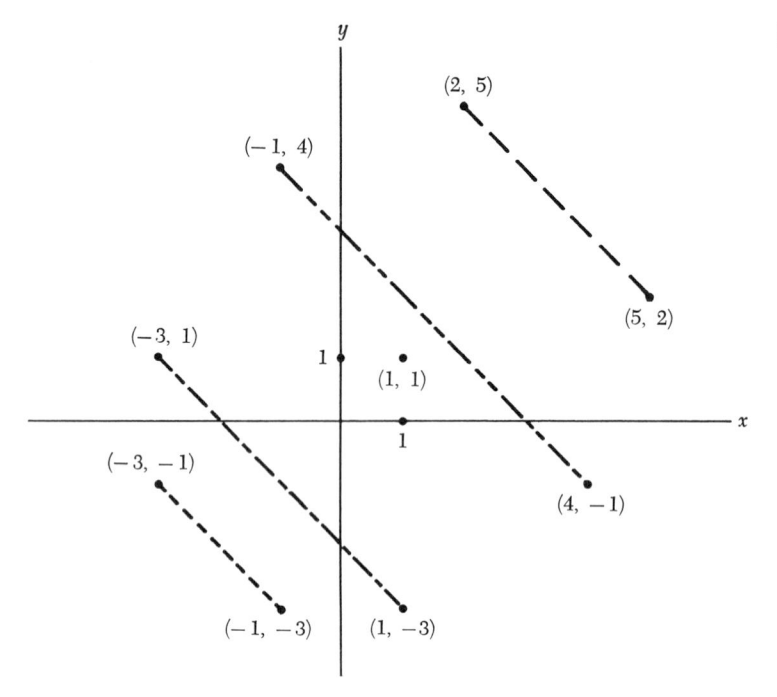

(2, 5)
(−1, 4)
(5, 2)
(−3, 1)
(1, 1)
(−3, −1)
(4, −1)
(−1, −3) (1, −3)

tions that have inverses occur in pairs, and each function in the pair is the inverse of the other. As noted in Example 5, there are functions which, in this situation, are paired with themselves, because they are self-inverse.

How does the graph of the inverse of a function compare with the graph of the function itself? Let F be the function and G be its inverse. A point belongs to the graph of G if and only if the graph of F contains the point obtained by an interchange of the coordinates of that point. Several pairs of points are shown in Fig. 4; in each pair, the points are related by the mutual interchange of coordinates. Also shown is a point that is unaltered by interchanging its coordinates.

The geometric effect of an interchange of coordinates is to exchange the roles of the x-coordinate axis and the y-coordinate axis. A point on the graph of the identity function $\{(x, y) \mid y = x\}$ is unaffected by an interchange of coordinates. Any other point in the coordinate plane is converted, under a coordinate interchange, into another point that is its reflection in the line $y = x$. (Pictorially, imagine a reflection as if the line behaved like a mirror.)

The preceding discussion applies to every point on the graph of F. Thus, the graph of F and the graph of its inverse G are mirror reflections of each other in the line $y = x$. The two curves are congruent to one another. Physically speaking, we may obtain the graph of either one of the functions from the other by letting the line bisecting the positive quadrant serve as an axis of rotation (in three-dimensional space) and rotating halfway around. Figure 5 shows the three pairs of mutually inverse functions discussed in Examples 1, 2, 3. Each figure also shows the "mirror," the line $y = x$.

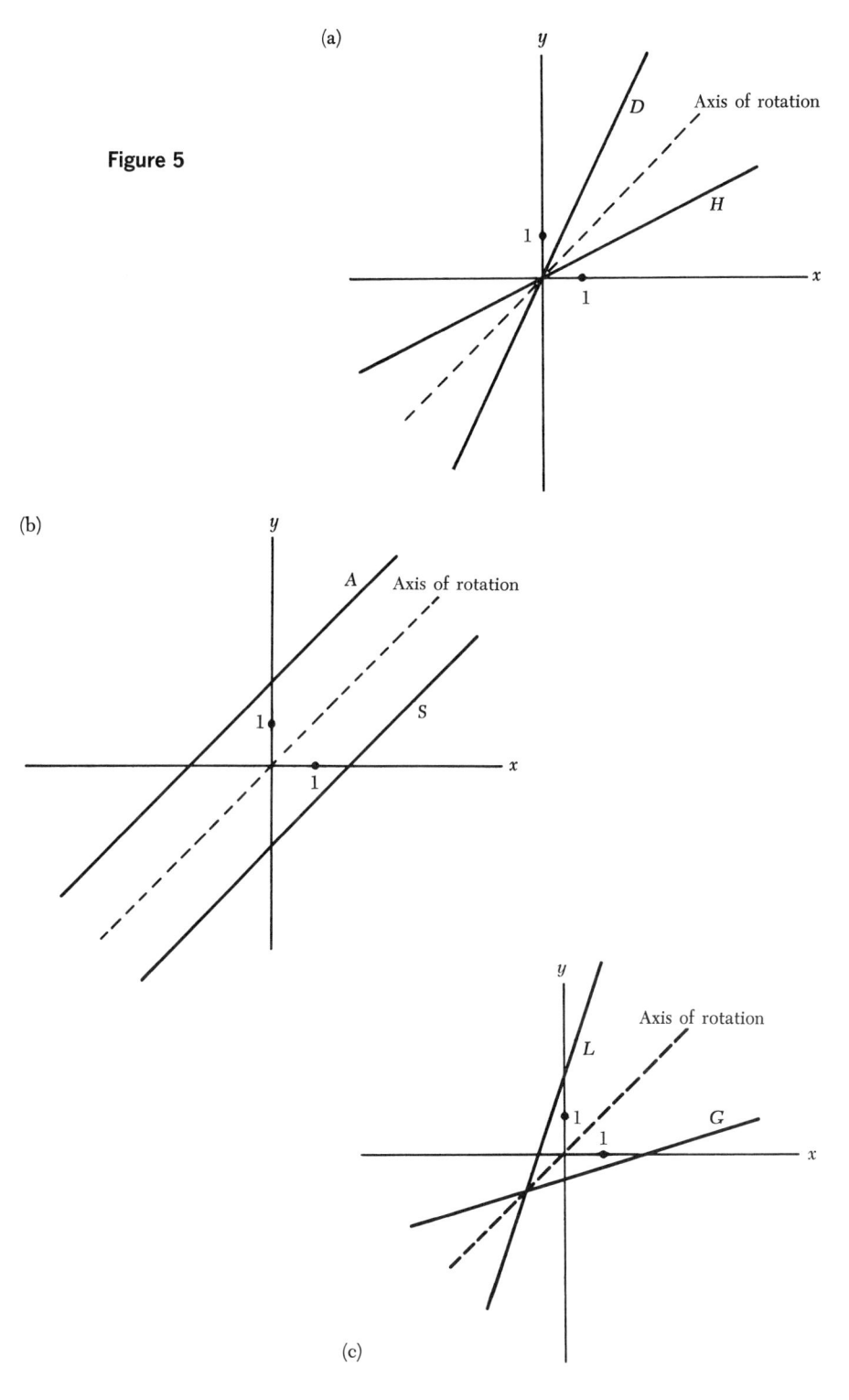

Figure 5

(a)

(b)

(c)

3 / Geometric interpretation

EXERCISES B

1. Let H be the third-power function given by the formula $H(x) = x^3$.
   (a) If $(c, d) \in H$, express d in terms of c and then express c in terms of d.
   (b) Give a formula for the inverse of H.

2. Let F be the function $\{(x, y) \mid y = 8(x + 4)^3\}$.
   (a) $(c, d) \in F$ if and only if $d = 8(c + 4)^3$. Solve the equation for c in terms of d.
   (b) Show that (d, c) is an element belonging to the inverse of F if and only if $c = \frac{1}{2}d^{1/3} - 4$.
   (c) Give a formula for the inverse of F.
   (d) What function is the inverse of the function whose formula was requested in part (c)?

3. Each of the following is a formula for a function. In each case find a formula for the inverse of the given function.
   (a) $\frac{1}{2}x^3$. (b) $x^3 + 1$. (c) $(x - 8)^{1/3}$.
   (d) $x^{1/3} - 8$. (e) $-x^3$. (f) x^{-3}.

4. Let F be a function that has an inverse, and let G be the inverse of F. Explain why all four of the following statements have the same meaning: $b = F(a)$; $(a, b) \in F$; $(b, a) \in G$; $a = G(b)$.

5. Let F and G be two functions that are inverses of each other.
   (a) If $F(3) = 5$, find $G(5)$.
   (b) If $G(-2) = 1$, evaluate $F(1)$.
   (c) If $(8, -6) \in F$, what is the image of -6 under G?
   (d) If $G(c) = d$, simplify $F(d)$.

6. Explain why each of the following functions fails to have an inverse.
   (a) $\{(x, y) \mid y = -x^2\}$.
   (b) $\{(x, y) \mid y = 5\}$.
   (c) $\{(x, y) \mid y = |x|\}$.
   (d) $\{(x, y) \mid [x < 0) \wedge (y = -1)] \vee [(x > 0) \wedge (y = 1)]\}$.

7. Explain why every quadratic function fails to have an inverse.

8. Explain why every function whose graph is a semicircle fails to have an inverse.

9. Find the inverse of each of the following functions.
   (a) $\{(x, y) \mid [(x, y) \geq (0, 0)] \wedge [x^2 + y^2 = 4]\}$.
   (b) $\{(x, y) \mid [(x, y) \geq (0, 0)] \wedge [x^2 + 9y^2 = 36]\}$.
   (c) $\{(x, y) \mid [(-x, y) \geq (0, 0)] \wedge [x^2 + y^2 = 9]\}$.

4 Inverses for certain functions

We now specialize our discussion to the case of linear functions. The two principal results are stated in the following theorem.

Theorem 1 Every linear function has an inverse, and the inverse of each linear function is a linear function.

PROOF Consider any linear function. If we call the function L, then L has a formula of the type $mx + b$, where m is a nonzero number and b is a number. If (c, d) is any member of L, then $mc + b = d$; we may rewrite this equation as $mc = d - b$ or as $c = \frac{1}{m}(d - b)$. In the latter step, we have taken advantage of the hypothesis that $m \neq 0$. Thus, the ordered pair (d, c) belongs to the linear function K given by the formula $K(x) = \frac{1}{m}x - \frac{b}{m}$. We observe that $(c, d) \in L$ if and only if $(d, c) \in K$. Thus, K is the inverse for L. We have shown that L has an inverse and that the inverse is itself a linear function.

The geometrical interpretation of Theorem 1 is significant. A linear function has as its graph a line that is not parallel to either coordinate axis. If we imagine the line $y = x$ as a "mirror," the reflection of this graph in the mirror is also a line that is neither horizontal nor vertical. Thus, the reflection is the graph of a function—indeed, of a linear function.

The criterion for a function to have an inverse is that no two of its members have the same right component. Geometrically, this criterion states that no two points belonging to the graph of the function have the same y-coordinate. In other words, no horizontal line intersects the graph at more than one point.

We recall that a nonempty set of ordered pairs is a function if and only if the graph of the set meets no vertical line at more than one point. Hence, combining results, we obtain the following theorem.

Theorem 2 A nonempty set of ordered pairs is a function possessing an inverse if and only if its graph and each line parallel to a coordinate axis have an intersection consisting of at most one point.

Example 6 The second-power function Q is the set $\{(x, y) \mid y = x^2\}$. Its graph is a parabola. There are many horizontal lines, each of which intersects the parabola in two points. Thus, we confirm, in geometrical language, the result noted in Example 4 that Q does not have an inverse.

An important subset of Q, however, does have an inverse. Let Q^* be the set of all ordered pairs such that the left component is a nonnegative number and the right component is the second power of the first component; in symbols,

$$Q^* = \{(x, y) \mid (x \geq 0) \wedge (y = x^2)\}.$$

Note that Q^* is the subset of Q obtained by deleting all ordered pairs whose first component is negative. We may also observe that Q^* is the intersection of Q and the right coordinate halfplane, because

$$\begin{aligned} Q^* &= \{(x, y) \mid (x \geq 0) \wedge (y = x^2)\} \\ &= \{(x, y) \mid y = x^2\} \cap \{(x, y) \mid x \geq 0\} \\ &= Q \cap \{(x, y) \mid x \geq 0\}. \end{aligned}$$

The graph of Q^* is a subset of the graph of Q, namely, the portion of the parabola that lies in the halfplane $\{(x, y) | x \geq 0\}$. We may speak of the curve as a semiparabola (Fig. 6).

This semiparabola has the property that no horizontal line intersects the curve at more than one point. The function Q^* has an inverse. If $(c, d) \in Q^*$, then $c \geq 0$ and $c^2 = d$; hence $c = d^{1/2}$. Consequently, (d, c) belongs to the set $\{(x, y) | y = x^{1/2}\}$. The inverse of the function Q^* is the square-root function given by the formula $x^{1/2}$. We recall (see Section 5 in Chapter 7) that, in our work, this formula is applicable only for nonnegative numbers. In other words, the domain of the square-root function consists of numbers that are either positive or zero.

As an application of our observation about the graphs of a pair of mutually inverse functions, the graph of the square-root function is obtainable from the graph of Q^* by a mirror reflection in the line $y = x$. Consequently, the graph of $\{(x, y) | y = x^{1/2}\}$ is a semiparabola, a subset of a parabola with vertex at the origin and with a vertical directrix. The curve is shown in Fig. 7.

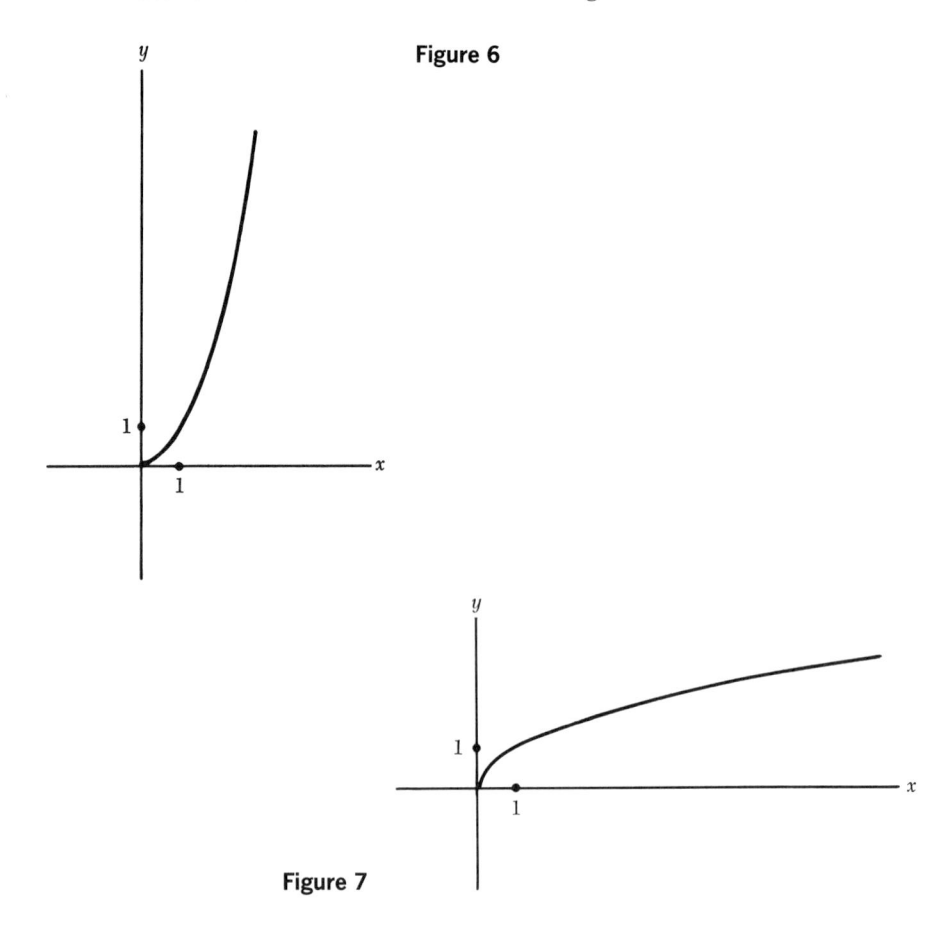

Figure 6

Figure 7

Inverse and Rational Functions

EXERCISES C

1. Find a formula for the inverse of each of the following functions.
   (a) $\{(x, y) | y = 3 - x\}$.
   (b) $\{(x, y) | y = -x/2\}$.
   (c) $\{(x, y) | y = -x\}$.
   (d) $\{(x, y) | y = x\}$.

2. Let F be the function given by the formula $F(x) = (x - 1)^2$.
   (a) Show that F has no inverse.
   (b) Show that if $c \geq 1$ and if $d = (c - 1)^2$, then $c = d^{1/2} + 1$.
   (c) Show that the function $F \cap \{(x, y) | x \geq 1\}$ has an inverse.
   (d) Show that $1 + x^{1/2}$ is a formula for the inverse mentioned in part (c).

3. Find a formula for the inverse of each of the following functions.
   (a) $\{(x, y) | (x \geq 0) \wedge (y = 3x^2 - 2)\}$.
   (b) $\{(x, y) | (x \geq -5) \wedge (y = x^2 + 10x + 29)\}$.

4. Sketch the semiparabola whose equation is $y = 2 + x^{1/2}$.

5. Sketch the graph of each of the following equations.
   (a) $y = (x - 3)^{1/2}$.
   (b) $y = 2x^{1/2} + 3$.
   (c) $y = -x^{1/2}$.
   (d) $y = (-x)^{1/2}$.

6. Let $Q^* = \{(x, y) | (x \geq 0) \wedge (y = x^2)\}$, and let H be the square-root function $\{(x, y) | (x \geq 0) \wedge (y = x^{1/2})\}$.
   (a) If $(5, q) \in Q^*$, find q.
   (b) If $(q, 4) \in H$, find q.
   (c) If $(5, q) \in H$, find q.
   (d) If $(1.6, r) \in Q^*$, find $H(r)$.
   (e) If $(1.6, r) \in H$, find $Q^*(r)$.
   (f) If $Q^*(0.3) = m$, find $H(m)$.
   (g) If $Q^*(s) = t$, what is $H(t)$?
   (h) If $H(s) = t$, what is $Q^*(t)$?

7. Find every linear function that is the inverse of itself.

8. For each of the following, tell whether the curve is the graph of a function. If it is, tell whether the function has an inverse. If the curve is the graph of a function with an inverse, sketch the graph of the inverse function. In every case, be prepared to justify your decision.

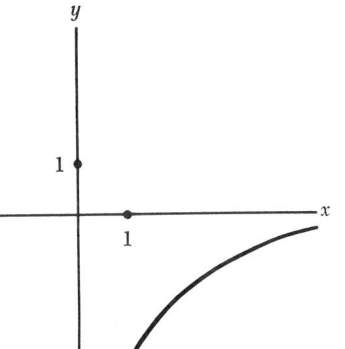

(a) Figure 8

(b) Figure 9

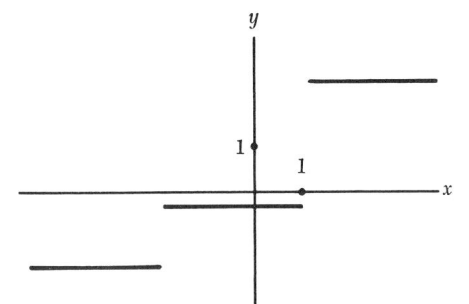

(c) Figure 10

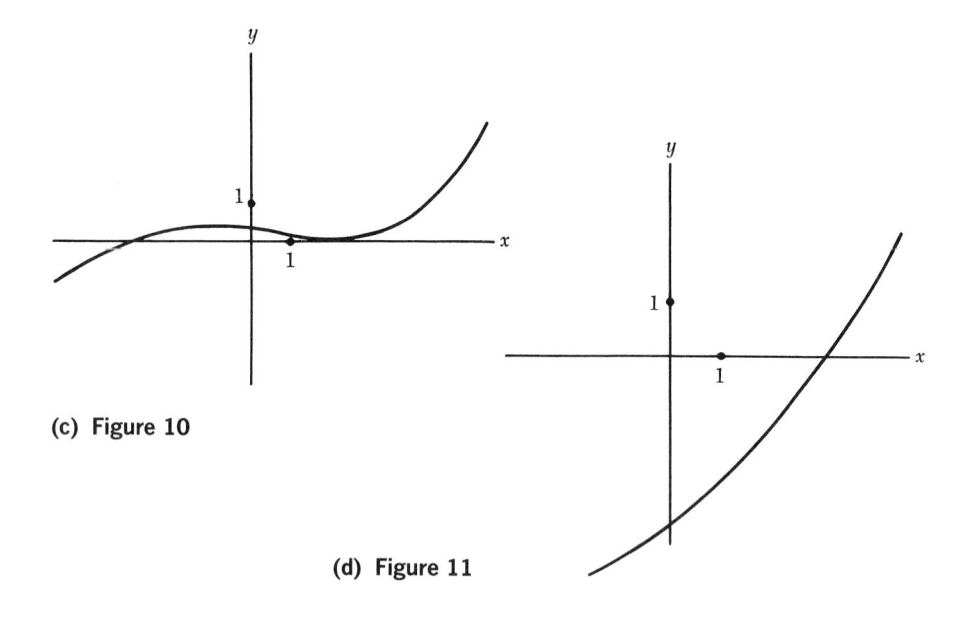

(d) Figure 11

5 Reflections

Let $V = (a, b)$ be a two-dimensional vector. In Section 3 we contrasted (from the geometrical viewpoint) the vector V and the vector (b, a) obtained from V by interchanging components. The vectors are reflections of each other in the "mirror" given by the equation $y = x$. (See Fig. 12.)

The distance of the point (a, b) from the origin is $(a^2 + b^2)^{1/2}$. This is the same as the distance of the point (b, a) from the origin. Both of the points (a, b) and (b, a) lie on a circle with its center at the origin. We wish to note some other points that also lie on this same circle and compare their positions with the location of the point (a, b).

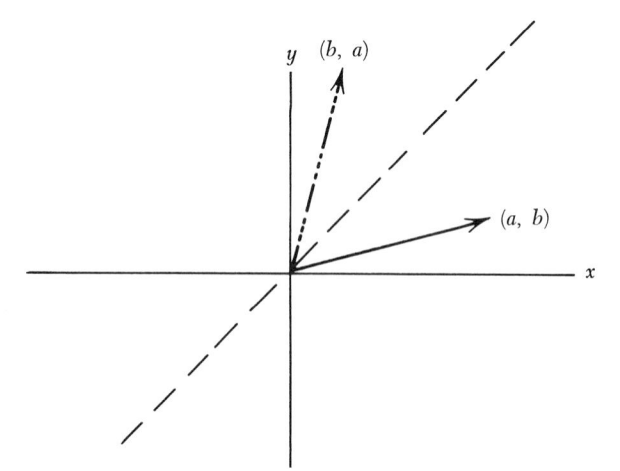

Figure 12

414   Inverse and Rational Functions

Figure 13a

(a, b)

(−a, −b)

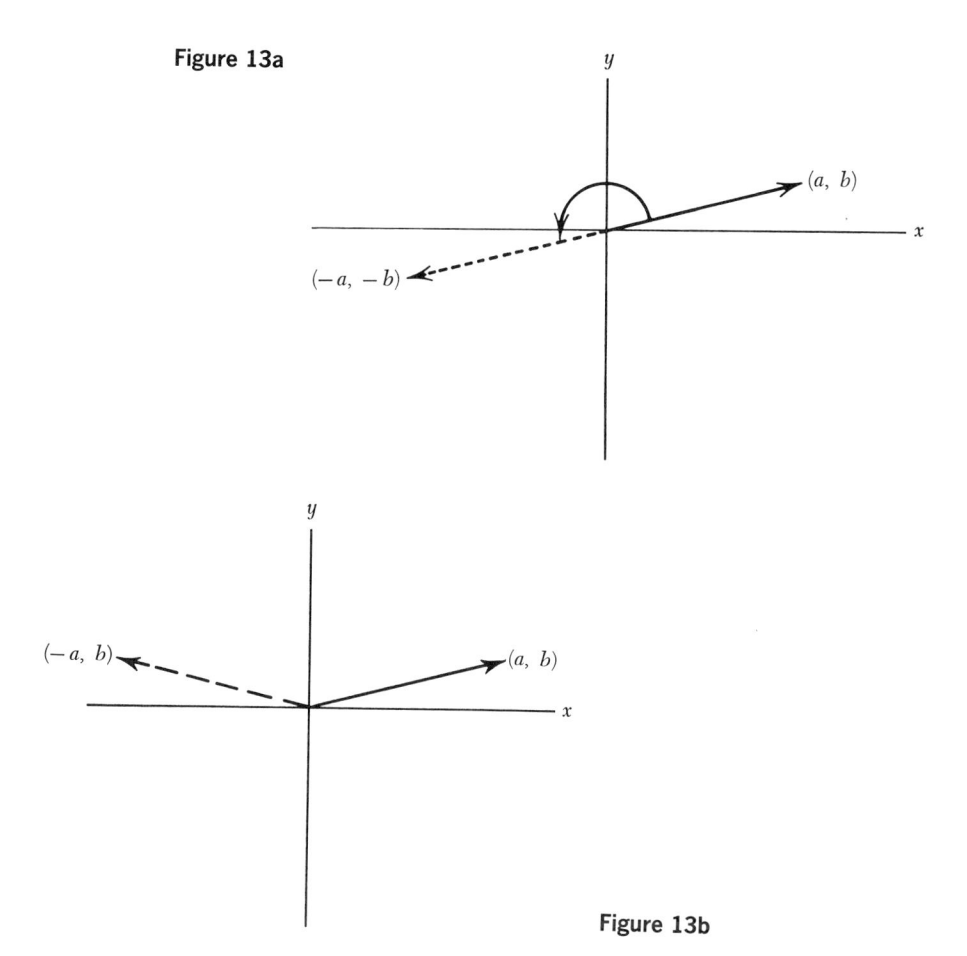

Figure 13b

(−a, b) (a, b)

First, consider the vector $-V$. Its components, namely, $(-a, -b)$, are the negatives of the respective components of V. Geometrically, $-V$ is like the vector V but in the opposite direction from the origin. (See Fig. 13a.) The origin is the midpoint of the segment joining the two points (a, b) and $(-a, -b)$. Physically, we may transform the vector V into the vector $-V$ by rotating it halfway around the origin; this viewpoint is suggested by the semicircular arrow in Fig. 13a.

Next, consider the vector $(-a, b)$. The right component of $(-a, b)$ is the same as the right component of (a, b), while the left components agree in absolute value but differ in sign. Geometrically, the vector $(-a, b)$ resembles the vector V except that the two vectors are oriented oppositely from the y-coordinate axis. (See Fig. 13b.) The points (a, b) and $(-a, b)$ are reflections of each other in the vertical coordinate axis, which may be imagined as a mirror. The axis is the perpendicular-bisector of the segment joining the two points.

Finally, consider the vector $(a, -b)$. This vector resembles V except that the two vectors are oriented oppositely from the x-coordinate axis. In this

5 / Reflections

Figure 13c

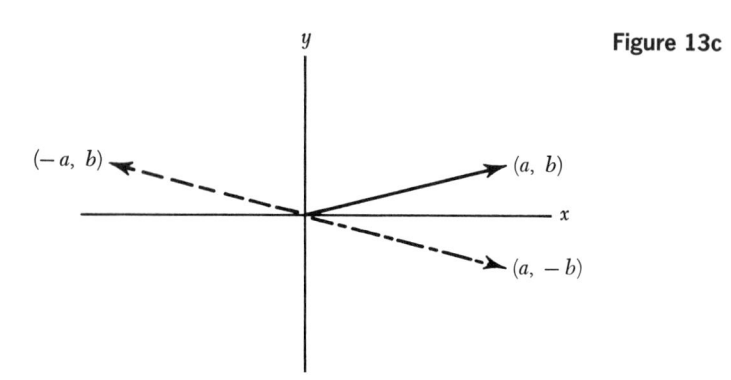

case, the horizontal coordinate axis serves as a mirror of reflection. We also note that the vector $(a, -b)$ is the negative of the vector $(-a, b)$ and that these two vectors are in opposite directions from the origin (see Fig. 13c).

All of the five points (a, b), (b, a), $(-a, -b)$, $(-a, b)$, $(a, -b)$, lie on a circle with center at the origin, namely, the circle whose equation is

$$x^2 + y^2 = a^2 + b^2.$$

Example 7 The five points for the case where $(a, b) = (3, 1)$ are shown in Fig. 14a. For the case $(a, b) = (2, 5)$, the five points are shown in Fig. 14b. Our discussion has not required that a and b be positive numbers. In case $(a, b) = (-1, 4)$, then $(b, a) = (4, -1)$, $(-a, -b) = (1, -4)$, $(-a, b) = (1, 4)$, and $(a, -b) = (-1, -4)$. These five points are shown in Fig. 14c.

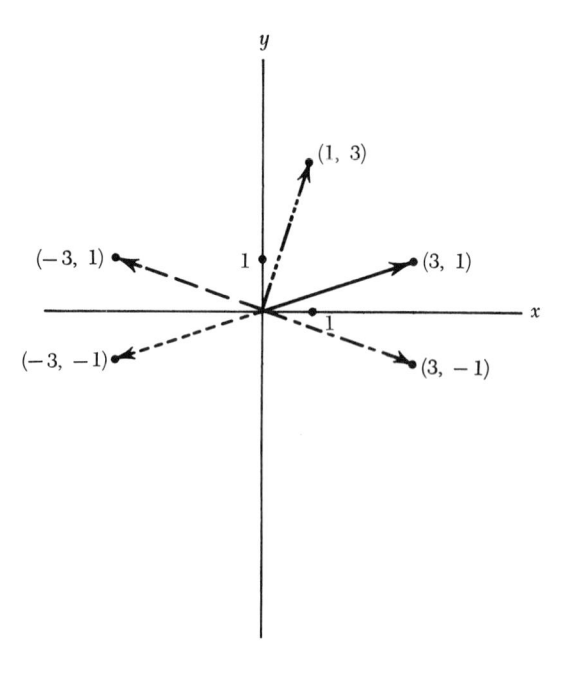

Figure 14a

416 Inverse and Rational Functions

Figure 14b

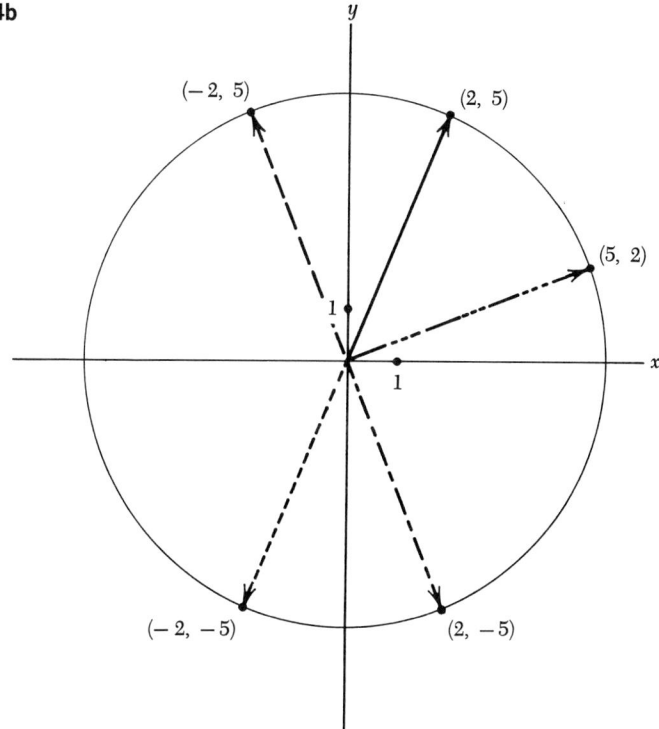

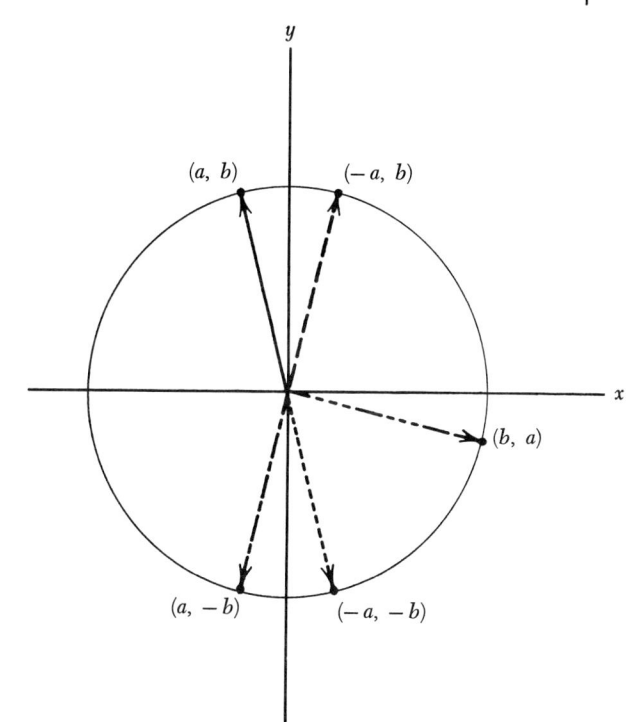

Figure 14c

5 / Reflections   **417**

We now apply our discussion to functions.

Example 8 Let L be the linear function given by the formula $2x - 3$. A point (a, b) belongs to the graph of L if and only if $b = 2a - 3$. The graph of L is a line.

Suppose that, in the equation $y = 2x - 3$ for this line, we replace x by $-x$. We obtain a new equation, namely, $y = -2x - 3$. Now $(a, b) \in L$ if and only if

$$b = 2a - 3,$$

hence, if and only if

$$b = -2(-a) - 3,$$

that is, if and only if the point $(-a, b)$ satisfies the new equation. But $(-a, b)$ is the reflection of (a, b) in the y-coordinate axis. Since the preceding discussion applies to every point, the entire graph of the equation

$$y = -2x - 3$$

is the reflection, in the vertical axis, of the graph of L. (See Fig. 15.) Physically, we may obtain the graph of $y = -2x - 3$ by rotating the graph of L halfway around the y-coordinate axis.

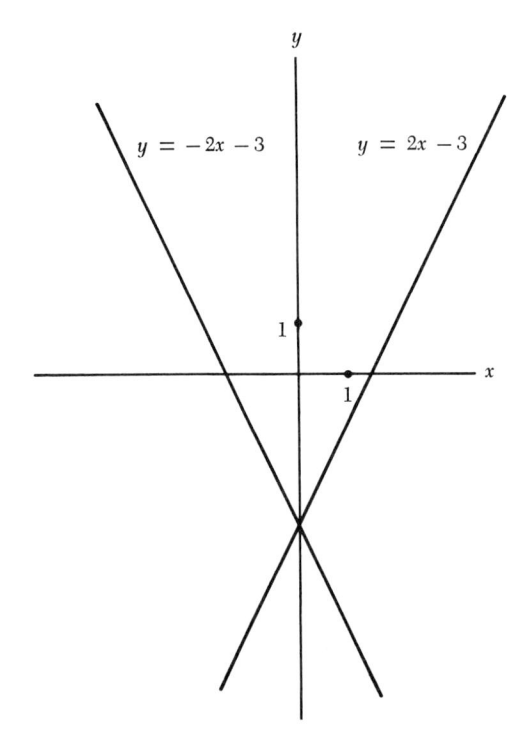

Figure 15

Inverse and Rational Functions

Now suppose that, in the equation $y = 2x - 3$, we replace y by $-y$. We obtain a new equation, namely, $-y = 2x - 3$. Again, $(a, b) \in L$ if and only if
$$b = 2a - 3,$$
hence, if and only if
$$-(-b) = 2a - 3,$$
that is, if and only if the point $(a, -b)$ satisfies the new equation. Since $(a, -b)$ is the reflection of (a, b) in the horizontal coordinate axis, the graph of the equation
$$-y = 2x - 3$$
is the reflection of the graph of L. Now $-y = 2x - 3$ is equivalent to $y = -2x + 3$. (See Fig. 16.) Note that $-2x + 3 = -(2x - 3) = -L(x)$. Physically, the graph of the function $-L$ is obtainable from the graph of L by a half rotation around the x-coordinate axis.

Finally, suppose that, in the equation $y = 2x - 3$, we simultaneously replace x by $-x$ and y by $-y$. This gives us still another equation, namely, $-y = -2x - 3$. The condition that (a, b) satisfies the original equation, namely, that
$$b = 2a - 3,$$
is the same as
$$-(-b) = -2(-a) - 3,$$

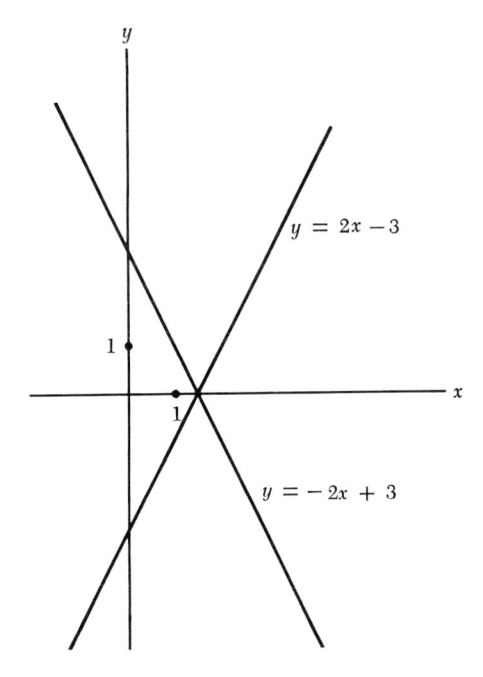

Figure 16

5 / Reflections

Figure 17

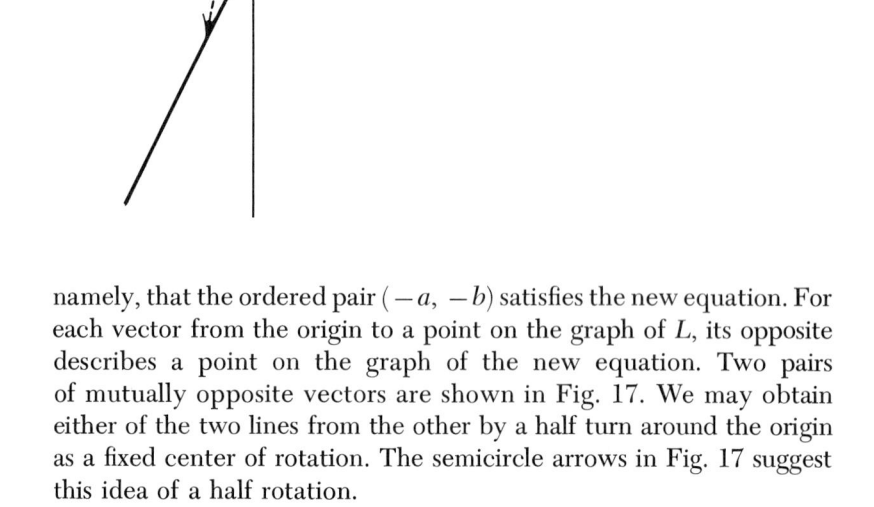

namely, that the ordered pair $(-a, -b)$ satisfies the new equation. For each vector from the origin to a point on the graph of L, its opposite describes a point on the graph of the new equation. Two pairs of mutually opposite vectors are shown in Fig. 17. We may obtain either of the two lines from the other by a half turn around the origin as a fixed center of rotation. The semicircle arrows in Fig. 17 suggest this idea of a half rotation.

The preceding example treated a linear function. The same type of discussion is applicable to any function f. We are thus able to contrast the graphs of the following equations:

$$y = f(x),$$
$$y = f(-x),$$
$$-y = f(x) \quad \text{or} \quad y = -f(x),$$
$$-y = f(-x) \quad \text{or} \quad y = -f(-x).$$

The type of argument employed in Example 8 establishes the following summary.

420   Inverse and Rational Functions

Theorem 3 For any function f, the graphs of all four equations $y = f(x)$, $y = f(-x)$, $y = -f(x)$, $y = -f(-x)$ are congruent to each other. More specifically:

(a) The graph of $y = f(-x)$ is the reflection of the graph of f in the y-coordinate axis.

(b) The graph of $y = -f(x)$ is the reflection of the graph of f in the x-coordinate axis.

(c) The graph of $y = -f(-x)$ is obtainable from the graph of f by a half rotation around the origin.

EXERCISES D

1. For each of the following choices of the point (a, b), plot on a single coordinate system the five points (a, b), $(-a, -b)$, $(-a, b)$, $(a, -b)$, (b, a).
    (a) $(a, b) = (1, 6)$.
    (b) $(a, b) = (-3, 1)$.
    (c) $(a, b) = (2, -\frac{1}{2})$.
    (d) $(a, b) = (-3, -5)$.
    (e) $(a, b) = (4, 0)$.
    (f) $(a, b) = (0, -\sqrt{2})$.
    (g) $(a, b) = (-3, 0)$.
    (h) $(a, b) = (0, 1)$.
    (i) $(a, b) = (0, 0)$.

2. In each of the following equations, replace x by $-x$ and simplify the equation obtained. (The letters a, b, c, m represent nonzero constants.)
    (a) $y = -3x + 4$.
    (b) $y = x$.
    (c) $y = mx + b$.
    (d) $y = x^2 - 2$.
    (e) $y = 1/x$.
    (f) $y = x^3 - 2x$.
    (g) $y = x^2/(1 - x^4)$.
    (h) $y = ax^2 + bx + c$.
    (i) $y = 0$.
    (j) $y = \begin{cases} 0 & \text{if } x < 0, \\ x & \text{if } x \geq 0. \end{cases}$
    (k) $x^2 + y^2 = 25$.
    (l) $y = x^{1/2}$.
    (m) $y = 1 + x^{1/2}$.

3. For each equation in Exercise 2, replace y by $-y$ and simplify the equation obtained.

4. For each equation in Exercise 2, replace both x and y by their respective negatives and simplify the resulting equation.

5. Let Q be the second-power function given by the formula $Q(x) = x^2$.
    (a) Sketch the graph of Q.
    (b) Using Theorem 3, compare the graph of the equation $y = Q(-x)$ with the graph of Q.
    (c) Using Theorem 3, sketch the graph of the equation $y = -Q(x)$.
    (d) Using Theorem 3, describe the graph of the equation $y = -Q(-x)$.

6. Let R be the reciprocal function $\{(x, y) \mid xy = 1\}$.
    (a) Sketch the graph of R.
    (b) Using Theorem 3, sketch the graph of the equation $y = R(-x)$.
    (c) Using Theorem 3, sketch the graph of the equation $y = -R(x)$.
    (d) How do the graphs in parts (b) and (c) compare?
    (e) Describe the graph of the equation $y = -R(-x)$.

7. Let m and b be nonzero real numbers. In each part of this Exercise, a pair of formulas is given. As an application of Theorem 3, compare the graphs of the linear functions given by the respective formulas in each pair.
    (a) $mx + b$ and $-mx + b$.
    (b) $mx + b$ and $-mx - b$.
    (c) $mx + b$ and $mx - b$.

8. Let a, b, c be nonzero real numbers. In each part below, a pair of formulas is given. Applying Theorem 3, compare the graphs of the quadratic functions given by the respective formulas in each pair.
   (a) $ax^2 + bx + c$ and $-ax^2 - bx - c$.
   (b) $ax^2 + bx + c$ and $-ax^2 + bx - c$.
   (c) $ax^2 + bx + c$ and $ax^2 - bx + c$.

9. The graph of the equation $y = x - x^{1/2}$ is reflected in the y-coordinate axis. Find an equation for the curve in its new position.

10. The graph of the equation $y = x(1 + x^2)^{-1}$ is rotated halfway around the origin. What is an equation for the curve in its new location?

11. For each of the following "mirrors" of reflection, find an equation for the curve obtained by reflecting the graph of the equation $y = 1/(1 + x^2)$ in the given axis.
    (a) The x-coordinate axis.   (b) The y-coordinate axis.

12. Repeat Exercise 11 for the original equation $(x - 1)^2 + (y + 2)^2 = 16$.

13. Each diagram below shows a curve. For each part, make four separate drawings as follows. First, sketch the curve obtained by reflecting the given curve in the y-coordinate axis. Second, sketch the curve that is the reflection of the given curve in the horizontal coordinate axis. Third, sketch the curve obtained by rotating the given curve halfway around the origin. Fourth, sketch the curve obtained by reflecting the given curve in the line $y = x$.

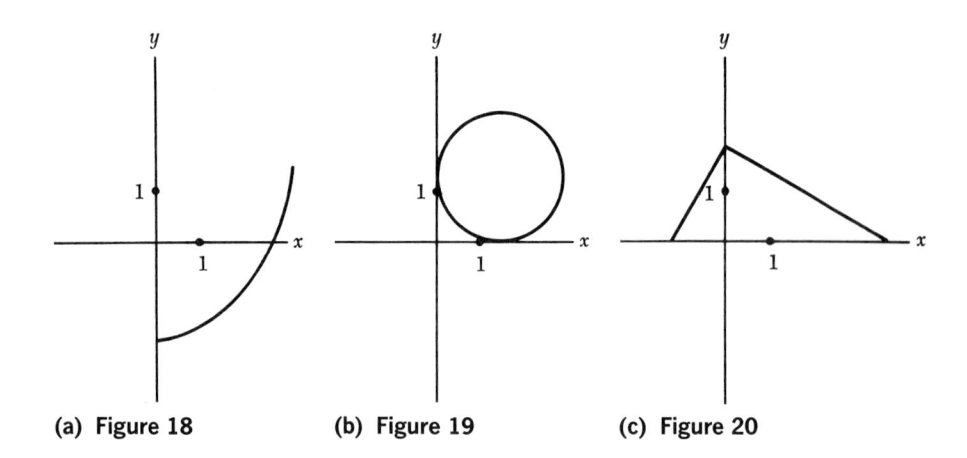

(a) Figure 18   (b) Figure 19   (c) Figure 20

6 Even rational functions

For any function, a discussion of its significant characteristics and an examination of its graph ought to be complementary aids to our understanding and appreciation of the function. A diagram showing the graph should reveal, as strikingly as a picture can, the important features of the function. On the other hand, a careful analysis of the properties of a function often can make far easier the task of drawing its graph. We will look at some examples of this type of assistance.

Example 9 Sketch the graph of the function given by the formula x^{-2}.

We begin by observing that the formula does not apply to the number 0, but is applicable to every other number. Thus, the domain of the function is the set of all nonzero numbers, that is, $\{x \mid x \neq 0\}$. Geometrically speaking, the graph does not meet the y-coordinate axis.

Next we note that, for the even exponent -2, the power of any number is positive. Consequently, the range of the function is contained in the set of all positive numbers, that is, it is a subset of $\{y \mid y > 0\}$. Geometrically speaking, the graph lies in the upper coordinate halfplane.

We exploit further the fact that the exponent -2 is even. For any number a, we have $1/a^2 = 1/(-a)^2$. Hence, two numbers that are negatives of each other have the same image. For example, 6 and -6 have the same image $\frac{1}{36}$; also, $-\frac{4}{3}$ and $\frac{4}{3}$ have the same number $\frac{9}{16}$ as image. Geometrically speaking, any two points on the graph that are equally distant from the y-axis (one on the left, and the other on the right of the coordinate axis) have the same height above the x-axis. Therefore, the graph is the reflection of itself in the vertical coordinate axis.

If we reflect the curve in the y-coordinate axis, we obtain the same curve again. This corroborates Theorem 3, because if we replace x by $-x$ in the formula x^{-2}, we obtain $(-x)^{-2} = x^{-2}$; the new function is the same as the original function.

If a and c are two positive numbers such that $c > a$, then $c^2 > a^2$ and $c^{-2} < a^{-2}$. That is, the greater of any two positive numbers has the smaller image under the function. The function is a decreasing function on the subset $\{x \mid x > 0\}$ of its domain. (Review Section 8 in Chapter 9.) Geometrically speaking, the portion of the graph in the right coordinate halfplane is a falling curve.

If c is a large positive number, then c^2 is very large and $c^{-2} = 1/c^2$ is very small. Consequently, far out to the right, the curve is near the horizontal axis. If the number a is small in absolute value, then a^2 is very small and $a^{-2} = 1/a^2$ is very large. Therefore, points on the curve near the vertical axis are very high up.

The various remarks we have made concerning the function $\{(x, y) \mid y = 1/x^2\}$ enable us to draw its graph very acceptably after plotting only a couple of individual points. We note that the equation $y = x^{-2}$ is satisfied by the points $(1, 1)$ and $(\frac{1}{2}, 4)$. The remainder of

Figure 21

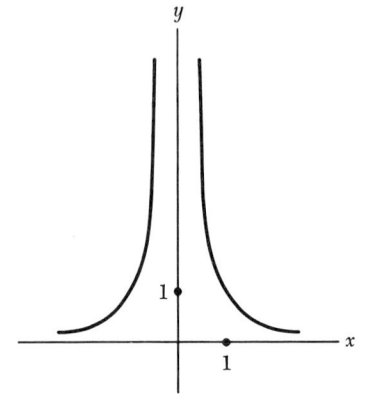

the picture in Fig. 21 is drawn on the basis of the above-mentioned properties of the function.

Note that although this function is a decreasing function on $\{x \mid x > 0\}$, it is not a "decreasing function." In fact, it is an increasing function on $\{x \mid x < 0\}$.

The power function $\{(x, y) \mid y = x^2\}$ and the power function $\{(x, y) \mid y = x^{-2}\}$, in which the exponents are 2 and -2 respectively, are members of a class of functions known as even functions. Although the name "even" function arises because some of the best illustrations are power functions with even exponents, an even function does not need to be a power function. In general, an even function is characterized by the fact that the image of a number under the function does not depend upon whether the number is positive or negative. In other words, any two numbers with the same absolute value, that is, any two numbers such as, say, c and $-c$, have the same image.

Definition A function F is called an *even function* provided, for any number c in its domain, $F(-c) = F(c)$.

An even function F is characterized as follows: if a number d is the image of a number c, then the same d is also the image of $-c$. Rephrased, if $(c, d) \in F$, then $(-c, d) \in F$. We now recall the preceding section. Geometrically speaking, a function is even if the reflection, in the y-coordinate axis, of every point on its graph is also a point belonging to the graph. Rephrased, a function is even if and only if the graph of the function is its own reflection in the vertical axis. We sometimes express this geometrical characterization by saying that the graph is *symmetric* with respect to the y-axis.

Example 10 Discuss the function given by the formula $\frac{1}{1 + x^2}$ and sketch its graph.

(a) Every real number belongs to the domain of the function. In-

424 Inverse and Rational Functions

deed, if c is any number, we may apply the prescribed rule and obtain the image $(1 + c^2)^{-1}$ for c.

(b) More specifically, given c, then $c^2 \geq 0$, hence $1 + c^2 \geq 1$. The division of 1 by a denominator that is at least 1 yields a positive quotient that is at most 1. That is, $0 < 1/(1 + c^2) \leq 1$ for every c. The image of every number belongs to $\{y \,|\, 0 < y \leq 1\}$. In other words, the range of the function is contained in the set $\{y \,|\, 0 < y \leq 1\}$. Geometrically, the graph lies in a horizontal strip one unit wide, extending from the x-coordinate axis to a height one unit above. The graph may touch the upper boundary of the strip, but not the lower boundary.

(c) Since

$$\frac{1}{1 + a^2} = \frac{1}{1 + (-a)^2} \qquad \text{for every number } a,$$

the function is an even function. This fact gives us the privilege of simplifying the subsequent discussion by confining our investigation to the right coordinate halfplane $\{(x, y) \,|\, x \geq 0\}$. This privilege arises because we know that the portion of the graph in the left halfplane is a reflection of the part in the right halfplane.

(d) The point $(0, 1)$ belongs to the graph.

(e) The function is decreasing on the set $\{x \,|\, x \geq 0\}$. For, if $c > a \geq 0$, then $c^2 > a^2$ and $1 + c^2 > 1 + a^2$ and therefore $(1 + c^2)^{-1} < (1 + a^2)^{-1}$; hence, the greater of two nonnegative numbers has the smaller image.

(f) If c is a large positive number, then $1 + c^2$ is very large and $1/(1 + c^2)$ is very small. Points on the curve that are far to the right are very near the horizontal axis.

After noting that $(1, \frac{1}{2})$ belongs to the function, we may sketch the graph with the aid of the above comments. The curve shown in Fig. 22 is called a *witch* or, more specifically, a *witch of Agnesi*.

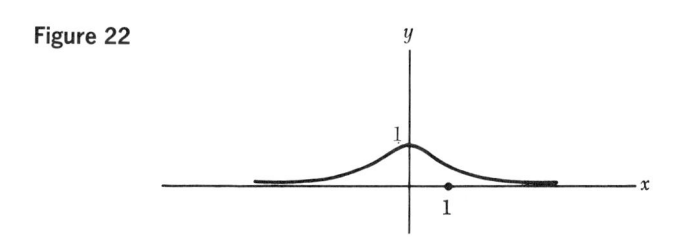

Figure 22

Example 11 Discuss the function $\{(x, y) \,|\, y = 1/(1 - x^2)\}$, and sketch its graph.

(a) The formula enables us to compute an image for every real number with only two exceptions. We must exclude each number for which the denominator $1 - x^2$ has the value zero. Thus, the domain of the function consists of all real numbers except 1 and -1.

(b) The function is an even function. Again, the symmetry of the graph permits us to confine further investigation to the right coordinate halfplane.

(c) The ordered pair (0, 1) belongs to the function.

(d) The function is increasing on $\{x \mid 0 \leq x < 1\}$. For suppose that $0 \leq a < c < 1$; then $0 \leq a^2 < c^2 < 1$, and $1 - a^2 > 1 - c^2 > 0$, and $(1 - a^2)^{-1} < (1 - c^2)^{-1}$. Thus, c, which is greater than a, has a greater image under the function than a has.

(e) The function is increasing on $\{x \mid x > 1\}$. For suppose that $1 < a < c$; then $1 < a^2 < c^2$, and $0 < a^2 - 1 < c^2 - 1$, and $(a^2 - 1)^{-1} > (c^2 - 1)^{-1}$; finally, multiplying each member of the latter inequality by -1, we obtain $(1 - a^2)^{-1} < (1 - c^2)^{-1}$. Thus, the greater of the two numbers a and c has the greater image.

(f) Points on the graph that are far to the right are very near, but below, the x-coordinate axis.

(g) Points on the graph with x-coordinate close to 1 are very far from the horizontal axis. For, if b is only slightly less or greater than 1, then $1 - b^2$ is nearly 0, and $(1 - b^2)^{-1}$ is a number that is large in absolute value.

The graph of the function is shown in Fig. 23. Note that although the function is increasing on the set $\{x \mid 0 \leq x < 1\}$ and is increasing on the set $\{x \mid x > 1\}$, the function does not have the property of being increasing on the set $\{x \mid x \geq 0\}$. Indeed, 2 is greater than $\frac{1}{2}$, but the image of 2 is less than the image of $\frac{1}{2}$.

The three functions discussed in Examples 9, 10, 11 have respective formulas $1/x^2$, $1/(1 + x^2)$, $1/(1 - x^2)$. In spite of the strong resemblance among the formulas, the respective graphs, as shown in Figs. 21, 22, 23, are quite dissimilar. The functions share the common property of being even functions. Each of their graphs is its own reflection in the y-coordinate axis.

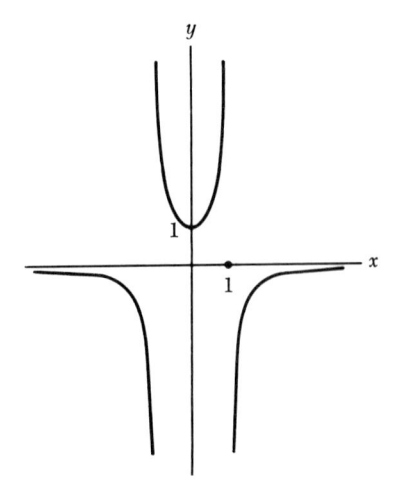

Figure 23

Inverse and Rational Functions

EXERCISES E

1. Show that the two functions in each of the following pairs have graphs that are congruent to one another.
   (a) x^{-2} and $(x+4)^{-2}$.
   (b) $\dfrac{3}{1+x^2}$ and $\dfrac{3}{1+(1-x)^2}$.
   (c) $(x-3)^{-2}$ and $x^{-2}-3$.
   (d) $\dfrac{1}{4+x^2}$ and $\dfrac{1}{5+2x+x^2}$.
   (e) $\dfrac{1}{1-x^2}+3$ and $\dfrac{1}{1-(x+\frac{1}{2})^2}$.

2. Each of the following formulas describes a function. For each function make an extensive list of its properties and sketch the graph of the function.
   (a) $\dfrac{4}{x^2}$.
   (b) $-\dfrac{1}{x^2}$.
   (c) $\dfrac{10}{2+x^2}$.
   (d) $\dfrac{4}{x^2}-1$.
   (e) $\dfrac{2}{1+x}$.
   (f) $\dfrac{3}{1-x}$.
   (g) $\dfrac{1}{9-x^2}$.
   (h) $\dfrac{9}{1-x^2}$.
   (i) $\dfrac{5}{(x-2)^2}$.
   (j) $\dfrac{2}{1+(x-3)^2}+1$.

3. For each of the following functions, decide whether it is an even function; justify your decision.
   (a) $\{(x,y) \mid y = x^4 - 2x^2 + 3\}$.
   (b) $\{(x,y) \mid y = 5\}$.
   (c) $\{(x,y) \mid y = |x|\}$.
   (d) $\{(x,y) \mid [(x \ne 0) \wedge (y=0)] \vee [(x,y) = (0,3)]\}$.
   (e) $\{(x,y) \mid y = x + x^{-1}\}$.

7 Other rational functions

In contrast with the idea of an even function is the concept of an odd function. Among the best illustrations of an odd function are the various power functions with an odd exponent.

Example 12 Consider the third-power function given by the formula x^3. Since the exponent 3 is odd, the function is an odd function. For any number c, we recall that $(-c)^3 = -c^3$. Thus, the third powers of c and of $-c$ are negatives of each other. That is, for any two numbers with the same absolute value, of which one is positive and the other is negative, their images under the third-power function are equal in absolute value but differ in sign. Rephrased, each image is the negative of the other.

The discussion in Example 12 suggests the following definition.

Figure 24

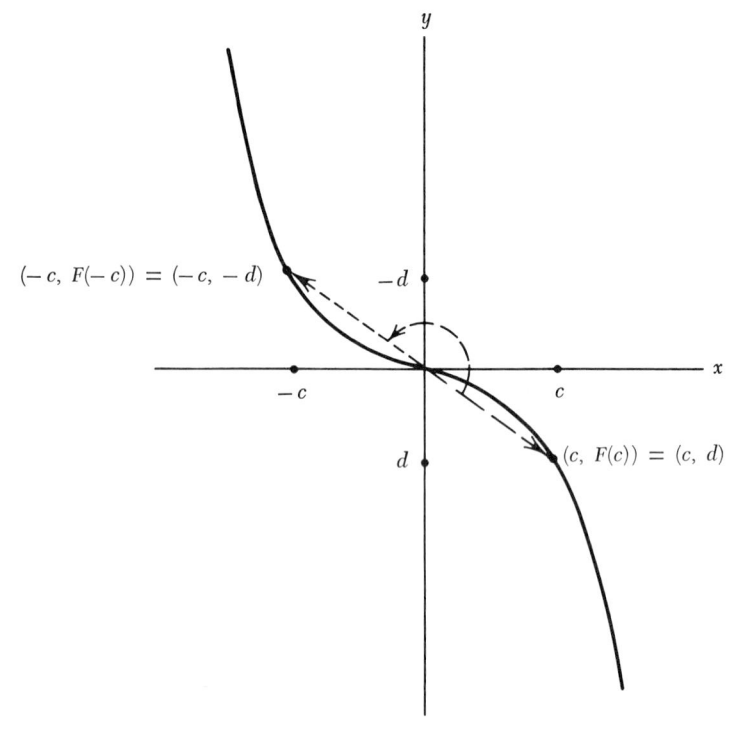

(−c, F(−c)) = (−c, −d)

(c, F(c)) = (c, d)

Definition A function F is called an *odd function* provided, for any number c in its domain, $F(-c) = -F(c)$.

Figure 24 illustrates an odd function. The definition characterizes an odd function F as follows: if a number d is the image of a number c [so that $d = F(c)$], then the image of $-c$ is $F(-c) = -F(c) = -d$. In other words, if $(c, d) \in F$, then $(-c, -d) \in F$. (See Fig. 24.) Recalling Section 5, we recognize the following interpretation of an odd function. For any point on the graph of the function, a half turn around the origin gives another point on the graph. In brief, a function is odd if and only if its graph is the same as the curve obtained by rotating its graph halfway around the origin. We sometimes say that the graph of an odd function is *symmetric* with respect to the origin.

A word of caution may be appropriate. The words "even" and "odd," as applied to functions, are not to be considered as direct opposites. Some functions are even, some functions are odd, many functions are neither even nor odd, and it is possible for a function to be both even and odd.

Example 13 Discuss the function

$$H = \{(x, y) \mid y = x^3 + x^2 - 2x\},$$

and sketch its graph.

428 Inverse and Rational Functions

Although the formula $x^3 + x^2 - 2x$ for H is often useful, another formula for the same function also is helpful, namely,

$$H(x) = x(x + 2)(x - 1).$$

From the factored version, we find that the equation $H(x) = 0$ has three roots, $-2, 0, 1$. Consequently, the graph of H and the x-coordinate axis have three points in common.

If x is a large positive number, then each of the factors $x, x + 2, x - 1$ is a large positive number and their product $H(x)$ is very large. If x is a negative number and large in absolute value, then each of the factors $x, x + 2, x - 1$ is a negative number large in absolute value; their product $H(x)$ is large numerically and, since the number of negative factors is three, the product is negative.

If $-2 < x < 0$, then the factors $x, x + 2, x - 1$ are respectively negative, positive, negative; the product $H(x)$ is therefore positive. The portion of the graph for $-2 < x < 0$ lies in the upper coordinate halfplane. If $0 < x < 1$, then the factors $x, x + 2, x - 1$ are respectively positive, positive, negative. The portion of the graph for $0 < x < 1$ lies below the x-coordinate axis.

The preceding two paragraphs are summarized in Table 1.

Table 1

	$x + 2$	x	$x - 1$	$H(x)$
x negative large	negative large	negative large	negative large	negative large
$-2 < x < 0$	positive	negative	negative	positive
$0 < x < 1$	positive	positive	negative	negative
x positive large	positive large	positive large	positive large	positive large

The graph of H is a cubical parabola. Its appearance (see Fig. 25) is typical of the graphs of many polynomial functions of the third degree.

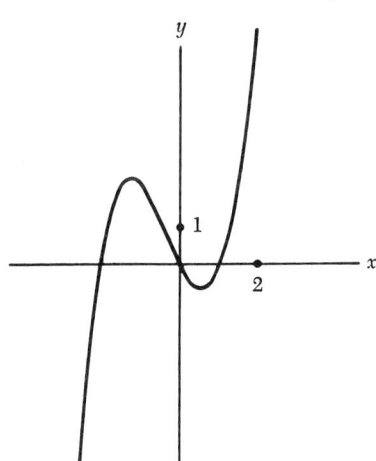

Figure 25

EXERCISES F

1. Consider the function F given by the formula $(x-1)(x-3)(x-6)$.
   (a) Show that if $c > 6$, then $F(c) > 0$.
   (b) Show that if $3 < c < 6$, then $F(c)$ is the product of three factors, of which one is negative and the other two are positive.
   (c) Show that if $1 < c < 3$, then $F(c)$ is positive because it is a product of two negative factors and one positive factor.
   (d) Show that $\{1, 3, 6\}$ is the set of roots of the equation $F(x) = 0$.
   (e) What is the algebraic sign of $F(c)$ in case c is less than the least root 1?
   (f) Evaluate $F(0)$.
   (g) Show that F is an increasing function on $\{x \mid x > 6\}$.
   (h) Sketch the graph of F.

2. Let G be the function given by the formula $x^2(x^2 - 4)$.
   (a) Find the set of roots of the equation $G(x) = 0$.
   (b) Find the set of roots of the inequality $G(x) < 0$.
   (c) Determine whether G is an odd function or an even function or neither.
   (d) Show that G is an increasing function on $\{x \mid x > 2\}$.
   (e) Show that $G(2^{1/2}) < G(1)$.
   (f) Sketch the graph of G.

3. Sketch the graph of each of the following functions. In each case use appropriate properties of the function to help in sketching.
   (a) $x^3 - 9x$.
   (b) $x^3 - 8$.
   (c) $(x+2)(x+1)(x-1)(x-3)$.
   (d) x^5.
   (e) $\dfrac{|x|}{x}$.
   (f) $\dfrac{1-x^2}{1+x^2}$.

4. For each of the following pairs of functions, explain why the graphs of the two functions are congruent to one another.
   (a) x^3 and $-x^3$.
   (b) x^3 and $1 - (x-1)^3$.
   (c) $\dfrac{2}{1+x^2}$ and $\dfrac{1-x^2}{1+x^2}$.
   (d) $\dfrac{1}{1-x}$ and $\dfrac{x}{1-x}$.

5. For each of the following functions, tell whether it is an even function or an odd function or neither. Give the reason in each case.
   (a) $x^3 - x$.
   (b) $x^3 - 1$.
   (c) $\dfrac{x^4+1}{x^2+1}$.
   (d) $x - |x|$.
   (e) $\{(x, y) \mid [(x < 0) \wedge (y = -1)] \vee [(x > 0) \wedge (y = 1)] \vee [(x, y) = (0, 0)]\}$.

6. Sketch the graph of each of the following equations.
   (a) $y = x + |x|$.
   (b) $y = |x|^{1/2}$.
   (c) $y = x^{-1/2}$.
   (d) $y = \dfrac{x}{1+x}$.

Transcendental Functions

1 Introduction

In the previous three chapters we have been studying algebraic functions. In this chapter we turn our attention to various transcendental functions. We begin with functions of the exponential type, then discuss functions involving logarithms, and later investigate trigonometric functions.

2 Functions of exponential type

In Chapter 7 we described what an exponent is in many important situations. In particular, we defined the power b^r, where the base b is a positive number and the exponent r is a rational number. As a brief review, if r is a natural number, then b^r is the product of r factors each of which is b; if the exponent is 0, then the power is 1; if r is the quotient s/t of natural numbers s and t, then b^r is the positive number whose tth power is b^s; finally, if r is any rational number, then $b^{-r} = 1/b^r$.

Suppose we select a positive number b as a base and keep it fixed. Then the power b^r is a number that is determined by r. The rule that assigns b^r as an image of r describes a function. Such a function may be called of exponential type. $f(x) = 2^x$ *plotted on next page.*

Example 1 Suppose the chosen base is 2, and consider the formula 2^x for a function. We may make a table showing several images under this function (Table 1a). Figure 1a shows the corresponding points plotted on a two-dimensional coordinate system. The formula 2^x is also meaningful for rational values of x that are not integers, and

Table 1a

x	3	2	1	0	-1	-2	-3
2^x	8	4	2	1	$\frac{1}{2}$	$\frac{1}{4}$	$\frac{1}{8}$

431

Table 1b

x	$\frac{5}{2}$	$\frac{5}{3}$	$\frac{4}{3}$	$\frac{1}{2}$	$\frac{1}{5}$	$-\frac{1}{4}$	$-\frac{1}{2}$	$-\frac{5}{4}$	$-\frac{7}{2}$
2^x	$4(2^{1/2})$	$2(4^{1/3})$	$2(2^{1/3})$	$2^{1/2}$	$2^{1/5}$	$\frac{1}{2^{1/4}}$	$\frac{1}{2^{1/2}}$	$\frac{1}{2(2^{1/4})}$	$\frac{1}{8(2^{1/2})}$
2^x (nearest hundredth)	5.66	3.17	2.52	1.41	1.15	0.84	0.71	0.42	0.09

we may continue Table 1a with the selected entries in Table 1b. By plotting the corresponding points on the diagram in Fig. 1a, we obtain Fig. 1b.

The points plotted in Fig. 1b vividly portray a significant feature: namely, of two plotted points, the point with the greater x-coordinate also has the greater y-coordinate. This characteristic corresponds to

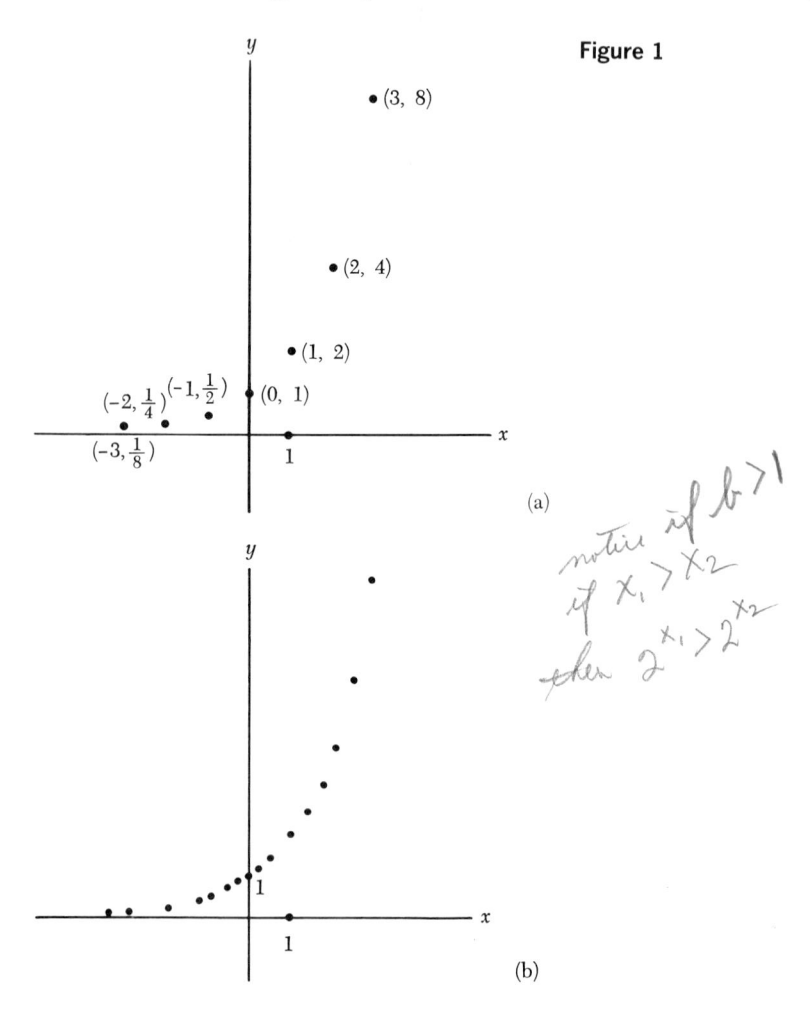

Figure 1

(a)

• (3, 8)

• (2, 4)

• (1, 2)

$(-2, \frac{1}{4})$ $(-1, \frac{1}{2})$ • (0, 1)

$(-3, \frac{1}{8})$

*notice if $b > 1$
if $x_1 > x_2$
then $2^{x_1} > 2^{x_2}$*

(b)

432   Transcendental Functions

an important property of exponents mentioned in Chapter 7. If q and r are rational numbers such that $q < r$, then $2^q < 2^r$. Of two different powers, the greater is the power given by the greater of the exponents. These observations remind us of the notion of an increasing function.

We have not yet described what is meant by the symbol 2^x in case x is an irrational number. Expressions such as $2^{\sqrt{2}}$ or 2^π have not been defined. We wish to extend the notion of an exponent so that the formula 2^x will give a function whose domain consists of *all* real numbers. This can be done in such a manner that the resulting function is an increasing function. Furthermore, if we impose the requirement that the function be increasing on the set of all real numbers, then there is only one way to accomplish the extension. These claims we do not attempt to prove, although they should appear plausible from a thoughtful inspection of Fig. 1b. We use these ideas as a basis for the next definition.

Definition Let b be a real number greater than 1. The powers b^x for irrational numbers x are defined in such a manner that the function $\{(x, y) \mid y = b^x\}$ is an increasing function on the set of all real numbers.

The situation of the above definition is reversed for the case of a base b satisfying the condition $0 < b < 1$; then the function given by the formula b^x is a decreasing function. Finally, if $b = 1$, then b^x is defined to be 1 for every real number x. Our study of mathematics will not require any further extension of the notion of a power.

On the basis of the above definition, we can now sketch the graph of the function given by the formula 2^x. The graph contains all the points plotted in Fig. 1b. Since the function is increasing, the graph is a rising curve. (See Fig. 2.)

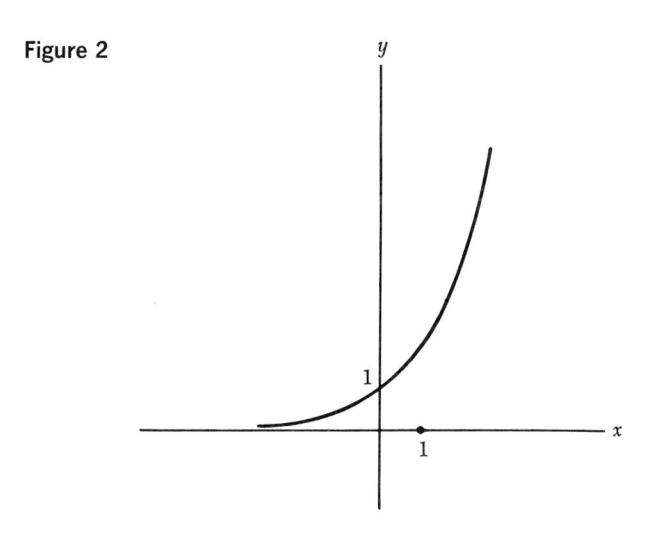

Figure 2

2 / Functions of exponential type

The familiar properties of exponents apply to all powers of a positive base. We list these properties in review:

[handwritten annotations:]
$b \cdot b = b^2$
$b \cdot b \cdot b = b^3$
$b^2 \cdot b^3 = b \cdot b \cdot b \cdot b \cdot b = b^5$

$$b^u \cdot b^v = b^{u+v}; \qquad \frac{b^u}{b^v} = b^{u-v}; \qquad (b^u)^v = b^{uv}.$$

EXERCISES A

1. Consider the function G given by $G(x) = 3^x$.
   (a) Find $G(0)$.
   (b) Show that if r and s are natural numbers such that $r < s$, then $3^r < 3^s$.
   (c) Show that if q and r are negative integers such that $q < r$, then $G(q) < G(r)$.
   (d) Write one or more significant inequalities about the image of a negative number n under the function G.
   (e) How many roots does the equation $G(x) = 0$ have?
   (f) Sketch the graph of the equation $y = G(x)$.

2. Let H be the function $\{(x, y) \mid y = 10^x\}$.
   (a) Sketch the graph of H.
   (b) Compare and contrast the graph of H and the graph of G in Exercise 1f. Mention ways in which the curves are alike, and tell how they are unlike.

3. Let F be the function $\{(x, y) \mid y = (\frac{1}{2})^x\}$.
   (a) Sketch the graph of F.
   (b) Compare and contrast the graph of F and the graph of G in Exercise 1f.
   (c) Explain why the graph of F is congruent to the graph of the equation $y = 2^x$. [HINT: See Theorem 3 in Section 5 of Chapter 11.]

4. For each of the following pairs, show that the graphs of the two equations in the pair are congruent to one another, and explain how the graph of the second equation is located on the coordinate system relative to the other graph.
   (a) $y = 2^x$ and $y = 2^x - 3$.
   (b) $y = 2^x$ and $y = 2^{x-3}$.
   (c) $y = 3^x - 4$ and $y = 3^{x+4}$.
   (d) $y = 2^x$ and $y = 8 \cdot 2^x$.
   (e) $y = 10^x$ and $y = 1 - 10^x$.
   (f) $y = 2^{1+2x}$ and $y = (\frac{1}{4})^x$.

5. Using the ideas developed in Exercise 4, sketch the graph of each of the following functions.
   (a) $\{(x, y) \mid y = 3 - 2^x\}$.
   (b) $\{(x, y) \mid y = 16 \cdot 2^x\}$.
   (c) $\{(x, y) \mid y = \frac{1}{2} \cdot 2^x + 5\}$.
   (d) $\{(x, y) \mid y = 3^{x/2}\}$.

6. Simplify each of the following numbers.
   (a) $(10^{\sqrt{2}})^2$.
   (b) $(10^{1/3})^\pi / (10^{1/6})^\pi$.
   (c) $4^{(1+\sqrt{2})/2}$.

3 The exponential function

For each positive number b greater than 1, we have an increasing function $\{(x, y) \mid y = b^x\}$. Different choices for b yield different functions. Numbers like 2 and 10 are sometimes useful choices for the base b. However, by far the most important base, both practically and theoretically, is a certain irrational number customarily denoted by e. The reasons for the preference for this particular number e can be best explained by the calculus and consequently will not be discussed here. Nevertheless, we shall henceforth pick almost exclusively the number e as our base b.

There are a variety of ways of identifying the number e. We may define e as follows: e is the number having the property that the graph of the set $\{(x, y) \mid (1 \leq x \leq e) \wedge (0 \leq y \leq x^{-1})\}$ has one unit of area. Let us examine this definition in more detail.

Consider the reciprocal function given by the formula $1/x$, and consider the portion of its graph in the positive quadrant. See Fig. 3. If we choose any

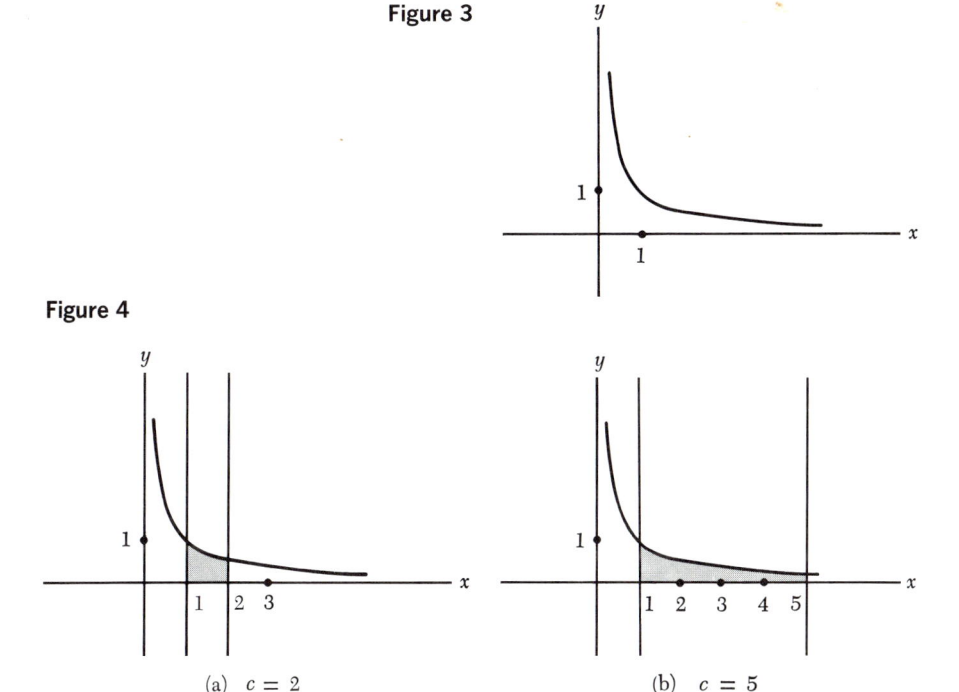

Figure 3

Figure 4

(a) $c = 2$

(b) $c = 5$

number c greater than 1, then we have determined a region of the plane whose boundary consists of parts of the following: the x-coordinate axis (at the bottom), the line $x = 1$ (at the left), the hyperbola $y = 1/x$ (at the top), and the line $x = c$ (at the right). Figures 4a and 4b show this region in a shaded fashion for two different values of the number c, namely, $c = 2$ and $c = 5$, respectively. The larger the number c we choose, the farther to the right the right-hand boundary of the region is, and consequently the greater the area of the region is. It appears that, if c is allowed to increase gradually, then the area also increases gradually. Now an inspection of Fig. 4a reveals that the area of the shaded region in the case $c = 2$ is less than one, while we find from Fig. 4b that in the case $c = 5$, there is more than one unit of area in the shaded region. It may seem reasonable that, by properly picking c, the region so determined will have exactly one unit of area. This can indeed be

3 / The exponential function

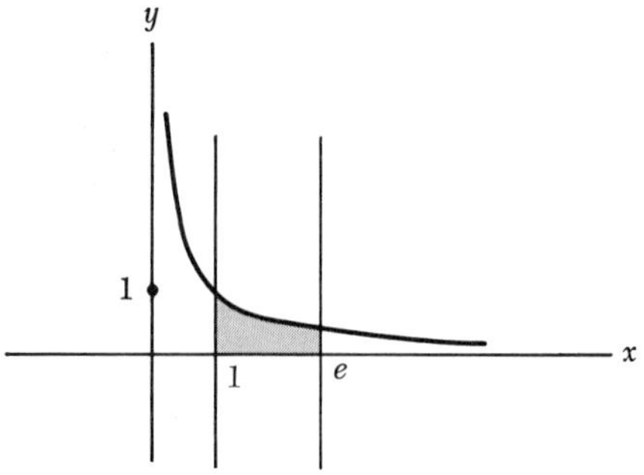

Figure 5

done, and the number picked to satisfy this requirement is the number called e. The shaded region in Fig. 5 has exactly one unit of area.

Now $e > 2$, as we have observed; furthermore, $e < 3$. Approximately, e is 2.7; more specifically, $2.718 < e < 2.719$. We repeat that e is an irrational number and therefore is not expressible exactly in the form of a terminating decimal or an ordinary rational fraction. For most of our purposes, whenever we need to approximate e, we shall be content to remember that e is a little more than 2.7.

Definition The function given by the formula e^x is called the *exponential function*.

Sometimes the exponential function is denoted by the symbol: exp. Thus, $\exp = \{(x, y) \mid y = e^x\}$.

We summarize several properties of the exponential function. Most of these have already been observed as applicable to any function having a formula of the type b^x for a fixed base b greater than 1.

1. The domain of the exponential function consists of all real numbers.
2. The range of the exponential function is the set of all positive numbers.
3. The exponential function is an increasing function.
4. $(0, 1) \in \exp$.
5. The image of a large positive number under the exponential function is a very large number; that is, if x is large, then e^x is very large.
6. The image of a large negative number is very near zero; that is, if $x < 0$ and if $|x|$ is large, then e^x is very small.

Referring to item 2 in the preceding list, it is important to keep in mind that since e is positive, every power of e is also positive. As illustrations of item 5, note that, for a number such as 10, the image e^{10} is greater than 22,000; also e^{100} exceeds

$$26{,}880{,}000{,}000{,}000{,}000{,}000{,}000{,}000{,}000{,}000{,}000{,}000{,}000{,}000;$$

and the image of a number like two million would require a good-sized book to print the part of the decimal expansion *before* the decimal point. On the other hand, as an illustration of item 6, $e^{-20} < 0.000\ 000\ 0021$.

436   Transcendental Functions

Figure 6

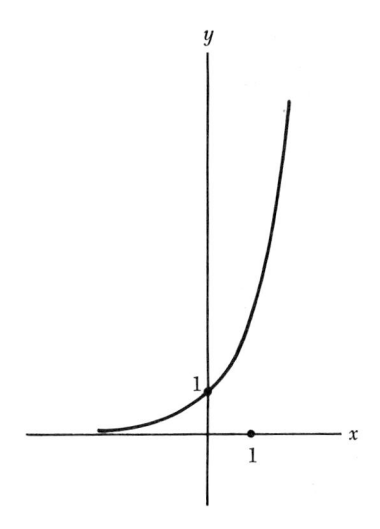

[Graph showing exponential function curve passing through (0,1), increasing rapidly]

Observe carefully how each of the above-mentioned properties is portrayed by the graph of the exponential function, as shown in Fig. 6.

The fact that exp is an increasing function means that, if a and c are real numbers such that $a < c$, then $\exp a < \exp c$, that is, $e^a < e^c$.

We apply to the exponential function some familiar laws for exponents:

$$\exp(u + v) = (\exp u)(\exp v);$$

$$\exp(u - v) = \frac{\exp u}{\exp v};$$

$$(\exp u)^v = \exp(uv).$$

$(e^u)^v = e^{uv}$

Several selected images under the exponential function are approximated, correct to three significant digits, in Table 2. A more extensive list of values appears in the Appendix.

Table 2

x	1	2	3	$\frac{1}{2}$	-1	-2
$\exp x$	2.72	7.39	20.1	1.65	0.368	0.135

EXERCISES B

1. In this exercise use only the appropriate values given in Table 2.
   (a) Using the entry in the table for exp 1, compute exp 2, and compare your result with the entry in the table.
   (b) Using the entry in the table for exp 1, compute e^{-1}, and compare your result with the entry in the table.
   (c) Use an appropriate entry from the table to compute e^4.
   (d) Calculate an approximation for $\exp \frac{3}{2}$.
   (e) Find an approximation for $(\exp 3)/(\exp 2)$.
   (f) Calculate an approximation for $\exp(-\frac{1}{2})$.

2. Refer to the information in Table 2. Which of the following four numbers is the greatest: 6 exp 1, 3 exp 2, 2 exp 3, exp 6?

3. If $t = 2$, evaluate $\exp(-t^2/2)$.

4. The function given by the formula $(\exp x)^2$ is also given by one of the following formulas: $\exp x^2$, $2 \exp x$, e^{2x}. Which one? Why?

5. The formula $[\exp(-x)]^2$ gives a function that is also given by one or more of the following formulas: $\exp x^2$, e^{2x}, $\exp(-2x)$, $2\exp(-x)$, e^{-x^2}. Which formula (or formulas)? Explain.

6. The graphs of the three functions below are all congruent to each other. Explain how the graph of each of them is located relative to the graph of the exponential function.
   (a) $\{(x, y) \mid y = e^x - 2\}$.   (b) $\{(x, y) \mid y = e^{x-2}\}$.   (c) $\{(x, y) \mid y = -e^x\}$.

7. Each of the following formulas describes a function. Give the range of each function.
   (a) $e^x - 4$.   (b) e^{x-4}.   (c) $4 \exp x$.
   (d) $4 - 3e^x$.   (e) e^{-x}.   (f) $\exp(1/x) - 1$.

8. Show, in detail, that the area of the region $\{(x, y) \mid (1 \leq x \leq 4) \wedge (0 \leq y \leq 1/x)\}$ is greater than one.

4 Applications

We often encounter functions that are obtained by combining the exponential function with one or more functions of other types. If h and k are nonzero numbers, then the formula he^{kx} gives a function of exponential type. More generally, we may consider a formula such as $he^{kx} + d$, where d is a number.

Example 2 Let F be the function given by the formula $3e^{x/2}$. The symbol exp is particularly useful in simplifying notation where complicated exponents may be involved; here we can write

$$F(x) = 3 \exp(x/2).$$

We discuss various properties of F and then use these in sketching the graph of F.

— x values

1. Since the formula for F is applicable to every real number, Dom F consists of all real numbers.
2. Since the image of a number under exp is positive, every number has a positive image under F; that is, Rg F consists of positive numbers only.
3. $F(0) = 3$.
4. Let a and c be numbers such that $a < c$. Then $a/2 < c/2$; since exp is an increasing function, $\exp(a/2) < \exp(c/2)$; finally then $3e^{a/2} < 3e^{c/2}$, that is, $F(a) < F(c)$. In words, F is an increasing function.

438   Transcendental Functions

5. If x is large, then $x/2$ is large, $e^{x/2}$ is very large, and $F(x)$ is very large.
6. If $x < 0$ and $|x|$ is large, then $x/2$ is a large negative number, $\exp(x/2)$ is very near zero, and $F(x)$ is very near zero.

The geometrical interpretations of these properties are as follows:

1. Every number is an x-coordinate for some point on the graph of F.
2. The graph of F is contained in the upper coordinate halfplane.
3. The graph intersects the y-coordinate axis three units above the origin.
4. The graph is a rising curve.
5. Points on the graph far to the right of the vertical axis are very far above the horizontal axis.
6. Points on the curve far to the left of the y-coordinate axis are very near the x-coordinate axis.

These observations enable us to sketch the curve rather easily. If we also plot the two points $(2, 3e)$ and $(-2, 3/e)$, which are approximately $(2, 8.1)$ and $(-2, 1.1)$, we obtain the graph of F shown in Fig. 7.

Example 3 Discuss the function given by the formula e^{x+2}, and sketch its graph.

The formula e^{x+2} is obtainable from the formula e^x by substituting $x + 2$ in place of x. As we learned in Section 7 in Chapter 10, this remark means that the graph of the equation $y = e^{x+2}$ is congruent to the graph of exp and is located two units to the left of the graph of exp. (See Fig. 8.)

Figure 7

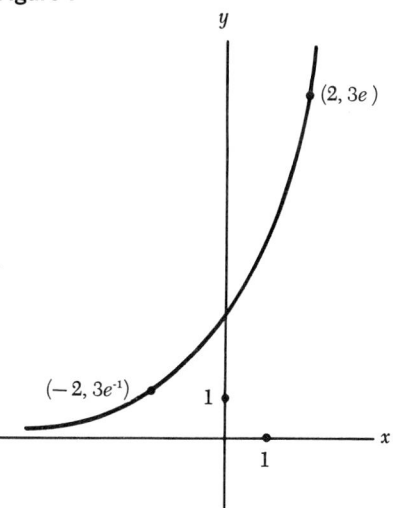

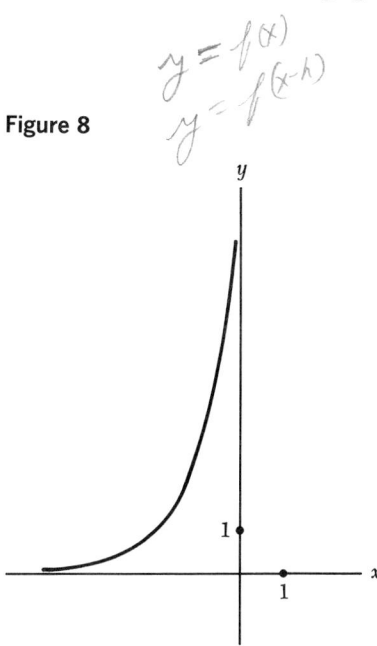

Figure 8

4 / Applications

Thus, the function given by e^{x+2} has many properties like those of the exponential function:

1. Its domain contains all numbers.
2. Its range is the set of all positive numbers.
3. It is an increasing function.
4. It assigns a very large image to each large number.
5. It assigns a very small image to each large negative number.

Since $(0, 1) \in \exp$, the ordered pair $(-2, 1)$ belongs to the function $\{(x, y) \mid y = \exp(x + 2)\}$. The latter function contains $(0, e^2)$. Since $e^{x+2} = e^2 e^x$, the function under consideration is of exponential type $h e^{kx}$, with $h = e^2$ and $k = 1$.

Example 4 Consider the function G given by $G(x) = 3 - e^x$. Since $G(x) = -e^x + 3$, the graph of G is congruent to, and three units above, the graph of the function $H = \{(x, y) \mid y = -e^x\}$. Since $H = -\exp$, the graph of H and the graph of the exponential function are congruent to each other, because each may be obtained from the other by a reflection in the x-coordinate axis. Thus, the graph of G is also congruent to the graph of $\exp$. (See Fig. 9.)

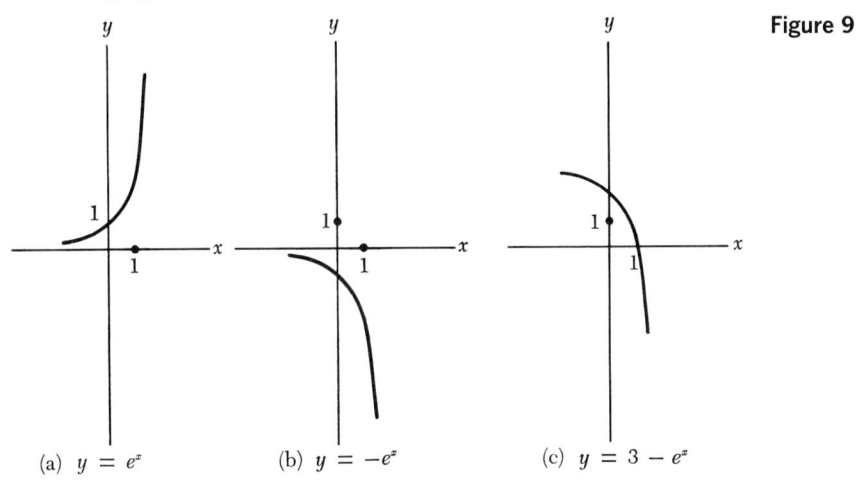

Figure 9

(a) $y = e^x$ (b) $y = -e^x$ (c) $y = 3 - e^x$

Since $\exp$ is an increasing function, its negative, namely H, is a decreasing function; consequently, G is also a decreasing function. Since $\text{Rg} \exp$ consists of all the positive numbers, $\text{Rg } H = \text{Rg}(-\exp)$ consists of all the negative numbers and therefore $\text{Rg } G = \{y \mid y < 3\}$. The domain of G consists of all real numbers. The graph of G intersects the y-coordinate axis at $(0, 2)$.

Example 5 Consider the function F given by $F(x) = \exp(-2x)$. The function F is of exponential type $h e^{kx}$ with $(h, k) = (1, -2)$.

Let a and c be two real numbers. If $a < c$, then $-2a > -2c$; therefore, since $\exp$ is an increasing function, $\exp(-2a) > \exp(-2c)$,

Figure 10

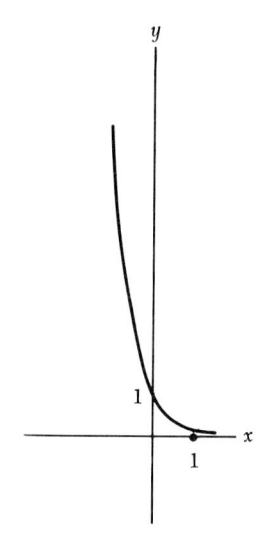

that is, $F(a) > F(c)$. Thus, F is a decreasing function. Since every power of e is positive, the range of F consists of positive numbers. If x is a large positive number, then $-2x$ is a large negative number, and $F(x)$ is very near zero. If x is a large negative number, then $-2x$ is a large positive number, and $F(x)$ is a very large number. Figure 10 shows the graph of the equation $y = e^{-2x}$.

Many practical problems involving the exponential function are concerned with growth or decay.

Example 6 Consider the dissolving of a solid substance in a solution, such as salt in water. The phenomenon proceeds gradually, with more of the substance being dissolved as more time elapses. If we let t be the number of units of time that have elapsed since the substance was placed in the liquid, then the proportion of the original substance that remains undissolved may be expressed by a formula of exponential type, say by the formula e^{-bt}, where b is a positive constant depending on the substance and the solvent.

If one-half of the substance dissolves in 5 minutes, what proportion remains after 10 minutes? At the time when half of the original substance is dissolved, then the proportion still undissolved is, of course, also one-half. Consequently, the given information yields the equation $e^{-5b} = \frac{1}{2}$. We are asked to evaluate e^{-10b}. Now $e^{-10b} = (e^{-5b})^2 = (\frac{1}{2})^2 = \frac{1}{4}$. Thus, after 10 minutes one-quarter of the substance has not dissolved.

Example 7 Suppose that an investment of A dollars is made in a savings account. Suppose that periodic payments of interest are

4 / Applications 441

credited to the account m times in a year. Suppose that the quoted yearly interest rate is r. Then the interest rate paid at each payment period is r/m. At the end of one year the value of the account (as we showed in Section 3 of Chapter 7) is $B = A[1 + (r/m)]^m$ dollars.

The accumulated amount B depends upon the number m of times interest is paid. The larger m is, the greater B is. One might guess that by compounding the interest often enough, the accumulated amount could become very large. This, however, is not the case. It can be shown that, no matter what m is, B cannot exceed $A \cdot e^r$.

As a numerical illustration, consider an investment of $1000 in an account quoting an interest rate of 6%. If interest is paid only once, then the value of the account at the end of the year is $1060.00. If interest is paid at the end of each month, then the accumulated amount is approximately $1061.68. The investor has earned an extra $1.68.

Now $1000e^{0.06} < 1061.84$. Consequently, regardless of how frequently interest is credited to the account, the year-end value cannot exceed $1061.84. We note that this amount is only 16 cents greater than monthly compounding gives. The careful investor may seriously evaluate the advertising glamor of institutions offering to compound interest daily or perhaps oftener.

EXERCISES C

1. Let H be the function $\{(x, y) | y = 2e^x\}$.
   (a) Calculate a decimal approximation for $H(2)$. [HINT: See Table 2 in Section 3.]
   (b) Compute, approximately, $H(-1)$.
   (c) Give the range of H.
   (d) Show that H is an increasing function.
   (e) Fill the blank with the best response: If $t > 0$, then $H(t) > $ _____ .
   (f) Fill the respective blanks with the best responses: If $t < 0$, then _____ $< H(t) <$ _____ .
   (g) Sketch the graph of H.

2. Using the result of Exercise 1g, sketch the graph of the function given by the formula $-2 \exp x$.

3. Let G be the function $\{(x, y) | y = 2e^{-x}\}$, and let H be the function discussed in Exercise 1.
   (a) Give the domain of G.
   (b) If (c, d) belongs to H, show that $(-c, d)$ belongs to G.
   (c) Show that if $(p, q) \in G$, then $(-p, q) \in H$.
   (d) What is the range of the function G?
   (e) Show that G is a decreasing function.
   (f) Sketch the graph of G.
   (g) Describe how the graph of G is related to the graph of H.

4. The function given by the formula e^{x-3} is the same as the function given by a formula of the type he^{kx}. Find h and k.

5. For each of the following formulas, decide whether the function given by the formula is increasing, or decreasing, or neither.
   (a) $5e^{3x}$.
   (b) $5e^{-3x} + 4$.
   (c) $-1/e^x$.
   (d) $-\exp x^2$.
   (e) $e^x + e^{-x}$.
   (f) $e^x - e^{-x}$.

6. For each of the following formulas, decide whether the given function is even, odd, or neither.
   (a) $(e^x)^2$.
   (b) e^{-x^2}.
   (c) $\dfrac{e^x + e^{-x}}{2}$.
   (d) $\dfrac{e^x - e^{-x}}{2}$.

7. For each of the following functions, discuss its important properties and then sketch its graph.
   (a) $\{(x, y) \mid y = 5 - \exp x\}$.
   (b) $\left\{(x, y) \,\Big|\, y = \dfrac{\exp x + \exp(-x)}{2}\right\}$.
   (c) $\left\{(x, y) \,\Big|\, y = \dfrac{\exp x - \exp(-x)}{2}\right\}$.
   (d) $\left\{(x, y) \,\Big|\, y = \dfrac{1}{e^{2x}}\right\}$.
   (e) $\{(x, y) \mid y = \exp(x^2)\}$.
   (f) $\{(x, y) \mid y = \exp(-x^2)\}$.
   (g) $\{(x, y) \mid y = \exp 2x - 1\}$.
   (h) $\{(x, y) \mid y = \exp(4x - 3)\}$.

8. In the study of certain physical changes, such as radioactive decay, we encounter a formula of the type $1 - \exp(-bt)$, where b is a positive constant. The formula may represent the probability that a particle of a radioactive substance will disintegrate within t units of time, measured from a suitable instant chosen as the time origin. The probability is thus a function of elapsed time. Suppose that, as a particular case, $b = \frac{1}{4}$.
   (a) Calculate the probability for $t = 2$. (See Table 2 in Section 3.)
   (b) Calculate the probability for $t = 8$.
   (c) What is the probability that the particle will not disintegrate before $t = 2$ but will disintegrate before $t = 8$?
   (d) What is the probability that the particle will disintegrate between $t = 4$ and $t = 8$?
   (e) What is the probability that the particle will not disintegrate until after $t = 12$?
   (f) Let c and d be positive numbers such that $c < d$. What is the probability that the disintegration will occur when $c \leq t < d$?
   (g) Calculate the probability for $t = 0$.
   (h) Sketch the graph of the function $\{(t, y) \mid [t \geq 0] \wedge [y = 1 - \exp(-t/4)]\}$.
   (i) What is the range of the function in part (h)?

9. As a consequence of an assumption frequently made in population studies, the number of organisms is related to time by a function of exponential type. Suppose that n is the number of bacteria in a certain culture at time t (measured in hours). Suppose that the culture of one million bacteria when $t = 0$ has doubled in numbers three hours later.
   (a) Show that $n = 1{,}000{,}000 e^{kt}$, where k is the number such that $e^{3k} = 2$.
   (b) How many bacteria are in the culture when $t = 6$?
   (c) Approximately how many bacteria are in the culture when $t = 4$? [HINT: See Table 1b in Section 2.]
   (d) What is the approximate value of t when the bacteria population is 3,170,000?

10. If $0 < r < 0.07$, then e^r is approximately the same as $1 + r + (r^2/2)$; the error of approximation is less than 0.0001. (The preceding result may be established by calculus techniques.) Use this information to find the maximum amount of interest that may be earned in a year on an investment of A dollars at a quoted yearly rate of r, in each of the following cases.
    (a) $A = \$1000$, $r = 3\%$.
    (b) $A = \$4000$, $r = 2\%$.
    (c) $A = \$9000$, $r = 0.0175$.

$y = a^x$ is an increasing function

5 Inverse of an increasing function

We now wish to observe that the exponential function has an inverse and to study the properties of the inverse of exp. Since exp is an increasing function, we may apply the two general results in the next theorem.

Theorem 1 Every increasing function has an inverse. Furthermore, the inverse of every increasing function is itself an increasing function.

Before we give the proof of Theorem 1, it will be worthwhile to interpret geometrically the two assertions. We are considering a function whose graph is a rising curve. For an illustration, see Fig. 11. We reflect the curve in the mirror $y = x$. Our claims are that the reflected curve is the graph of a function and that it is a rising curve. Figure 11 makes the two propositions plausible.

We now begin our proof of Theorem 1 by considering the first of the propositions. Let F be any increasing function. In order to show that F has an inverse, we need to show that any two ordered pairs belonging to F have

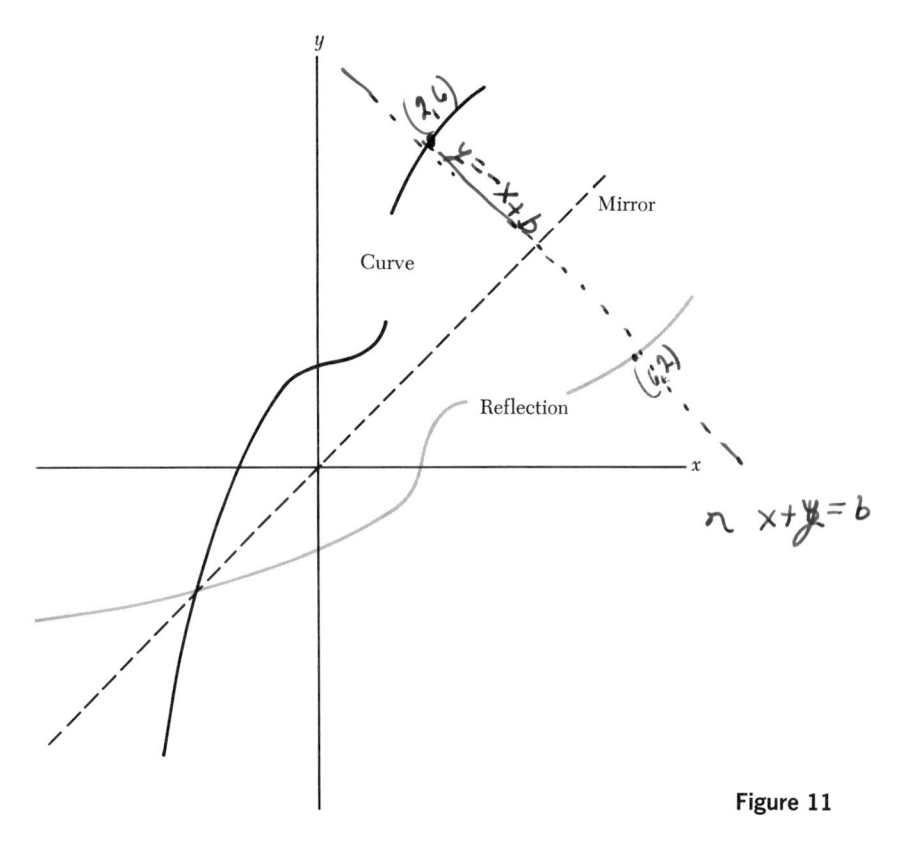

(2,6)

$y = -x + b$

Mirror

Curve

Reflection

(6,2)

$x + y = b$

Figure 11

444 Transcendental Functions

*for ordered pairs (x, y)
if $x_1 > x_2$ then $y_1 > y_2$*

different right components. Consider then any two ordered pairs in F. Since F is a function, these ordered pairs have different left components. Thus, one of the left components is greater than the other. Since F is increasing, the pair with the greater left component has the greater right component. Thus, the two ordered pairs have different right components. Since this applies to any two ordered pairs in F, the function F has an inverse.

We now establish the second assertion of the theorem. As before, let F be any increasing function. We now know that F has an inverse; denote the inverse by G. We claim that G is an increasing function. Let p and q be any two numbers in the domain of G such that $p < q$. Let their images under G be denoted by r and s, respectively; thus, $(p, r) \in G$ and $(q, s) \in G$. Since $p < q$, our claim will be established if we can show that $r < s$.

Since G is the inverse of F, the function F contains the ordered pairs (r, p) and (s, q) obtained from (p, r) and (q, s), respectively, by interchanging components. Three possibilities are conceivable: either $r = s$ or $r > s$ or $r < s$. We shall show that neither the first nor the second of these possibilities is compatible with our hypotheses. If $r = s$, then the hypothesis that F is a function guarantees that $p = F(r) = F(s) = q$, a contradiction to the condition $p < q$. If $r > s$, then the hypothesis that F is increasing guarantees that $p = F(r) > F(s) = q$, a contradiction to the condition $p < q$. The sole remaining alternative is that $r < s$. Since $G(p) < G(q)$ whenever $p < q$, the function G is increasing. The proof of the theorem is complete.

6 The logarithmic function

The exponential function is an increasing function. Therefore, by Theorem 1, the exponential function has an inverse. We have a special name for this inverse.

Definition The inverse of the exponential function e^x is called the *logarithmic function*.

The logarithmic function is sometimes denoted by the symbol log and sometimes by the symbol ln (which arises from the initial letters of the Latin phrase translated by "logarithm natural"). We shall regularly use "ln," since "log" often represents a different function than the inverse of exp.

By the second part of Theorem 1, ln is an increasing function.

From the definition of the inverse for a function, we observe that an ordered pair (t, u) belongs to the logarithmic function if and only if the ordered pair (u, t)—obtained by interchanging components—belongs to the exponential function. That is,

$$(t, u) \in \ln \quad \text{if and only if} \quad (u, t) \in \exp.$$

Expressed in different notation, this biconditional statement becomes

$$u = \ln t \quad \text{if and only if} \quad t = e^u.$$

by definition

In the preceding paragraph, t, being a power of e, is necessarily positive, while u is not restricted. The domain of the logarithmic function is the same as the range of the exponential function, for a left component of an ordered pair in ln is a right component of an ordered pair in exp. Thus, Dom ln $= \{x \mid x > 0\}$. By a similar discussion, Rg ln $=$ Dom exp hence the range of the logarithmic function contains all real numbers.

We repeat. We consider the logarithm of a number only in case the number is positive.

Since the equations $u = \ln t$ and $t = e^u$ are logically equivalent, we may combine them by eliminating t. Replacing t in the first equation by e^u, we obtain

$$\boxed{u = \ln e^u} \quad \text{for every real number } u.$$

The same pair of equivalent equations may be combined by replacing u in the second equation by $\ln t$; we obtain

$$\boxed{t = e^{\ln t}} \quad \text{for every positive number } t.$$

Since $(0, 1) \in \exp$, we conclude that $(1, 0) \in \ln$.

Since the logarithmic function and the exponential function are inverses of each other, their graphs are congruent to one another. The graph of one of them may be obtained from the graph of the other by a reflection in the line whose equation is $y = x$. (See Section 3 of Chapter 11 for a comparison of the graphs of mutually inverse functions.) The graph of exp, shown in Fig. 6, may be reflected to yield the graph of ln, shown in Fig. 12.

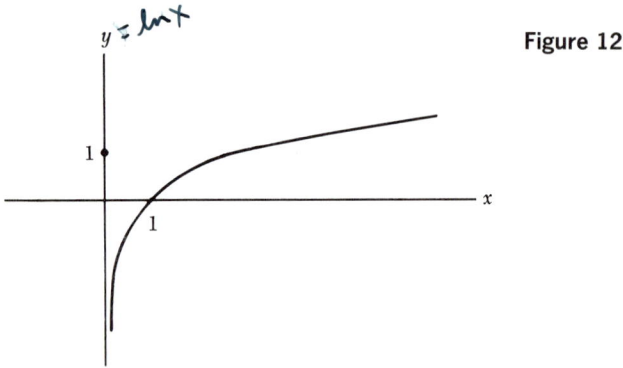

y = ln x

Figure 12

Observe how the graph of ln reveals some of the important properties of the function.

1. The graph lies in the right coordinate halfplane because the domain of ln consists of positive numbers. *Range is reals.*

446   Transcendental Functions

2. Every number is the y-coordinate of a point on the graph, since Rg ln contains all real numbers.
3. The only intersection of the curve and a coordinate axis is $(1, 0)$.
4. The graph is a rising curve because ln is an increasing function.
5. The image of a very large number is large.
6. The image of a very small (positive) number is a large negative number.

Furthermore, we note that if $0 < x < 1$, then $\ln x < 0$; while if $x > 1$, then $\ln x > 0$.

EXERCISES D

1. Simplify each of the following.
   (a) $e^{\ln 3}$.
   (b) $\ln \exp(1 + \sqrt{2})$.
   (c) $\exp \ln \frac{3}{4}$.
   (d) $e^2 e^{\ln 5}$.
   (e) $\exp(7 + \ln 8)$.
   (f) $\exp \ln 5 - \ln e^3$.

2. For each of the following, give the domain of the function given by the formula.
   (a) $\ln x$.
   (b) $\ln x - 3$.
   (c) $\ln(x - 3)$.
   (d) $\ln(-x)$.
   (e) $\ln x^2$.

3. If c is a number such that $e^c < 4$, show that $\ln e^c < \ln 4$.

4. If d is a number such that $\ln d < 3$, show that $e^{\ln d} < e^3$ and that $0 < d < e^3$.

5. Find the set of roots for each of the following statements.
   (a) $\ln x = 1$.
   (b) $e^x = \frac{3}{7}$.
   (c) $\ln x > 1$.
   (d) $e^x > \frac{3}{7}$.
   (e) $3 < \ln x < 7$.
   (f) $5 < e^x < 8$.
   (g) $-\frac{1}{2} \leq \ln x \leq \frac{1}{2}$.
   (h) $-\frac{1}{2} \leq e^x \leq \frac{1}{2}$.
   (i) $\ln x = e$.
   (j) $e^x = \ln 2$.
   (k) $e^x = \ln \frac{1}{2}$.

6. Let Q be the quadratic function $\{(x, y) \mid y = (1 + x)(2 - x)\}$.
   (a) What numbers are roots of the equation $Q(x) = 0$?
   (b) Sketch the graph of Q.
   (c) What numbers are roots of the inequality $Q(x) > 0$?
   (d) What is the domain of the function given by $\ln[(1 + x)(2 - x)]$?
   (e) What is the vertex of the parabola that is the graph of Q?
   (f) What is the maximum of the set Rg Q?
   (g) What is the maximum value of the function $\{(x, y) \mid y = \ln[Q(x)]\}$?

7. Each of the following formulas describes a function. In each case, find a formula for the inverse of the given function.
   (a) $e^{x/2}$.
   (b) $-e^x$.
   (c) $\ln(x/2)$.
   (d) $\ln(-x)$.

8. Let F be the function given by the formula $\ln(-x)$.
   (a) What is the domain of F?
   (b) What is the range of F?
   (c) Show that the inverse of F (see Exercise 7d) is a decreasing function.
   (d) Show that F is a decreasing function.
   (e) Sketch the graph of the inverse of F.
   (f) Sketch the graph of F.

9. Prove each of the following propositions.
   (a) Every decreasing function has an inverse.
   (b) The inverse of every decreasing function is also a decreasing function.

7 Applications

The exponential and logarithmic functions are inverses of each other. Consequently, to each property of one of them there corresponds a related property of the other. Related to the following three properties of the exponential function,

$$\exp(u + v) = (\exp u)(\exp v),$$
$$(\exp u)^v = \exp(uv),$$
$$\exp(u - v) = \frac{\exp u}{\exp v},$$

are, respectively, the following important properties of the logarithmic function:

$$\ln(ab) = \ln a + \ln b \qquad \text{if } a > 0 \text{ and } b > 0,$$
$$\ln a^q = q \ln a \qquad \text{if } a > 0,$$
$$\ln \frac{a}{b} = \ln a - \ln b \qquad \text{if } a > 0 \text{ and } b > 0.$$

We may establish these results if we keep in mind that $t = e^{\ln t}$ for every positive t and $u = \ln e^u$ for every number u. Suppose that a and b are positive numbers. Then

proof.

$$\ln(a \cdot b) = \ln(e^{\ln a} \cdot e^{\ln b}) = \ln(e^{\ln a + \ln b}) = \ln a + \ln b.$$

Also

proof.

$$\ln(a^q) = \ln[(e^{\ln a})^q] = \ln[e^{q \ln a}] = q \ln a.$$

Finally,

$$\ln \frac{a}{b} = \ln(ab^{-1}) = \ln a + \ln b^{-1} = \ln a + (-1) \ln b = \ln a - \ln b.$$

Example 8 If we are given the information that $(\ln 3, \ln 7) = (1.099, 1.946)$, approximately, then we may calculate decimal approximations for $\ln 21$ and $\ln \frac{7}{9}$ as follows:

$$\ln 21 = \ln(3 \cdot 7) = \ln 3 + \ln 7 = 1.099 + 1.946 = 3.045;$$
$$\ln \frac{7}{9} = \ln 7 - \ln 9 = \ln 7 - \ln 3^2 = \ln 7 - 2 \ln 3$$
$$= 1.946 - 2(1.099) = -0.252.$$

As a partial check, we observe that since $\frac{7}{9}$ is less than 1, its logarithm is negative.

Example 9 Find the inverse of the function F given by the formula $F(x) = 5e^{-x/4}$.

448 Transcendental Functions

We note that F is a decreasing function and therefore has an inverse; suppose we call the inverse H. An ordered pair (c, d) belongs to H if and only if $(d, c) \in F$, that is, if and only if $c = 5e^{-d/4}$. We solve this equation for d in terms of c by the following steps:

let $F(x) = c$, $x = d$

$$\frac{c}{5} = e^{-d/4},$$

$$\ln \frac{c}{5} = -\frac{d}{4},$$

$$d = -4 \ln \frac{c}{5}.$$

inverse $G(x) = d$, $x = c$

The key step is the second where we applied the principle that a statement of the form $t = e^u$ is logically equivalent to the statement that $\ln t = u$. We have found that $(c, d) \in H$ if and only if $d = -4 \ln (c/5)$. Thus, a formula for the inverse of F is given by

$$H(x) = -4 \ln \frac{x}{5}.$$

Example 10 Two functions are given by the respective formulas $1 + \ln (3x - 3)$ and $\ln (2x + 5)$. Show that the graphs of these two functions are congruent to one another.

We express each of the given functions by a different formula. Now

$$1 + \ln (3x - 3) = 1 + \ln 3(x - 1) = 1 + \ln 3 + \ln (x - 1);$$

the graph of

$$y = \ln (x - 1) + (1 + \ln 3)$$

is congruent to the graph of the logarithmic function and is obtainable from the graph of $\ln$ by translating one unit to the right and $(1 + \ln 3)$ units up. Next,

$$\ln (2x + 5) = \ln [2 \cdot (x + \tfrac{5}{2})] = \ln (x + \tfrac{5}{2}) + \ln 2;$$

thus, the graph of

$$y = \ln (2x + 5)$$

is also congruent to the graph of the logarithmic function and is obtainable from the latter graph by a translation of $\tfrac{5}{2}$ units to the left and $\ln 2$ units up. Since the graphs of the given functions are each congruent to the graph of $\ln$, they are congruent to one another.

Example 11 Sketch the graph of the equation $y = \ln (2 - x)$. A number having a logarithm is positive. Consequently, we are

7 / Applications 449

interested only in values of x such that $2 - x > 0$. The graph lies in the halfplane $\{(x, y) \mid x < 2\}$. The graph of

$$y = \ln(2 - x) = \ln[-(x - 2)]$$

is congruent to, and two units to the right of, the graph of $y = \ln(-x)$. The latter graph is the reflection, in the y-coordinate axis, of the graph of $y = \ln x$. Thus, we may begin with the known graph of ln, reflect it, and then translate it in order to obtain the desired graph. See Fig. 13.

Figure 13

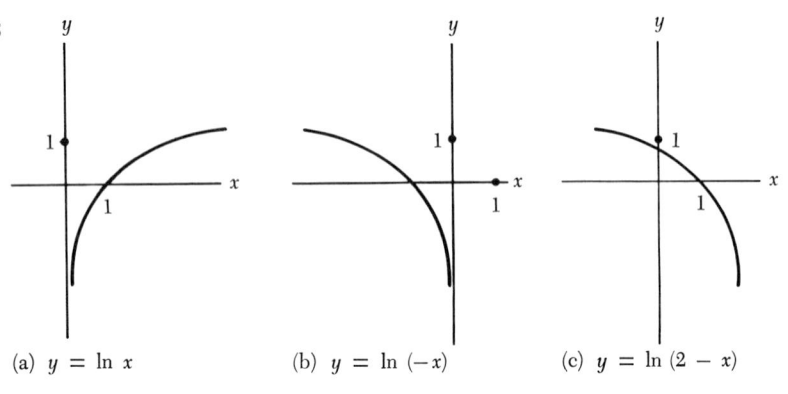

(a) $y = \ln x$ (b) $y = \ln(-x)$ (c) $y = \ln(2 - x)$

We observe that the graph of the equation $y = \ln(2 - x)$ intersects the x-coordinate axis at $(1, 0)$ and the y-coordinate axis at $(0, \ln 2)$. We also observe that the function given by the formula $\ln(2 - x)$ is a decreasing function: indeed, if $a < c < 2$, then $2 - a > 2 - c > 0$ and $\ln(2 - a) > \ln(2 - c)$.

Example 12 Find each root of the equation $e^{7x-1} = 4$.

The given equation is equivalent to $7x - 1 = \ln 4$ and therefore to $7x = 1 + \ln 4$. The equation has one root, namely, $(1 + \ln 4)/7$.

Example 13 Find each root of the equation $\ln(5 + 2x) = 3 + \ln 2$.

As in the preceding example, we exploit the definition of a logarithm, namely, that the statement $\ln t = u$ is logically equivalent to $t = e^u$. The equation given in this problem is equivalent to $5 + 2x = e^{3 + \ln 2}$. The latter equation may be solved by the following sequence of steps:

$$5 + 2x = e^{3 + \ln 2} = e^{\ln 2} \cdot e^3 = 2e^3,$$
$$2x = 2e^3 - 5,$$
$$x = e^3 - \tfrac{5}{2}.$$

Thus, the given equation has one root, namely, $e^3 - \tfrac{5}{2}$.

450 Transcendental Functions

Example 14 Solve the inequality $\exp(-x + 4) > \frac{1}{3}$.
Since ln is an increasing function,
$$\ln[\exp(-x + 4)] > \ln \tfrac{1}{3}.$$
In the latter inequality, the left member is
$$\ln[\exp(-x + 4)] = -x + 4$$
because ln and exp are inverses of each other, while the right member is
$$\ln \tfrac{1}{3} = \ln 3^{-1} = -\ln 3.$$
Thus, the inequality becomes $-x + 4 > -\ln 3$, which is equivalent to $4 + \ln 3 > x$. The set of roots of the original inequality is the set of numbers x such that
$$x < 4 + \ln 3.$$

Example 15 Let the sequence g defined on a set J of integers be a geometric sequence. (Recall Section 8 of Chapter 7.) There are nonzero constants a and r such that the sequence g is given by the formula
$$g(k) = ar^k \qquad \text{for each } k \in J.$$
Suppose that each of the numbers a and r is positive. We may form a new function f defined by
$$f(k) = \ln g(k) \qquad \text{for each } k \in J.$$
Utilizing the properties of the logarithmic function and the hypotheses that $a > 0$ and $r > 0$, we obtain
$$f(k) = \ln(ar^k) = \ln r^k + \ln a = k \ln r + \ln a.$$
Since $\ln r$ and $\ln a$ are constants, the condition that
$$f(k) = (\ln r)k + (\ln a) \qquad \text{for each } k \in J$$
shows that f is an arithmetic sequence with domain J.

Example 16 Suppose that in a physical experiment a scientist is studying the relationship between two variables x and y. The experimental results yield the data in Table 3. To gain further insight into the relationship, the scientist plots the corresponding points on a

Table 3

x	3.0	3.5	4.0	4.5	5.0	5.5	6.0
y	5.7	6.9	8.2	10.3	12.7	15.4	18.7

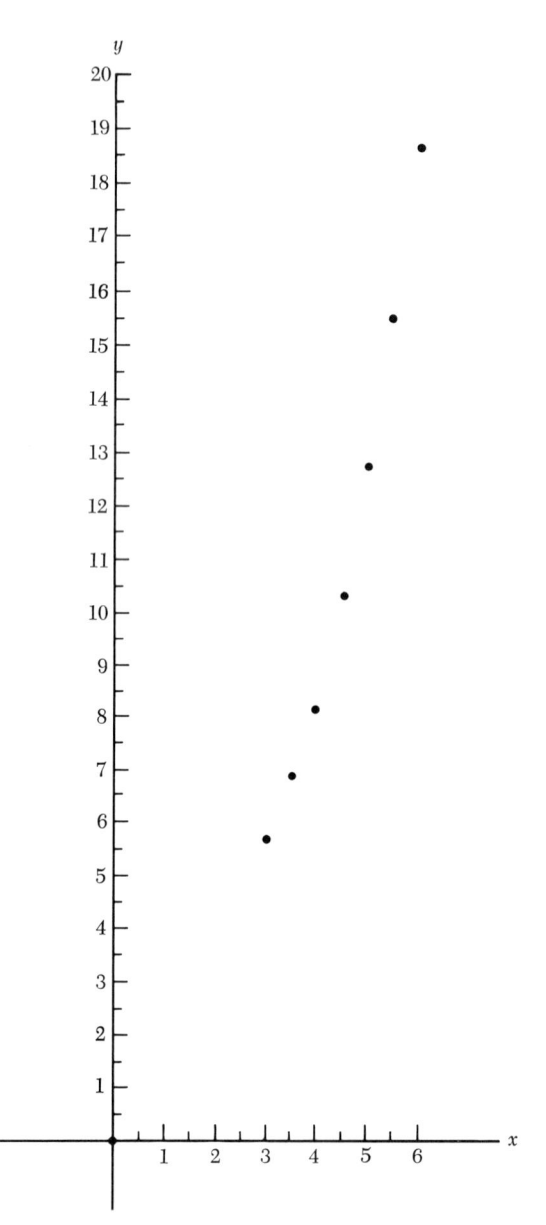

Figure 14

coordinate system. By inspection of Fig. 14, or possibly by a careful examination of the data, or perhaps on the basis of relevant theory pertaining to the experimental situation, the researcher conjectures that the dependency of y on x is a function of exponential type. That is, he guesses that there are constants h and k such that $y = he^{kx}$.

One approach toward finding such constants is to introduce the new variable w given by $w = \ln y$. Then

$$w = \ln(he^{kx}) = \ln e^{kx} + \ln h = kx + \ln h.$$

Since k and $\ln h$ are constants, we note that w is a linear function of x. In other words, if we plot ordered pairs $(x, w) = (x, \ln y)$, the points should be on a line.

We find the logarithms of the observed values of y and construct Table 4. Using the seven ordered pairs (x, w) of experimental data,

Table 4

x	3.0	3.5	4.0	4.5	5.0	5.5	6.0
$w = \ln y$	1.74	1.93	2.10	2.33	2.54	2.73	2.93

we draw Fig. 15a. The points appear to be approximately collinear. Because of limitations on physical equipment and measurement technique, the scientist would not anticipate his data to fit perfectly on a line. There are advanced methods for choosing a best line to fit points which are approximately collinear. The seven points and a good fitting line are shown in Fig. 15b. This line has slope 0.4, and an equation for the line is

$$w = 0.4x + 0.53.$$

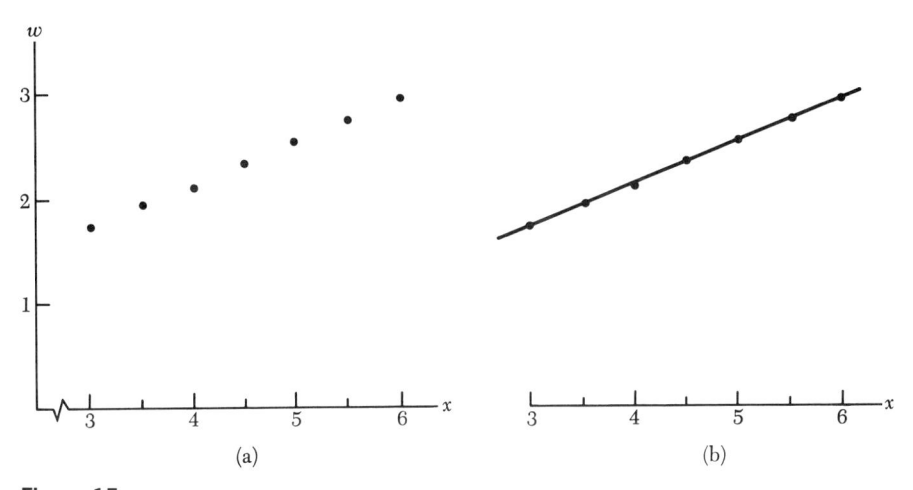

Figure 15

Thus, $y = \exp w = \exp(0.4x + 0.53) = (\exp 0.53)(\exp 0.4x)$. Since $\exp 0.53$ is approximately 1.7, a formula that is compatible with the given experimental data is

$$y = 1.7e^{0.4x}.$$

The computational and graphical labor in practical problems such as Example 16 and Exercises 22 and 23 below may be reduced with the aid of specially designed graph papers.

7 / Applications

Table 5

x	1.5	2	2.5	7	0.8	0.6	0.2
ln x	0.41	0.69	0.92	1.95	−0.22	−0.51	−1.61

Table 5 gives the logarithm, correct to the nearest hundredth, for several selected numbers. A more extensive list of values appears in the Appendix.

EXERCISES E

1. Use the values in Table 5 to compute an approximation to each of the following.
   (a) ln 14. (b) ln 49. (c) ln $\frac{7}{2}$. (d) ln 0.48.
   (e) ln $\frac{2}{5}$. (f) ln $\frac{5}{3}$. (g) ln 5. (h) ln 16.
   (i) ln 3. (j) ln $\sqrt{3}$. (k) ln 10.

2. Use Table 2 (in Section 3) and/or Table 5 to find each of the following, approximately.
   (a) ln 20.1. (b) ln 0.368. (c) exp 0.92.
   (d) $e^{1.95}$. (e) $e^{0.22}$. (f) exp (−0.41).

3. If ln a and ln b are known, how may you find ln $(1/ab)$?

4. If ln c is known, give ln $(1/\sqrt{c})$.

5. If ln $a = b$, find a simpler expression for $e^{b/2}$.

6. Find a different formula for the function given by ln $(1/x)$.

7. If exp $c = d$, find a simpler expression for ln $d^{1/3}$.

8. Each of the following formulas describes a function. In each case, find a formula for the inverse of the given function.
   (a) $7e^{-4x}$. (b) exp $(3x - 5)$. (c) $-e^{2x/3}$. (d) $2/e^x$.
   (e) ln $(4 + 3x)$. (f) $\frac{1}{2}$ ln $(x - 8)$. (g) $-\frac{3}{4}$ ln $(5 - 2x)$. (h) $-\ln \frac{1}{1+x}$.

9. If t is a positive number, find a simpler expression for exp $(-\ln t)$.

10. (a) Show that, if x is positive, then 2 ln x = ln x^2.
    (b) Explain why the two functions given by the respective formulas 2 ln x and ln x^2 are not the same.
    (c) Explain why the function given by ln x^2 is the same as the function given by the formula 2 ln $|x|$.

11. What is the set of roots for the equation 3 ln x = ln x^3?

12. Let a, b, c be positive numbers. Solve each of the following equations.
    (a) ln x = ln a − ln b + 2 ln c. (b) ln x = 5 ln a + $\frac{1}{2}$ ln b.
    (c) ln x = ln $(a + b)$ + 2c. (d) ln $\frac{1}{x}$ = ln $\frac{2}{a}$ − 3 ln c.

13. Find the root of each of the following equations.
    (a) $e^x = 10e^7$. (b) ln x = 5 − ln 4.
    (c) $e^{2x} = 3e^4$. (d) ln $(1 - x) = 6 + $ ln $\frac{1}{2}$.

14. Sketch the graph of each of the following equations.
    (a) $y = -\ln x$. (b) $y = \ln 2x$. (c) $y = 0.7 + \ln x$. (d) $y = \ln \frac{1}{x}$.
    (e) $y = \ln \frac{x}{3}$. (f) $y = \frac{\ln x}{\ln 3}$. (g) $y = \ln x - 1.1$. (h) $y = \ln (x - 3)$.

454 Transcendental Functions

15. Find the domain of each of the following functions.
    (a) $\{(x, y) | y = \ln(x + 3)^2\}$.
    (b) $\{(x, y) | y = \ln(2x - 3)^2\}$.
    (c) $\{(x, y) | y = \ln[(2x - 3)(3x - 4)]\}$.
    (d) $\{(x, y) | y = \ln(x^2 - x)\}$.
    (e) $\{(x, y) | y = \ln x + \ln(x - 1)\}$.

16. Solve each of the following inequalities.
    (a) $e^x > e^{-1/2}$.
    (b) $e^{-x} > \exp \frac{2}{5}$.
    (c) $2e^x < 3$.
    (d) $\ln(x + 1) \geq \ln 6$.
    (e) $\ln(1 - x) + 1 \geq 0$.
    (f) $1 + \exp\left(-\frac{2x}{3} + \frac{1}{6}\right) > 0$.
    (g) $\ln x^2 \geq 4$.
    (h) $\frac{1}{2e^{x/3}} \leq 6e^3$.
    (i) $\ln(3x + 2) > 5$.
    (j) $\ln 2x < -\frac{1}{4}$.

17. (a) On a single coordinate system, draw carefully the graph of exp and the graph of the function $\{(x, y) | y = 1 + x\}$.
    (b) Using your diagram, solve the inequality $e^x > 1 + x$.

18. (a) Using Exercise 17, solve the inequality $\ln(1 + x) < x$.
    (b) On a single coordinate system, draw carefully the graphs of the two functions given by the respective formulas $\ln(1 + x)$ and x.
    (c) Explain how your drawing in part (b) confirms your answer in part (a).

19. Let f be an arithmetic sequence defined on a set J of integers. Show that the function g, given by
    $$g(n) = \exp[f(n)] \quad \text{for each } n \in J,$$
    is a geometric sequence.

20. If a function y maps each number x in its domain onto a positive number, if $w = \ln y$, and if w is a linear function of x, show that y is a function of exponential type.

21. Suppose that, for the function y of x, every value of x is positive and every value of y is positive. Let $w = \ln y$, and let $z = \ln x$. Suppose that w is a linear function of z. With the given information, express y as a function of x.

22. The following table gives experimentally measured data concerning ordered pairs (x, y).

x	2.0	2.2	2.4	2.7	3.0	3.5
y	12.5	14.5	17.0	20.5	24.5	32.0

    If $z = \ln x$ and $w = \ln y$, then the corresponding ordered pairs (z, w) are given in the following table.

z	0.69	0.79	0.88	0.99	1.10	1.25
w	2.53	2.67	2.83	3.02	3.20	3.47

    (a) On a large diagram, plot (with great care) the six experimental points (z, w).
    (b) By laying a straightedge on your figure, observe that the six points are approximately collinear, but that the line through the first two points does not fit the data satisfactorily.
    (c) Show that the linear equation $w = 1.7z + 1.33$ fits the data reasonably well.
    (d) Use the approximate relationship in part (c) to express y as a function of x. (For your convenience, exp 1.33 is approximately 3.8.)

7 / Applications   455

23. Experimental data concerning ordered pairs (x, y) are presented in the accompanying table. Corresponding values of $z = \ln x$ and $w = \ln y$ are also tabulated.
    (a) On a coordinate system, plot (with great care) the seven observed points (x, y).
    (b) On a large diagram plot (with great care) the seven points (z, w).
    (c) On another large diagram plot (with great care) the seven points (x, w).
    (d) In which of the three diagrams does it appear that (subject to reasonable tolerance for experimental error in observed data) the plotted points are nearly collinear?
    (e) Find an equation expressing y as a function of x that reasonably fits the observed data.

x	10	12	15	18	22	30	40
y	40	39	38	36	35	31	28
z	2.30	2.48	2.71	2.89	3.09	3.40	3.69
w	3.69	3.66	3.64	3.58	3.56	3.43	3.33

8 Normal probability distribution

In our study of random variables in Chapter 8, we restricted attention to the case of an experiment having only finitely many possible outcomes. Consequently, a random variable in this situation has a limited number of possible values, because a random variable is essentially a function that assigns a number to every possible outcome. Very often a random variable may have a great many different possible values—hundreds or thousands of them—and calculations of probabilities associated with such random variables may be quite laborious. We may prefer to approximate such a probability differently.

We may extend the notion of a random variable to allow the set of possible values for the random variable to be the set of all real numbers. A full discussion of the extension and its associated properties involves the methods of the calculus. Without undertaking in this book the task of establishing all the calculus background, we nevertheless wish to gain at least a partial understanding of how a random variable in the extended sense may assist in estimating probabilities involving familiar random variables.

We devote this and the next sections to introducing the *standard normal variable*, a random variable that has paramount importance in probability and statistics. The normal distribution has tremendously broad applications; so widespread is its use that some persons with a little statistical training have acquired the erroneous belief that the normal distribution handles all problems and thereby abuse it by attempting to exploit it in situations where its use is not justified.

The description of the standard normal variable begins with a function whose formula involves two important irrational numbers, the number e which we have been studying as the base of the exponential function and the number π which is first met as the ratio of the circumference of a circle to its own diameter. We recall from geometry that π is somewhat larger than 3, being 3.14 to the nearest hundredth.

Definition The *normal probability density function* is the function given by the formula $\dfrac{1}{\sqrt{2\pi}} e^{-x^2/2}$.

We often denote this function by ϕ (the lower case Greek letter "phi"). Thus,

$$\phi(x) = (2\pi)^{-1/2} \exp\left(-\frac{x^2}{2}\right) \quad \text{for every real number } x.$$

The domain of ϕ consists of all real numbers. For any number c, we have $-c^2/2 \leq 0$; consequently, $0 < \exp(-c^2/2) \leq 1$, and therefore $0 < \phi(c) \leq (2\pi)^{-1/2}$. That is, the range of ϕ consists of positive numbers that do not exceed $1/\sqrt{2\pi}$. Now a decimal approximation for the number $1/\sqrt{2\pi}$ is 0.399; thus, the image of every number under ϕ is between zero and 0.4.

The function ϕ is an even function, since the only occurrence of x in the formula is in the expression x^2. The function ϕ is a decreasing function on the set $\{x \mid x \geq 0\}$. Indeed, let a and c be numbers such that $0 \leq a < c$; then $a^2/2 < c^2/2$, hence $-a^2/2 > -c^2/2$; since exp is an increasing function, $\exp(-a^2/2) > \exp(-c^2/2)$, and finally $\phi(a) > \phi(c)$. An even function that is a decreasing function on the set of positive numbers is of course an increasing function on the set of negative numbers.

The remarks of the two preceding paragraphs have geometrical significance and will aid us in sketching the graph of the normal density function. The graph lies in a rather narrow horizontal strip bounded below by the x-coordinate axis and having width $(2\pi)^{-1/2}$. The graph touches the upper edge of the strip at the point $(0, [2\pi]^{-1/2})$ but does not meet the lower boundary. The point $(0, [2\pi]^{-1/2})$ is therefore the highest point on the curve, as the graph is a rising curve on the left and is a falling curve on the right of this maximum point. The portion of the curve on one side of the y-coordinate axis is the reflection, in the axis, of the portion on the other side. (See Fig. 16.)

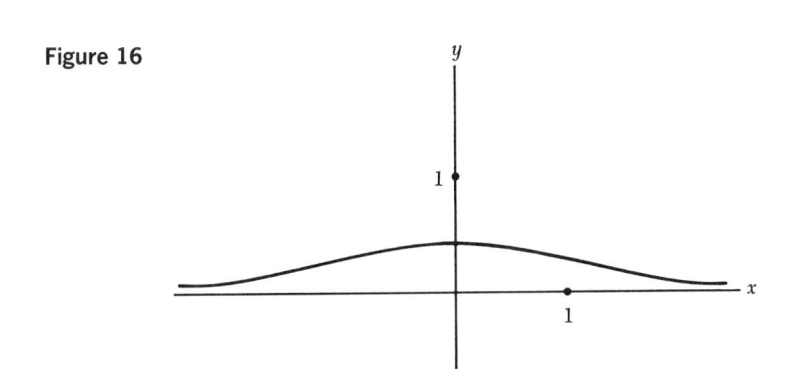

Figure 16

Figure 17

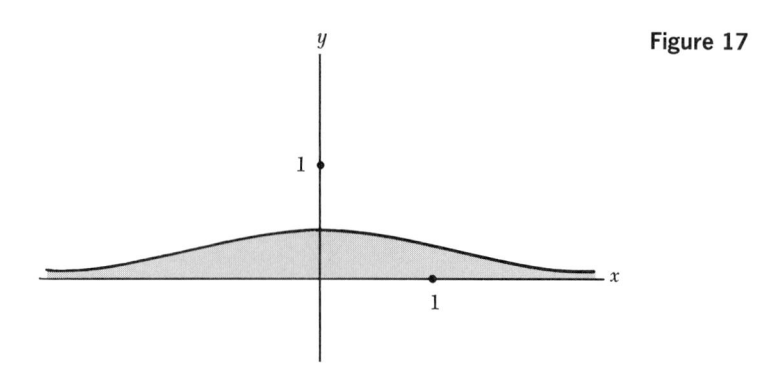

One further property of the function ϕ has particular importance in probability. This property we state and will use, although the calculus is required in order to show that it is meaningful and applicable. The region in the coordinate plane between the x-coordinate axis and the graph of ϕ has exactly one unit of area. This region may be expressed as the graph of the set $\{(x, y) \mid 0 \leq y \leq (2\pi)^{-1/2} \exp(-x^2/2)\}$. It is the region which is shaded in Fig. 17. The property involving area explains why the rather unusual numerical factor $1/\sqrt{2\pi}$ appears in the definition of the normal density function ϕ. If this factor were omitted from the formula for simplicity, or if it were replaced by some other number, then the area of the region above the axis and below the graph of the modified function would not be one. We need this number one for probability reasons.

Table 6 gives several selected images under the function ϕ, each expressed approximately to the nearest thousandth.

Table 6

x	0	$\frac{1}{2}$	1	$\frac{3}{2}$	2	$\frac{5}{2}$	3	$\frac{7}{2}$
$\phi(x)$	0.399	0.352	0.242	0.130	0.054	0.018	0.004	0.001

We use the function ϕ in order to describe the normal probability distribution.

Definition The *normal probability distribution function* is the rule assigning to each real number c an image that is the area of the graph of the set

$$\{(x, y) \mid [x \leq c] \wedge [0 \leq y \leq \phi(x)]\}.$$

Let us denote the normal distribution function by F. Then $F(1)$ is the area of the shaded region in Fig. 18a; $F(\frac{1}{2})$ is the area of the shaded region in Fig. 18b; $F(-\frac{1}{2})$ is the area of the region shown in Fig. 18c. In each of these diagrams the upper boundary of the region is a portion of the graph of the equation $y = \phi(x)$.

Figure 18

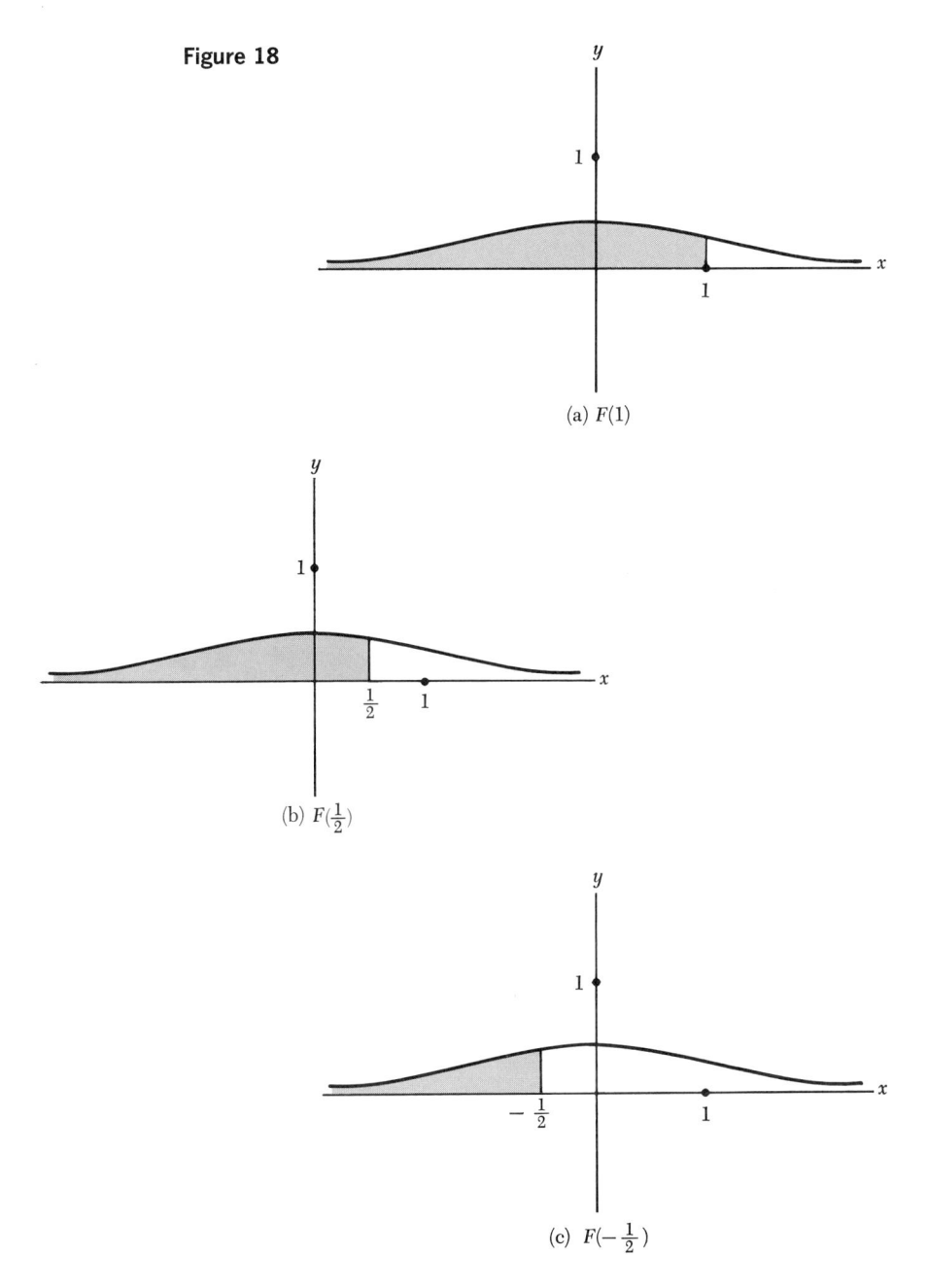

(a) $F(1)$

(b) $F(\frac{1}{2})$

(c) $F(-\frac{1}{2})$

In general, for each c, the number $F(c)$ tells the number of units of area in the region of the coordinate plane that is bounded below by a halfline of the x-coordinate axis, bounded above by a portion of the graph of ϕ, and bounded on the right by a segment of the line whose equation is $x = c$.

Every real number has an image under F; and the image is positive, since it is the area of a region.

8 / Normal probability distribution

An inspection of Fig. 18 reveals that $F(\frac{1}{2}) < F(1)$, because the shaded region in Fig. 18b is a subset of the region represented in Fig. 18a. Likewise $F(-\frac{1}{2}) < F(\frac{1}{2})$. These remarks suggest that F is an increasing function: the greater the number c, the farther toward the right the right-hand edge of the region is, and consequently the greater the area of the region is.

On the other hand, every region involved in the definition of F is a subset of the set $\{(x, y) \mid 0 \leq y \leq \phi(x)\}$ portrayed in Fig. 17. Since the area of the graph of the latter set is one, every image under F is less than one. Combining results, we conclude that $\operatorname{Rg} F = \{t \mid 0 < t < 1\}$.

Since the graph of ϕ is symmetric with respect to the y-coordinate axis, the two subregions in Fig. 17 on either side of the axis are congruent and have the same area. The area of the subregion in the left coordinate half-plane, namely $\frac{1}{2}$, is (by definition) $F(0)$.

Now consider the images of two numbers, say c and $-c$, which are negatives of each other. Their respective images are the areas of the shaded regions B and A in Figs. 19b and 19a. Since ϕ is an even function, its graph is symmetric about the y-coordinate axis. This symmetry implies that the region D denoted by the darker tint in Fig. 19b is congruent to region A in Fig. 19a, since each region is the reflection of the other in the y-coordinate axis. Region D therefore has the same area as region A, namely $F(-c)$. The area of region B is $F(c)$. Since the area of the union of the nonoverlapping regions B and D is one, we conclude that $F(c) + F(-c) = 1$.

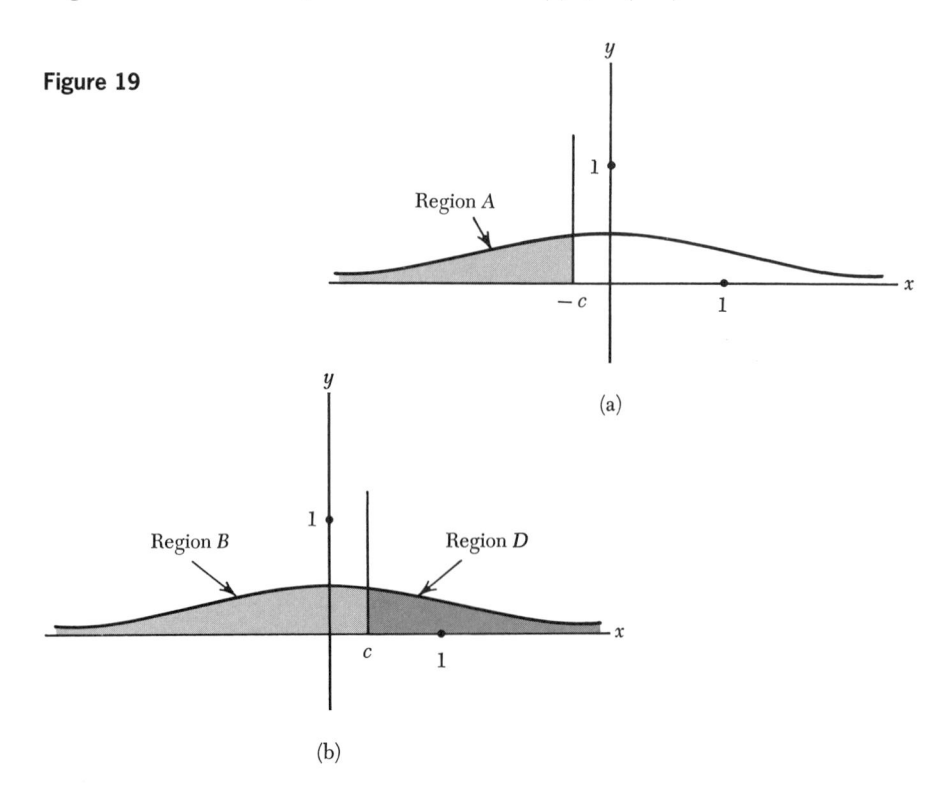

Figure 19

(a)

(b)

460 Transcendental Functions

Figure 20

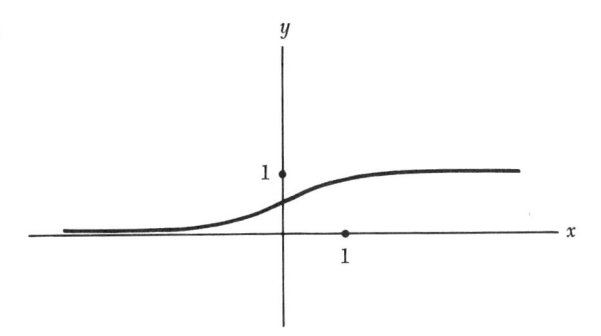

We summarize the properties of the normal distribution function F:

1. Its domain consists of all real numbers.
2. Its range consists of numbers between 0 and 1.
3. It is an increasing function.
4. $F(0) = \frac{1}{2}$.
5. Every number c satisfies the condition $F(c) + F(-c) = 1$.

Geometrically the graph of F is a gradually rising curve each of whose points has a y-coordinate between 0 and 1. See Fig. 20. Table 7 gives approximations for selected ordered pairs belonging to F. Three-decimal-place entries are approximations to the nearest thousandth.

Table 7

c	$\frac{1}{4}$	$\frac{1}{2}$	0.674	$\frac{3}{4}$	1	1.282	$\frac{3}{2}$	1.645	1.960	2	$\frac{5}{2}$	3
$F(c)$	0.599	0.691	$\frac{3}{4}$	0.773	0.841	$\frac{9}{10}$	0.933	$\frac{19}{20}$	$\frac{39}{40}$	0.977	0.994	0.999

c	1.10	1.11	1.12	1.13	1.14	1.15	1.16	1.17	1.18	1.19	1.20
$F(c)$	0.864	0.867	0.869	0.871	0.873	0.875	0.877	0.879	0.881	0.883	0.885

Example 17 Since $F(-2) = 1 - F(2)$, we find from Table 7 that $F(-2) = 1 - 0.977 = 0.023$, approximately.

EXERCISES F

[NOTE: In these exercises, let ϕ and F be the normal probability density and distribution functions, respectively.]

1. Evaluate, approximately, each of the following.
   (a) $\phi(-1)$.   (b) $\phi(-2)$.

2. Using an appropriate entry from Table 2 (in Section 3), verify the approximate value of $\phi(2)$ given in Table 6 in this section.

3. Explain how the data concerning ϕ reveal that $0.018 < \phi(2.1) < 0.054$.

4. Estimate $\phi(\frac{1}{4})$ in decimal form.

5. Estimate $\phi(4)$.

6. Evaluate, approximately, each of the following.
   (a) $F(0.75)$.
   (b) $F(1.15) - F(0.25)$.
   (c) $1 - F(\frac{1}{2})$.
   (d) $F(-1)$.
   (e) $F(-\frac{1}{4}) - F(-\frac{3}{4})$.
   (f) $F(1.10) - F(-1.10)$.
   (g) $F(1.13) - F(0)$.
   (h) $F(0) - F(-\frac{3}{2})$.

7. Find the root of each of the following equations.
   (a) $F(x) = 0.933$.
   (b) $F(x) = 0.067$.
   (c) $F(x) - F(0) = 0.4$.
   (d) $1 - F(x) = 0.227$.
   (e) $1 - F(x) = 0.691$.
   (f) $F(0) - F(x) = \frac{1}{4}$.
   (g) $F(x) - F(-x) = 0.866$.
   (h) $F(x) - F(-x) = \frac{1}{2}$.

8. Show that, $F(c) - F(-c) = 2F(c) - 1$ for every number c.

9. Discuss the properties of the function $\{(x, y) \mid y = 1 - F(x)\}$, and then sketch its graph.

10. Find approximately the area of the region $\{(x, y) \mid [-1 \leq x \leq 2] \wedge [0 \leq y \leq \phi(x)]\}$.

9 Standard normal variable

In the extended sense, a random variable may have any real number as a value. We are primarily concerned with a random variable that has a normal probability distribution.

Definition A random variable T is called a *standard normal variable* provided that, for every real number c, the probability of the statement $T \leq c$ and the probability of the statement $T < c$ are each equal to the number $F(c)$, where F is the normal probability distribution function.

A standard normal variable is sometimes said to have a standard normal probability distribution.

In symbols, T is a standard normal variable if and only if
$$\Pr[T \leq c] = F(c) = \Pr[T < c] \quad \text{for every } c.$$

Example 18 If T is a standard normal variable, compute each of the following probabilities approximately: $\Pr[T \leq 1.20]$, $\Pr[T < -\frac{1}{2}]$, $\Pr[T \geq 1]$, $\Pr[\frac{1}{4} \leq T \leq \frac{3}{4}]$.

Using Table 7 for the function F, we find that $\Pr[T \leq 1.20] = F(1.20) = 0.885$.

We use the table along with the property $F(c) + F(-c) = 1$ to obtain $\Pr[T < -\frac{1}{2}] = F(-\frac{1}{2}) = 1 - F(\frac{1}{2}) = 1 - 0.691 = 0.309$.

The statement $T \geq 1$ is the negation of the statement $T < 1$. We apply the principle that $\Pr[p] = 1 - \Pr[\sim p]$ to obtain $\Pr[T \geq 1] = 1 - \Pr[T < 1] = 1 - 0.841 = 0.159$.

Finally, the statement $\frac{1}{4} \leq T \leq \frac{3}{4}$ and the statement $T < \frac{1}{4}$ form a logically inconsistent pair of statements whose disjunction is equivalent to $T \leq \frac{3}{4}$. Hence, the probability of the latter statement is the sum of the probabilities of the two former statements; that is,

$$\Pr[T \leq \tfrac{3}{4}] = \Pr[\tfrac{1}{4} \leq T \leq \tfrac{3}{4}] + \Pr[T < \tfrac{1}{4}].$$

Solving for the desired number, we find

$$\Pr[\tfrac{1}{4} \leq T \leq \tfrac{3}{4}] = \Pr[T \leq \tfrac{3}{4}] - \Pr[T < \tfrac{1}{4}] = 0.773 - 0.599 = 0.174.$$

The preceding examples have illustrated two relationships which we now develop.

Theorem 2 Let T be a standard normal variable. If c is a real number, then $\Pr[T > c] = 1 - F(c)$.

PROOF Indeed, the statement $T > c$ is the negation of the statement $T \leq c$. Hence,

$$\Pr[T > c] = 1 - \Pr[T \leq c] = 1 - F(c).$$

Theorem 3 Let T be a standard normal variable. If c and d are real numbers such that $c < d$, then $\Pr[c \leq T \leq d] = F(d) - F(c)$.

PROOF The statement $T \leq d$ is logically equivalent to the disjunction of the two statements $T < c$ and $c \leq T \leq d$ which form a logically inconsistent pair. Therefore,

$$\Pr[T \leq d] = \Pr[T < c] + \Pr[c \leq T \leq d].$$

Rephrased, this equation becomes

$$\Pr[c \leq T \leq d] = \Pr[T \leq d] - \Pr[T < c] = F(d) - F(c).$$

EXERCISES G

1. Let T be a standard normal variable. Find, approximately, each of the following probabilities.
   (a) $\Pr[T \leq 0.5]$.
   (b) $\Pr[0 \leq T < \tfrac{1}{2}]$.
   (c) $\Pr[\tfrac{1}{4} \leq T \leq 2]$.
   (d) $\Pr[T > 1.5]$.
   (e) $\Pr[T \leq -\tfrac{1}{2}]$.
   (f) $\Pr[T > -1]$.
   (g) $\Pr[-1 < T < 2]$.
   (h) $\Pr[T < 0.5 \mid T < 1.5]$.
   (i) $\Pr[-\tfrac{1}{4} \leq T \leq \tfrac{1}{4} \mid -1 \leq T \leq 1]$.
   (j) $\Pr[|T| \leq 1]$.
   (k) $\Pr[|T| > 2]$.

2. Solve each of the following equations for the unknown number u.
   (a) $\Pr[T \leq u] = 0.9$.
   (b) $\Pr[0 \leq T \leq u] = \tfrac{1}{4}$.
   (c) $\Pr[T > u] = 0.95$.
   (d) $\Pr[|T| < u] = \tfrac{1}{2}$.

10 Binomial and normal distributions

One of the many uses of the normal probability distribution is to approximate probabilities that arise from binomial distributions. Consider a Bernoulli experiment consisting of a sequence of independent trials of some activity. Suppose, as usual, that we are interested in a certain statement pertaining to the outcome of a trial, so that each trial may be deemed to yield either a "success" or a "failure" according as its outcome does or does not belong to the truth set of the statement. Let S be the number of successes in the experiment. As we have learned in Chapter 8, the random variable S has a binomial distribution that depends upon two parameters, the number n of trials and the probability p of success on each trial. In fact, for each $k \in \{0, 1, \ldots, n\}$,

$$\Pr[S = k] = \binom{n}{k}(1-p)^{n-k}p^k.$$

If a and b are two integers such that $0 \leq a < b \leq n$, then

$$\Pr[a \leq S \leq b] = \sum_{k=a}^{b} \binom{n}{k}(1-p)^{n-k}p^k.$$

For a small value of n, the calculation of a probability by one of these formulas is a reasonable task. But for moderate or large values of n, the binomial coefficients and high exponents involve considerable computational labor. The difficulty may be removed if we have access to adequate tables. However, sufficiently extensive tables for the binomial distributions are not as readily available as tables for the standard normal distribution. In many practical situations we may employ the normal distribution as an approximation to a given binomial distribution and obtain numerically satisfactory results with little effort from the tables for the normal distribution function F.

From Section 24 of Chapter 8 we recall that each (nonconstant) random variable X has associated with it a standard variable $(X - \mu)/\sigma$, where μ and σ are the mean and the standard deviation of X. Thus, $\mu = E[X]$ and $\sigma = (\text{Var}[X])^{1/2}$. The mean and the standard deviation for the standard random variable $(X - \mu)/\sigma$ are 0 and 1, respectively.

For the case of the above Bernoulli experiment and the random variable S giving the number of successes, the mean and variance of S are (see Section 25 in Chapter 8) np and $np(1-p)$, respectively. Hence, the standard random variable associated with S is $(S - np)/(npq)^{1/2}$, where $q = 1 - p$.

Generally speaking, the standard random variable $(S - np)/(npq)^{1/2}$ and the standard normal variable T have probability distributions that are approximately the same. Each of these "standard" variables has mean 0 and standard deviation 1. In case n is large and p is not near to either 0 or 1, then the probability of a statement about one of these variables is approximately equal to the probability of the corresponding statement about the other. The larger n is, the better the approximation is. A convenient "rule of thumb" to apply is the following: in case $5 \leq np \leq n - 5$, then the approximation is reasonably good in most practical situations.

Example 19 Let the random variable S have a binomial distribution in which $(n, p) = (400, \frac{4}{5})$. Consider the practical statement $310 < S < 330$, and call it r. We wish to evaluate $\Pr[r]$. Since each value of S is necessarily an integer, r means that S is one of the 19 members of the set $\{311, 312, \ldots, 329\}$. In converting r to a statement about the normal variable T whose values are not restricted to integers, we obtain better results if we express r in the fashion $310.5 \leq S \leq 329.5$. We notice that the integers satisfying this latter chain inequality are precisely the 19 members of the set $\{311, 312, \ldots, 329\}$, and we also notice that the distance between the endpoints $310\frac{1}{2}$ and $329\frac{1}{2}$ is precisely 19, the same as the number of integral values of S involved.

Now $E[S] = np = 400 \cdot \frac{4}{5} = 320$ and $\text{Var}[S] = np(1 - p) = 400 \cdot \frac{4}{5} \cdot \frac{1}{5} = 64$. Consequently, the standard random variable associated with S is $(S - 320)/8$. The statement r is equivalent to

$$-9.5 \leq S - 320 \leq 9.5$$

or to

$$-\frac{9.5}{8} \leq \frac{S - 320}{8} \leq \frac{9.5}{8}.$$

We now make the approximating step. The desired number $\Pr[r]$ is approximately

$$\Pr\left[-\frac{9.5}{8} \leq T \leq \frac{9.5}{8}\right],$$

where T is the standard normal variable. Since $9.5/8$ is 1.19 to the nearest hundredth, we may use Table 7 (in Section 8) to calculate the following approximation:

$$\Pr[r] = \Pr[-1.19 \leq T \leq 1.19] = F(1.19) - F(-1.19)$$
$$= (0.883) - (1 - 0.883) = 0.766.$$

Example 20 An experiment consists in tossing a true coin 625 times in succession. If X is the number of heads that appear, then the random variable X has a binomial distribution with the ordered pair of parameters $(n, p) = (625, \frac{1}{2})$. The mean and the standard deviation of X are $\mu = 312.5$ and $\sigma = (625 \cdot \frac{1}{2} \cdot \frac{1}{2})^{1/2} = \frac{25}{2}$. Find an integer m such that $\Pr[\mu \leq X \leq m]$ is approximately $\frac{1}{4}$.

Since μ is midway between consecutive integers, the statement $\mu \leq X \leq m$ may be interpreted as $\mu \leq X \leq m + \frac{1}{2}$, and the latter is equivalent to

$$0 \leq \frac{X - \mu}{\sigma} \leq \frac{m + \frac{1}{2} - \mu}{\sigma}.$$

10 / Binomial and normal distributions 465

Now
$$\frac{m + \frac{1}{2} - \mu}{\sigma} = \frac{m - 312}{12.5}.$$

Converting from the statement expressed in terms of the standard random variable associated with X over to a statement in terms of the standard normal variable T, we wish to find an integer m such that

$$\Pr\left[0 \leq T \leq \frac{m - 312}{12.5}\right] = \frac{1}{4}.$$

From the table for the normal distribution function F (Table 7), we find that $(m - 312)/12.5 = 0.674$. Solving for m, we obtain $m = 320.4$. Since we required m to be an integer, we may pick 320 as the best choice.

Actually, $\Pr[\mu \leq X \leq 320]$ is a little less than $\frac{1}{4}$, being about 0.239. By comparison, $\Pr[\mu \leq X \leq 321]$ is approximately 0.264, and we have by reference to more extensive tables confirmed that the best choice for an integer m to meet the condition is 320.

Thus, the chances are approximately 1 out of 4 that in 625 coin tosses, heads will appear between 313 and 320 times inclusive.

EXERCISES H

1. Let X be a random variable with a binomial distribution in which $n = 800$ and $p = \frac{1}{3}$.
   (a) Consider the statement $X < 270$. Interpreting the given statement as $X \leq 269.5$, show that $\Pr[X < 270]$ is approximately $F(0.21)$.
   (b) Consider the statement $X > 260$. Show that the probability of this statement is approximately $1 - F(-0.46)$.
   (c) Show that $\Pr[X > 260]$ is approximately $F(0.46)$.
   (d) Show that $\Pr[250 \leq X \leq 300]$ is approximately $F(2.54) - F(-1.29)$.
   (e) Express $\Pr[X \geq 265]$, approximately, in terms of an image under the function F.
   (f) Express $\Pr[260 \leq X \leq 261]$, approximately, in terms of one or more images under F.
   (g) Likewise, express $\Pr[X = 270]$.

2. Let X be a random variable with a binomial distribution in which $(n, p) = (10000, \frac{1}{10})$. Suppose that a complete table of values of the normal distribution function F were available. Express each of the following in terms of numbers that can be approximately evaluated by referring to the table.
   (a) $\Pr[X > 1050]$.
   (b) $\Pr[X \leq 900]$.
   (c) $\Pr[|X - 1000| > 20]$.
   (d) $\Pr[960 < X < 1000]$.
   (e) $\Pr[(|X - 1000| > 50)|(|X - 1000| > 20)]$.

3. Experience has shown that 10% of the items produced by a certain machine are defective. If a lot of 10,000 of the outputs from this machine are selected at random, consider the probability that the number of defectives in this lot is more than 977 but not more than 1035.
   (a) Write an exact expression for the desired probability.
   (b) Then use the normal distribution to calculate an approximation for the number.

4. In a Bernoulli experiment consisting of 40,000 independent trials, the probability of success on each trial is 2%. Use the normal distribution to calculate an approximation to the probability that the number of successes in the experiment is between 730 and 767 inclusive.

5. Let S be the number of successes in a Bernoulli experiment consisting of n independent trials such that p and $q = 1 - p$ are the probabilities for success and failure, respectively, on each individual trial. Let $\mu = E[S]$ and $\sigma = (\text{Var}[S])^{1/2}$. Suppose that n is a large number.
   (a) Evaluate $\Pr[\mu - \sigma \leq S \leq \mu + \sigma]$.
   (b) Evaluate $\Pr[np - 2(npq)^{1/2} \leq S \leq np + 2(npq)^{1/2}]$.
   (c) Evaluate $\Pr[S \leq \mu]$.
   (d) Evaluate $\Pr[|S - \mu| < 3\sigma]$.
   (e) Evaluate $\Pr[\mu \leq S \leq \mu + 2\sigma]$.
   (f) Find the root k of the equation $\Pr[\mu - k\sigma \leq S \leq \mu + k\sigma] = 0.9$.
   (g) Find the root k of the equation $\Pr[\mu - k\sigma \leq S \leq \mu + k\sigma] = 95\%$.
   (h) Find the root k of the equation $\Pr\left[\left|\dfrac{S - \mu}{\sigma}\right| > k\right] = \dfrac{1}{2}$.

6. Suppose that a true coin is tossed one million times. The expected value of the number of heads is 500,000.
   (a) Show that the chances are more than 4 out of 5 that the observed number of heads is *not* within 100 of the "expected" number 500,000.
   (b) The preceding statement might be interpreted as saying, "In one sense, there is not much chance that the observed number of heads will be 'close' to the expected number of heads." Comment on this interpretation.

7. Consider an experiment consisting of 80 fair rolls of a pair of manufactured dice.
   (a) Under the assumption that the dice are true, write an expression for the probability that a total of seven spots occurs (exactly) 16 times during the 80 rolls. (The calculated value of this number, to three decimal places, is 0.082.)
   (b) Use the normal distribution in order to find an approximation for the probability mentioned in part (a).
   (c) Compare your approximation in part (b) with the quoted value in part (a).
   (d) Still under the assumption that the dice are true, use the normal distribution to estimate the probability that the number of occurrences of seven spots is more than 24.
   (e) Comment on an experiment, of the prescribed type, in which a total of seven spots actually appears 25 times.

11 Winding function

The name "trigonometry" is derived from a combination of words meaning the measurement of triangles. Although the trigonometric functions still retain considerable importance in geometrical measurement and its applications, the functions now have a much broader sphere of influence. A large part of the significance of the trigonometric functions is their characteristic of being *periodic* functions. This viewpoint is clearer if we approach the trigonometric functions by way of a circle rather than in terms of triangles. In fact, the trigonometric functions are now sometimes referred to as circular functions.

Figure 21

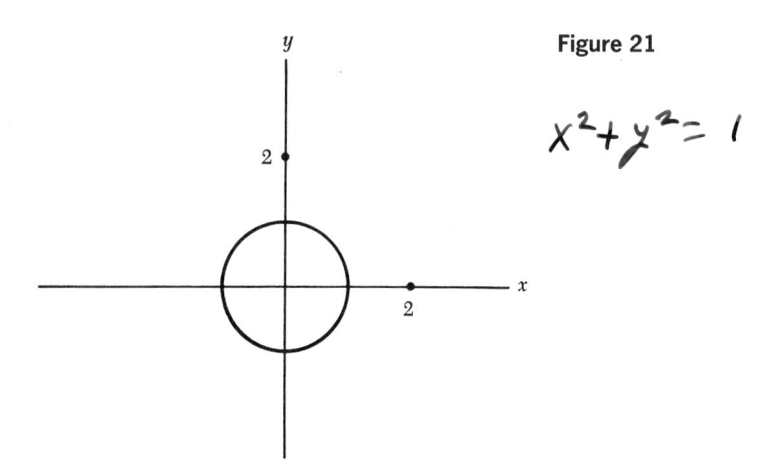

$x^2 + y^2 = 1$

Among the several trigonometric functions, the sine and the cosine are most worthy of our attention. These we shall introduce by a study of the unit circle and the winding function. Later we shall briefly encounter the tangent function.

On a two-dimensional coordinate system, the circle with the origin as center and with radius one is often called *the unit circle.* In symbols the unit circle is the graph of the set $\{(x, y) \mid x^2 + y^2 = 1\}$. (See Fig. 21.) Since one is the radius of the unit circle, we recall from geometry that the length (or circumference) of the unit circle is 2π.

Suppose we denote by U the point $(1, 0)$. Then U is the point of intersection of the unit circle and the right coordinate halfplane and the horizontal coordinate axis.

We consider a rule that assigns to each real number a point on the unit circle. This rule is a function—a function of a somewhat different type than the functions we have been studying, for, although its domain is the set of all real numbers, its range consists of points rather than numbers. In fact, the range consists of precisely the points belonging to the unit circle.

> **Definition** The *winding function* is the rule that associates to each real number r the point on the unit circle determined as follows: starting from the point U on the circle, wind around the circle, winding counterclockwise or clockwise according as r is positive or negative; the point reached by winding in the specified direction a distance of $|r|$ units is the image of r under the winding function.

> **Example 21** Find the points on the unit circle that are the respective images under the winding function of the numbers $\pi/2$, 2π, $-9\pi/4$.
>
> To find the image of $\pi/2$, we wind counterclockwise. Since $\pi/2$ is one-fourth of 2π, the distance we wind is a quarter of the length

468 Transcendental Functions

Figure 22

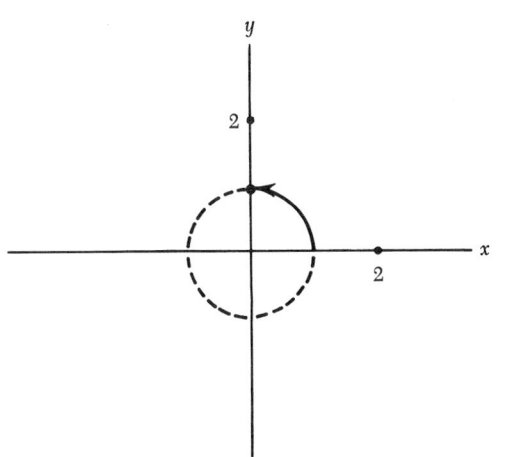

r is a measure of circum.

Tie together r and θ in radians + degrees.

of the circle. Consequently, the winding is along the circular arc from U to the point (0, 1). (See Fig. 22.)

The point associated with 2π is obtained by winding 2π units counterclockwise. Since 2π is precisely the length of the circle, the winding brings us back again to the point U. The image of 2π is the point (1,0).

For the negative number $-9\pi/4$, we wind clockwise a distance of $9\pi/4$ units. Since $9\pi/4 = 2\pi + (\pi/4) = 2\pi(1 + \frac{1}{8})$, the winding process requires a complete revolution and one-eighth of a second revolution. Intermediate stages during the process are suggested in Figs. 23a and 23b; the end result is pictured in Fig. 23c, where part of the unit circle has been traced twice. The point we reach is midway through the lower right quadrant. It lies on the line that bisects the quadrant, namely, the line whose equation is $y = -x$. It is the sole point belonging to the set

$$\{(x, y) \mid (x > 0) \wedge (y = -x) \wedge (x^2 + y^2 = 1)\}.$$

+ real   *(-45°)*   *on circle.*

Figure 23

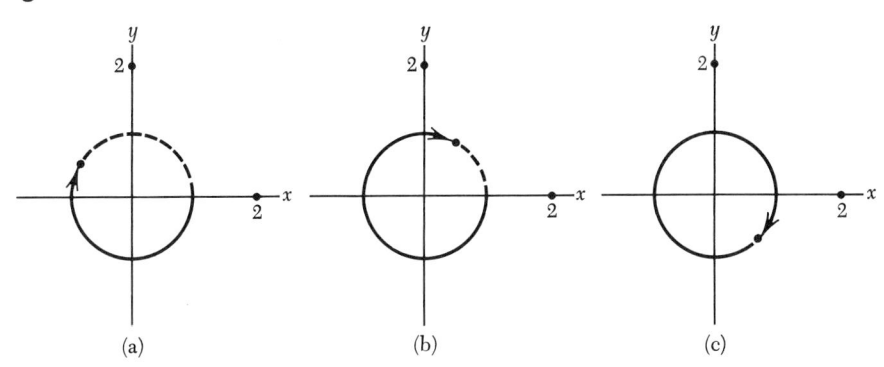

(a)   (b)   (c)

The image of $-9\pi/4$ is the point $(2^{-1/2}, -2^{-1/2})$.

The same point is also reached after just one-eighth of a revolution. Thus, the same point is associated with the number $-\pi/4$.

Example 22 Find the image under the winding function for the number 0, for the number 1, and for the number 2.

The process of winding a distance of 0 leaves us still at the starting point U. Thus, the image of 0 is the point $(1, 0)$.

Figure 24 shows a circular arc of length 1, the same length as the radius of the unit circle. The image of 1 under the winding function is the point A. If we draw a diagram on a suitably large scale, we may measure the coordinates of A and find that, approximated to the nearest hundredth, they are $(0.54, 0.84)$.

Figure 25 shows a circular arc of length 2, one of whose endpoints is U. Point B is the image of 2. On a suitably large diagram, we may find by measurement that B has coordinates $(-0.42, 0.91)$, expressed to the nearest hundredth.

Figure 24

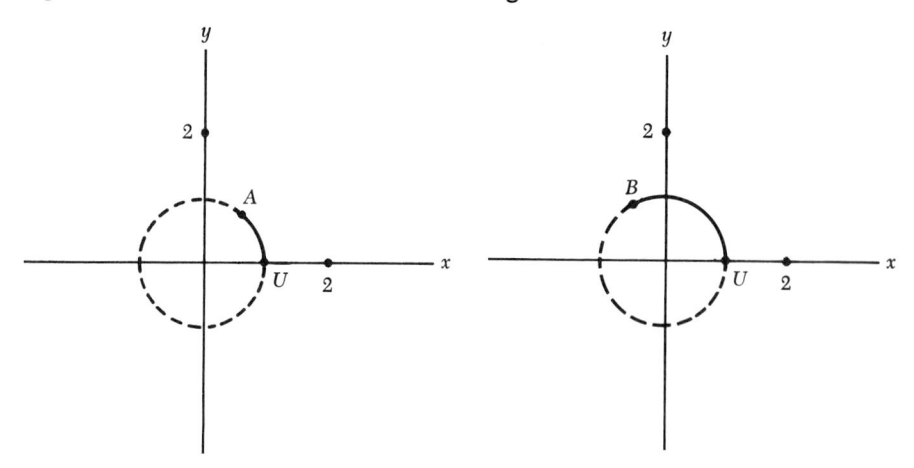

Figure 25

EXERCISES I

1. Let Q be a point in the positive quadrant, on the unit circle, and with x-coordinate 0.8. What is the y-coordinate of Q?

2. Let A be a point on the unit circle with second coordinate $\frac{5}{13}$. What are the possibilities for the first coordinate of A?

3. Let B be a point on the unit circle with negative y-coordinate and with x-coordinate $\frac{1}{3}$. What is the y-coordinate of B?

4. Let G be the point in the positive quadrant, on the unit circle, and with x-coordinate $\frac{1}{2}$.
   (a) What is the y-coordinate of G?
   (b) Find the distance between G and U.

(c) What special kind of a triangle is the triangle whose vertices are G, U, and the origin?
(d) The length of the circular arc from U to G is what fractional part of the circumference of the unit circle?
(e) Show that the point G is the image of the number $\pi/3$ under the winding function.

5. Let H be the point in the positive quadrant, on the unit circle, and with y-coordinate $\frac{1}{2}$.
   (a) What is the x-coordinate of H?
   (b) Using a development like that in Exercise 4, show that the winding function assigns the point H to the number $\pi/6$.

6. Let G be the point described in Exercise 4. Let J be the reflection of G in the y-coordinate axis. Let K be the reflection of J in the x-coordinate axis.
   (a) What are the coordinates of the point J?
   (b) Find the coordinates of the point K.
   (c) Find the distance between the points J and U.
   (d) How far apart are the points J and K?
   (e) What special kind of a triangle is the triangle with vertices U, J, K?
   (f) Find a number whose image under the winding function is the point J.
   (g) Find a number that is mapped by the winding function onto the point K.

7. Referring to Exercises 4, 5, and 6, find the point that is the image under the winding function for each of the following numbers. *(angle)*
   (a) $2\pi/3$.   (b) $-\pi/3$.   (c) $-5\pi/6$.   (d) $11\pi/6$.   (e) $4\pi/3$.   (f) $10\pi/3$.

8. For each of the following pairs of numbers, describe the relationship between their images under the winding function.
   (a) $\pi/4$ and $9\pi/4$.   (b) 2 and $2 + 2\pi$.   (c) -5 and $2\pi - 5$.
   (d) $\frac{3}{4}$ and $\frac{3}{4} - 2\pi$.   (e) $2\pi/3$ and $5\pi/3$.   (f) 1 and $1 + \pi$.
   (g) 3 and $3 - \pi$.   (h) 8 and $8 - 5\pi$.   (i) $3\pi/4$ and $-3\pi/4$.
   (j) 1 and -1.   (k) $\frac{7}{3}$ and $-\frac{7}{3}$.

9. If r is any real number, how does the point assigned to r by the winding function compare with the image of each of the following?
   (a) $r + 2\pi$.   (b) $r + \pi$.   (c) $-r$.

12   Sine and cosine functions

We have seen, in examples and exercises, that two different numbers may have the same image under the winding function. Thus, the point U is the image of the number 0 and also of 2π; furthermore, U is the image of 4π and the image of -2π, as well as the image of every number that is the product of an integer and 2π. Likewise, the point (0, 1) is the image of $\pi/2$, the image of $(\pi/2) + 2\pi = 5\pi/2$, the image of $(\pi/2) - 2\pi = -3\pi/2$, and the image of many other numbers.

If we apply the winding process to find the image of any number r, then one extra counterclockwise revolution will give the image of the number $r + 2\pi$. But the point reached in each case is the same point on the unit circle. Thus, the image of $r + 2\pi$ is the same as the image of r, for every number r.

We may describe the preceding property of the winding function by saying that it is *periodic*, that is, the images recur, over and over again, in a regular or periodic fashion. A change of 2π in the value of a member of the domain does not affect the image. Since 2π is the smallest positive number having this characteristic, we call the number 2π the *period* of the winding function.

Definition A function F is called a *periodic function* provided there is a smallest positive number p such that, for every number r in the domain of F, the numbers r and $r + p$ have the same image under F. The number p is called the *period* of the function F.

The winding function is periodic with period 2π. The trigonometric functions which we are now ready to introduce are also periodic.

Definition The *cosine* function is the rule that associates to every real number r the first coordinate of the point that is the image of r under the winding function.

The cosine function is abbreviated cos; and the image of the number r under cos may be denoted by $\cos(r)$, or more simply by $\cos r$ in case no confusion can arise.

Example 23 We found in Example 21 that the point associated with $\pi/2$ by the winding function is $(0, 1)$. The x-coordinate of this point, namely 0, is the cosine of $\pi/2$. We write: $\cos \frac{\pi}{2} = 0$.

We also found in Example 21 that $(2^{-1/2}, -2^{-1/2})$ is the image of $-9\pi/4$ under the winding function. Since $2^{-1/2}$ is the first coordinate of this point, we conclude that $\cos(-9\pi/4) = 2^{-1/2}$. Notice that here we retain the parentheses in writing $\cos(-9\pi/4)$.

We may use the images under the winding function which we found in Example 22 to deduce that $\cos 0 = 1$ and that $\cos 1$ is approximately 0.54.

Definition The *sine* function is the rule that associates to each real number r the second coordinate of the point that is the image of r under the winding function.

The sine function is abbreviated sin; the image of the number r may be written $\sin(r)$, although commonly the parentheses are suppressed in case no confusion can arise by using the simpler notation $\sin r$.

Example 24 In Examples 21 and 22 we found the points associated with several numbers under the winding function. In each case the y-coordinate of the point is the sine of the number. Thus,

Figure 26

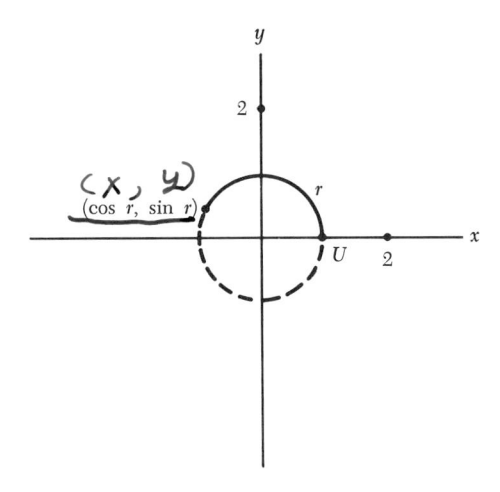

(x, y)
(cos r, sin r)

$\sin(\pi/2) = 1$, $\sin 2\pi = 0$, $\sin(-9\pi/4) = -1/\sqrt{2}$, $\sin 0 = 0$; also, we have the approximations $\sin 1 = 0.84$ and $\sin 2 = 0.91$.

Every real number has an image under the winding function, and the image is a point, which of course has a first and a second coordinate. Thus, every real number belongs to the domain of the cosine function and also belongs to Dom sin.

Since the first and second coordinates of the point assigned to r by the winding function are $\cos r$ and $\sin r$, the point is ($\cos r$, $\sin r$). (See Fig. 26.) According to the winding process, we reach the point ($\cos r$, $\sin r$) by winding r units around the unit circle from U. Since the point ($\cos r$, $\sin r$) lies on the unit circle, its coordinates satisfy the equation of the circle, namely, $x^2 + y^2 = 1$. Therefore, $(\cos r)^2 + (\sin r)^2 = 1$. Since it is customary to abbreviate $(\cos r)^2$ by $\cos^2 r$ and similarly for $(\sin r)^2$, we obtain the extremely important result that

$$\sin^2 r + \cos^2 r = 1 \qquad \text{for every real number } r.$$

identity

Since the range of the winding function is the set of all points on the unit circle, the range of cos is the set of all first coordinates of points on the unit circle, namely, the set of all numbers between -1 and 1 inclusive. Likewise, the range of the sine function, the set of all y-coordinates of points on the unit circle, consists of all numbers between -1 and 1 inclusive. Thus, in symbols, we have

$$\text{Rg cos} = \{t \mid -1 \leq t \leq 1\} = \text{Rg sin}.$$

We recall that the winding function is periodic with period 2π. Given any number r, the numbers r and $r + 2\pi$ have the same point as image under the winding function. The first coordinate of this point is the cosine of r and also the cosine of $r + 2\pi$. Consequently,

$$\cos(r + 2\pi) = \cos r \qquad \text{for every number } r.$$

12 / Sine and cosine functions

A similar argument applied to the second coordinate shows that
$$\sin(r + 2\pi) = \sin r \quad \text{for every number } r.$$
Now 2π is the smallest positive number p satisfying either of the conditions
$$\cos(0 + p) = \cos 0 = 1 \quad \text{or} \quad \sin\left(\frac{\pi}{2} + p\right) = \sin\frac{\pi}{2} = 1.$$
Hence, each of the functions, cos and sin, is periodic with period 2π.

EXERCISES J

1. Find the image of each of the following numbers under the sine function.
   (a) π.   (b) $3\pi/4$.   (c) 0.   (d) $5\pi/4$.
   (e) $\pi/6$.   (f) $-3\pi/2$.   (g) $\pi/3$.   (h) $2\pi/3$.
   (i) $-5\pi/6$. (j) $23\pi/6$.   (k) $-19\pi/4$.

2. Find the cosine of each of the following numbers.
   (a) 5π.   (b) $-3\pi/4$.   (c) $-\pi/4$.   (d) 0.
   (e) $11\pi/2$.   (f) $\pi/3$.   (g) $4\pi/3$.   (h) $\pi/6$.
   (i) $-7\pi/6$.   (j) $11\pi/6$.   (k) $19\pi/6$.

3. Let r be a number such that $\sin r = \frac{2}{3}$.
   (a) In this case what are the possibilities for the number $\cos^2 r$?
   (b) What are the possibilities for the number $\cos r$?

4. If $\cos t = -\frac{3}{4}$, what are the possibilities for the number $\sin t$?

5. "Let s be a number such that $\cos s = \pi/3$." Criticize, or discuss, the preceding statement.

6. Find three different numbers x such that $\cos x = 1$.

7. (a) Find two different points on the unit circle each of which has y-coordinate $1/\sqrt{2}$.
   (b) Find three different numbers s such that $\sin s = 2^{-1/2}$.

8. Find five different roots for each of the following equations.
   (a) $\sin x = \frac{1}{2}$.   (b) $\sin x = 0$.   (c) $\sin x = -1$.
   (d) $\cos x = -2^{-1/2}$.   (e) $\cos x = \frac{1}{2}$.   (f) $\sin x = -\frac{1}{2} \cdot 3^{1/2}$.
   (g) $\cos x = \sqrt{3}/2$.   (h) $2 \sin x - 1 = 0$.   (i) $\cos(2x) = -1$.
   (j) $\sin(3x) = -\frac{1}{2}$.   (k) $\cos(\frac{1}{2}x) = 0$.

9. Compare the cosines of the following four numbers: 1, $1 + 4\pi$, $1 - 6\pi$, $1 + 10000\pi$.

10. Compare the following three numbers: $\sin 2$, $\sin(2 - 2\pi)$, $\sin(2 + 8\pi)$.

13 Properties of sine and cosine

Consider two numbers that are negatives of each other, say r and $-r$. We wish to compare the images of these two numbers under the winding function. For each of the numbers the process is to wind $|r|$ units around the unit circle

Figure 27

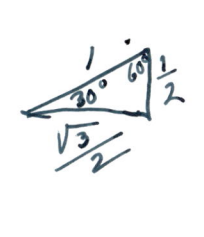

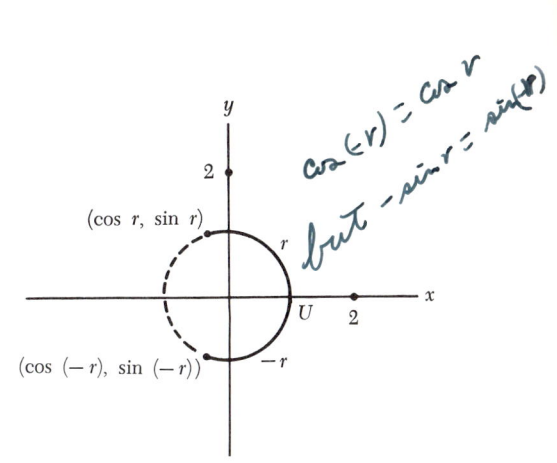

starting from the point U. The distinction is that in one case we wind counterclockwise and in the other case clockwise. Since the distance is the same in each case, the respective points we reach are reflections of each other in the horizontal coordinate axis. One of the points is $(\cos r, \sin r)$ and the other is $(\cos(-r), \sin(-r))$. (See Fig. 27.) Since each of the points $(\cos r, \sin r)$ and $(\cos(-r), \sin(-r))$ is the reflection of the other in the x-coordinate axis, the points have the same first coordinate, and the second coordinates of the two points are negatives of each other. Therefore we have the following relationships:

$$\cos(-r) = \cos r \quad \text{for every number } r;$$
$$\sin(-r) = -\sin r \quad \text{for every number } r.$$

The first of these relationships means that cos is an even function; the second means that the sine function is an odd function.

We assert that cos is a decreasing function on the set $\{r \mid 0 \leq r \leq \pi\}$. Consider any two numbers a and c such that $0 \leq a < c \leq \pi$. To find the images of a and c under the cosine function, we start at the point U and wind around the upper semicircle of the unit circle. The winding process applied to c takes us farther than the point corresponding to a. (See Fig. 28.) Consequently, the

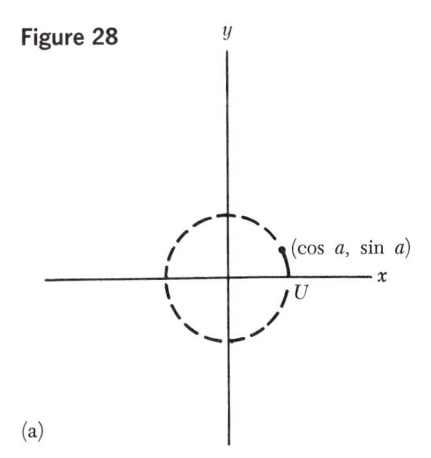

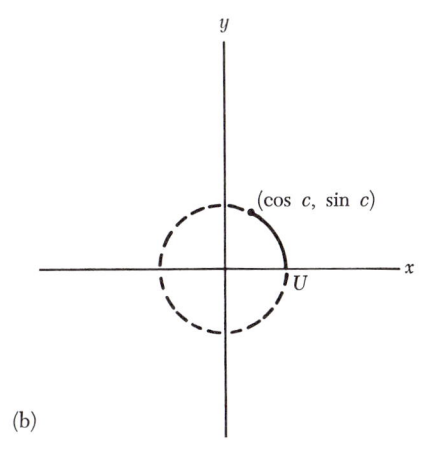

Figure 28

(a)

(b)

13 / Properties of sine and cosine

point associated with c has a smaller x-coordinate than the point corresponding to a. In other words, cos c < cos a. We have verified the assertion that cos is a decreasing function on $\{r \mid 0 \leq r \leq \pi\}$.

A similar discussion applied to the y-coordinates reveals that the sine function is an increasing function on the set $\{r \mid 0 \leq r \leq \pi/2\}$ but is a decreasing function on the set $\{r \mid \pi/2 \leq r \leq \pi\}$.

We now summarize some of the important characteristics of the cosine function. These we will exploit in sketching the graph of the function.

1. Dom cos consists of all real numbers. ~values for r
2. Rg cos = $\{t \mid -1 \leq t \leq 1\}$. values for cos r
3. cos is a periodic function with period 2π.
4. cos is an even function. $\cos(-r) = \cos r$
5. cos is a decreasing function on $\{r \mid 0 \leq r \leq \pi\}$.
6. $(0, 1) \in$ cos.

The preceding properties may be used in sketching the graph of the equation $y = \cos x$ according to the following steps. The curve contains the point $(0, 1)$ and is a falling curve for values of x between 0 and π (see Fig. 29a). Since Rg cos contains no number less than -1, the curve does not fall more than one unit below the horizontal axis; in fact, $(\pi, -1)$ belongs to the graph (see Fig. 29b). Since cos is an even function, its graph is symmetric with respect to the vertical coordinate axis; consequently, the reflection in the y-axis of any portion of the curve in the right coordinate halfplane is also a part of the curve (see Fig. 29c, which shows the portion of the curve pictured in Fig.

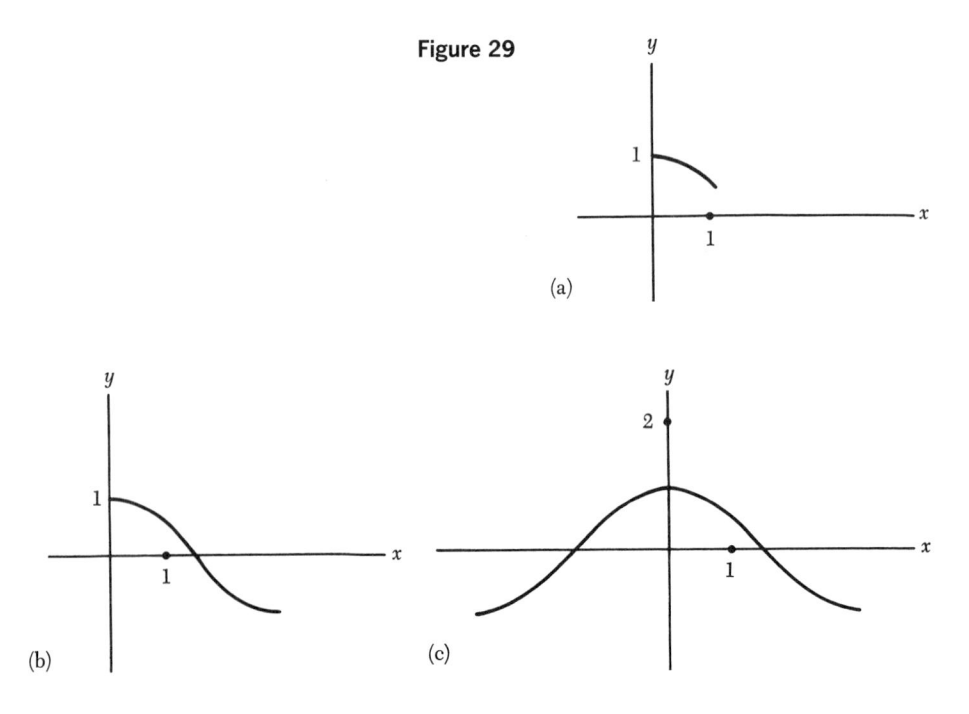

Figure 29

(a)

(b)

(c)

476   Transcendental Functions

Figure 30

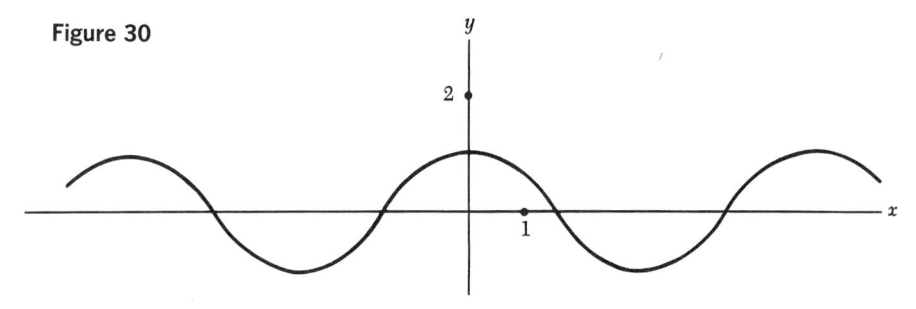

29b and its reflection). Finally, we make use of the periodic characteristic. A displacement of 2π units in the horizontal direction reproduces the same height on the curve; thus, the entire curve may be obtained by repeating, over and over again, any arc that extends 2π units horizontally. Such an arc is drawn in Fig. 29c; a recurring wave of which Fig. 29c shows one cycle is the graph of the equation $y = \cos x$ (see Fig. 30).

Notice how each of the six properties cited above is depicted on the graph of the cosine function.

An analogous discussion may be made pertaining to the sine function. Some noteworthy characteristics of the function are the following.

1. Dom sin is the set of all real numbers.
2. Rg sin = $\{t \mid -1 \leq t \leq 1\}$.
3. sin is a periodic function with period 2π.
4. sin is an odd function.
5. sin is an increasing function on $\{r \mid 0 \leq r \leq \pi/2\}$ and is a decreasing function on $\{r \mid \pi/2 \leq r \leq \pi\}$.
6. $(0, 0) \in $ sin.

We may begin tracing the sine curve from the origin into the positive quadrant. The curve attains a maximum point at $(\pi/2, 1)$, and beyond that it falls to $(\pi, 0)$ (see Fig. 31a). For each point $(r, \sin r)$ on the graph, the point $-(r, \sin r)$ obtained by rotating $(r, \sin r)$ halfway around the origin also belongs to the curve, because $-(r, \sin r) = (-r, -\sin r) = (-r, \sin(-r))$. The curve is symmetric with respect to the origin. This observation enables us to extend the portion sketched in Fig. 31a, for a half rotation of this around the origin yields another arc of the curve; together they form the portion of the graph pictured in Fig. 31b. The sine function is periodic. The graph of

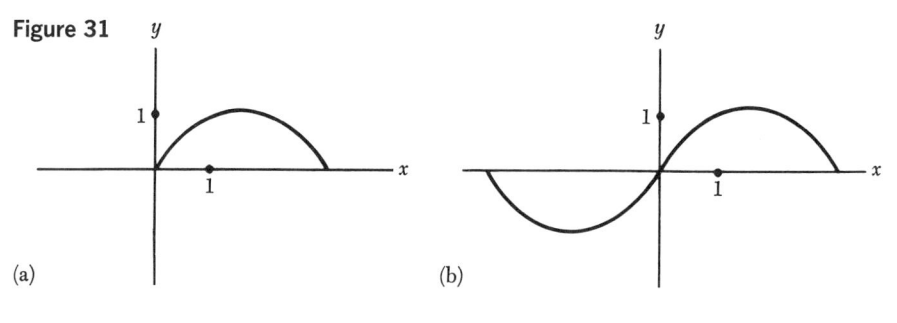

Figure 31

(a)   (b)

13 / Properties of sine and cosine

Figure 32

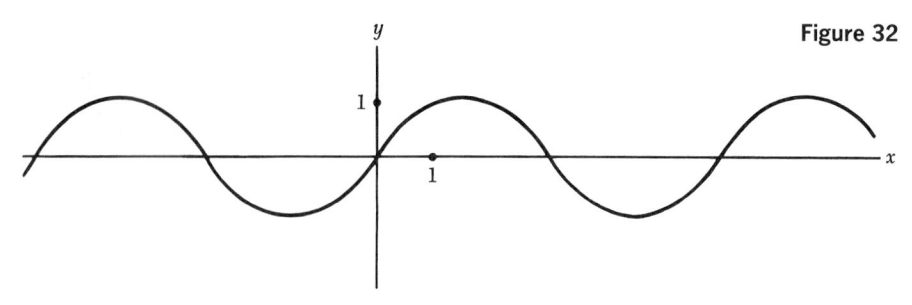

the equation $y = \sin x$ is the wavy curve suggested in Fig. 32 and obtained by repeating, again and again, the single cycle represented in Fig. 31b.

A horizontal translation of the sine wave by 2π units shifts the curve into itself. The curve is the union of arcs, all of which are congruent to one another and each of which has endpoints 2π units apart.

EXERCISES K

work

1. Approximately, $\cos 1 = 0.54$ and $(0.71)^{1/2} = 0.84$. Use this information to find each of the following, to the nearest hundredth.
   (a) $\cos(-1)$. $= .54$ (b) $-\cos 1$.   (c) $\cos(1 + 2\pi)$.
   (d) $\cos(-1 + 2\pi) = .54$ (e) $\cos^2 1. = .29$ (f) $\sin^2 1. = .71$
   (g) $\sin 1. \sqrt{1-\cos^2 1} = .84$ (h) $\sin(-1)$.

2. Approximately, $\sin 2 = 0.91$ and $(0.17)^{1/2} = 0.42$. Use this information to find each of the following, to the nearest hundredth.
   (a) $\sin(2 + 4\pi)$.   (b) $\sin(-2)$.   (c) $\sin(10\pi - 2)$.   (d) $\sin^2 2$.
   (e) $\cos^2 2$.   (f) $\cos 2$.   (g) $\cos(-2)$.   (h) $\cos(2\pi - 2)$.

3. Refer to Exercise I-9b, and let r be a real number.
   (a) How does $\cos(r + \pi)$ compare with $\cos r$?
   (b) How is $\sin(r + \pi)$ related to $\sin r$?

4. Approximately, $\cos 6 = 0.96$ and $(0.08)^{1/2} = 0.28$. Use this information to find each of the following, to the nearest hundredth.
   (a) $\cos(6 - 2\pi)$.   (b) $\cos(-6)$.   (c) $\cos(6 + \pi)$.
   (d) $\sin^2 6$.   (e) $\sin 6$.   (f) $\sin(-6)$.
   (g) $\sin(\pi - 6)$.

5. Directly from the definition of the specified function, show each of the following.
   (a) The sine function is increasing on the set $\{r | -\pi/2 < r < \pi/2\}$.
   (b) The sine function is decreasing on the set $\{r | \pi/2 < r < 3\pi/2\}$.
   (c) $\cos$ is increasing on the set $\{r | -\pi \leq r \leq 0\}$.

6. Sketch the graph of each of the following equations.
   (a) $y = \sin x$.   (b) $y = -\sin x$.
   (c) $y = \sin x + 3$.   (d) $y = \sin(x - 2)$.

7. Sketch the graph of each of the following functions.
   (a) $\{(x, y) | y = \cos(-x)\}$.   (b) $\{(x, y) | y = -\cos x\}$.
   (c) $\{(x, y) | y = \cos x - 1\}$.   (d) $\{(x, y) | y = \cos(x + 2)\}$.

8. Find the distance between the two points in each of the following pairs.
   (a) The point U with coordinates $(1, 0)$, and the point on the unit circle that is assigned to $\pi/2$ by the winding function.

(b) The image of $\pi/3$ under the winding function, and U.
(c) The images of $8\pi/3$ and $5\pi/4$ under the winding function.
(d) The points $\left(\cos\dfrac{5\pi}{6}, \sin\dfrac{5\pi}{6}\right)$ and $\left(\cos\dfrac{3\pi}{2}, \sin\dfrac{3\pi}{2}\right)$.
(e) The point $(\cos r, \sin r)$, and U.
(f) The point $(\cos r, \sin r)$, and the point $(\cos(-r), \sin(-r))$.

14 Addition formulas

Let r be a real number. The point P that is the image of r under the winding function has coordinates $(\cos r, \sin r)$. We reach P by winding $|r|$ units around the unit circle from the point U with coordinates $(1, 0)$. We inquire how far apart the points U and P are. (See Fig. 33.) The formula for the distance between two points, as developed in Section 3 of Chapter 10, tells us that the distance between U and P is $[(\cos r - 1)^2 + (\sin r)^2]^{1/2}$. Now we may use the binomial theorem and the relationship $\sin^2 r + \cos^2 r = 1$ in order to obtain a simpler formula for the distance:

$$[(\cos r - 1)^2 + (\sin r)^2]^{1/2} = [\cos^2 r - 2\cos r + 1 + \sin^2 r]^{1/2}$$
$$= (2 - 2\cos r)^{1/2}.$$

The formula just developed is also applicable for the distance between any two points on the unit circle such that one may be reached from the other by starting at the latter and winding $|r|$ units around the circle. Indeed, if J and K are two such points, then the segment joining J and K (see Fig. 34) is congruent to the segment joining U and P (see Fig. 33). Hence, the distance between J and K is the same as the distance between U and P, namely, $(2 - 2\cos r)^{1/2}$.

Figure 33

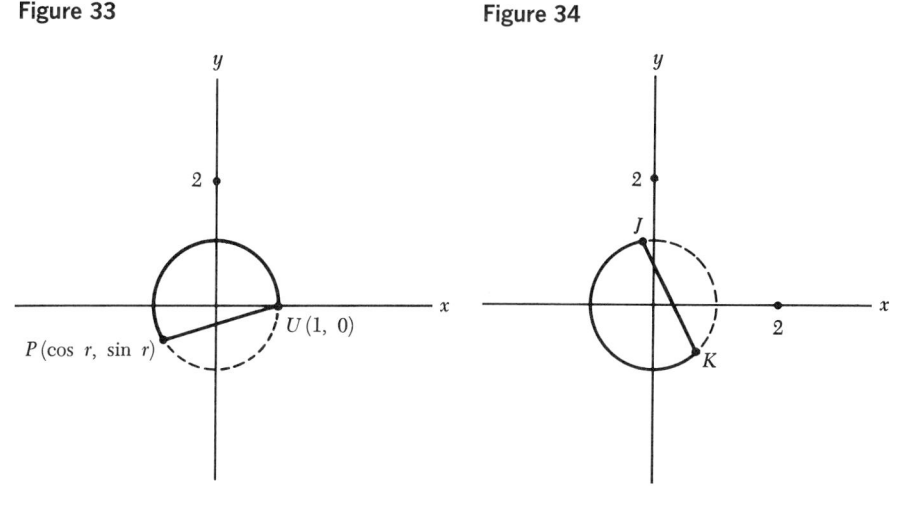

Figure 34

We are ready to develop several very important results concerning the sine or cosine of a sum or difference of numbers. The first of these results is the following statement.

$$\cos(t - r) = \cos t \cos r + \sin t \sin r \qquad \text{for all ordered pairs } (t, r).$$

Suppose that r and t are any real numbers. If we start at U and apply the winding process in the usual manner, we find that the point P on the unit circle associated with the number r is $(\cos r, \sin r)$. If afterwards we continue from the point P already reached and wind an additional $t - r$ units around the circle, then we reach the same point Q as if we begin at U and locate the image of $r + (t - r) = t$ under the winding function; this image Q is $(\cos t, \sin t)$. (See Fig. 35.)

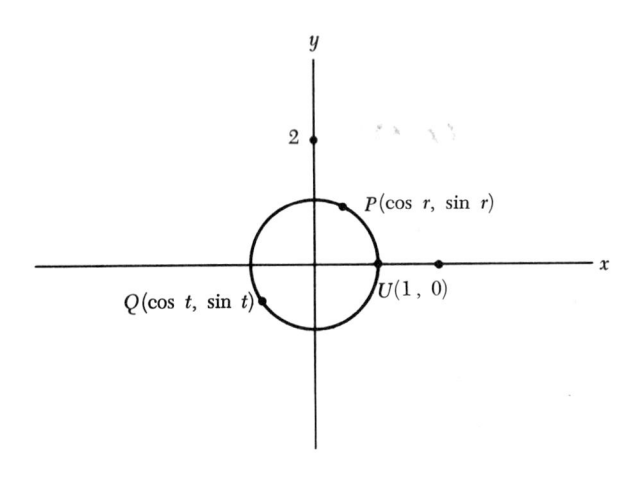

Figure 35

Since Q may be reached from P by winding a distance of $|t - r|$ units around the unit circle, the above development reveals that the distance between P and Q is ~~angle is (t-r)~~ *same as previous page*

$$[2 - 2\cos(t - r)]^{1/2}.$$

On the other hand, the distance between P and Q with respective coordinates $(\cos r, \sin r)$ and $(\cos t, \sin t)$ is given by

$$[(\cos t - \cos r)^2 + (\sin t - \sin r)^2]^{1/2}$$

brute expansion
$$= [\cos^2 t - 2\cos t \cos r + \cos^2 r + \sin^2 t - 2\sin t \sin r + \sin^2 r]^{1/2}$$
$$= [2 - 2\cos t \cos r - 2\sin t \sin r]^{1/2}.$$

Since any two expressions for the distance between P and Q must agree, we obtain

$$[2 - 2\cos(t - r)]^{1/2} = [2 - 2\cos t \cos r - 2\sin t \sin r]^{1/2}.$$

Hence,

$$2 - 2\cos(t - r) = 2 - 2\cos t \cos r - 2\sin t \sin r,$$

which may be simplified to

$$\cos(t - r) = \cos t \cos r + \sin t \sin r.$$ *proved.*

In words, the cosine of the difference between two numbers is the sum of the product of their cosines and the product of their sines.

Example 25 Verify the above formula for the ordered pair $(t, r) = (\pi/2, \pi/3)$.

On the one hand,

$$t - r = \frac{\pi}{2} - \frac{\pi}{3} = \frac{\pi}{6} \quad \text{and} \quad \cos\frac{\pi}{6} = \frac{\sqrt{3}}{2}.$$

On the other hand,

$$\cos t \cos r + \sin t \sin r = \left(\cos\frac{\pi}{2}\right)\left(\cos\frac{\pi}{3}\right) + \left(\sin\frac{\pi}{2}\right)\left(\sin\frac{\pi}{3}\right)$$

$$= 0 \cdot \frac{1}{2} + 1 \cdot \frac{\sqrt{3}}{2} = \frac{\sqrt{3}}{2}.$$

work

Example 26 Calculate $\cos(\pi/12)$.

Since $\frac{1}{12} = \frac{1}{3} - \frac{1}{4}$, we find that

$$\cos\frac{\pi}{12} = \cos\left(\frac{\pi}{3} - \frac{\pi}{4}\right)$$

$$= \cos\frac{\pi}{3}\cos\frac{\pi}{4} + \sin\frac{\pi}{3}\sin\frac{\pi}{4}$$

$$= \frac{1}{2} \cdot \frac{\sqrt{2}}{2} + \frac{\sqrt{3}}{2} \cdot \frac{\sqrt{2}}{2}$$

$$= \frac{\sqrt{2} + \sqrt{6}}{4}.$$

As a decimal approximation, $\cos(\pi/12) = 0.966$.

Example 27 An important special case of the above formula occurs when $t = \pi/2$. Then the right member of the equation is

$$\cos\frac{\pi}{2}\cos r + \sin\frac{\pi}{2}\sin r = 0(\cos r) + 1(\sin r)$$

$$= \sin r.$$

Hence,

$$\cos\left(\frac{\pi}{2} - r\right) = \sin r \quad \text{for every number } r.$$

In words, the sine of any number is the cosine of the difference between $\pi/2$ and the number.

Example 28 We apply the preceding sentence to the number $(\pi/2) - r$. Then
$$\sin\left(\frac{\pi}{2} - r\right) = \cos\left[\frac{\pi}{2} - \left(\frac{\pi}{2} - r\right)\right]$$
$$= \cos r.$$

Thus,
$$\cos r = \sin\left(\frac{\pi}{2} - r\right) \quad \text{for every number } r.$$

This relation enables us to explain the similarity in names between the *sine* function and the *cosine* function. In the application to trigonometry, the difference $(\pi/2) - r$ may be called the complement of r. The two initial letters of the word *complement* combine with *sine* to give *cosine*, for a number's cosine is the number's complement's sine.

Fully as important as the formula for the cosine of the difference of two numbers is the expansion of the image of the sum of two numbers under cos. We recall that cos is an even function, that the sine function is odd, and that $t + r = t - (-r)$. Then
$$\cos[t + r] = \cos[t - (-r)]$$
$$= \cos t \cos(-r) + \sin t \sin(-r)$$
$$= \cos t \cos r + \sin t (-\sin r).$$

We have proved the following theorem.

Theorem 4 $\cos(t + r) = \cos t \cos r - \sin t \sin r$ for all ordered pairs (t, r).

We combine recent results in order to prove the next theorem.

Theorem 5 $\sin(t + r) = \sin t \cos r + \sin r \cos t$ for all ordered pairs (t, r).

The various steps in the proof use the results concerning $\cos\left(\frac{\pi}{2} - r\right)$, $\cos(t - r)$, and $\sin\left(\frac{\pi}{2} - r\right)$. We have
$$\sin[t + r] = \cos\left[\frac{\pi}{2} - (t + r)\right]$$
$$= \cos\left[\left(\frac{\pi}{2} - t\right) - r\right]$$
$$= \cos\left(\frac{\pi}{2} - t\right)\cos r + \sin\left(\frac{\pi}{2} - t\right)\sin r$$
$$= \sin t \cos r + \cos t \sin r.$$

482   Transcendental Functions

Since important ideas are often easier to apply when expressed in words rather than in symbols, we restate Theorems 4 and 5. The cosine of a sum of two numbers is the difference between the product of their cosines and the product of their sines. The sine of the sum of two numbers is the sum of the two products obtained by multiplying the sine of each of the numbers by the cosine of the other number.

Example 29 According to Example 22, approximations for (cos 1, sin 1) and (cos 2, sin 2) are (0.54, 0.84) and (−0.42, 0.91), respectively. Consequently, we may calculate an approximation for (cos 3, sin 3) as follows:

$$\begin{aligned}
\cos 3 &= \cos(2+1) \\
&= (\cos 2)(\cos 1) - (\sin 2)(\sin 1) \\
&= (-0.42)(0.54) - (0.91)(0.84) \\
&= -0.99; \\
\sin 3 &= \sin(2+1) \\
&= (\sin 2)(\cos 1) + (\sin 1)(\cos 2) \\
&= (0.91)(0.54) + (0.84)(-0.42) \\
&= 0.14.
\end{aligned}$$

We may also approximate cos 4 in the same manner:

$$\begin{aligned}
\cos 4 &= \cos(2+2) \\
&= (\cos 2)(\cos 2) - (\sin 2)(\sin 2) \\
&= (-0.42)^2 - (0.91)^2 \\
&= -0.65.
\end{aligned}$$

We again recall that the sine and cosine functions are respectively odd and even. Consequently,

$$\begin{aligned}
\sin[t-r] &= \sin[t+(-r)] \\
&= \sin t \cos(-r) + \sin(-r) \cos t \\
&= \sin t \cos r + (-\sin r) \cos t.
\end{aligned}$$

Therefore, $\sin(t-r) = \sin t \cos r - \sin r \cos t$ for all ordered pairs (t, r).

Example 30 Evaluate $\sin(7\pi/12)$.

work Since $\frac{7}{12} = \frac{1}{3} + \frac{1}{4}$, we find that

$$\begin{aligned}
\sin \frac{7\pi}{12} &= \sin\left(\frac{\pi}{3} + \frac{\pi}{4}\right) \\
&= \sin \frac{\pi}{3} \cos \frac{\pi}{4} + \sin \frac{\pi}{4} \cos \frac{\pi}{3} \\
&= \frac{\sqrt{3}}{2} \cdot \frac{\sqrt{2}}{2} + \frac{\sqrt{2}}{2} \cdot \frac{1}{2} \\
&= \frac{\sqrt{6} + \sqrt{2}}{4}.
\end{aligned}$$

As an alternative verification,

$$\sin \frac{7\pi}{12} = \sin \left(\frac{\pi}{2} + \frac{\pi}{12}\right)$$

$$= \sin \frac{\pi}{2} \cos \frac{\pi}{12} + \sin \frac{\pi}{12} \cos \frac{\pi}{2}$$

$$= 1\left(\cos \frac{\pi}{12}\right) + \left(\sin \frac{\pi}{12}\right)0$$

$$= \cos \frac{\pi}{12},$$

and $\cos (\pi/12)$ we evaluated in Example 26.

Example 31 Simplify $\sin \left(\frac{\pi}{2} + r\right)$ and $\cos \left(\frac{\pi}{2} + r\right)$.

First,

$$\sin \left(\frac{\pi}{2} + r\right) = \left(\sin \frac{\pi}{2}\right)(\cos r) + (\sin r)\left(\cos \frac{\pi}{2}\right) = \cos r,$$

since $\left(\cos \frac{\pi}{2}, \sin \frac{\pi}{2}\right) = (0, 1)$. The special case in which $\pi/12$ plays the role of the number r was encountered in Example 30.

Second,

$$\cos \left(\frac{\pi}{2} + r\right) = \left(\cos \frac{\pi}{2}\right)(\cos r) - \left(\sin \frac{\pi}{2}\right)(\sin r) = -\sin r,$$

again since $\left(\cos \frac{\pi}{2}, \sin \frac{\pi}{2}\right) = (0, 1)$.

EXERCISES L

1. Verify that $\cos [(\pi/2) - r] = \sin r$ for each of the following values of r.
   (a) $r = \pi/3$. [HINT: Calculate $\sin (\pi/3)$ and $\cos (\pi/6)$, and then compare the two numbers.]
   (b) $r = 3\pi/4$.   (c) $r = -\pi/6$.   (d) $r = 7\pi/2$.

2. Given that $\cos 1 = 0.54$ approximately, find $\sin 0.57$ approximately.

3. Given that $\sin 2 = 0.91$ approximately, find $\cos (-0.43)$ approximately.

4. Verify that $\cos (t + r) = \cos t \cos r - \sin t \sin r$ for each of the following pairs of numbers. That is, calculate each of the two members of the asserted equation, and then compare the results.
   (a) $(t, r) = \left(\frac{\pi}{6}, \frac{2\pi}{3}\right)$.   (b) $(t, r) = \left(\frac{3\pi}{4}, -\frac{5\pi}{4}\right)$.

5. Apply the formula for $\cos (t + r)$ to each of the following pairs of numbers.
   (a) $(t, r) = \left(\frac{\pi}{4}, \frac{\pi}{3}\right)$.   (b) $(t, r) = \left(\frac{5\pi}{4}, -\frac{\pi}{6}\right)$.

484   Transcendental Functions

6. Calculate each of the following.
   (a) $\sin \frac{5\pi}{12}$.
   (b) $\sin \frac{\pi}{12}$.
   (c) $\sin \frac{17}{12}\pi$.

7. Use the information in Example 29 to calculate an approximation for each of the following numbers.
   (a) $\sin 4$.
   (b) $\sin(-3)$.
   (c) $\sin(\pi - 3)$.
   (d) $\cos 5$.
   (e) $\sin 6$.
   (f) $\sin(2\pi - 5)$.

8. Given the following approximations: $\sin \frac{1}{4} = 0.25$, $\cos \frac{1}{4} = 0.97$, $\sin \frac{1}{5} = 0.20$, $\cos \frac{1}{5} = 0.98$. Calculate each of the following approximately.
   (a) $\sin 0.45$.
   (b) $\cos 0.45$.
   (c) $\sin 1.12$.
   (d) $\cos 0.70$.
   (e) $(\sin 0.82)^2 + (\cos 0.82)^2$.

9. Simplify each of the following.
   (a) $\sin(\pi - t)$.
   (b) $\cos(\pi - t)$.
   (c) $\sin(\pi + t)$.

10. For each of the following formulas, what is the range of the function given by the formula?
    (a) $\sin 5x$.
    (b) $5 \cos x$.
    (c) $\sin x + 3$.
    (d) $\cos(x - 7)$.
    (e) $5 - 2 \cos x$.
    (f) $3 + 4 \sin(x - 8)$.

11. If $0 < r < 0.2$, then $\cos r$ is approximately the same as $1 - (r^2/2)$; the error of approximation is less than 0.0001. (The preceding result may be established by calculus techniques.)
    (a) Use this information to calculate $\cos 0.1$ to four decimal places.
    (b) Use the information to calculate $\cos(-0.08)$ to four decimal places.
    (c) Use the information to calculate an approximation for $\sin 1.4508$.

12. If $0 < r < 0.4$, then $\sin r$ is approximately the same as $r - (r^3/6)$; the error of approximation is less than 0.0001. (The preceding result may be established by calculus techniques.) Use this information to evaluate the following.
    (a) $\sin \frac{1}{8}$, correct to four decimal places.
    (b) $\sin(-\frac{3}{20})$, correct to the nearest ten-thousandth.
    (c) $\cos 1.6708$.

15 Double

A special case of the formula for the sine of a sum of numbers arises if the numbers are the same. We have

$$\sin(r + r) = \sin r \cos r + \sin r \cos r.$$

Since the left member of the equation is $\sin(2r)$ and the right member is $2 \sin r \cos r$, we have derived the result that

$$\sin 2r = 2 \sin r \cos r \quad \text{for every number } r.$$

We may, in a like fashion, specialize the formula for the cosine of a sum. We obtain

$$\cos(r + r) = \cos r \cos r - \sin r \sin r;$$

thus,

$$\cos 2r = \cos^2 r - \sin^2 r \quad \text{for every number } r.$$

If we subtract each member of the preceding equation from 1, the difference is

$$1 - \cos 2r = (\cos^2 r + \sin^2 r) - (\cos^2 r - \sin^2 r) = 2 \sin^2 r.$$

Hence,

$$\sin^2 r = \frac{1 - \cos 2r}{2} \quad \text{for every number } r.$$

If we add 1 to $\cos 2r$, the sum is

$$1 + \cos 2r = (\cos^2 r + \sin^2 r) + (\cos^2 r - \sin^2 r) = 2 \cos^2 r.$$

Therefore,

$$\cos^2 r = \frac{1 + \cos 2r}{2} \quad \text{for every number } r.$$

Example 32 Evaluate $\cos(3\pi/8)$.

We apply the preceding equation and obtain

$$\cos^2\left(\frac{3\pi}{8}\right) = \frac{1 + \cos(2 \cdot \frac{3}{8}\pi)}{2} = \frac{1}{2}\left(1 + \cos\frac{3\pi}{4}\right) = \frac{1}{2}(1 - 2^{-1/2}).$$

Now $\cos(3\pi/8)$ is positive, since $0 < 3\pi/8 < \pi/2$. Thus,

$$\cos\frac{3\pi}{8} = \left(\frac{1 - 2^{-1/2}}{2}\right)^{1/2}.$$

As a decimal approximation,

$$\sqrt{\frac{1}{2}\left(1 - \frac{1}{\sqrt{2}}\right)} = 0.38,$$

to the nearest hundredth.

Example 33 Evaluate $\sin(17\pi/12)$.

We first calculate the second power of the desired number:

$$\left(\sin\frac{17\pi}{12}\right)^2 = \frac{1 - \cos(2 \cdot \frac{17}{12}\pi)}{2} = \frac{1}{2}\left(1 - \cos\frac{5\pi}{6}\right)$$

$$= \frac{1}{2}\left(1 + \frac{\sqrt{3}}{2}\right) = \frac{2 + \sqrt{3}}{4}.$$

Since $\pi < 17\pi/12 < 2\pi$, we note that we are seeking a negative number. Hence,

$$\sin\frac{17\pi}{12} = -\left(\frac{2 + 3^{1/2}}{4}\right)^{1/2} = -\frac{(2 + 3^{1/2})^{1/2}}{2}.$$

To the nearest hundredth, the desired number is -0.97.

Transcendental Functions

In the preceding paragraph, we utilized the periodicity of the cosine function, where we replaced

$$\cos\left(2 \cdot \frac{17\pi}{12}\right) = \cos \frac{17\pi}{6} = \cos\left(2\pi + \frac{5\pi}{6}\right)$$

by

$$\cos \frac{5\pi}{6}.$$

EXERCISES M

1. Simplify each of the following.
   (a) sin 2 cos 5 + sin 5 cos 2.
   (b) $\cos \frac{1}{2} \cos \frac{1}{4} - \sin \frac{1}{2} \sin \frac{1}{4}$.
   (c) cos 1.8 cos 1.2 + sin 1.8 sin 1.2.
   (d) $\sin \frac{3}{2} \cos \frac{1}{2} - \sin \frac{1}{2} \cos \frac{3}{2}$.

2. Simplify each of the following.
   (a) 2 sin 3 cos 3.   (b) $\cos^2\left(\frac{1}{3}\right) - \sin^2\left(\frac{1}{3}\right)$.
   (c) $1 - 2\sin^2(a/2)$.   (d) $2\cos^2(b/2) - 1$.

3. Express each of the following in a different fashion.
   (a) $(1 - \cos 12)/2$.   (b) $1 + \cos 50$.   (c) $2\sin^2 \frac{3}{4}$.
   (d) $2\cos^2\left(-\frac{3}{5}\right)$. (e) $\sin^2 a$.   (f) $\cos^2(b/2)$.

4. Compute each of the following.
   (a) $\sin \frac{5\pi}{8}$.   (b) $\cos \frac{5\pi}{8}$.   (c) $\sin \frac{1}{2}$.   (d) $\cos \frac{1}{2}$.

5. Suppose that t is a number such that $\pi < t < 2\pi$ and $\cos t = \frac{5}{13}$.
   (a) Evaluate sin t.
   (b) Evaluate sin $2t$.
   (c) Evaluate sin $\frac{1}{2}t$.
   (d) Evaluate cos $\frac{1}{2}t$.

6. Let the number w satisfy the conditions that $-4 < w < -2$ and $\sin w = \frac{1}{3}$. Find each of the following numbers.
   (a) cos w.   (b) sin $2w$.   (c) cos $2w$.
   (d) sin $3w$.   (e) cos $\frac{1}{2}w$.   (f) $\sin \frac{w}{2}$.
   (g) $\cos \frac{w}{4}$.

7. On a single coordinate system, plot and label each of the following points: (cos 2, sin 2); (3 cos 2, 3 sin 2); ($\frac{1}{2}$ cos 2, $\frac{1}{2}$ sin 2); (cos (−1), sin (−1)); (5 cos (−1), 5 sin (−1)).

16 Applications of the addition formulas

On a two-dimensional coordinate system, let Q be a point different from the origin. The closed halfline that contains Q and whose endpoint is the origin is sometimes called the *ray from the origin through Q*. (See Fig. 36.) This ray

Figure 36

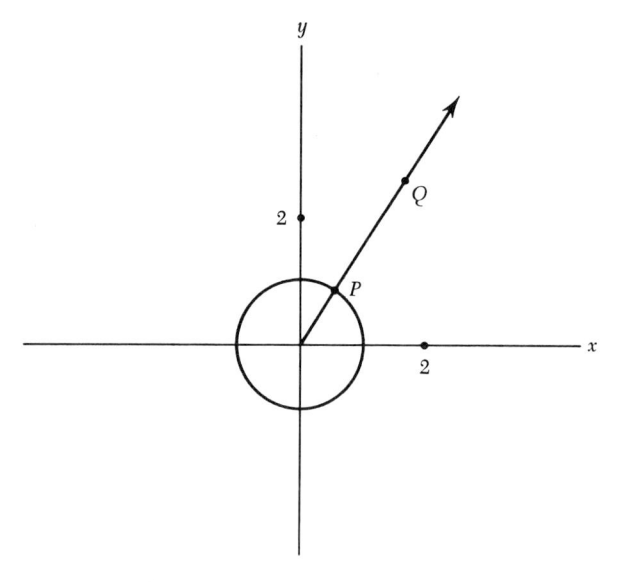

intersects the unit circle at one point, which we may call P. Since P is on the unit circle, its coordinates are (cos r, sin r), where r is a number whose image under the winding function is P.

Suppose that d is the distance between the origin O and the point Q. Then the vector from O to Q is d times the vector from O to P. Thus, the vector associated with the point Q is

$$d \cdot (\cos r, \sin r) = (d \cos r, d \sin r).$$

The components of this vector are respectively the coordinates of the point Q.

We have shown that the coordinates of any point Q other than the origin are expressible in the form

$$(d \cos r, d \sin r),$$

where d is the distance between the origin and Q and where the ray from the origin through Q intersects the unit circle at the image of r under the winding function.

Example 34 The point $(-4, 3)$ is $[(-4)^2 + (3)^2]^{1/2} = 5$ units from the origin. See Fig. 37. We may express $(-4, 3)$ as $5(\cos t, \sin t)$ or $(5 \cos t, 5 \sin t)$, where the image of t under the winding function is the intersection of the unit circle and the ray from O through Q.

Example 35 Consider the number $8 \cos 3 - 15 \sin 3$. The distance between the origin and the point $(-15, 8)$ is 17. Consequently, there is a number c such that the ray from $(0, 0)$ through $(-15, 8)$ intersects the unit circle at $(\cos c, \sin c)$. In terms of this number c, we have

$$(-15, 8) = 17 (\cos c, \sin c);$$

Figure 37

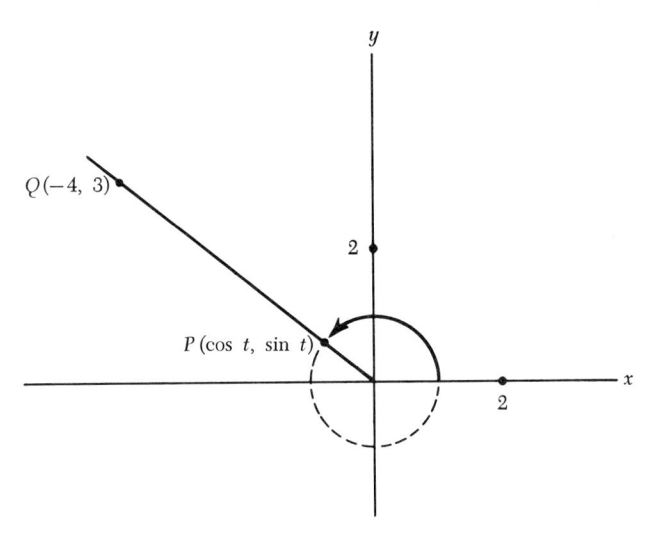

in other words, $-15 = 17 \cos c$ and $8 = 17 \sin c$. Therefore,

$$8 \cos 3 - 15 \sin 3 = (17 \sin c) \cos 3 + (17 \cos c) \sin 3$$
$$= 17 (\sin c \cos 3 + \sin 3 \cos c)$$
$$= 17 \sin (3 + c).$$

Example 36 Discuss the function given by the formula $\sin x + 2 \cos x$, and sketch its graph.

The coefficients of $\sin x$ and of $\cos x$ are 1 and 2, respectively. The point $(1, 2)$ is $\sqrt{5}$ units from the origin. Thus,

$$5^{-1/2} \cdot (1, 2) = \left(\frac{1}{5^{1/2}}, \frac{2}{5^{1/2}}\right)$$

is a point on the unit circle and therefore is the image of a number, say b, under the winding function. Since $(1, 2)$ belongs to the positive quadrant, we may choose b to satisfy the condition[1] $0 < b < \pi/2$. Now, $5^{-1/2} \cdot (1, 2) = (\cos b, \sin b)$, and we may express $(1, 2)$ as $(\sqrt{5} \cos b, \sqrt{5} \sin b)$. Hence,

$$\sin x + 2 \cos x = (\sqrt{5} \cos b) \sin x + (\sqrt{5} \sin b) \cos x$$
$$= 5^{1/2} (\sin x \cos b + \sin b \cos x)$$
$$= 5^{1/2} \sin (x + b).$$

The graph of the equation $y = \sin (x + b)$ is congruent to the graph of the sine function, because the former curve is obtainable from the latter by a horizontal translation of b units to the left. For the function given by the formula $\sin (x + b)$, the domain consists of all real numbers, the range consists of all numbers between -1 and 1 inclusive, and the period is 2π.

---
[1] By drawing a large diagram and measuring, or by referring to a table, we may find that the number b, subject to this condition, is approximately 1.107.

The function given by sin x + 2 cos x has the following properties.
1. Its domain consists of all real numbers.
2. Its range is the set $\{t \mid |t| \leq 5^{1/2}\}$.
3. It is periodic with period 2π.
4. It contains the ordered pair (0, 2).

To sketch the graph, we visualize the graph of the sine function, we shift the sine wave b units to the left, and then we multiply each y-coordinate by the amplitude factor $5^{1/2}$. See Fig. 38.

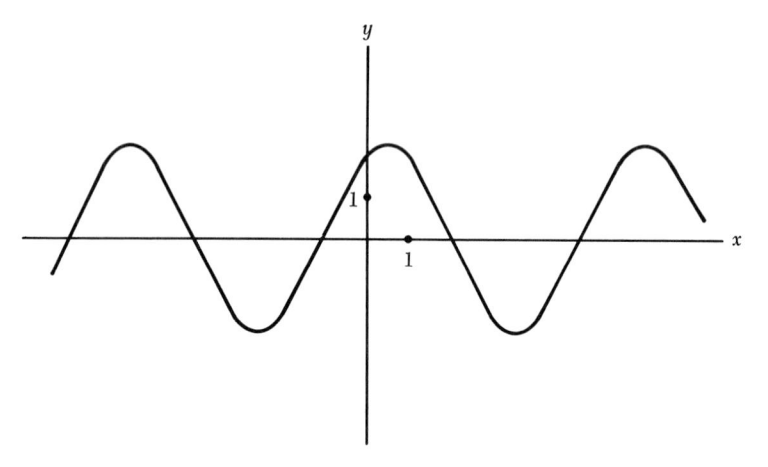

Figure 38

Example 37 Find each member of the set
$$\{x \mid (0 < x < 2\pi) \wedge (4 \sin x \cos x = 1)\}.$$

Since $2 \sin x \cos x = \sin 2x$, the condition $4 \sin x \cos x = 1$ is equivalent to $2 \sin 2x = 1$ and hence to
$$\sin 2x = \tfrac{1}{2}.$$

Since $0 < x < 2\pi$, we are interested in numbers $2x$ between 0 and 4π. A number between 0 and 4π that has the image $\tfrac{1}{2}$ under the sine function is

$$2x = \frac{\pi}{6} \quad \text{or} \quad 2x = \frac{5\pi}{6}$$

$$\text{or} \quad 2x = \frac{\pi}{6} + 2\pi = \frac{13\pi}{6} \quad \text{or} \quad 2x = \frac{5\pi}{6} + 2\pi = \frac{17\pi}{6}.$$

Thus, there are four members of the given set, namely,

$$\frac{\pi}{12}, \frac{5\pi}{12}, \frac{13\pi}{12}, \frac{17\pi}{12}.$$

Transcendental Functions

EXERCISES N

1. Let b be a real number such that $(\cos b, \sin b) = (-\frac{12}{13}, \frac{5}{13})$. Find another expression for the number $-12 \cos 4 + 5 \sin 4$.

2. For each of the following formulas, find another formula that represents the same function as the function prescribed by the given formula.
   (a) $\sin x + \cos x$.
   (b) $\sin x - \cos x$.
   (c) $2 \cos x - \sin x$.

3. Show that the function given by the formula $\sin 2x$ is periodic with period π. [HINT: Two facts must be established. First, $\sin 2(r + \pi) = \sin 2r$ for every number r. Second, no positive number p less than π has the property that $\sin 2(r + p) = \sin 2r$ for every r.]

4. Show that the function $\{(x, y) \mid y = \cos(x/3)\}$ is periodic with period 6π.

5. Show that the period of the periodic function $\{(x, y) \mid y = \sin 2\pi x\}$ is one.

6. Let k be a positive number.
   (a) Show that the function given by the formula $\cos kx$ is periodic.
   (b) What is the period of this function?
   (c) Justify your answer to part (b).

7. Let k be a positive number, let h be a nonzero number, let m be a number, and let G be the function $\{(x, y) \mid y = h \sin kx + m\}$.
   (a) Show that G is periodic.
   (b) What is the period of G? Justify your answer.
   (c) What is the domain of G?
   (d) Give Rg G.

8. For each of the following, make a list of the important analytical properties of the function given by the formula. Discuss such topics as the domain, the range, the period (if any), whether the function is odd or even, where the function is increasing, where decreasing, the image of the number 0.
   (a) $4 \sin x$.
   (b) $2 \cos x$.
   (c) $-3 \cos x$.
   (d) $2 \sin \left(x - \frac{\pi}{3}\right)$.
   (e) $\sin 2\pi x$.
   (f) $6 \cos \pi x$.
   (g) $\sin x + \cos x$.
   (h) $\cos x - \sin x$.
   (i) $\sqrt{3} \cos x + \sin x$.
   (j) $1 - 3 \cos 4x$.
   (k) $\frac{1}{2} + \frac{1}{2} \cos 2x$.
   (l) $\sin^2 x$.
   (m) $4 \sin x \cos x$.

9. Using the results of Exercise 8, sketch the graph of each of the following equations.
   (a) $y = 4 \sin x$.
   (b) $y = 2 \cos x$.
   (c) $y = -3 \cos x$.
   (d) $y = 2 \sin \left(x - \frac{\pi}{3}\right)$.
   (e) $y = \sin 2\pi x$.
   (f) $y = 6 \cos \pi x$.
   (g) $y = \sin x + \cos x$.
   (h) $y = \cos x - \sin x$.
   (i) $y = \sqrt{3} \cos x + \sin x$.
   (j) $y = 1 - 3 \cos 4x$.
   (k) $y = \frac{1}{2} + \frac{1}{2} \cos 2x$.
   (l) $y = \sin^2 x$.
   (m) $y = 4 \sin x \cos x$.

10. Show that the graph of sin is obtainable from the graph of cos by translating (or shifting) horizontally, and tell how far and in what direction the cosine curve should be translated to obtain the sine curve.

17 Tangent function

Let r be a real number. The image of r under the winding function is a point on the unit circle, which we may call P; its coordinates are $(\cos r, \sin r)$. Suppose that P does not lie on a coordinate axis. Then the line joining P and the origin has slope $(\sin r)/(\cos r)$, and an equation for the line is

$$y = \frac{\sin r}{\cos r} x.$$

(See Fig. 39.)

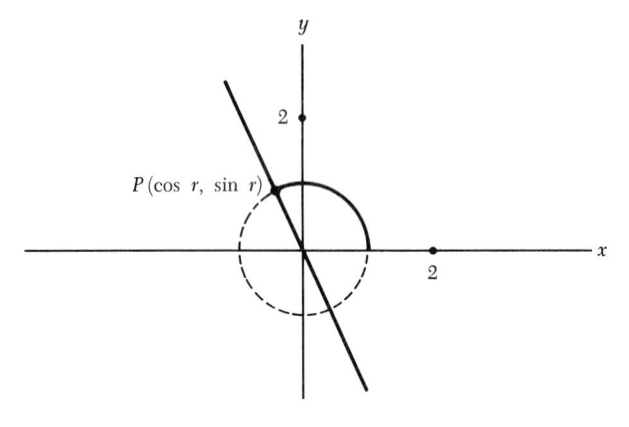

Figure 39

The slope of this line is determined by the number r. This suggests a new function. We proceed to introduce the tangent function.

Definition For each real number r such that $\cos r \neq 0$, the *tangent* of the number r is the quotient $(\sin r)/(\cos r)$.

If we abbreviate the tangent function by tan, then

$$\tan r = \frac{\sin r}{\cos r}.$$

Example 38 Since the image of $3\pi/4$ under the winding function is the point $\left(\cos \frac{3\pi}{4}, \sin \frac{3\pi}{4}\right) = (-2^{-1/2}, 2^{-1/2})$, we calculate

$$\tan \frac{3\pi}{4} = \frac{2^{-1/2}}{-2^{-1/2}} = -1.$$

The point $(1, 0)$ is associated with the number 0 by the winding process; consequently,

$$\tan 0 = \frac{0}{1} = 0.$$

By contrast, $\pi/2$ does not belong to the domain of the tangent function, since $\pi/2$ is not a root of the inequality $\cos r \neq 0$.

Transcendental Functions

Example 39 Consider the ray from the origin through the point Q with coordinates $(-1, -4)$. Suppose this ray intersects the unit circle at the image of the number t under the winding function. Since the distance between Q and the origin is $d = \sqrt{1^2 + 4^2} = \sqrt{17}$, the point Q has coordinates $(\sqrt{17} \cos t, \sqrt{17} \sin t)$. Thus,

$$\tan t = \frac{\sin t}{\cos t} = \frac{\sqrt{17} \sin t}{\sqrt{17} \cos t} = \frac{-4}{-1} = 4.$$

Geometrically speaking, the tangent of a number r is the slope of the line joining the origin and the point on the unit circle that is matched with r by the winding process. In case the line is vertical, it has no slope, and the corresponding number r does not belong to Dom tan. In case the line is horizontal, the slope may be considered to be zero, and the corresponding number r has the property that $\tan r = 0$.

The domain of the tangent function contains all numbers r except those for which $\cos r = 0$. These exceptional values are $\pi/2, -\pi/2, 3\pi/2, -3\pi/2, 5\pi/2$, etc., namely, the multiples of $\pi/2$ in which the multiplier is an odd integer. Since any number may be the slope of a line, the range of tan contains all numbers.

Like the sine and cosine functions, their quotient is also periodic. However, the period of tan is not 2π but π, as the following relations suggest:

$$\tan (r + \pi) = \frac{\sin (r + \pi)}{\cos (r + \pi)} = \frac{\sin r \cos \pi + \sin \pi \cos r}{\cos r \cos \pi - \sin \pi \sin r} = \frac{-\sin r}{-\cos r} = \tan r$$

for all numbers $r \in$ Dom tan.

Since the sine and cosine functions are odd and even, respectively, their quotient is an odd function. We may verify this claim as follows:

$$\tan (-r) = \frac{\sin (-r)}{\cos (-r)} = \frac{-\sin r}{\cos r} = -\tan r$$

for every $r \in$ Dom tan.

The tangent function is increasing on the set $\{r \mid 0 < r < \pi/2\}$. To establish this assertion, consider two numbers a, c such that $0 < a < c < \pi/2$. Then (see Fig. 28) the slope of the line joining $(\cos c, \sin c)$ and the origin is greater than the slope of the line joining $(\cos a, \sin a)$ and the origin. That is, $\tan a < \tan c$.

We summarize several significant properties.

1. Dom tan consists of all numbers except the odd integral multiples of $\pi/2$.
2. Rg tan consists of all real numbers.
3. The tangent function is periodic with period π.
4. The tangent function is an odd function.
5. The tangent function is increasing on the set $\{r \mid 0 < r < \pi/2\}$.
6. $(0, 0) \in$ tan.

These properties enable us to sketch easily the graph of the equation $y = \tan x$. By property 3, the curve is the union of branches, all of which are congruent to one another and each of which lies in a vertical strip π units wide. By property 1, one of these branches lies in the region $\{(x, y) \mid -\pi/2 < x < \pi/2\}$; after drawing this branch, others may be obtained by translation. By property 4, the semibranch in the region $\{(x, y) \mid -\pi/2 < x < 0\}$ is obtainable from the semibranch lying in the region $\{(x, y) \mid 0 < x < \pi/2\}$ by rotating halfway around the origin. By properties 6 and 5, together with the fact that $(\pi/4, 1) \in \tan$, we obtain the partial graph shown in Fig. 40a. By property 2, we then obtain the semibranch in Fig. 40b. Rotation finishes the full branch pictured in Fig. 40c. Finally, periodicity enables us to complete the sketch, as shown in Fig. 41.

Figure 40

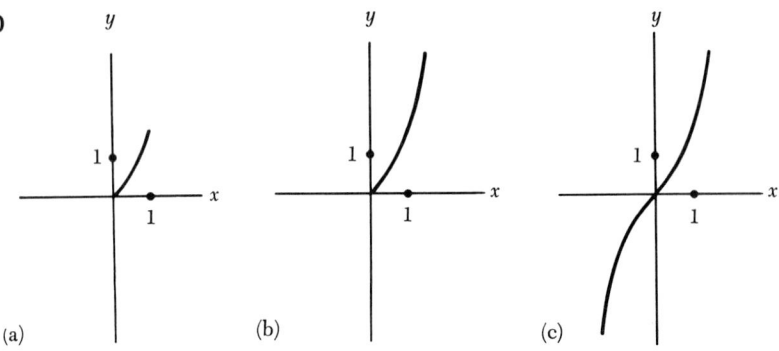

(a) (b) (c)

Figure 41

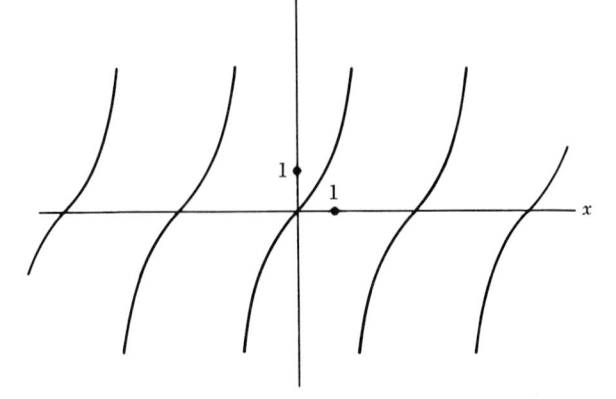

EXERCISES O

1. Evaluate each of the following.

    (a) $\tan \frac{\pi}{4}$.   (b) $\tan \pi$.   (c) $\tan \frac{2\pi}{3}$.   (d) $\tan\left(-\frac{\pi}{6}\right)$.   (e) $\tan\left(-\frac{11}{4}\pi\right)$.

2. If b is a number such that $\sin b = \frac{2}{3}$ and $\cos b < 0$, find $\tan b$.

3. If c is a number such that $\pi < c < 2\pi$ and $\cos c = \frac{3}{4}$, find $\tan c$.

4. Find an equation for the line that contains the point $(2, -1)$ and whose slope is $\tan 3$.

5. The ray from the origin through the point $(3, -8)$ intersects the unit circle at the point that matches the number q under the winding function.
   (a) Find $\tan q$.   (b) Find $\sin q$.

6. The line through the origin and the point $(-2, 3)$ intersects the unit circle at the point $(\cos q, \sin q)$.
   (a) Evaluate $\tan q$.
   (b) Explain why the information is not sufficient to determine $\cos q$.
   (c) List all the logically possible values for $\cos q$ in this situation.
   (d) List all the logical possibilities for the number $\sin q$.

7. For the set in each of the following parts, list all the members of the set.
   (a) $\{x \mid (0 < x < 2\pi) \wedge (\sin x = \frac{1}{2})\}$.
   (b) $\{x \mid (0 < x < 2\pi) \wedge (\tan x = 1)\}$.
   (c) $\{x \mid 0 < x < \pi\} \cap \{x \mid \cos x = -2^{-1/2}\}$.
   (d) $\{x \mid (-\pi < x < \pi) \wedge (\cos 2x = 3^{1/2}/2)\}$.
   (e) $\{x \mid -\pi < x < \pi\} \cap \{x \mid 4 \sin x \cos x = -3^{1/2}\}$.

8. Find three roots of the equation $2 \cos x \cos 2 - 2 \sin x \sin 2 = 1$.

9. Solve each of the following systems of inequalities.
   (a) $0 < x < \pi/2$, $\sin x > \frac{1}{2}$.
   (b) $0 < x < 2\pi$, $\sin x < -\frac{1}{2}$.
   (c) $-\pi/2 < x < \pi/2$, $\tan^2 x < 1$.
   (d) $0 < x < 2\pi$, $\cos^2 2x - \sin^2 2x > 2^{-1/2}$.

10. Let p be a positive real number, and consider the statement that
    $$\tan (r + p) = \tan r \quad \text{for every number } r \in \text{Dom tan}.$$
    We have seen that the statement becomes true when p is replaced by π.
    (a) Show that the statement is true when 2π is substituted for p.
    (b) Find still another value of p for which the statement is true.
    (c) Show that the sentence becomes a false statement if $\pi/2$ is substituted for p.
    (d) Complete the proof of the claim that the period of $\tan$ is π by showing that π is the *smallest* of the positive numbers p for which the statement is true.

11. Show that the function given by the formula $\tan 2x$ is periodic with period $\pi/2$.

12. For what positive number k does the function $\{(x, y) \mid y = \tan kx\}$ have period one?

13. For each of the following formulas, discuss the function given by the formula, and then sketch the graph of the function.
    (a) $\tan 2x$.   (b) $-\tan \frac{1}{2}x$.   (c) $\tan x + 1$.   (d) $3 \tan \left(x - \frac{\pi}{4}\right)$.

14. Let r and t be real numbers such that neither r nor t nor the sum $r + t$ is an odd multiple of $\pi/2$. Prove that
    $$\tan (t + r) = \frac{\tan t + \tan r}{1 - \tan t \tan r}.$$
    [HINT: Transform the right member into an expression involving images under the sine and cosine functions, and then simplify to obtain the left member.]

18 Summary

We collect together some important identities involving the trigonometric functions. The following equations are true for all real numbers r, t (except where otherwise indicated).

1. $\sin^2 r + \cos^2 r = 1$.
2. $\sin(t + r) = \sin t \cos r + \sin r \cos t$.
2a. $\sin(r + 2\pi) = \sin r$.
2b. $\sin 2r = 2 \sin r \cos r$.
3. $\sin(t - r) = \sin t \cos r - \sin r \cos t$.
3a. $\sin(-r) = -\sin r$.
3b. $\sin\left(\dfrac{\pi}{2} - r\right) = \cos r$.
4. $\cos(t + r) = \cos t \cos r - \sin t \sin r$.
4a. $\cos(r + 2\pi) = \cos r$.
4b. $\cos 2r = \cos^2 r - \sin^2 r$.
5. $\cos(t - r) = \cos t \cos r + \sin t \sin r$.
5a. $\cos(-r) = \cos r$.
5b. $\cos\left(\dfrac{\pi}{2} - r\right) = \sin r$.
6. $1 + \cos 2r = 2 \cos^2 r$.
7. $1 - \cos 2r = 2 \sin^2 r$.
8. $\tan r = (\sin r)/(\cos r)$ if r is not an odd multiple of $\pi/2$.
9. $\tan(r + \pi) = \tan r$ if r is not an odd multiple of $\pi/2$.

19 Angle

There are several worthwhile notions of an angle. One of these may be associated with the winding process. On a two-dimensional coordinate system, consider the ray from the origin O through the point U with coordinates $(1, 0)$. Suppose a real number r is given. As we apply the winding process, winding r units counterclockwise around the unit circle from U to determine the point P that matches r, we may imagine the ray as rotating about its endpoint along with the winding. The terminal position of the ray contains P. (See Fig. 42.)

This rotation may be considered to generate the angle UOP. The number r serves as a measure of the angle. We call r the *radian measure* of the angle.

In case $r = \pi/2$, we wind around one quadrant of the unit circle and reach the point $(0, 1)$. The angle generated is a right angle. Thus, the radian measure and the degree measure of an angle are related by the statement

$$\frac{\pi}{2} \text{ radians} = 90 \text{ degrees} = \frac{1}{4} \text{ revolution}.$$

Transcendental Functions

Figure 42

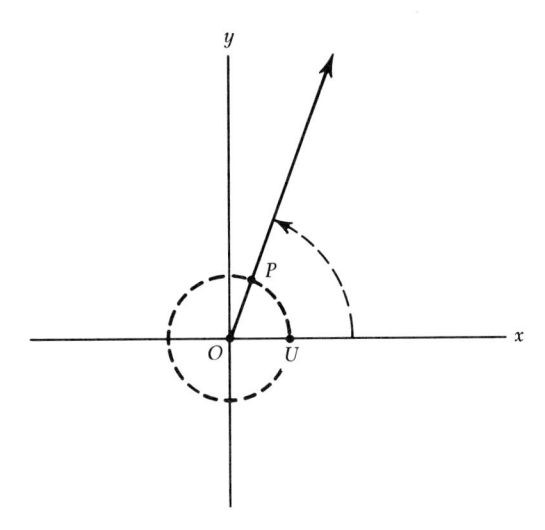

As consequences, we find that 1 degree is $\pi/180$ radians and that 1 radian is $180/\pi$ degrees. Since $180/\pi$ is approximately 57.3, we conclude that 1 radian is nearly $57°18'$.

In many applications we speak of the "sine of an angle." The sine of an angle is a number. If a given angle has the number r as its radian measure, then the sine of the angle is the same as the sine of the number r. We restate the two preceding sentences as a definition.

Definition The *sine of an angle* is the same number as the sine of the radian measure of the angle.

Analogous definitions describe the *cosine of an angle* or the *tangent of an angle*; we merely replace each occurrence of the word "sine" in the above definition by "cosine" or by "tangent," respectively.

Example 40 Since $60° = \frac{2}{3} \cdot 90°$ corresponds to $\frac{2}{3} \cdot (\pi/2) = \pi/3$ radians,

$$\sin 60° = \sin \frac{\pi}{3} = \frac{\sqrt{3}}{2}, \qquad \cos 60° = \cos \frac{\pi}{3} = \frac{1}{2},$$

and $\quad \tan 60° = \tan \dfrac{\pi}{3} = 3^{1/2}.$

Since $225° = \frac{5}{2} \cdot 90°$ corresponds to $\frac{5}{2} \cdot (\pi/2) = \frac{5}{4}\pi$ radians,

$$\sin 225° = \sin \frac{5\pi}{4} = -\frac{\sqrt{2}}{2}, \qquad \cos 225° = \cos \frac{5\pi}{4} = -\frac{\sqrt{2}}{2},$$

and $\quad \tan 225° = \tan \dfrac{5\pi}{4} = 1.$

19 / Angle

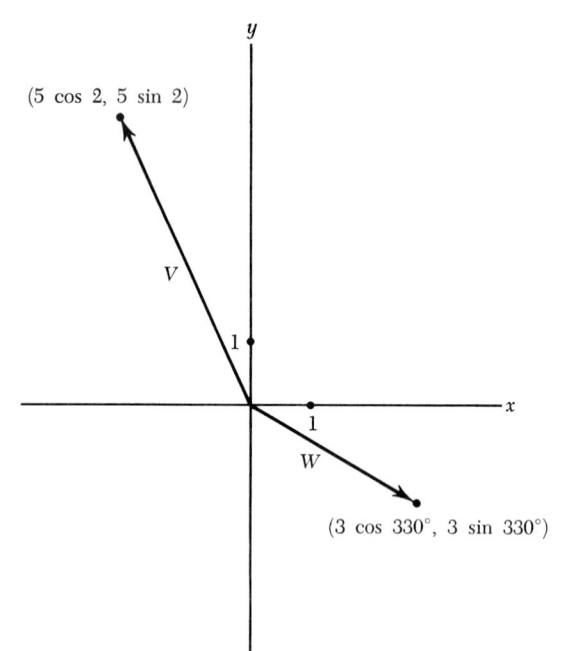

Figure 43

(5 cos 2, 5 sin 2)

V

(3 cos 330°, 3 sin 330°)

W

Example 41 Let V be the vector 5 (cos 2, sin 2); let W be the vector with initial point at the origin and with terminal point three units away along the ray that makes an angle of 330° with the ray from O through U. (See Fig. 43.) Find the sum $V + W$.

Since $W = 3\,(\cos 330°, \sin 330°)$, the sum $V + W$ is

$$5\,(\cos 2, \sin 2) + 3\,(\cos 330°, \sin 330°)$$
$$= \left(5 \cos 2 + 3 \cos \frac{11\pi}{6},\ 5 \sin 2 + 3 \sin \frac{11\pi}{6}\right).$$

If we prefer a decimal approximation for the components of $V + W$, we may calculate

$$5\,(\cos 2, \sin 2) + 3(\cos 330°, \sin 330°)$$
$$= 5(-0.42, 0.91) + 3(0.87, -0.50)$$
$$= (0.51, 3.05).$$

20 Rotation matrix

Consider the rule that assigns to each real number p the matrix

$$\begin{bmatrix} \cos p & -\sin p \\ \sin p & \cos p \end{bmatrix}.$$

498 Transcendental Functions

This rule describes a function whose domain is the set of all real numbers and whose range is a set of matrices rather than a set of numbers. If we denote the image of the number p under this function by $M(p)$, then

$$M(p) = \begin{bmatrix} \cos p & -\sin p \\ \sin p & \cos p \end{bmatrix}.$$

We observe that the image of the number 0 is the identity matrix, because

$$M(0) = \begin{bmatrix} \cos 0 & -\sin 0 \\ \sin 0 & \cos 0 \end{bmatrix} = \begin{bmatrix} 1 & 0 \\ 0 & 1 \end{bmatrix} = I.$$

Since each of the functions cos, sin, $-$sin appearing in the matrix $M(p)$ has period 2π, the function M also is periodic with period 2π.

On a two-dimensional coordinate system, let Q be a point other than the origin. The ray from the origin through Q makes an angle of, say, r radians with the ray through U. If the positive number d is the distance between Q and the origin, then (see Section 16) the point Q may be identified by the vector $\begin{bmatrix} d \cos r \\ d \sin r \end{bmatrix}$. We have here chosen to write a column vector in order to fit with our work in Chapter 3.

For a fixed number p, the matrix $M(p)$ is associated with a linear transformation on a two-dimensional vector space. The image of the vector

$$V = \begin{bmatrix} d \cos r \\ d \sin r \end{bmatrix}$$

under the linear transformation is the product $M(p) \cdot V$. In order to evaluate this product, we use Theorems 4 and 5 in Section 14.

$$M(p) \cdot V = \begin{bmatrix} \cos p & -\sin p \\ \sin p & \cos p \end{bmatrix} \cdot \begin{bmatrix} d \cos r \\ d \sin r \end{bmatrix}$$

$$= \begin{bmatrix} d(\cos p \cos r - \sin p \sin r) \\ d(\sin p \cos r + \cos p \sin r) \end{bmatrix}$$

$$= \begin{bmatrix} d \cos (r + p) \\ d \sin (r + p) \end{bmatrix}.$$

The point Q' represented by the product is also d units from the origin. The ray from the origin through Q' makes an angle of $r + p$ radians with the ray through U. Thus (see Fig. 44), the ray from the origin through Q' makes an angle of p radians with the ray through Q. The geometric effect of the linear transformation on the vector V is to rotate it around the origin through an angle of p radians. The matrix $M(p)$ may be considered as a rotation matrix. We have established the following theorem.

Figure 44

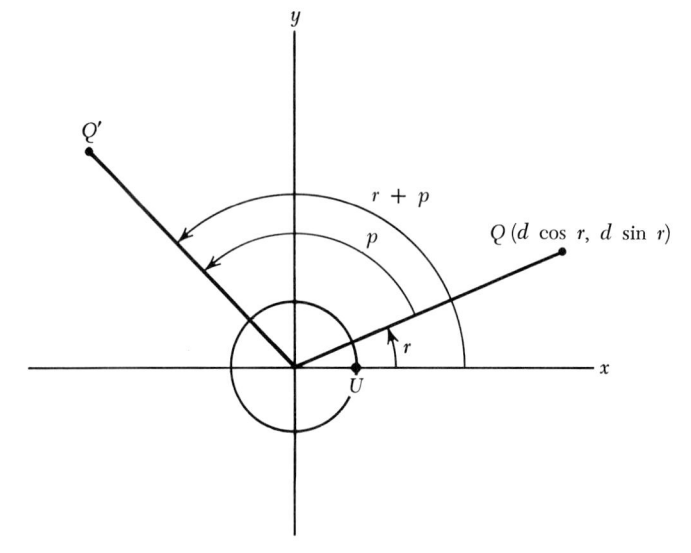

Theorem 6 For each real number p, the linear transformation associated with the matrix $M(p)$ is a counterclockwise rotation of the vector space around the origin through an angle of p radians.

Example 42 The matrix associated with $p = \dfrac{\pi}{4}$ is

$$\begin{bmatrix} 2^{-1/2} & -2^{-1/2} \\ 2^{-1/2} & 2^{-1/2} \end{bmatrix}.$$

If V is the vector

$$\begin{bmatrix} 4\cos\dfrac{3\pi}{4} \\ 4\sin\dfrac{3\pi}{4} \end{bmatrix} = \begin{bmatrix} -2^{3/2} \\ 2^{3/2} \end{bmatrix},$$

then the image of V is

$$\begin{bmatrix} 2^{-1/2} & -2^{-1/2} \\ 2^{-1/2} & 2^{-1/2} \end{bmatrix} \cdot \begin{bmatrix} -2^{3/2} \\ 2^{3/2} \end{bmatrix} = \begin{bmatrix} -4 \\ 0 \end{bmatrix}.$$

Since

$$\begin{bmatrix} -4 \\ 0 \end{bmatrix} = \begin{bmatrix} 4\cos\pi \\ 4\sin\pi \end{bmatrix},$$

we verify that the vector has been rotated $\tfrac{1}{8}$ of a revolution around the origin. (See Fig. 45.)

500   Transcendental Functions

Figure 45

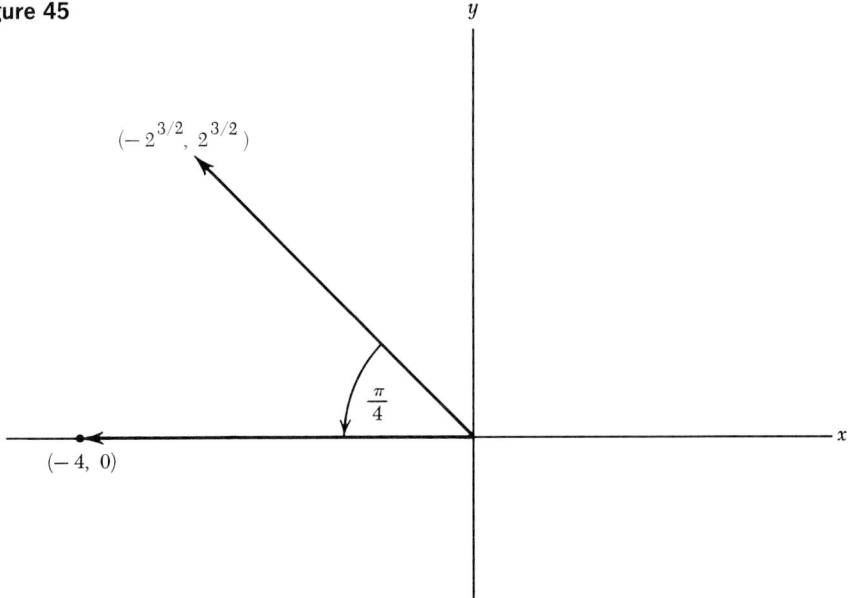

(−2^{3/2}, 2^{3/2})

(−4, 0)

π/4

y

x

EXERCISES P

1. Several angle measures are given below in either degrees or radians. Give the equivalent in the other unit of measurement.
   (a) 60 degrees
   (b) $5\pi/6$ radians.
   (c) $\frac{1}{2}$ radian.
   (d) 9000 degrees.
   (e) 22°30′.
   (f) 30 radians.
   (g) $(\frac{1}{2})$°.
   (h) $4\pi/3$ radians.
   (i) $\pi/4$ radians.
   (j) 1′.
   (k) (0.01)°.
   (l) 0.01 radian.

2. Evaluate each of the following.
   (a) sin 30°.
   (b) cos 90°.
   (c) tan 45°.
   (d) cos 120°.
   (e) tan 120°.
   (f) sin (−90°).
   (g) cos (−540°).
   (h) tan (−150°).
   (i) sin (−225°).
   (j) cos (−60°).
   (k) tan 180°.
   (l) cos 15°.
   (m) sin (67.5)°.
   (n) tan 112°30′.
   (o) sin 46° cos 29° + sin 29° cos 46°.
   (p) $(\sin 18°)^2 + (\cos 18°)^2$.

3. Evaluate the rotation matrix $M(\pi/2)$.

4. If the point $(-1, 1)$ is rotated 1 radian counterclockwise around the origin, what are the coordinates of the image point?

5. Let a point in two-dimensional space be represented by the column vector $\begin{bmatrix} a \\ b \end{bmatrix}$.
   (a) Evaluate the rotation matrix $M(\pi)$.
   (b) Find the image of $\begin{bmatrix} a \\ b \end{bmatrix}$ under the rotation matrix $M(\pi)$.
   (c) Describe the geometrical interpretation of the linear transformation associated with $M(\pi)$.
   (d) Compare the above results with Theorem 3c in Section 5 of Chapter 11.

6. Let p and t be real numbers.
   (a) Multiply the rotation matrices $M(p)$ and $M(t)$ together, and show that the product is the matrix $M(p + t)$.
   (b) Interpret the result in part (a) from the viewpoint of linear transformations.

7. Let p be a real number.
   (a) Show algebraically that the rotation matrix $M(-p)$ is the inverse of the matrix $M(p)$.
   (b) Explain geometrically why the inverse of $M(p)$ should be $M(-p)$.

8. Let N be the matrix $\begin{bmatrix} -1 & 0 \\ 0 & 1 \end{bmatrix}$, and consider the rotation matrix $M(\pi)$ discussed in Exercise 5.
   (a) Calculate the image of a vector $\begin{bmatrix} a \\ b \end{bmatrix}$ under the linear transformation associated with the matrix N.
   (b) Describe the geometrical effect of the linear transformation associated with N. [HINT: Compare Theorem 3a in Section 5 of Chapter 11.]
   (c) Show that $N \cdot M(\pi) = M(\pi) \cdot N$.
   (d) Describe geometrically the linear transformation associated with the product $N \cdot M(\pi)$.

21 Triangles and trigonometry

The next two examples illustrate some applications of the trigonometric functions to the measurement of triangles.

Example 43 Let the vertices of a certain triangle be denoted A, B, C. Suppose that the angle at A is measured as $67°$ and that the distances between A and B and between A and C are measured as 32 meters and 18 meters, respectively.

We introduce a two-dimensional coordinate system in the plane containing the triangle: we choose A as the origin, we choose B to lie on the horizontal axis in the right coordinate halfplane, and we choose C to lie in the upper coordinate halfplane. (See Fig. 46.) Then the coordinates of B are $(32, 0)$. Since $67°$ is the measure of the angle BAC, the coordinates of the point C which is 18 meters from A are $(18 \cos 67°, 18 \sin 67°)$. Thus, the distance between B and C is

$$[(18 \cos 67° - 32)^2 + (18 \sin 67°)^2]^{1/2}$$
$$= [(18 \cos 67°)^2 - 2(32)(18 \cos 67°) + (32)^2 + (18 \sin 67°)^2]^{1/2}$$
$$= [18^2 + 32^2 - 2(32)(18) \cos 67°]^{1/2},$$

since $(\cos 67°)^2 + (\sin 67°)^2 = 1$. The expression for the distance may be simplified further to be

$$(1348 - 1152 \cos 67°)^{1/2}.$$

Figure 46

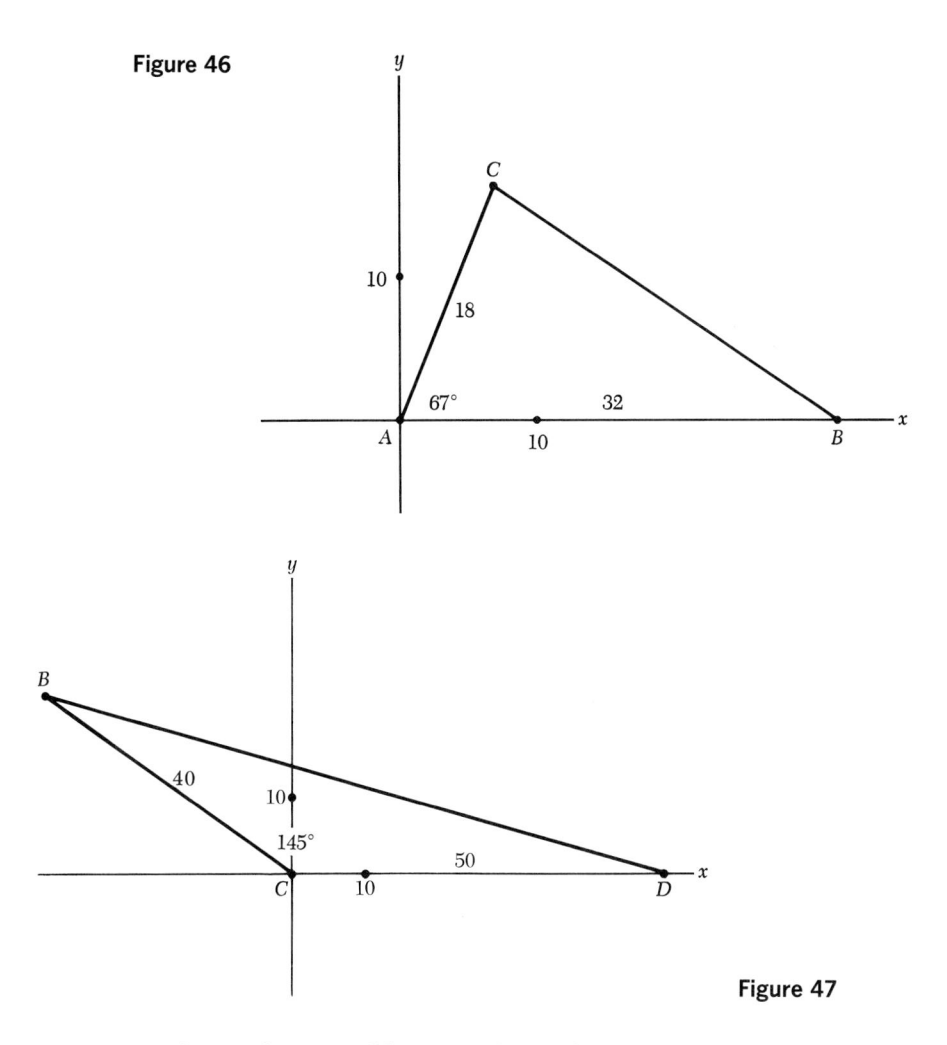

Figure 47

If we refer to a table giving decimal approximations, we find that $\cos 67°$ is 0.391 to the nearest thousandth. Hence, we may calculate an approximation as follows:

$$(1348 - 1152 \cos 67°)^{1/2} = [1348 - 1152(0.391)]^{1/2} = (898)^{1/2}.$$

The distance between B and C is nearly 30 meters.

Example 44 Let the vertices of a certain triangle be called B, C, D. Suppose that B and D are, respectively, 40 meters and 50 meters distant from the point C; also suppose that the measure of the angle BCD is 145°. Calculate the area of the interior of the triangle.

In the plane that contains the triangle we choose a coordinate system: let C be the origin, let D be on the horizontal coordinate axis in the right coordinate halfplane, and let B be in the upper coordinate halfplane. (See Fig. 47.)

21 / Triangles and trigonometry   503

Since the distance from C to D is 50, the point D has coordinates (50, 0). The coordinates of the point B are (40 cos 145°, 40 sin 145°). If we consider the segment joining vertices C and D as the base of the triangle, then the altitude of the triangle is the height of B above the x-coordinate axis, namely, the y-coordinate of B. The usual formula for the area of a triangular region (one-half the product of the base and the altitude) gives in this case an area $\frac{1}{2}(50)(40 \sin 145°)$. After simplification, we obtain 1000 sin 145°. According to tables, sin 145° is 0.574 to the nearest thousandth. The area of the region enclosed by the triangle is approximately 574 square meters.

EXERCISES Q

1. A triangle has vertices at the points A, B, C with respective coordinates (0, 0), (3, 0), (−1, 6).
   (a) Calculate the tangent of the angle BAC.
   (b) Find the distance b between the points A and C.
   (c) Calculate the sine of the angle BAC.
   (d) Find the distance c between A and B.
   (e) Calculate the cosine of the angle BAC.
   (f) If a is the distance between B and C, find a^2.
   (g) Using the results of parts (b), (d), (e), calculate the number
      $b^2 + c^2 - 2bc \cos (\angle BAC)$.
   (h) Compare your answers in part (f) and part (g).

2. Repeat Exercise 1 for the triangle whose vertices A, B, C are the points (0, 0), (5, 0), (7, 1), respectively.

3. A triangle has vertices K, L, M. The distance between K and L is 4 meters, the distance between M and L is 7 meters, and the sine of the acute angle MLK is 2/3.
   (a) Establish a convenient coordinate system in the plane containing the triangle.
   (b) Find the coordinates of each of the three points.
   (c) Find the distance between K and M.

4. Repeat Exercise 3, with the following modification: replace the word "acute" by the word "obtuse."

5. A triangle has vertices F, G, H. The cosine of the angle FGH is $-\frac{1}{4}$. The distance between G and F is 20 feet; the distance between G and H is 30 feet.
   (a) Establish a coordinate system in the plane containing the triangle, and then find the coordinates of each of the vertices.
   (b) Find the slope of each line that contains a side of the triangle.
   (c) Find the midpoint of each side of the triangle.
   (d) Find the distance between each vertex and the midpoint of the opposite side of the triangle.
   (e) Find the length of the altitude of the triangle from vertex H.
   (f) Find the area of the interior of the triangle.

6. Repeat Exercise 5, with the following modified data. The distance between G and F is 60 meters. The distance between G and H is 10 meters. The cosine of the angle FGH is $\frac{3}{5}$.

7. Find the sum of the following three vectors: the vector $(2, -9)$, the vector that is 5 times the vector $(\cos(\pi/3), \sin(\pi/3))$, and the vector from the origin to a point seven units away in the direction of the ray that makes an angle of $210°$ with the ray from O through U.

8. One formula states that the area of a triangular region is one-half the product of the lengths of any two sides of the triangle and the sine of the angle included by these two sides. With the aid of a suitable coordinate system, prove that this formula is correct.

9. Let the lengths of the sides of a triangle be the numbers a, b, c, respectively. Let A be the vertex of the triangle opposite the side of length a. An important generalization of the theorem of Pythagoras is expressed by the equation

$$a^2 = b^2 + c^2 - 2bc\cos(\angle A).$$

With the aid of a suitable coordinate system (and some ideas suggested by Example 43 and Exercise 1), prove that this equation is true for every triangle.

Appendix

A Table of Functional Values

The values in columns 2 through 7 of the following Table of Functional Values have been extracted from the following books, which were prepared by the Federal Works Agency, Works Projects Administration for the City of New York, under the sponsorship of the National Bureau of Standards: *Tables of the Exponential Function e^x*, 1939; *Table of Natural Logarithms*, volume III, 1941; *Tables of Probability Functions*, volume II, 1942; *Tables of Sines and Cosines for Radian Arguments*, 1940. The values in column 8 are from *Table of Circular and Hyperbolic Tangents and Cotangents for Radian Arguments*, prepared by the Mathematical Tables Project, National Bureau of Standards (New York: Columbia University Press, 1947).

A Table of Functional Values*

x	e^x	$\ln x$	$\phi(x)$	$F(x)$	$\sin x$	$\cos x$	$\tan x$
0.01	1.010	−4.605	0.399	0.504	0.010	1.000	0.010
0.02	1.020	−3.912	0.399	0.508	0.020	1.000	0.020
0.03	1.030	−3.507	0.399	0.512	0.030	1.000	0.030
0.04	1.041	−3.219	0.399	0.516	0.040	0.999	0.040
0.05	1.051	−2.996	0.398	0.520	0.050	0.999	0.050
0.06	1.062	−2.813	0.398	0.524	0.060	0.998	0.060
0.07	1.073	−2.659	0.398	0.528	0.070	0.998	0.070
0.08	1.083	−2.526	0.398	0.532	0.080	0.997	0.080
0.09	1.094	−2.408	0.397	0.536	0.090	0.996	0.090
0.10	1.105	−2.303	0.397	0.540	0.100	0.995	0.100
0.11	1.116	−2.207	0.397	0.544	0.110	0.994	0.110
0.12	1.127	−2.120	0.396	0.548	0.120	0.993	0.121
0.13	1.139	−2.040	0.396	0.552	0.130	0.992	0.131
0.14	1.150	−1.966	0.395	0.556	0.140	0.990	0.141
0.15	1.162	−1.897	0.394	0.560	0.149	0.989	0.151
0.16	1.174	−1.833	0.394	0.564	0.159	0.987	0.161
0.17	1.185	−1.772	0.393	0.567	0.169	0.986	0.172
0.18	1.197	−1.715	0.393	0.571	0.179	0.984	0.182
0.19	1.209	−1.661	0.392	0.575	0.189	0.982	0.192
0.20	1.221	−1.609	0.391	0.579	0.199	0.980	0.203
0.21	1.234	−1.561	0.390	0.583	0.208	0.978	0.213
0.22	1.246	−1.514	0.389	0.587	0.218	0.976	0.224
0.23	1.259	−1.470	0.389	0.591	0.228	0.974	0.234
0.24	1.271	−1.427	0.388	0.595	0.238	0.971	0.245
0.25	1.284	−1.386	0.387	0.599	0.247	0.969	0.255
0.26	1.297	−1.347	0.386	0.603	0.257	0.966	0.266
0.27	1.310	−1.309	0.385	0.606	0.267	0.964	0.277
0.28	1.323	−1.273	0.384	0.610	0.276	0.961	0.288
0.29	1.336	−1.238	0.383	0.614	0.286	0.958	0.298
0.30	1.350	−1.204	0.381	0.618	0.296	0.955	0.309
0.31	1.363	−1.171	0.380	0.622	0.305	0.952	0.320
0.32	1.377	−1.139	0.379	0.626	0.315	0.949	0.331
0.33	1.391	−1.109	0.378	0.629	0.324	0.946	0.343
0.34	1.405	−1.079	0.377	0.633	0.333	0.943	0.354
0.35	1.419	−1.050	0.375	0.637	0.343	0.939	0.365
0.36	1.433	−1.022	0.374	0.641	0.352	0.936	0.376
0.37	1.448	−0.994	0.373	0.644	0.362	0.932	0.388
0.38	1.462	−0.968	0.371	0.648	0.371	0.929	0.399
0.39	1.477	−0.942	0.370	0.652	0.380	0.925	0.411
0.40	1.492	−0.916	0.368	0.655	0.389	0.921	0.423
0.41	1.507	−0.892	0.367	0.659	0.399	0.917	0.435
0.42	1.522	−0.868	0.365	0.663	0.408	0.913	0.447
0.43	1.537	−0.844	0.364	0.666	0.417	0.909	0.459
0.44	1.553	−0.821	0.362	0.670	0.426	0.905	0.471
0.45	1.568	−0.799	0.361	0.674	0.435	0.900	0.483
0.46	1.584	−0.777	0.359	0.677	0.444	0.896	0.495
0.47	1.600	−0.755	0.357	0.681	0.453	0.892	0.508
0.48	1.616	−0.734	0.356	0.684	0.462	0.887	0.521
0.49	1.632	−0.713	0.354	0.688	0.471	0.882	0.533
0.50	1.649	−0.693	0.352	0.691	0.479	0.878	0.546

*The tabulated functions ϕ and F are given as follows: $\phi(x) = (2\pi)^{-1/2} \exp(-x^2/2)$; for every real number c, $F(c)$ is the area of the graph of the set $\{(x, y) \mid [x \leq c] \wedge [0 \leq y \leq \phi(x)]\}$.

A Table of Functional Values (continued)

x	e^x	$\ln x$	$\phi(x)$	$F(x)$	$\sin x$	$\cos x$	$\tan x$
0.51	1.665	−0.673	0.350	0.694	0.488	0.873	0.559
0.52	1.682	−0.654	0.348	0.698	0.497	0.868	0.573
0.53	1.699	−0.635	0.347	0.702	0.506	0.863	0.586
0.54	1.716	−0.616	0.345	0.705	0.514	0.858	0.599
0.55	1.733	−0.598	0.343	0.709	0.523	0.853	0.613
0.56	1.751	−0.580	0.341	0.712	0.531	0.847	0.627
0.57	1.768	−0.562	0.339	0.716	0.540	0.842	0.641
0.58	1.786	−0.545	0.337	0.719	0.548	0.836	0.655
0.59	1.804	−0.528	0.335	0.722	0.556	0.831	0.670
0.60	1.822	−0.511	0.333	0.726	0.565	0.825	0.684
0.61	1.840	−0.494	0.331	0.729	0.573	0.820	0.699
0.62	1.859	−0.478	0.329	0.732	0.581	0.814	0.714
0.63	1.878	−0.462	0.327	0.736	0.589	0.808	0.729
0.64	1.896	−0.446	0.325	0.739	0.597	0.802	0.745
0.65	1.916	−0.431	0.323	0.742	0.605	0.796	0.760
0.66	1.935	−0.416	0.321	0.745	0.613	0.790	0.776
0.67	1.954	−0.400	0.319	0.749	0.621	0.784	0.792
0.68	1.974	−0.386	0.317	0.752	0.629	0.778	0.809
0.69	1.994	−0.371	0.314	0.755	0.637	0.771	0.825
0.70	2.014	−0.357	0.312	0.758	0.644	0.765	0.842
0.71	2.034	−0.342	0.310	0.761	0.652	0.758	0.860
0.72	2.054	−0.329	0.308	0.764	0.659	0.752	0.877
0.73	2.075	−0.315	0.306	0.767	0.667	0.745	0.895
0.74	2.096	−0.301	0.303	0.770	0.674	0.738	0.913
0.75	2.117	−0.288	0.301	0.773	0.682	0.732	0.932
0.76	2.138	−0.274	0.299	0.776	0.689	0.725	0.950
0.77	2.160	−0.261	0.297	0.779	0.696	0.718	0.970
0.78	2.181	−0.248	0.294	0.782	0.703	0.711	0.989
0.79	2.203	−0.236	0.292	0.785	0.710	0.704	1.009
0.80	2.226	−0.223	0.290	0.788	0.717	0.697	1.030
0.81	2.248	−0.211	0.287	0.791	0.724	0.689	1.050
0.82	2.270	−0.198	0.285	0.794	0.731	0.682	1.072
0.83	2.293	−0.186	0.283	0.797	0.738	0.675	1.093
0.84	2.316	−0.174	0.280	0.800	0.745	0.667	1.116
0.85	2.340	−0.163	0.278	0.802	0.751	0.660	1.138
0.86	2.363	−0.151	0.276	0.805	0.758	0.652	1.162
0.87	2.387	−0.139	0.273	0.808	0.764	0.645	1.185
0.88	2.411	−0.128	0.271	0.811	0.771	0.637	1.210
0.89	2.435	−0.117	0.268	0.813	0.777	0.629	1.235
0.90	2.460	−0.105	0.266	0.816	0.783	0.622	1.260
0.91	2.484	−0.094	0.264	0.819	0.790	0.614	1.286
0.92	2.509	−0.083	0.261	0.821	0.796	0.606	1.313
0.93	2.535	−0.073	0.259	0.824	0.802	0.598	1.341
0.94	2.560	−0.062	0.256	0.826	0.808	0.590	1.369
0.95	2.586	−0.051	0.254	0.829	0.813	0.582	1.398
0.96	2.612	−0.041	0.252	0.831	0.819	0.574	1.428
0.97	2.638	−0.030	0.249	0.834	0.825	0.565	1.459
0.98	2.664	−0.020	0.247	0.836	0.830	0.557	1.491
0.99	2.691	−0.010	0.244	0.839	0.836	0.549	1.524
1.00	2.718	0.000	0.242	0.841	0.841	0.540	1.557

A Table of Functional Values (continued)

x	e^x	$\ln x$	$\phi(x)$	$F(x)$	$\sin x$	$\cos x$	$\tan x$
1.01	2.746	0.010	0.240	0.844	0.847	0.532	1.592
1.02	2.773	0.020	0.237	0.846	0.852	0.523	1.628
1.03	2.801	0.030	0.235	0.848	0.857	0.515	1.665
1.04	2.829	0.039	0.232	0.851	0.862	0.506	1.704
1.05	2.858	0.049	0.230	0.853	0.867	0.498	1.743
1.06	2.886	0.058	0.227	0.855	0.872	0.489	1.784
1.07	2.915	0.068	0.225	0.858	0.877	0.480	1.827
1.08	2.945	0.077	0.223	0.860	0.882	0.471	1.871
1.09	2.974	0.086	0.220	0.862	0.887	0.462	1.917
1.10	3.004	0.095	0.218	0.864	0.891	0.454	1.965
1.11	3.034	0.104	0.215	0.867	0.896	0.445	2.014
1.12	3.065	0.113	0.213	0.869	0.900	0.436	2.066
1.13	3.096	0.122	0.211	0.871	0.904	0.427	2.120
1.14	3.127	0.131	0.208	0.873	0.909	0.418	2.176
1.15	3.158	0.140	0.206	0.875	0.913	0.408	2.234
1.16	3.190	0.148	0.204	0.877	0.917	0.399	2.296
1.17	3.222	0.157	0.201	0.879	0.921	0.390	2.360
1.18	3.254	0.166	0.199	0.881	0.925	0.381	2.427
1.19	3.287	0.174	0.197	0.883	0.928	0.372	2.498
1.20	3.320	0.182	0.194	0.885	0.932	0.362	2.572
1.21	3.353	0.191	0.192	0.887	0.936	0.353	2.650
1.22	3.387	0.199	0.190	0.889	0.939	0.344	2.733
1.23	3.421	0.207	0.187	0.891	0.942	0.334	2.820
1.24	3.456	0.215	0.185	0.893	0.946	0.325	2.912
1.25	3.490	0.223	0.183	0.894	0.949	0.315	3.010
1.26	3.525	0.231	0.180	0.896	0.952	0.306	3.113
1.27	3.561	0.239	0.178	0.898	0.955	0.296	3.224
1.28	3.597	0.247	0.176	0.900	0.958	0.287	3.341
1.29	3.633	0.255	0.174	0.901	0.961	0.277	3.467
1.30	3.669	0.262	0.171	0.903	0.964	0.267	3.602
1.31	3.706	0.270	0.169	0.905	0.966	0.258	3.747
1.32	3.743	0.278	0.167	0.907	0.969	0.248	3.903
1.33	3.781	0.285	0.165	0.908	0.971	0.238	4.072
1.34	3.819	0.293	0.163	0.910	0.973	0.229	4.256
1.35	3.857	0.300	0.160	0.911	0.976	0.219	4.455
1.36	3.896	0.307	0.158	0.913	0.978	0.209	4.673
1.37	3.935	0.315	0.156	0.915	0.980	0.199	4.913
1.38	3.975	0.322	0.154	0.916	0.982	0.190	5.177
1.39	4.015	0.329	0.152	0.918	0.984	0.180	5.471
1.40	4.055	0.336	0.150	0.919	0.985	0.170	5.798
1.41	4.096	0.344	0.148	0.921	0.987	0.160	6.165
1.42	4.137	0.351	0.146	0.922	0.989	0.150	6.581
1.43	4.179	0.358	0.144	0.924	0.990	0.140	7.055
1.44	4.221	0.365	0.141	0.925	0.991	0.130	7.602
1.45	4.263	0.372	0.139	0.926	0.993	0.121	8.238
1.46	4.306	0.378	0.137	0.928	0.994	0.111	8.989
1.47	4.349	0.385	0.135	0.929	0.995	0.101	9.887
1.48	4.393	0.392	0.133	0.931	0.996	0.091	10.983
1.49	4.437	0.399	0.131	0.932	0.997	0.081	12.350
1.50	4.482	0.405	0.130	0.933	0.997	0.071	14.101

A Table of Functional Values (continued)

x	e^x	$\ln x$	$\phi(x)$	$F(x)$	$\sin x$	$\cos x$	$\tan x$
1.55	4.711	0.438	0.120	0.939	1.000	0.021	48.078
1.60	4.953	0.470	0.111	0.945	1.000	−0.029	−34.233
1.65	5.207	0.501	0.102	0.951	0.997	−0.079	−12.599
1.70	5.474	0.531	0.094	0.955	0.992	−0.129	−7.697
1.75	5.755	0.560	0.086	0.960	0.984	−0.178	−5.520
1.80	6.050	0.588	0.079	0.964	0.974	−0.227	−4.286
1.85	6.360	0.615	0.072	0.968	0.961	−0.276	−3.488
1.90	6.686	0.642	0.066	0.971	0.946	−0.323	−2.927
1.95	7.029	0.668	0.060	0.974	0.929	−0.370	−2.509
2.00	7.389	0.693	0.054	0.977	0.909	−0.416	−2.185
2.05	7.768	0.718	0.049	0.980	0.887	−0.461	−1.925
2.10	8.166	0.742	0.044	0.982	0.863	−0.505	−1.710
2.15	8.585	0.765	0.040	0.984	0.837	−0.547	−1.529
2.20	9.025	0.788	0.035	0.986	0.808	−0.589	−1.374
2.25	9.488	0.811	0.032	0.988	0.778	−0.628	−1.239
2.30	9.974	0.833	0.028	0.989	0.746	−0.666	−1.119
2.35	10.486	0.854	0.025	0.991	0.711	−0.703	−1.012
2.40	11.023	0.875	0.022	0.992	0.675	−0.737	−0.916
2.45	11.588	0.896	0.020	0.993	0.638	−0.770	−0.828
2.50	12.182	0.916	0.018	0.994	0.598	−0.801	−0.747
2.55	12.807	0.936	0.015	0.995	0.558	−0.830	−0.672
2.60	13.464	0.956	0.014	0.995	0.516	−0.857	−0.602
2.65	14.154	0.975	0.012	0.996	0.472	−0.882	−0.535
2.70	14.880	0.993	0.010	0.997	0.427	−0.904	−0.473
2.75	15.643	1.012	0.009	0.997	0.382	−0.924	−0.413
2.80	16.445	1.030	0.008	0.997	0.335	−0.942	−0.356
2.85	17.288	1.047	0.007	0.998	0.287	−0.958	−0.300
2.90	18.174	1.065	0.006	0.998	0.239	−0.971	−0.246
2.95	19.106	1.082	0.005	0.998	0.190	−0.982	−0.194
3.00	20.086	1.099	0.004	0.999	0.141	−0.990	−0.143
3.05	21.115	1.115	0.004	0.999	0.091	−0.996	−0.092
3.10	22.198	1.131	0.003	0.999	0.042	−0.999	−0.042
3.15	23.336	1.147	0.003	0.999	−0.008	−1.000	0.008
3.20	24.533	1.163	0.002	0.999	−0.058	−0.998	0.058
3.25	25.790	1.179	0.002	0.999	−0.108	−0.994	0.109
3.30	27.113	1.194	0.002	1.000	−0.158	−0.987	0.160
3.35	28.503	1.209	0.001	1.000	−0.207	−0.978	0.211
3.40	29.964	1.224	0.001	1.000	−0.256	−0.967	0.264
3.45	31.500	1.238	0.001	1.000	−0.304	−0.953	0.319
3.50	33.115	1.253	0.001	1.000	−0.351	−0.936	0.375
3.55	34.813	1.267	0.001	1.000	−0.397	−0.918	0.433
3.60	36.598	1.281	0.001	1.000	−0.443	−0.897	0.493
3.65	38.475	1.295	0.001	1.000	−0.487	−0.874	0.557
3.70	40.447	1.308	0.000	1.000	−0.530	−0.848	0.625
3.75	42.521	1.322	0.000	1.000	−0.572	−0.821	0.697
3.80	44.701	1.335	0.000	1.000	−0.612	−0.791	0.774
3.85	46.993	1.348	0.000	1.000	−0.651	−0.759	0.857
3.90	49.402	1.361	0.000	1.000	−0.688	−0.726	0.947
3.95	51.935	1.374	0.000	1.000	−0.723	−0.691	1.047
4.00	54.598	1.386	0.000	1.000	−0.757	−0.654	1.158

Answers to Selected Exercises

Chapter 1

Linear Functions

Exercises A **1.** (a) 2, (b) -8. **2.** (a) 8. **3.** (b) 0. **4.** All except $(-1, 5)$.

Exercises B **6.** $m + b$. **9.** (c) m. **10.** (b) $-\frac{1}{4}$, (f) -1. **12.** -3.

Exercises C **2.** $c = -\frac{2}{3}, d = \frac{1}{2}$. **4.** (b) $-\frac{1}{3}$, (c) 0. **6.** $1 + (2a/b)$.

Exercises D **2.** -5. **4.** $\frac{39}{4}$. **6.** (a) Linear function given by formula $-\frac{2}{3}x + \frac{4}{3}$, (c) Linear function given by formula $-\frac{32}{21}x$. **7.** (b) $\frac{3}{4}$, (d) $\frac{6}{5}$. **8.** $-b/a$. **9.** -1.

Exercises E **2.** $\frac{2}{3}$. **4.** $m = -3$, $b = 10$. **6.** $L(x) = 3x + 4$. **8.** $9x + 8y = 0$. **10.** $-\frac{3}{2}$. **15.** -1. **18.** $q = (mc - kb)/(m - k)$.

Exercises F **1.** (a) 100, (b) 0. **5.** Constant of proportionality is $\frac{1}{4}$. **8.** Constant of proportionality is $\frac{1}{90}$. **9.** (a) Approximately 10.76. **12.** (b) 105%. **13.** (a) $L(u) = \frac{5}{9}u - \frac{160}{9}$. **14.** (c) Decrease to $\frac{3}{4}$ times the original number.

Exercises G **1.** (c) 10, (e) 2, (f) $5^{1/2}$. **3.** (b) 6, (c) 9, (e) $a - b$. **4.** $-10d$. **5.** Vertices are $(-2, 1), (-1, -6), (1, 4)$.

Exercises H **1.** (b) Parallel, (c) Perpendicular. **4.** $x + 3y = 3$. **5.** $-a$. **6.** $5x + 2y + 14 = 0$. **10.** (b) $x = 4$, (e) $y = 1$, (g) $y = b$, (i) $ay = bx$. **11.** (b) 30, (e) $g = 30t + 7500$, (g) 375.

Chapter 2

Systems of Linear Equations

Exercises A **1.** (a) $(x, y) = (5, -2)$, (c) $(x, y, z) = (1, -6, 2)$, (e) $(w, x, y, z) = (1, -1, 0, 2)$. **2.** $(29, -11)$. **4.** $(a + b, a - b)$. **7.** The respective graphs of the two equations are a pair of parallel lines.

513

Exercises B **2. (b)** The points form the vertical line one unit to the right of the z-coordinate axis. **(e)** The points form the yz-coordinate plane. **4.** One unit behind the yz-coordinate plane. **7.** $y = 4$. **9.** $3x + 5 = 0$.

Exercises C **1. (b)** $(0.66, -0.49)$, **(c)** $(3, -1, 15)$. **2. (c)** $(4, 3, -6, 10)$. **5.** $(300, 105, 67, 164)$. **7.** $(270, 120, 48, 126)$. **8. (a)** $(23, -6)$, **(c)** No, **(e)** $(4, \frac{11}{2})$. **10.** $(1, \frac{3}{2}, -4)$. **13. (d)** $(0, 0, \ldots, 0)$.

Exercises D **1. (b)** 5, **(d)** 2. **3. (a)** $\begin{bmatrix} -5 & 15 & 10 \\ 0 & 20 & -25 \end{bmatrix}$, **(e)** $\begin{bmatrix} 9 & 5 & 20 \\ 2 & 30 & 1 \end{bmatrix}$.

5. $\begin{bmatrix} -1 & 2 \\ -1 & 1 \\ 1 & \frac{1}{3} \end{bmatrix}$. **6. (b)** $b_{ij} + a_{ij}$, **(d)** $(a_{ij} + b_{ij}) + c_{ij}$. **7. (b)** $(\frac{7}{2}, 5, \frac{5}{2})$.

Exercises E **1. (b)** -4, **(c)** $2a - b$. **2. (b)** $\begin{bmatrix} 3 & -2 & 4 \end{bmatrix} \begin{bmatrix} x \\ y \\ z \end{bmatrix} = 12$. **4.** $-a/b$.

6. (a) $ac + bd = 0$.

7. (a) $\begin{bmatrix} 41 & 43 \\ 50 & 22 \end{bmatrix}$, **(c)** $\begin{bmatrix} 2 & -5 & -11 \\ -8 & 5 & 23 \\ 14 & -5 & -35 \end{bmatrix}$, **(f)** HINT: The product in (f) is a 2×2 matrix.

8. (b) $\begin{bmatrix} 0 & 0 \\ 0 & 0 \end{bmatrix}$. **10. (a)** $\begin{bmatrix} 1 & -2 \\ 3 & 1 \\ -4 & -3 \\ -1 & 5 \end{bmatrix} \begin{bmatrix} x \\ y \end{bmatrix}$.

Exercises F **1.** $x + 2y - z = 9$, $4x - y + 2z = 21$.

4. (b) $\begin{bmatrix} 6 \\ -2 \\ -1 \end{bmatrix}$, **(d)** $\begin{bmatrix} 1 \\ 2 \\ -3 \\ -1 \end{bmatrix}$.

5. (b) $\begin{bmatrix} -\frac{4}{3} \\ \frac{5}{3} \end{bmatrix}$, **(d)** $\begin{bmatrix} (11a - 4b)/3 \\ (-13a + 5b)/3 \end{bmatrix}$.

8. (b) $\begin{bmatrix} ap + bq \\ cp + dq \end{bmatrix}$, **(e)** $\begin{bmatrix} a(h + p) + b(k + q) \\ c(h + p) + d(k + q) \end{bmatrix}$.

9. (a) $r_1 g_1 + r_2 g_2 + r_3 g_3 + r_4 g_4$,

(d) $u_1 g_1 + u_2 g_2 + u_3 g_3 + u_4 g_4 + k(r_1 g_1 + r_2 g_2 + r_3 g_3 + r_4 g_4)$.

Exercises G

1.

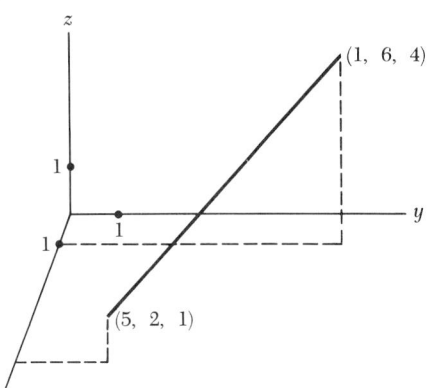

(1, 6, 4)

(5, 2, 1)

5. (c)

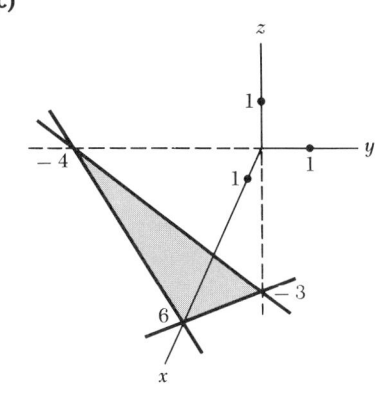

7. $2x + y + 5z = 10,$
$2x - y + 5z = 10$

(a), (b), (c)

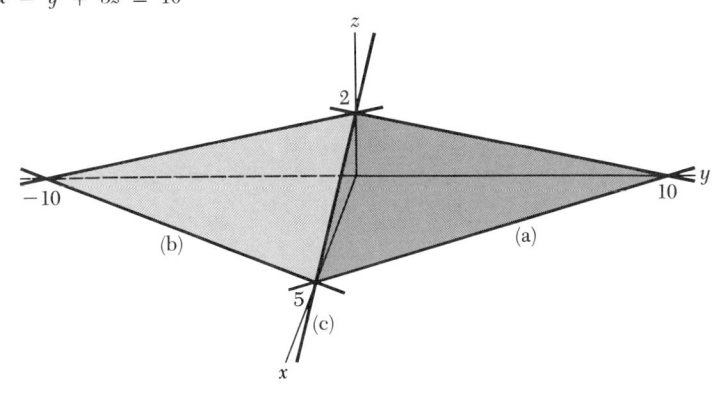

Exercises H **1. (a)** $x + z = 3$, **(d)** $16x + 3y + 12z = 24$.
3. $bcx + cay + abz = 2abc$. **5. (b)** $5x - 4y + 2 = 0$.

Exercises I **1. (a)** HINT: The set of vectors $(8 - \frac{2}{3}p, 1 - \frac{1}{3}p, p)$ for all numbers p is the same as the set of vectors $(6 + 2k, k, 3 - 3k)$ for all numbers k, as we may verify by letting $p = 3 - 3k$. **(c)** HINT: After solving the system, try letting $p = 6 - 3h$.
3. (c) $(3, 2, 0) + p(-1, -2, 1)$ for all numbers p. **4. (b)** Equation has no root, because the third row states that $[0 \ 0 \ 0 \ 0]X = 1$, whereas every four-dimensional column vector C satisfies the condition $[0 \ 0 \ 0 \ 0]C = 0$.

(c) $\begin{bmatrix} 0 \\ 0 \\ 0 \end{bmatrix}$, **(e)** $\begin{bmatrix} 5 \\ 0 \\ -2 \end{bmatrix} + p \begin{bmatrix} 3 \\ 1 \\ 0 \end{bmatrix}$ for all numbers p,

Answers to Selected Exercises 515

(g) $\begin{bmatrix} 5 \\ -2 \\ 0 \\ 6 \\ 0 \end{bmatrix} + p \begin{bmatrix} 1 \\ -2 \\ 1 \\ 0 \\ 0 \end{bmatrix} + q \begin{bmatrix} -3 \\ 1 \\ 0 \\ -2 \\ 1 \end{bmatrix}$ for every real number p and every real number q.

5. (c) A line, **(d)** HINT: Introduce $p = k - 3$. **8.** d/c. **10.** A line.

Exercises J

1. (c) $\begin{bmatrix} 0 \\ 1 \\ 1 \\ 1 \end{bmatrix} + p \begin{bmatrix} -7 \\ 6 \\ 4 \\ 7 \end{bmatrix}$ for all numbers p, **(d)** $\begin{bmatrix} 5 \\ 7 \end{bmatrix}$, **(h)** No root,

(i) $\begin{bmatrix} 0 \\ 1 \\ 0 \\ 0 \\ 0 \end{bmatrix} + p \begin{bmatrix} 1 \\ -2 \\ 1 \\ 0 \\ 0 \end{bmatrix} + q \begin{bmatrix} 3 \\ 0 \\ 0 \\ 1 \\ 0 \end{bmatrix} + r \begin{bmatrix} -2 \\ 1 \\ 0 \\ 0 \\ 1 \end{bmatrix}$ for all numbers p, q, r.

2. $5x - y + 3z = 10$. **4.** There are many planes containing all four points; one of them has the equation $2x - z = 17$.

Chapter 3
Linear Transformations

Exercises A 1. (a) $\begin{bmatrix} -2 \\ 7 \end{bmatrix}$, **(f)** $\begin{bmatrix} -\frac{5}{4} \\ \frac{7}{2} \end{bmatrix}$. **2. (b)** $\begin{bmatrix} 11 \\ 7 \\ 4 \end{bmatrix}$, **(f)** $\mathfrak{T}(4U) = \begin{bmatrix} 16 \\ -4 \\ 16 \end{bmatrix}$. **5. (b)** $\mathfrak{T}(Y)$ is the second column of M, namely, $\begin{bmatrix} 2 \\ -1 \\ 16 \end{bmatrix}$. **6. (a)** $\mathfrak{T}(X)$ is the first column of N, namely, $\begin{bmatrix} a \\ d \\ g \end{bmatrix}$.

Exercises B 1. (a) $\begin{bmatrix} 9 \\ 16 \end{bmatrix}$, **(c)** $\begin{bmatrix} 1 \\ 2 \end{bmatrix}$, **(g)** $\begin{bmatrix} 3 & 1 \\ 5 & 2 \end{bmatrix} \begin{bmatrix} a \\ b \end{bmatrix}$. **2. (a)** $\begin{bmatrix} -1 \\ -7 \\ 13 \end{bmatrix}$,

(d) $\begin{bmatrix} -a + b + c \\ -7a + 6b + 9c \\ 13a - 11b - 16c \end{bmatrix}$. 3. (b) $\begin{bmatrix} 1 \\ \frac{1}{2} \\ 0 \end{bmatrix}$, (f) $\begin{bmatrix} -8 & 1 & -5 \\ -\frac{1}{2} & \frac{1}{2} & -\frac{1}{2} \\ 2 & 0 & 1 \end{bmatrix} \begin{bmatrix} a \\ b \\ c \end{bmatrix}$.

Exercises C 1. (a) $\begin{bmatrix} 9 \\ 19 \end{bmatrix}$, (b) $\begin{bmatrix} 8 \\ -7 \end{bmatrix}$, (c) $\begin{bmatrix} -7 \\ 8 \end{bmatrix}$. 2. (d) $\begin{bmatrix} 4 & 0 \\ 2 & 0 \end{bmatrix}$.

3. (a) $S(T(U)) = \begin{bmatrix} 4 \\ -7 \\ 1 \end{bmatrix}$, (b) $T(S(U)) = \begin{bmatrix} -5 \\ 2 \\ 8 \end{bmatrix}$, (d) $\begin{bmatrix} -10 \\ 30 \\ 18 \end{bmatrix}$, (e) $\begin{bmatrix} 7 & -1 & -4 \\ 11 & 7 & -11 \\ -11 & 5 & 5 \end{bmatrix}$.

5. $\begin{bmatrix} \frac{17}{5} & -\frac{13}{5} \\ \frac{3}{5} & \frac{8}{5} \end{bmatrix}$.

Exercises D

1. (a) $\begin{bmatrix} 4a + 5b \\ a + 2b \end{bmatrix}$, (d) $\begin{bmatrix} 1 & 0 \\ 0 & 1 \end{bmatrix}$. 2. (a) $\begin{bmatrix} 1 & 1 & 0 \\ 1 & 1 & 1 \\ 0 & 1 & 1 \end{bmatrix}$, (d) $\begin{bmatrix} 0 & 1 & -1 \\ 1 & -1 & 1 \\ -1 & 1 & 0 \end{bmatrix}$.

3. (b) $\begin{bmatrix} 2 & -2 & 1 \\ -2 & 3 & -2 \\ 1 & -2 & 2 \end{bmatrix}$, (c) $\begin{bmatrix} 2 & 2 & 1 \\ 2 & 3 & 2 \\ 1 & 2 & 2 \end{bmatrix}$. 5. (b) $\begin{bmatrix} (c - r - 3s)/2 \\ (2a - 3c + 7r + s)/10 \\ r \\ s \end{bmatrix}$ for all real numbers r and s.

Exercises E

1. (b) $\begin{bmatrix} -1 \\ 2 \end{bmatrix}$. 3. (b) $\begin{bmatrix} \frac{9}{80} & -\frac{1}{80} \\ -\frac{1}{80} & \frac{9}{80} \end{bmatrix}$, (c) $\begin{bmatrix} -\frac{3}{4} & -\frac{1}{2} & -\frac{1}{4} \\ -\frac{1}{2} & -1 & -\frac{1}{2} \\ -\frac{1}{4} & -\frac{1}{2} & -\frac{3}{4} \end{bmatrix}$,

(f) $\begin{bmatrix} -16 & 2 & 5 & 4 \\ 12 & -2 & -2 & -3 \\ -5 & 1 & 1 & 1 \\ -4 & 1 & 0 & 1 \end{bmatrix}$.

5. (b) There are many roots; one choice for U is the vector $\begin{bmatrix} 9 \\ -1 \\ 7 \end{bmatrix}$. 6. (a) HINT: Application of the method yields a tableau whose bottom row is [0 0 0 ┆ 3 −2 1].
7. (a) BB, (e) BF.

Chapter 4

Linear Inequalities

Exercises A **1.** (a) 4, (c) 8, (e) -6. **3.** (f) HINT: The graph is the same as that in part (d). **4.** (a) The roots are the numbers less than $2 - \frac{3}{2}a$. **6.** (a) All except $(4, 0)$ and $(1, -2)$, (b) -5.

Exercises B

3. (b)

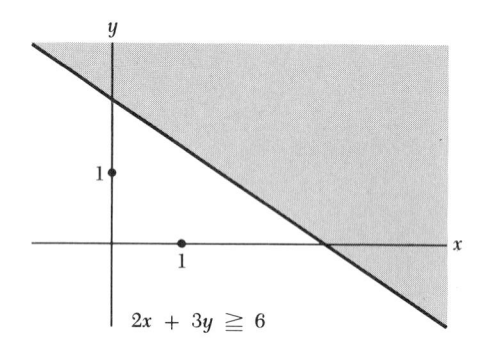

$2x + 3y \geq 6$

5. $2x - 7y \leq 3$. **6.** $(1, 4)$. **8.** (d) The graph consists of all points not lying on the xz-coordinate plane.

Exercises C

2.

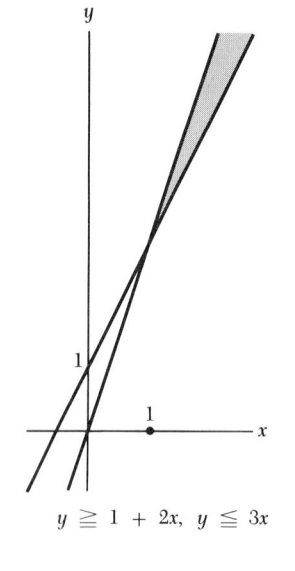

$y \geq 1 + 2x, \ y \leq 3x$

3.

$x \geq 0, \ y \geq 0, \ -3x + 2y \geq -6,$
$x - 5y \geq -10$

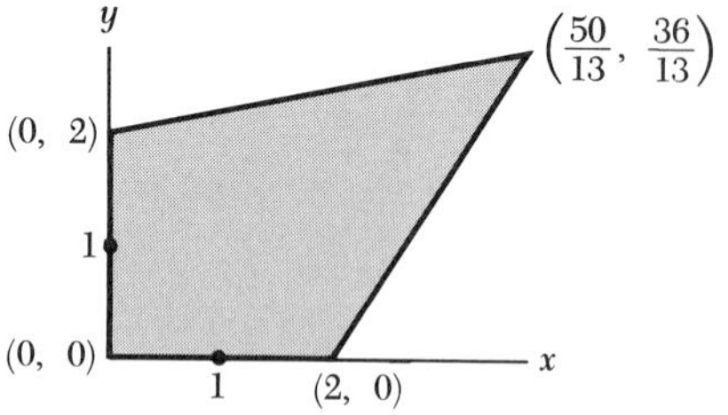

$\left(\frac{50}{13}, \frac{36}{13}\right)$

$(0, 2)$

$(0, 0) \quad (2, 0)$

4. (b) $\begin{bmatrix} 2 & -1 \\ -1 & 3 \\ -5 & 2 \end{bmatrix} \begin{bmatrix} x \\ y \end{bmatrix} \geq \begin{bmatrix} 4 \\ -1 \\ -12 \end{bmatrix}$. **5.** (a) $x + 3y \geq 13$. **6.** (c) $\begin{bmatrix} 1 & 2 \\ -7 & 1 \\ 1 & -3 \end{bmatrix} X \leq \begin{bmatrix} 12 \\ 6 \\ 2 \end{bmatrix}$.

518   Appendix

8. (a) The graph consists of a right angle and its interior; the vertex of the angle is the point (3, 2); one side extends horizontally toward the right from the vertex, and the other side extends vertically upward from the vertex. The interior includes all points that are above the horizontal side and to the right of the vertical side.

Exercises D **1.** $5x - 2y + 10z \geq 10$. **4.** The roots form a solid box, four units from back to front, one unit from left to right, and two units from bottom to top. The back, the front, the left side, the right side, the bottom, and the top lie in the respective planes $x = 2$, $x = 6$, $y = 4$, $y = 5$, $z = -1$, and $z = 1$. In geometrical language the set is a rectangular parallelepiped and its interior. **6.** Many answers are acceptable; one choice is given by $A = \begin{bmatrix} 1 & -2 \\ 1 & 2 \\ -1 & 3 \end{bmatrix}$ and $B = \begin{bmatrix} -1 \\ 3 \\ -8 \end{bmatrix}$. **7. (b)** Many answers are acceptable; one choice is $A = \begin{bmatrix} 1 & 2 \\ 4 & -1 \\ -1 & -1 \\ -3 & -2 \\ -6 & 1 \end{bmatrix}$, $B = \begin{bmatrix} -1 \\ -4 \\ -9 \\ -20 \\ -20 \end{bmatrix}$.

Exercises E **2. (b)** $(1, 7)$, $(1, -2)$, $(5, -2)$, **(d)** $(2, 0, 1)$, $(1, -1, 2)$, $(1, 1, 0)$, $(0, 2, 3)$.
3. (d) $(3, 2)$. **4.** $(11, 0)$. **6. (a)** $\frac{43}{11}$, **(c)** -2. **8.** Each point of the segment joining $(0, 7)$ and $(4, 5)$.

Exercises F

1. (a) $\begin{bmatrix} 1 & 0 \\ 0 & 1 \\ 1 & 3 \\ 4 & 5 \\ 4 & 1 \end{bmatrix} \begin{bmatrix} 40 \\ 15 \end{bmatrix} = \begin{bmatrix} 40 \\ 15 \\ 85 \\ 235 \\ 175 \end{bmatrix}$ is equal to or greater than $\begin{bmatrix} 0 \\ 0 \\ 80 \\ 230 \\ 110 \end{bmatrix}$; $[30 \; 20] \begin{bmatrix} 40 \\ 15 \end{bmatrix} = 1500$ is

greater than 1200. **3. (a)** Older machine, 6 days; newer machine, 4 days. **4.** $\frac{1}{7}$ pound Slix, $\frac{3}{7}$ pound Krix. **5.** 0 liter first solution, $\frac{2}{7}$ liter second solution, $\frac{20}{7}$ liter third solution. **6. (b)** $624, **(c)** $660.

Chapter 5
Sets and Logic

Exercises A **2. (a)** $\{f, g\}$, $\{f, h\}$, $\{g, h\}$, **(c)** T. **3.** All except (b) and (e). **5. (a)** 7.
6. (b) k. **7. (a)** Closed halfline, **(c)** Plane, **(e)** Point, **(h)** Point, **(j)** Closed halfplane.
8. All except (d), (f), and (g).

Exercises B **1. (a)** $\{a, f, d, j, g\}$, **(d)** $\{f\}$, **(e)** $\{b, d, h, j\}$. **2.** All except (a) and (c).
4. (b) $Q \cup \tilde{P}$, **(c)** $(P \cap \tilde{Q}) \cup (\tilde{P} \cap Q)$.
6. (c) $(P \cap \tilde{Q} \cap \tilde{R}) \cup (\tilde{P} \cap Q \cap \tilde{R}) \cup (\tilde{P} \cap \tilde{Q} \cap R)$. **7. (b)** T, **(e)** R, **(h)** R,

(j) R ∪ S, **(l)** S.   **9.** Each element belonging to P does not belong to Q (because P and Q are disjoint) and therefore belongs to Q̃; thus P ⊂ Q̃. (Here we do not present the argument showing that if P and Q are disjoint, then Q ⊂ P̃.)

Exercises C   **1.** Let c be any element of B̃. Then c does not belong to B. Hence c does not belong to the subset A of B. That is, c belongs to Ã. Since every element of B̃ is in Ã, B̃ ⊂ Ã.   **2. (c)** 23.   **4.** All except (c) and (d).   **5. (b)** 41, **(d)** 19, **(f)** 45.   **6. (b)** 78, **(f)** 0.

Exercises D   **2. (c)**

p	p ∨ p
T	T
F	F

3. (c)

p	u	p ∨ u
T	T	T
F	T	T

4. (b) w.   **5. (d)** False, **(f)** It is not the case that the earth is a planet or that water is not a chemical element.

Exercises E

1. (a)

p	q	~q	p ∨ ~q
T	T	F	T
T	F	T	T
F	T	F	F
F	F	T	T

(d)

p	q	(p ∨ q) ∧ (~p ∨ ~q)
T	T	F
T	F	T
F	T	T
F	F	F

3. (b) s ∧ ~r, **(c)** ~s ∧ ~r.

5. (b)

p	q	r	~p ∨ (q ∧ ~r)
T	T	T	F
T	T	F	T
T	F	T	F
T	F	F	F
F	T	T	T
F	T	F	T
F	F	T	T
F	F	F	T

7. (a)

p	~p	~(~p)
T	F	T
F	T	F

8. (b) p,   **(c)**

p	q	(p ∨ q) ∧ q
T	T	T
T	F	F
F	T	T
F	F	F

9. (d) The disjunction $q \lor (r \lor s)$ is false in case each of the statements $q, r \lor s$ is false; in every other case the disjunction is true. Now $r \lor s$ is false in case each of the statements r, s is false; otherwise $r \lor s$ is true. Thus $q \lor (r \lor s)$ is false only in case each of the three individual statements q, r, s is false; in every other case $q \lor (r \lor s)$ is true.

10. (b)

p	q	r	s	$\sim p \lor \sim q$	$q \lor (r \lor s)$	$[\sim p \lor \sim q] \land [q \lor (r \lor s)]$
T	T	T	T	F	T	F
T	T	T	F	F	T	F
T	T	F	T	F	T	F
T	T	F	F	F	T	F
T	F	T	T	T	T	T
T	F	T	F	T	T	T
T	F	F	T	T	T	T
T	F	F	F	T	F	F
F	T	T	T	T	T	T
F	T	T	F	T	T	T
F	T	F	T	T	T	T
F	T	F	F	T	T	T
F	F	T	T	T	T	T
F	F	T	F	T	T	T
F	F	F	T	T	T	T
F	F	F	F	T	F	F

Exercises F

1. (b)

p	q	$\sim p \rightarrow q$
T	T	T
T	F	T
F	T	T
F	F	F

(d)

p	q	$\sim q \lor p$	$q \rightarrow p$	$(\sim q \lor p) \rightarrow (q \rightarrow p)$
T	T	T	T	T
T	F	T	T	T
F	T	F	F	T
F	F	T	T	T

2. (b) $\sim h \leftrightarrow r$, **(c)** $(r \rightarrow \sim s) \land (s \rightarrow \sim r)$.

3. (b)

p	q	$p \land \sim q$	$(p \land \sim q) \rightarrow p$
T	T	F	T
T	F	T	T
F	T	F	T
F	F	F	T

(g)

p	q	r	$p \to q$	$q \to r$	$(p \to q) \wedge (q \to r)$	$p \to r$	$[(p \to q) \wedge (q \to r)] \to [p \to r]$
T	T	T	T	T	T	T	T
T	T	F	T	F	F	F	T
T	F	T	F	T	F	T	T
T	F	F	F	T	F	F	T
F	T	T	T	T	T	T	T
F	T	F	T	F	F	T	T
F	F	T	T	T	T	T	T
F	F	F	T	T	T	T	T

Exercises G **1. (c)** $(1, 3), (2, 1), (2, 3), (2, 4), (3, 1), (4, 1), (4, 2), (4, 3)$. **2. (a)** HINT: The number of ordered pairs is nine. **(c)** All except $(2, 3)$. **3. (a)** Red and red, red and green, red and blue, green and blue, **(c)** All.

Exercises H **1. (a)** $(P \cup \tilde{Q}) \cap (\tilde{P} \cup Q)$, **(b)** $P \cup \tilde{Q}$, **(c)** T. **2.** All answers affirmative except (b). **3.** $\{1, 2, 3, 4, 5\}$. **4.** $\left\{\dfrac{d-b}{m}\right\}$. **6. (b)** $\left\{\begin{bmatrix} 2 \\ -1 \end{bmatrix}\right\}$.

Exercises I **1. (c)** 0, **(d)** 1, **(e)** 0. **2. (b)** 2, 7, **(d)** 5, **(e)** $\{0, 1, 2, 3, 4, 5, 6, 7\}$. **4. (a)** $(0 + 1)(0 + 0) = 1 \cdot 0 = 0$ and $0 + 1 \cdot 0 = 0 + 0 = 0$. **5. (b)** $\{(1, 1)\}$.

Exercises J

4. (a) $0 + a = a + 0$ [because addition is commutative]
         $= a$ [because 0 is the additive identity].

(e) $(u + v)w = w(u + v)$ [because multiplication is commutative]
       $= wu + wv$ [because multiplication is distributive with respect to addition]
       $= uw + vw$ [because multiplication is commutative].

(g) $(a + b + c)(a + b + c) = a + b + c$ [because every element is idempotent under multiplication].

5. (c) c, **(d)** 1, **(e)** a, **(h)** b, **(i)** g. **6. (b)** The union of any subset of T and T itself is T. **(d)** For any subsets P, Q, R of T, $(P \cup Q) \cap R = (P \cap R) \cup (Q \cap R)$. **7. (a)** The negation of a logically true statement is logically false. **(d)** For any statement p, the disjunction of a logically false statement and p is logically equivalent to the statement p. **9. (b)** HINT: $cd' = (cd)d' = c(dd') = c \cdot 0 = 0$; supply the missing reasons.

Exercises K **2. (a)** HINT: Let c and d be elements such that $c + d = 0$. Then $c = c \cdot (c + d) = c \cdot 0 = 0$. Supply the missing reason for each of the three steps in the preceding sentence. Afterward show that $d = 0$. **3. (b)** For any subsets P, Q of T, $P \cap (\tilde{P} \cup Q) = P \cap Q$. **(c)** If the intersection of a pair of subsets of T is T itself, then each of the subsets is T. **4. (a)** For every triple (p, q, r) of statements pertaining to the outcome of an experiment, the compound statement $p \vee (q \wedge r)$ and the compound statement $(p \vee q) \wedge (p \vee r)$ form a logically equivalent pair. **6.** HINT: The proposition is stated in Example 41 (in Section 14).

7. (a) $(a + b)(a' + b) = (b + a)(b + a')$ [because addition is commutative]
$\qquad\qquad\qquad\quad = b + aa'$ [because addition is distributive with respect to multiplication]
$\qquad\qquad\qquad\quad = b + 0$ [because the product of an element and its complement is 0]
$\qquad\qquad\qquad\quad = b$ [because 0 is the additive identity].
(c) HINT: Simplification, with appropriate reasons, shows that the given element is equal to 1.

Exercises L

3. (b) $abc + ab'c + ab'c' + abc'$
$\qquad = ac(b + b') + ac'(b' + b)$ [because addition is associative, multiplication is associative and commutative, and multiplication is distributive over addition]
$\qquad = ac + ac'$ [because addition is commutative, the sum of an element and its complement is 1, and 1 is the multiplicative identity]
$\qquad = a(c + c')$ [because multiplication is distributive over addition]
$\qquad = a \cdot 1$ [because the sum of an element and its complement is 1]
$\qquad = a$ [because 1 is the multiplicative identity].

5. $xy + x'y' = (x + y')(y + x')$. 6. $xy'w + x'yz = (x' + y')(xw + yz)$.
9. (b) First, $[(xy)' + x'z]' = [x' + y' + x'z]'$ [by Theorem 8a]
$\qquad\qquad\qquad\qquad\quad = [x' + y']'$ [by Theorem 3a, applied to $x' + x'z$]
$\qquad\qquad\qquad\qquad\quad = (x')'(y')'$ [by Theorem 8b]
$\qquad\qquad\qquad\qquad\quad = xy$ [two applications of Theorem 7].
Hence, $[x + y][(xy)' + x'z] + z = [x + y]xy + z$ [by the above]
$\qquad\qquad\qquad\qquad\qquad\qquad = xy + z$ [by Theorem 3b, applied to $x(x + y)$].
(c) HINT: Simplification, with appropriate reasons, shows that the given element is equal to $uv'w$.

Chapter 6

Counting

Exercises A 3. 81. 5. 192. 7. hk. 9. (b) Every element of $A \cap P$ belongs to A; thus $A \cap P$ is a subset of A. Likewise $A \cap Q$ is a subset of A. For each element d in A, d belongs to W and hence belongs to exactly one of the subsets P, Q; thus d in A belongs to exactly one of the intersections $A \cap P$, $A \cap Q$. Since $A \cap P$ has 12 elements, it is not empty. Since $A \cap P$ has fewer members than A, there are elements of A that are not in $A \cap P$; thus $A \cap Q$ is not empty. Hence $\{A \cap P, A \cap Q\}$ is a partition of A.
(c) 14, (f) 8. 10. (a) 120, (d) 60. 12. 30. 13. (b) 30.

Exercises B 1. (a) 16, (c) 28. 2. (a) 19, (c) 38, (f) 12. 3. (b) 23, (e) 34.

Exercises C 2. (a) 13. 4. (a) 12. 5. (b) 338, (d) 325. 6. (a) 16, (c) 44. 7. (a) 26, (c) 51. 8. (b) $(x, y), (y, x)$. 9. (b) 12, (c) 16. 10. 12.

Answers to Selected Exercises 523

Exercises D **1. (b)** $13 \cdot 13 \cdot 12$, **(e)** $13 \cdot 12 \cdot 11 + 13 \cdot 13 \cdot 12$, or $13 \cdot 12 \cdot 24$, **(f)** $13 \cdot 12 \cdot 13 + 13 \cdot 13 \cdot 12$, or $13 \cdot 13 \cdot 24$, **(g)** 4, **(i)** $1 \cdot 11 \cdot 1 + 1 \cdot 12 \cdot 3 + 3 \cdot 12 \cdot 1 + 3 \cdot 13 \cdot 3 = 200$. **2. (b)** 6, **(e)** 15, **(f)** 10. **3. (a)** 125. **4.** 112. **5.** 720. **6.** 30.

Exercises E **1. (a)** 240, **(c)** 126, **(e)** 17280, **(g)** 35. **3.** Multiply 100! by 101. **5. (b)** n, **(c)** $n(n+1)$. **7.** $64 \cdot 63 \cdot 62 \cdot 61 \cdot 60 \cdot 59 \cdot 58 \cdot 57 \cdot 56$. **10.** 7! **11.** $(5!)(4!)$. **12.** 187200.

Exercises F **1. (b)** 28, **(c)** 220, **(f)** 15504. **3.** $\binom{52}{13} = 635{,}013{,}559{,}600$.

Exercises G **1.** $\binom{18}{6}$. **3. (a)** $\binom{13}{4}\binom{13}{4}\binom{13}{1}\binom{13}{4}$, **(d)** $2\binom{13}{6}2\binom{13}{5}\binom{13}{2}$, **(e)** $4\binom{13}{6}3\binom{13}{5}2\binom{13}{2}$, **(f)** 4^{13}, **(h)** $\binom{32}{13}$, **(i)** $\binom{2}{1}\binom{24}{10} + \binom{2}{1}\binom{24}{1}\binom{24}{9}$. **4.** $\binom{50}{15} \cdot 2^{15}$. **5. (a)** $\binom{1200}{30}$, **(b)** $\binom{18}{3}\binom{1182}{27}$. **6. (d)** $\binom{3}{1}\binom{81}{6}$. **8. (a)** 2520.

Exercises H **5. (a)** By hypothesis, $(n-k)/(k+1) > 1$ and $k \geq 0$. We multiply each member of the given inequality $(n-k)/(k+1) > 1$ by the positive number $k+1$ and obtain $n - k > k + 1$. To each member of the latter inequality we add $k - 1$, and we obtain $n - 1 > 2k$. Multiplication of each member of the latter inequality by $\frac{1}{2}$ yields $k < (n-1)/2$. **(d)** By Theorem 7, $\binom{n}{k+1} = \binom{n}{k} \cdot \frac{n-k}{k+1}$. Consequently the binomial coefficients $\binom{n}{k+1}$ and $\binom{n}{k}$ are the same if and only if the fraction $\frac{n-k}{k+1}$ is 1. A fraction is 1 if and only if the numerator and denominator of the fraction are the same. Thus $\binom{n}{k+1} = \binom{n}{k}$ if and only if $n - k = k + 1$; the latter equation is equivalent to $n - 1 = 2k$, which is satisfied if and only if $n - 1$ is even (that is, n is odd) and $k = (n-1)/2$.

Exercises I **1. (a)** $18 \cdot 17 \cdot 16 \cdot 15 \cdot 14$, or $\binom{18}{5} \cdot 5!$, **(c)** $5 \cdot 17 \cdot 16 \cdot 15 \cdot 14$, or $\binom{17}{4} \cdot 5!$, **(f)** 40320. **2.** 56. **3. (a)** 48, **(d)** 3744, **(f)** $\binom{4}{2}12\binom{4}{2}11\binom{4}{1} = 19008$, **(g)** 123552, **(h)** $\binom{13}{5}$, **(k)** 1020. **4. (b)** $\binom{n}{0} + \binom{n}{0+1} = 1 + n$ and $\binom{n+1}{0+1} = n + 1$.

Chapter 7

Powers and Sequences

Exercises A **1. (b)** $a^2 + 4ay + 4y^2$, **(f)** $v^2 - 2vw + w^2$, **(h)** $1 - 10x^2y^2 + 25x^4y^4$. **2. (b)** $b^3 - 3b^2w + 3bw^2 - w^3$, **(c)** $c^3 + 15c^2d + 75cd^2 + 125d^3$, **(e)** $b^4 + 4b^3y + 6b^2y^2 + 4by^3 + y^4$. **5.** $(x, y) = (80, 32)$. **6. (b)** 20. **8.** $\binom{72}{36} x^{36}(5z)^{36}$. **12. (a)** $(3c - 1)^2$, **(c)** $(1 + 4y)^3$. **13.** 1. **14. (a)** By hypothesis, $x + y = z$. Hence $z^2 = (x+y)^2 = x^2 + 2xy + y^2$. By hypothesis, x and y are positive. Hence $2xy > 0$. Therefore $z^2 = (x^2 + y^2) + 2xy > x^2 + y^2$.

Exercises B **1. (a)** $c^4 - 4c^3w + 6c^2w^2 - 4cw^3 + w^4$, **(c)** $16m^4 + 160m^3r + 600m^2r^2 + 1000mr^3 + 625r^4$, **(f)** $h^7 + 7h^6q^2 + 21h^5q^4 + 35h^4q^6 + 35h^3q^8 + 21h^2q^{10} + 7hq^{12}$

+ q^{14}. **2. (d)** $1^4 + 4 \cdot 1^3(0.1) + 6 \cdot 1^2(0.1)^2 + 4 \cdot 1 \cdot (0.1)^3 + (0.1)^4 = 1.4641$, **(f)** $(\sqrt{5})^3 - 3(\sqrt{5})^2 + 3\sqrt{5} - 1 = 8(\sqrt{5} - 2)$. **3. (b)** $(1 - 2y)^5$. **5.** $1 - p^6$. **7. (a)** $2318.55, **(b)** $2318.54. **10.** $7431.74. **12. (b)** Yes. Since x is positive by hypothesis, $x^2 > 0$. Thus $(1 - x)^2 = 1 - 2x + x^2 > 1 - 2x$; that is, $(1 - x)^m > 1 - mx$ for the case where $m = 2$. **(c)** HINT: In order to justify the decision that the statement is false, give a specific positive number x and a specific natural number m greater than 1 such that $(1 - x)^m$ fails to be greater than $1 - mx$.

Exercises C **1. (a)** a^{23}, **(d)** a^{11}, **(g)** $a^8 b^{19}$. **2. (b)** $b^4/c^7 d$, **(c)** 3, **(e)** 2. **3. (b)** -2.
5. (b) 2^{nk^2}. **7.** 741. **8. (a)** 4,360,000,000, **(c)** 251,000,000,000,000,000,000.
9. (c) Since $r > s$ by hypothesis, the difference $r - s$ is a natural number. By part (a), $b^{r-s} > 1$. We multiply each member of the latter inequality by the positive number b^s and obtain $b^{r-s} \cdot b^s > 1 \cdot b^s$, or $b^r > b^s$. **(e)** If $b = 0$, then $b^r = 0^r = 0$ and $b^s = 0^s = 0$. If $b = 1$, then $b^r = 1^r = 1$ and $b^s = 1^s = 1$.

Exercises D **1. (a)** 4, **(d)** $-\frac{1}{36}$, **(f)** 35, **(h)** 2, **(j)** 0. **2. (b)** 0.00012. **3. (b)** $b^{1/2}$, **(c)** $d^{-1/2}$, **(e)** $a^3 c^{-6}$, **(g)** $27g^{1/3}$, **(j)** $b^{-3} c^9$. **4. (b)** $1 - 6x^{1/3} + 12x^{2/3} - 8x$. **5. (c)** $1 - 2p$. **8. (a)** $2 \cdot (1.414) = 2.828$, **(b)** $1.732/3 = 0.577$ (to three decimal places). **10.** $\{t \mid t > 0\}$. **12.** $\{x \mid x \geq 0\}$. **14. (a)** $x > -5$, **(c)** $x \geq -\frac{5}{2}$. **15. (a)** By hypothesis, $r < s$. Consequently $s - r$ is a positive rational number. Since 2 is greater than 1, we may apply Theorem 2 to obtain $1 < 2^{s-r}$. Multiplication of each member of the latter inequality by the positive number 2^r yields $2^r < 2^{s-r} \cdot 2^r = 2^s$. **16. (b)** Concerning a pair of numbers (r, s), there are three possibilities: either $r < s$ or $r = s$ or $r > s$. We shall show that each of the first two alternatives contradicts the given hypothesis that $(\frac{1}{2})^r < (\frac{1}{2})^s$. Indeed, if $r < s$, then $(\frac{1}{2})^r > (\frac{1}{2})^s$, by Exercise 15c. Next, if $r = s$, then $(\frac{1}{2})^r = (\frac{1}{2})^s$. The only alternative compatible with the hypothesis is $r > s$.
17. (b) $\{2\}$, **(c)** $\{x \mid x < -12\}$.

Exercises E **1. (b)** 17, **(c)** 18, **(d)** 29. **2. (b)** 88, **(d)** 25, **(e)** 17. **3. (b)** 256. **5.** The images of 0, 1, 2, 3, 4, 5, 6 are 1, 0, 1, 0, 1, 0, 1, respectively.

Exercises F **1. (b)** $2 \cdot 3^{40}$, **(c)** 5, **(e)** $g(3 + 4) = 2 \cdot 3^7$ and $g(3) \cdot g(4) = (2 \cdot 3^3)(2 \cdot 3^4) = 2 \cdot 2 \cdot 3^7$; thus the latter number is double the former and is the greater of the numbers.
3. (a) $3n - 2$, **(b)** $3n + 1$. **5. (a)** $23 - 6n$, **(b)** 18. **7. (a)** $20(\frac{1}{2})^n$. **9. (a)** 100^{n-5}, **(b)** 26. **11. (a)** $u + n(v - u)$, **(d)** $\dfrac{2u + n(w - u)}{2}$. **12. (b)** $\dfrac{v^2}{w}\left(\dfrac{w}{v}\right)^n$.

Exercises G **1. (a)** 1, **(b)** 28, **(d)** $\frac{137}{60}$, **(g)** 132. **2. (a)** 64, **(c)** 0.999 999 999.
3. (b) 2^n, **(c)** 0. **5. (a)** $\frac{14}{3}$, **(d)** $18 \cdot 2^8 = 4608$.

Exercises H **1.** 500,000,500,000. **4. (a)** 555, **(b)** -855. **6.** 10760. **9.** 656.
11. $2^6 - 2^{-5}$. **13.** $[1 + q^{-1}][1 - (1 + q)^{-44}]$. **14. (b)** 0.3737373737373737.
17. (b) $\frac{1}{999}[1 - 10^{-1419}]$. **20. (a)** $\frac{62}{99}$, **(c)** $\frac{445}{333}$. **22. (b)** 98596, **(c)** 1798, **(d)** 46109.

Chapter 8

Probability

Exercises A **1. (a)** ONE, TWO, THREE, FOUR, FIVE, SIX, **(b)** $\frac{1}{6}$, **(c)** $\frac{1}{3}$. **2. (a)** (HEAD, HEAD), (HEAD, TAIL), (TAIL, HEAD), (TAIL, TAIL), **(c)** $\frac{3}{4}$. **4. (c)** $\frac{2}{13}$, **(d)** $\frac{10}{13}$. **5. (b)** $\frac{1}{4}$, **(d)** $\frac{4}{7}$. **6. (b)** $\frac{1}{12}$, **(d)** $\frac{1}{4}$, **(f)** $\frac{1}{2}$. **7. (b)** $\frac{9}{26}$, **(c)** $\frac{4}{13}$. **9.** $\frac{1}{8}$.

Exercises B   **1. (c)** $\frac{15}{34}$, **(f)** $\frac{1}{442}$, **(h)** $\frac{345}{442}$.   **2. (b)** $\frac{7}{36}$, **(d)** $\frac{5}{9}$.   **3. (c)** 0.027, **(d)** 0.006, **(g)** 0.18, **(h)** 0.488, **(j)** 0.054.   **4. (b)** $\frac{7}{9}$, **(c)** $\frac{17}{45}$.   **5. (d)** $\binom{488}{8}$, **(f)** $\binom{12}{2}\binom{488}{18} / \binom{500}{20}$, **(g)** $1 - \binom{488}{20} / \binom{500}{20}$.   **6. (a)** 4950, **(e)** 776/825, **(f)** 1/1650. **7. (b)** $\binom{24}{4} / \binom{30}{10} = \binom{10}{6} / \binom{30}{6}$, **(d)** $3! \binom{10}{3}\binom{10}{2}\binom{10}{1} / \binom{30}{6}$.

Exercises C   **1.** All except (b) and (d).   **3. (b)** $\frac{1}{4}$, **(d)** $\frac{5}{12}$, **(f)** $\frac{11}{12}$.   **4. (b)** $\frac{1}{6}$, **(e)** $\frac{4}{9}$. **5. (a)** HINT: See Exercise 5E-6b. **(c)** In case p is false.

Exercises D   **1. (c)** $\frac{2}{5}$, **(e)** $\frac{4}{5}$.   **2. (a)** $\frac{3}{10}$, **(c)** $\frac{1}{5}$, **(f)** $\frac{2}{5}$.   **3. (c)** $\frac{4}{5}$, **(e)** $\frac{3}{5}$.   **4. (b)** $\frac{3}{5}$, **(e)** $\frac{1}{25}$, **(g)** $\frac{11}{25}$.

Exercises E   **1. (a)** $\frac{1}{14}$, **(b)** $\frac{3}{7}$.   **3. (b)** For each $h \in \{1, 2, 3, 4, 5\}$, $\Pr[Y = h] = (6 - h)/15$.

5. (a)

k	0	1	2	3	4
Pr[W = k]	$\frac{81}{625}$	$\frac{2}{5}$	$\frac{6}{25}$	$\frac{18}{125}$	$\frac{54}{625}$

6. (a) 77, **(d)** $\frac{11}{25}$, **(e)** $\frac{181}{375}$.

Exercises F   **1. (d)** $\frac{5}{12}$, **(f)** $\frac{8}{21}$, **(g)** $\frac{5}{19}$, **(h)** $\frac{4}{7}$.   **2. (b)** $\frac{5}{12}$, **(c)** $\frac{5}{33}$, **(f)** $\frac{2}{5}$, **(h)** $\frac{11}{14}$.   **3. (c)** 4, **(g)** 2, **(k)** $\frac{1}{4}$, **(o)** $\frac{1}{2}$.   **5. (c)** Let d be the statement that at least one diamond is drawn, and let k be the statement that at least one king is drawn. The product of $\Pr[d] = \frac{15}{34}$ and $\Pr[k] = \frac{33}{221}$ is not equal to $\Pr[d \wedge k] = \frac{29}{442}$.   **6. (a)** $\frac{13}{15}$, **(c)** $\frac{4}{5}$, **(e)** $\frac{1}{45}$, **(g)** $\frac{1}{9}$, **(i)** $\frac{8}{15}$, **(k)** $\frac{9}{17}$.   **7. (b)** $\frac{11}{210}$, **(d)** $\frac{1}{3}$, **(e)** $\frac{33}{910}$, **(i)** $\frac{1}{5005}$, **(j)** $\frac{1}{3}$, **(m)** $\frac{17}{28}$, **(o)** $\frac{11}{105}$, **(q)** $\frac{1}{70}$.

Exercises G   **1.** 448/6561.

2.

k	0	1	2	3	4	5
Pr[X = k]	$\frac{243}{1024}$	$\frac{405}{1024}$	$\frac{270}{1024}$	$\frac{90}{1024}$	$\frac{15}{1024}$	$\frac{1}{1024}$

8. (a) 18232/59049, **(b)** 74/2279.   **9. (b)** $\frac{15}{28}$, **(c)** $\frac{3}{11}$, **(f)** $\frac{3}{10}$.   **10. (a)** 375/1024, **(d)** 583/1536, **(g)** 343/2332.   **11. (a)** 1859/5184, **(c)** 256/1859.

Exercises H   **2. (b)** $\frac{1}{221}$, **(e)** $\frac{1}{13}$, **(f)** $\frac{6}{25}$, **(h)** $\frac{25}{102}$, **(i)** $\frac{6}{425}$.

4.

h	0	1	2	3	4	5
Pr[Y = h]	$\frac{10}{27}$	$\frac{11}{27}$	$\frac{193}{1080}$	$\frac{211}{5400}$	$\frac{23}{5400}$	$\frac{1}{5400}$

5.

k	0	1	2	3
Pr[Z = k]	$\frac{5}{8}$	$\frac{15}{56}$	$\frac{5}{56}$	$\frac{1}{56}$

6. (c) $\frac{3}{8}$, **(e)** $\frac{3}{8}$.   **7. (b)** $\frac{5}{16}$, **(f)** $\frac{3}{8}$.   **8. (d)** $\frac{1}{4}$, **(e)** $\frac{3}{8}$.   **10.** $\frac{4}{11}$.   **12. (b)** $\frac{37}{81}$, **(c)** $\frac{8}{27}$, **(f)** $\frac{2}{9}$.

Exercises I   **1.** 1.   **3.** $\frac{3}{2}$.   **5.** $3p$.   **7. (a)** $\frac{3}{2}$, **(d)** $\frac{7}{3}$, **(f)** 9, **(j)** $\frac{2}{5}$.   **8. (b)** 0, **(d)** $\frac{3}{2}$, **(g)** $\frac{3}{2}$, **(i)** $\frac{1}{2}$.

Exercises J **3.** r^2. **6.** $\frac{35}{6}$. **7. (b)** $3. **8.** $\frac{3}{2}$.

Exercises K **1. (a)** $\frac{3}{4}$, **(c)** $\sqrt{71}/8$. **2. (b)** 2, **(d)** $\frac{21}{4}$.

Exercises L **1. (a)** r, **(c)** $\sqrt{5}/2$, **(f)** $\sqrt{3}/2$, **(i)** $\sqrt{35}/2$. **2. (a)** $\frac{15}{28}$, **(c)** $\frac{35}{12}$, **(f)** 3, **(h)** $\frac{3}{5}$, **(j)** 803/32400. **3.** $[(m^2 - 1)/12]^{1/2}$.

Exercises M **1. (b)** The pair (X, Z) is not an independent pair of random variables. Since $\Pr[(X, Z) = (1, 1)] = 0$ is different from $\Pr[X = 1] \cdot \Pr[Z = 1] = \frac{1}{2} \cdot \frac{1}{2}$, the statements $X = 1$ and $Z = 1$ do not form an independent pair of statements. [NOTE: This pair (X, Z) is an example of the fact that $E[XZ] = E[X] \cdot E[Z]$ can be true for a pair of random variables that is not an independent pair.] **(d)** The pair (X, Y) is not an independent pair of random variables. Since $\Pr[X = 9 \wedge Y = 2] = \frac{1}{10}$ is not the same as $\Pr[X = 9] \cdot \Pr[Y = 2] = 0.06$, the pair $(X = 9, Y = 2)$ is not an independent pair of statements. **2. (c)** 0.64, **(e)** $\Pr[1.4 \leq Y \leq 4.6] = \Pr[Y \in \{2, 3, 4\}] = 0.96$. **3. (a)** $\frac{3}{5}$, **(c)** $\frac{1}{3}$, **(d)** 1.

Exercises N **1. (a)** $\mu = 10$, $\sigma = 2\sqrt{2}$, **(d)** $(\mu, \sigma) = (\frac{200}{3}, \frac{20}{3})$. **2. (b)** 1. **3. (a)** W/r, **(c)** $(2Y - 5)/\sqrt{5}$. **4. (b)** 5, **(d)** 4, **(f)** 77, **(i)** 408, **(k)** 18, **(l)** 75. **5. (f)** $\frac{2}{3}$, **(h)** $\frac{1}{2}$. **6. (b)** $\frac{1}{4}$, **(c)** $\frac{3}{64}$, **(d)** $\frac{9}{64}$. **7. (f)** $\frac{1}{4}$, **(g)** $\frac{1}{12}$, **(i)** No.

Exercises O **1.** 7.8, 4.16, 2.04. **3.** 13, 298.4, 17.3.

Exercises P **1.** 7, 5, $\frac{179}{25}$. **4. (c)** $\frac{3}{68}$, **(d)** $\frac{39}{272}$.

5. (a)

k	0	$\frac{1}{4}$	$\frac{1}{2}$
$\Pr[\bar{X} = k]$	$\frac{188}{221}$	$\frac{32}{221}$	$\frac{1}{221}$

(c) $\frac{25}{2873}$, **(d)** $\frac{25}{884}$.

Chapter 9

Functions and Their Graphs

Exercises A **2. (c)** $y = 3 - x^2$

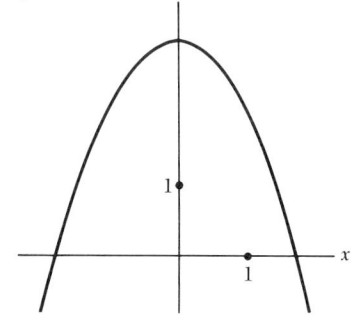

3. (b) The curves are congruent to one another. The graph of the equation $y = x^2 + 3$ is obtainable from the graph of $y = x^2$ by sliding the latter three units up on the coordinate system. **5. (a)** The curves are congruent to each other. A horizontal shift of three units toward the left carries the graph of $y = x^2$ into the graph of the equation $y = (x + 3)^2$. **10. (b)** $y = (x + 10)^2$, **(c)** $y = (x - 1)^3$, **(d)** $y = -x^4$.

Exercises B **2. (a)** $y = x^{-1} - 2$, **(d)** $y = (x + 5)^{-1}$. **7. (a)** The two curves are similar to each other. The graph of the equation $y = 4x^{-1}$ is obtainable from the graph of $y = x^{-1}$ by an expansion about the origin with a magnification factor of 2. A point belongs to the former graph if and only if its respective coordinates are double the coordinates of a point on the latter graph. **(b)** The two curves are congruent to one another. Each is the reflection of the other in the x-coordinate axis.

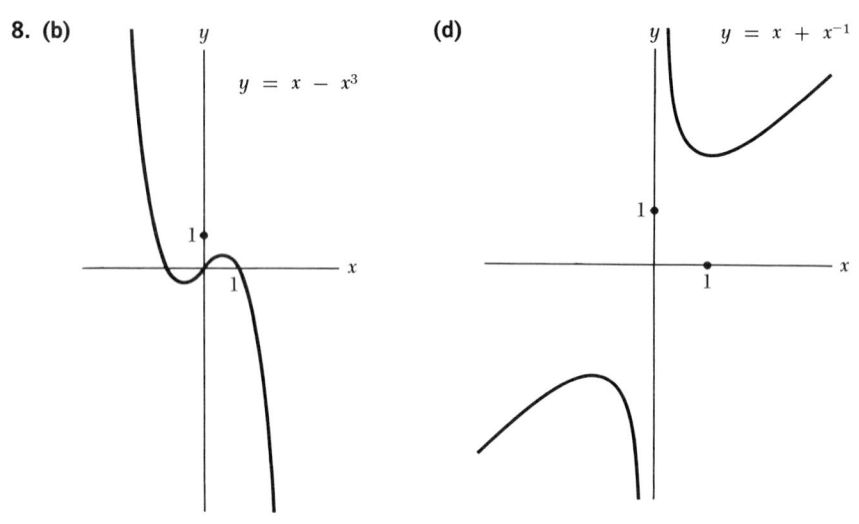

8. (b)

$y = x - x^3$

(d)

$y = x + x^{-1}$

Exercises C **1. (a)** $2x^2 + 3$, **(b)** $4 - x$. **2. (b)** For the exponent -2, each nonzero real number c has a uniquely determined power c^{-2}, and the number 0 has no power. Thus the set is a function. **(d)** The set contains two different ordered pairs with first component 1, namely, (1, 2) and (1, 3); therefore the set is not a function. **(g)** The set is not a function because it contains more than one ordered pair with left component 0; indeed, it contains (0, 1) and (0, 2) (as well as many others with first component 0). **(j)** Since each positive number has a unique fourth power, no number serves as the left component of more than one ordered pair in the set. The set is a function.

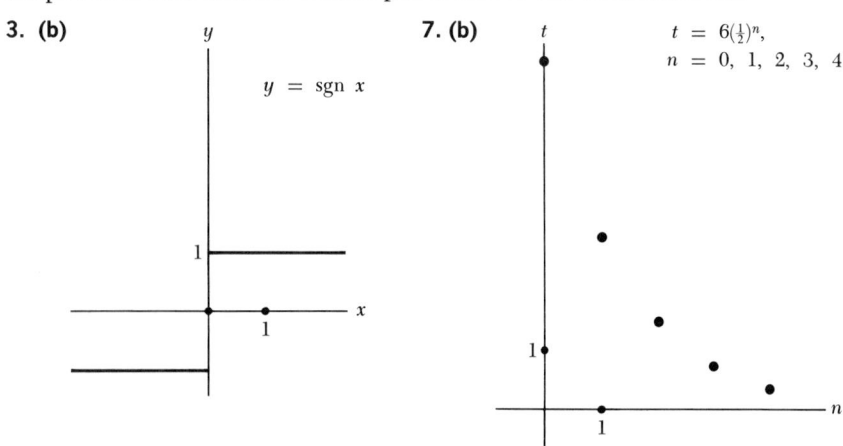

3. (b)

$y = \text{sgn } x$

7. (b)

$t = 6(\frac{1}{2})^n$,
$n = 0, 1, 2, 3, 4$

Exercises D 1. (c) The set J serves as the domain, and the sequence assigns to each member of J one number as its image. 2. (b) The set of all real numbers, (c) The set of all nonnegative real numbers, (e) $\{0, 2, 3, 6\}$. 3. (a) $\{1, 5\}$, (c) $\{-1, 1\}$, (e) $\{t \mid t \neq 0\}$, (g) $\{y \mid y > -2\}$. 4. (a) Six, (c) $\{\frac{1}{243}, \frac{10}{243}, \frac{32}{243}, \frac{40}{243}, \frac{80}{243}\}$,
(d) $t = \left(\frac{2}{3}\right)^5 \left(\frac{5}{k}\right) 2^{-k}$. 6. (b) $\{0, 1, 2, 3, 4, 5, 6, 7\}$, (d) $t = 10^{-7} \cdot \binom{7}{k} \cdot 9^{7-k}$.

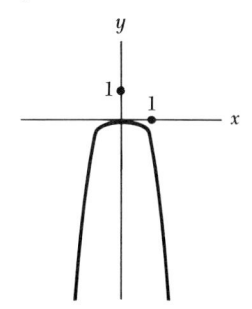

8. (b) $\quad y = -x^4$

Exercises E 1. (b) 8. 3. (a) 7.08, (c) 0. 5. (b) The absolute value of a negative number is defined as the additive inverse of the number. The additive inverse of a negative number is a positive number. 7. (b) $\{x \mid x \leq 0\}$. 9. (c) 4, (d) $b - a$.
10. (a)

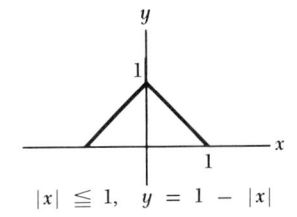

$|x| \leq 1, \quad y = 1 - |x|$

Exercises F 2. (b) Let a, c be numbers such that $a < c < 0$. Then ac is positive, and therefore $(ac)^{-1}$ is positive. By Section 3 of Chapter 4, $a(ac)^{-1} < c(ac)^{-1}$. That is, $a^{-1} > c^{-1}$. The reciprocal function is decreasing on the set $\{x \mid x < 0\}$. (c) Although $(-2) < 3$, it is not true that $1/(-2) > \frac{1}{3}$. (f) Let a, c be positive numbers such that $a < c$. Since the second-power function is increasing on the set of positive numbers, $a^2 < c^2$. Since a^2 and c^2 are positive and since the second-power function is increasing on $\{x \mid x > 0\}$, $(a^2)^2 < (c^2)^2$. That is, $a^4 < c^4$. Hence the fourth-power function is increasing on the set of all positive numbers. 4. (a) The domain of f is the set of all real numbers; the range of f consists of all numbers between 0 and 3; the function f is an increasing function; the image of 0 is $\frac{3}{2}$.

Chapter 10

Distance

Exercises A 2. $|b - a|$. 3. (b) 5, (d) $d - c$. 4. (a) 4, (c) $|p - b|$, (e) $|r|$.
6. (b) Since n is a negative number, $|n| = -n$. Hence $|n| \cdot |n| = (-n)(-n) = n \cdot n$.
7. (b) Since $y + 3$ is a real number, the equality is an application of the result established in Exercise 6c.

Answers to Selected Exercises   529

Exercises B 1. (c) $\sqrt{2}$, (e) $(a^2 + b^2)^{1/2}$, (f) 5, (h) $\frac{13}{3}$. 2. (a) $5 + 3\sqrt{17}$. 4. (c) $4x + 3y + 5 = 0$, (e) 5. 5. (c) $x + 2y + 2 = 0$, (f) $2\sqrt{5}$. 6. (b) $|y - 3| = 5$, (c) $[(x + 2)^2 + (y - 1)^2]^{1/2} = [(x - 1)^2 + (y + 3)^2]^{1/2}$, or $6x - 8y = 5$. 7. (c) 8, (d) $(2, 6, 0)$, (e) $\sqrt{17}$. 8. (c) $|c|$, (g) Since the segment joining Q and S is perpendicular to the plane $z = 0$, it is perpendicular to every segment that lies in the plane and contains S. Since one such segment is the segment joining S and the origin, the triangle with the given vertices has a right angle at vertex S.

Exercises C 1. (b) 13, (c) $2 \cdot 3^{1/2}$. 2. (b) $|c - f|$, (c) $(a^2 + c^2)^{1/2}$. 3. (b) $(2, 5, 1)$. 4. (b) 3, (c) $|a - r|$. 5. 2.

Exercises D 1. Let A be any point in the plane and not in the halfplane containing the focus F of the parabola. Then the segment joining A and F intersects the boundary of the halfplane, namely, the directrix of the parabola, at some point, which we may call B. The distance between A and F is greater than the distance between A and B. Consequently the distance between A and the focus is greater than the distance between A and the directrix; thus A does not belong to the parabola. Since the complement of the halfplane contains no point of the parabola, the parabola is a subset of the halfplane. 3. (b) Let (x, y) belong to the parabola. By Exercise 1, the point lies in the right halfplane whose boundary is the directrix; hence $x > -1$. The point (x, y) satisfies the criterion for membership in the parabola: its distance from the focus, namely, $[(x - 1)^2 + y^2]^{1/2}$, is equal to its distance from the directrix, namely, $|x - (-1)| = x + 1$.

Exercises E 3. (a) $(0, 0)$, (b) $(0, 5)$. 4. (b) $(0, -9)$, (c) $y = 9$. 5. (c) $y = -\frac{1}{12}$. 7. Vertex $(0, -2)$, focus $(0, -\frac{31}{16})$, directrix $16y + 33 = 0$. 9. (b) $4y = (x - 4)^2$. (c) Each distance is 5. 10. $4|p|$.

Exercises F 1. (a) Vertex $(1, 0)$, focus $(1, 2)$, directrix $y = -2$, (d) Vertex $(0, 40)$, focus $(0, 1601/40)$, directrix $40y = 1599$, (e) Vertex $(1, 5)$, focus $(1, \frac{61}{12})$, directrix $12y = 59$. 2. $8y = (x - 9)^2$. 4. $x^2 - 6x + 12y = 51$. 6. $y = -2x^2 - 4x - 5$. 8. $y = \frac{1}{8}x^2 - \frac{1}{4}x + \frac{25}{8}$.

Exercises G 2. (a) Vertex $(-3, 2)$, focus $(-3, \frac{9}{4})$, (c) Vertex $(\frac{5}{2}, \frac{25}{4})$, focus $(\frac{5}{2}, 6)$. 4. $\{(x, y) \mid y = 3x^2 - x - 4\}$. 5. $\{(x, y) \mid y = 2x^2 + 5\}$. 6. (d) 9, (f) $\frac{121}{4}$.

Exercises H 1. (b) $(a, h, k) = (1, \frac{3}{2}, \frac{7}{4})$, (c) $(a, h, k) = (2, -5, 7)$, (f) $(a, h, k) = (4, \frac{5}{2}, 9)$, (i) $(a, h, k) = (-3, -2, 13)$, (k) $(a, h, k) = (-\frac{1}{12}, 3, -\frac{1}{4})$, (n) $(a, h, k) = (-1, 0, 1)$. 2. (a) Vertex $(-5, -9)$, focus $(-5, -\frac{35}{4})$, directrix $4y + 37 = 0$, (d) Vertex $(1, -5)$, focus $(1, -\frac{39}{8})$, directrix $y = -\frac{41}{8}$, (j) Vertex $(-4, 5)$, focus $(-4, 4)$, directrix $y = 6$, (m) Vertex $(\frac{1}{2}, \frac{1}{2})$, focus $(\frac{1}{2}, \frac{11}{24})$, directrix $24y = 13$. 4. (c) The function has property (i) only. (d) The function has none of the properties. (g) The function has properties (ii) and (iii) only. (j) The function has property (ii) only. 5. 4. 7. (a) $x^2 + (1 - s)x + sy = s$, (c) $\left(\frac{s - 1}{2}, \frac{1}{2} + \frac{1}{4s}\right)$. 8. $4k(y - k) = (x - h)^2$. 10. $x^2 = 4p(y + p)$. 11. (a) $-b/2a$. 12. (a) The directrix of R is the line $y = -p$, and the extreme point of S is the point $(0, -p)$. The coordinates of the point $(0, -p)$ satisfy the equation of the line $y = -p$.

Exercises I 1. (a) 11.9, (d) $(-1, 3)$, (g) $\left(3a, 0, \frac{c^2 + 1}{2c}\right)$. 2. (b) 0, (d) $\frac{1}{2}$, (f) The point is between P and Q and is twice as far from Q as it is from P. 3. (c) The segment joining the points P and Q. 4. (b) $5x - 2y = 7$, (e) $2x + 5y + 3 = 0$. 5. (c) $[(x - 3)^2 + (y - 4)^2]^{1/2} = [(x + 1)^2 + (y + 6)^2]^{1/2}$. 6. (c) $12x - 2y - 4z + 19 = 0$, (e) $(-1, -\frac{1}{2}, 2)$.

530   Appendix

Exercises J   **2.** $(x + 5)^2 + (y - 7)^2 = 9$.   **4. (a)** Circle with center $(0, 3)$ and radius 4, **(c)** HINT: The curve is not a circle. **(d)** Empty set.   **5. (a)** $(x - 6)^2 + y^2 = 41$, **(c)** $x^2 + y^2 - 8x + 8y + 30 = 0$, **(f)** $x^2 + y^2 + 4x - 2y - 60 = 0$.
8. (c) $y = (8 - x^2)^{1/2}$ **(f)** Upper left quartercircle without endpoints, center $(0, 0)$, radius of 4.

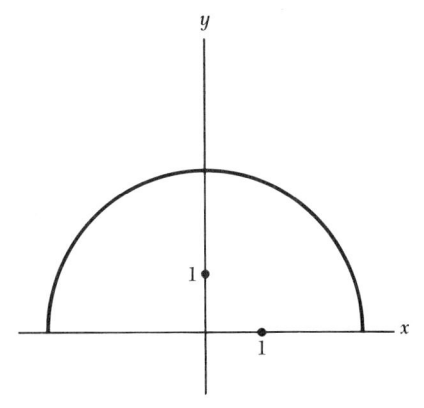

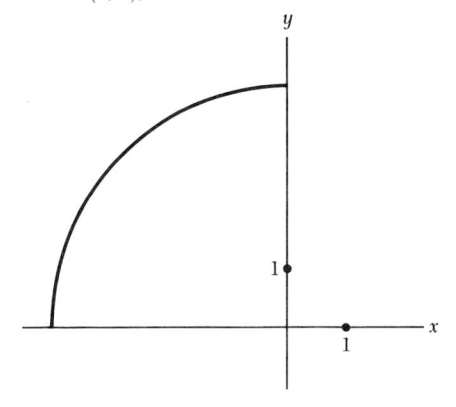

Exercises K   **1.** NOTE: In each part there are many acceptable answers for the point on the circle; we give one of the possible answers. **(c)** Center $(1, 0)$, radius $2^{1/2}$, point $(0, 1)$, **(f)** Center $(-\frac{5}{2}, \frac{7}{2})$, radius $\frac{1}{2}$, point $(-2, \frac{7}{2})$, **(h)** Center $(\frac{1}{3}, -\frac{4}{3})$, radius $3^{1/2}$, point $\left(\frac{1 + \sqrt{11}}{3}, 0\right)$, **(j)** Center $(-30, 40)$, radius 50, point $(0, 0)$.   **3. (a)** $x - 3y = 0$, **(c)** $x - y = 2$, **(d)** $(3, 1)$, **(h)** $x^2 + y^2 - 6x - 2y - 55 = 0$.   **4.** $13(x^2 + y^2) + 37x + 27y = 166$.   **5.** $5x^2 + 5y^2 - 22x + 8y - 61 = 0$.

Exercises L   **1. (a)** $(x - 3)^2 + (y + 4)^2 + (z - 1)^2 = 36$, **(d)** $(x - 2)^2 + (y - 2)^2 + (z - 5)^2 = 29$.   **2.** NOTE: In each part there are many acceptable answers for the point on the sphere; we give one. **(a)** Center $(4, -2, 0)$, radius 3, point $(1, -2, 0)$, **(c)** Center $(0, -\frac{3}{2}, -\frac{5}{2})$, radius $\frac{1}{2}$, point $(0, -\frac{3}{2}, 1)$.   **3.** $x^2 + y^2 + z^2 - 4x - 2y + 2z - 39 = 0$.   **6.** $(x - 1)^2 + (y + 5)^2 + (z + 4)^2 = 62$.   **7. (a)** 0, **(c)** 3, **(e)** 2, **(g)** $(3, \frac{1}{2}, \frac{1}{2})$, **(h)** $(9, 5, -7)$.   **9. (a)** Interior of circle with center at origin and radius $10^{1/2}$, **(c)** Union of the circle with center $(-\frac{3}{5}, \frac{4}{5})$ and radius $\sqrt{6}$ and the interior of this circle, **(f)** Interior of sphere with center $(0, -\frac{7}{4}, \frac{9}{2})$ and radius $\frac{5}{4}$.   **10. (a)** $(3, -2, 1)$, **(b)** $2\sqrt{6}$, **(g)** $x^2 + y^2 + z^2 - 6x + 4y - 2z \geq 10$.

11. (a) $x^2 + y^2 < 1$

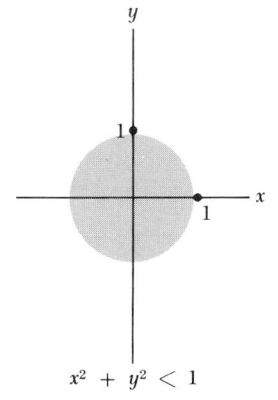

(d) $-1 < x < 1$

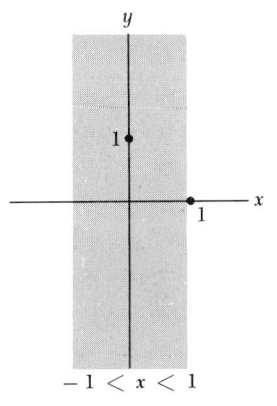

(f)

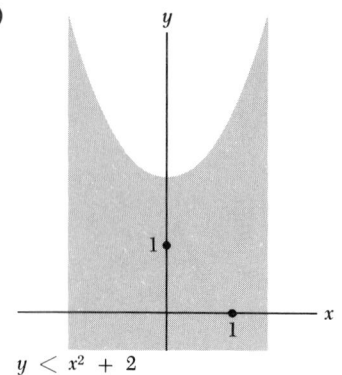

$y < x^2 + 2$

12. (b)

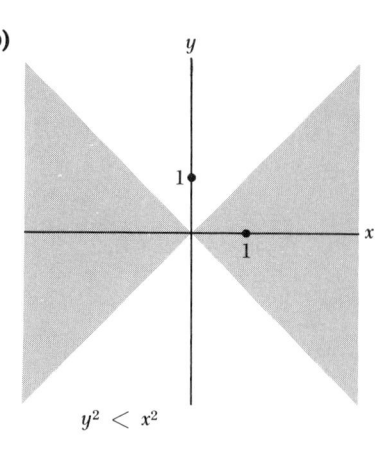

$y^2 < x^2$

Chapter 11

Inverse and Rational Functions

Exercises A **1. (c)** The inverse of L contains $(-2, -\frac{1}{2})$ because L contains $(-\frac{1}{2}, -2)$. **(f)** $\frac{1}{4}$. **(g)** The two slopes are reciprocals of each other. **2. (d)** $x - 3$. **3. (a)** $c = (d + 1)/3$. **4. (c)** $-\frac{1}{2}x + \frac{5}{2}$, **(d)** G. **5. (a)** $2x$, **(c)** $(x - 1)/9$, **(e)** $-x + 12$. **6. (b)** $\{(7, 2)\}$, **(c)** $\{(x, y) \,|\, [(x \leq -1) \wedge (y = -4x - 3)] \vee [(x > -1) \wedge (y = -x)]\}$. **8. (a)** An ordered pair belongs to H if and only if the ordered pair obtained from it by interchanging components belongs to G. Therefore a number is the left component of an ordered pair in H if and only if it is the right component of an ordered pair in G. Hence the set of all left components of ordered pairs belonging to H is the same as the set of all right components of ordered pairs belonging to G. In other words, Dom $H = $ Rg G. **(d)** Inside the left-hand oval.

Exercises B **1. (a)** $d = c^3$, $c = d^{1/3}$, **(b)** $x^{1/3}$. **3. (b)** $(x - 1)^{1/3}$, **(c)** $x^3 + 8$, **(e)** $-x^{1/3}$. **5. (b)** -2, **(c)** 8, **(d)** c. **6. (b)** The function contains the ordered pairs $(0, 5)$ and $(1, 5)$; these two (different) ordered pairs have the same right component 5. **(c)** The two ordered pairs $(-1, 1)$ and $(1, 1)$ have the same second component, and they both belong to the absolute-value function. **8.** There is a point on the semicircle between the endpoint and the midpoint of the arc. By the symmetry of the curve, there is another point on the semicircle such that the two points have the same second coordinate. Hence the function whose graph is the semicircle fails to qualify for having an inverse. **9. (a)** The function is the inverse of itself. **(b)** $\{(x, y) \,|\, [(x, y) \geq (0, 0)] \wedge [9x^2 + y^2 = 36]\}$.

Exercises C **1. (a)** $3 - x$, **(b)** $-2x$, **(d)** x. **3. (a)** $[(x + 2)/3]^{1/2}$, **(b)** $-5 + (x - 4)^{1/2}$. **5. (d)** $y = (-x)^{1/2}$

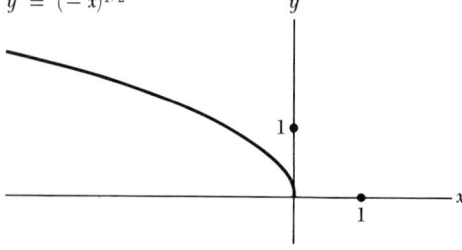

6. (a) 25, (b) 16, (d) 1.6, (g) s. **8.** HINT: All four curves are graphs of functions, but only two of the functions have inverses.

Exercises D **2.** (a) $y = 3x + 4$, (d) $y = x^2 - 2$, (h) $y = ax^2 - bx + c$, (i) $y = 0$, (l) $y = (-x)^{1/2}$. **3.** (a) $y = 3x - 4$, (d) $y = 2 - x^2$, (e) $y = -1/x$, (g) $y = x^2/(x^4 - 1)$, (k) $x^2 + y^2 = 25$, (m) $y = -1 - x^{1/2}$. **4.** (b) $y = x$, (d) $y = 2 - x^2$, (f) $y = x^3 - 2x$, (h) $y = -ax^2 + bx - c$, (i) $y = 0$, (l) $y = -(-x)^{1/2}$. **5.** (b) The graphs are the same. (d) The graph of the equation $y = -Q(-x)$ is the reflection of the graph of Q in the x-axis. **6.** (d) The graphs coincide. **7.** (a) The two graphs are reflections of each other in the y-coordinate axis. (c) Each of the graphs is obtainable from the other by a half rotation around the origin. **8.** (a) The two graphs are reflections of each other in the x-axis. **9.** $y = -x - (-x)^{1/2}$. **10.** $y = x(1 + x^2)^{-1}$. **12.** (a) $(x - 1)^2 + (y - 2)^2 = 16$, (b) $(x + 1)^2 + (y + 2)^2 = 16$.

Exercises E **1.** (a) The graph of the equation $y = (x + 4)^{-2}$ is obtainable by translating the graph of $y = x^{-2}$ four units to the left; the graphs are congruent to each other. (c) The graphs of the equations $y = (x - 3)^{-2}$ and $y = x^{-2} - 3$ are obtainable from the graph of the function x^{-2} by translating three units to the right and by translating three units down, respectively. Since each given curve is congruent to the graph of x^{-2}, the given graphs are congruent to each other. (d) Since $[5 + 2x + x^2]^{-1} = [4 + (x + 1)^2]^{-1}$, the graph of this function is congruent to the graph of the function $(4 + x^2)^{-1}$ and is located one unit toward the left from the latter curve. **2.** (b) The domain consists of all numbers except zero; the range consists of all negative numbers; the function is an even function; the function is increasing on the set of positive numbers; the image of a large number is very near zero; the image of a number near zero is very large in absolute value. (The graph, not shown here, is the reflection in the horizontal coordinate axis of the curve shown in Fig. 21 in Section 6 of Chapter 11.) (e) The domain consists of all numbers except -1; the range consists of all numbers except 0; the function is decreasing on $\{x | x > -1\}$ and is decreasing on $\{x | x < -1\}$; the image of a large number is near zero; the image of a number near -1 is large in absolute value; the image of 0 is 2. (The figure is not shown here.) (g) The domain consists of all numbers except 3 and -3; the range is $\{y | (y \geq \tfrac{1}{9}) \vee (y < 0)\}$; the function is an even function; the function is increasing on $\{x | 0 \leq x < 3\}$ and is increasing on $\{x | x > 3\}$; the image of a large number is a negative number near zero; the image of a number c near 3 is large in absolute value, either positive or negative according as $c < 3$ or $c > 3$; the image of 0 is $\tfrac{1}{9}$.

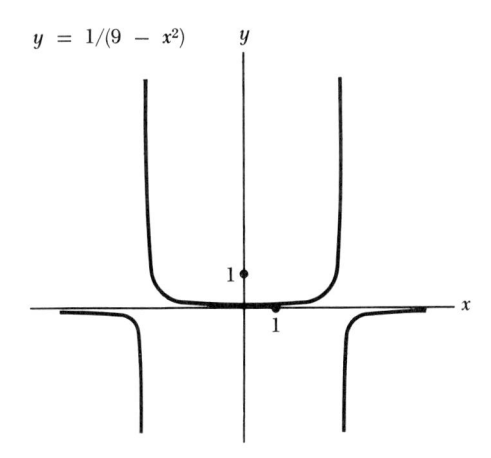

$y = 1/(9 - x^2)$

3. (b) For every real number c, the image of $-c$ under the function is 5 and is the same as the image of c under the function. The function is an even function. (e) The images of 1 and −1 under the function, namely, 2 and −2, respectively, are different numbers. The function is not an even function.

Exercises F **2. (a)** $\{-2, 0, 2\}$, **(b)** $\{x \,|\, 0 < |x| < 2\}$, **(c)** Even function, **(d)** Let a, c be numbers such that $2 < a < c$. Then $4 < a^2 < c^2$ and $0 < a^2 - 4 < c^2 - 4$. Hence $G(a) = a^2(a^2 - 4) < c^2(a^2 - 4) < c^2(c^2 - 4) = G(c)$. Thus G is increasing on $\{x \,|\, x > 2\}$. **4. (b)** The graph of the function $1 - (x - 1)^3$ is congruent to, and one unit above, the graph of the function $-(x - 1)^3$; the latter curve is congruent to (because it is the reflection in the x-axis of) the graph of the function $(x - 1)^3$; the latter curve is congruent to, and one unit to the right of, the graph of the function x^3. **(d)** Since $x/(1 - x) = 1/(1 - x) - 1$, the graph of this function is obtainable from the graph of $1/(1 - x)$ by translating one unit down. Hence the two graphs are congruent to each other. **5. (a)** The function is an odd function, because $(-c)^3 - (-c) = -(c^3 - c)$ for every real number c. **(d)** The function is neither even nor odd, because the respective images of 1 and −1 are 0 and −2, a pair of numbers which are neither equal to one another nor negatives of one another.

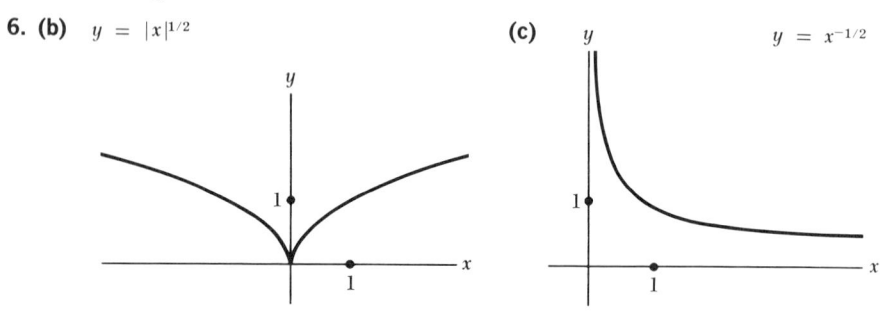

6. (b) $y = |x|^{1/2}$ **(c)** $y = x^{-1/2}$

Chapter 12

Transcendental Functions

Exercises A **1. (c)** Since $q < r < 0$, it follows that $0 < -r < -q$. Applying part (b), we obtain $3^{-r} < 3^{-q}$. We multiply each member of the latter inequality by the positive number 3^{q+r} and obtain $3^{q+r} \cdot 3^{-r} < 3^{q+r} \cdot 3^{-q}$, or $G(q) = 3^q < 3^r = G(r)$. **(e)** None. **2. (b)** The curves have the same general appearance: each of them lies in the upper coordinate halfplane; each is a rising curve, rising rather gradually in the left coordinate halfplane and steeply on the right of the y-axis; each intersects the y-coordinate axis at the point $(0, 1)$. However the graph of H rises more steeply than the graph of G, as evidenced, for example, by the fact that, although $H(0) = G(0)$, the images of 5 under G and H are 243 and 100,000, respectively. **3. (c)** Since $(\frac{1}{2})^x = 2^{-x}$, the graph of F and the graph of $y = 2^x$ are reflections of each other in the vertical coordinate axis. Consequently the curves are congruent to one another. **4. (b)** The graph of the equation $y = 2^{x-3}$ is congruent to the graph of $y = 2^x$ because it is obtainable from the latter curve by a translation of three units to the right. **(c)** The graph of the equation $y = 3^x - 4$ is obtainable from the graph of the equation $y = 3^x$ by a vertical translation of four units down. The graph of the equation $y = 3^{x+4}$ is also obtainable by translating the graph of $y = 3^x$, namely, by a translation of four units to the left. Since each of the two given curves is congruent to the graph of $y = 3^x$, they are congruent to one another. The graph of $y = 3^{x+4}$ is located four units above and four units to the left from the graph of $y = 3^x - 4$.

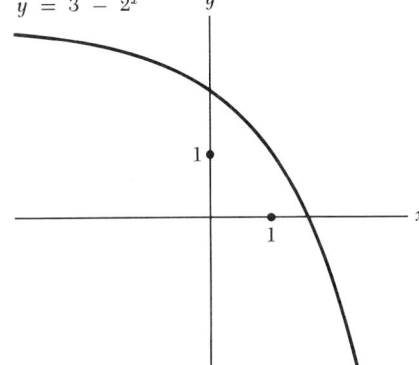

5. (a) $y = 3 - 2^x$ **6. (b)** $10^{\pi/6}$.

Exercises B   **1. (d)** 4.49 [using $\exp \frac{3}{2} = e \cdot e^{1/2}$], **(f)** 0.606 [using $\exp(-\frac{1}{2}) = 1/(\exp \frac{1}{2})$]. **4.** $(\exp x)^2 = (e^x)^2 = e^{2x}$.   **6. (a)** The curve is two units below the graph of exp. **(c)** The curve is the reflection of the graph of exp in the horizontal coordinate axis. **7. (a)** $\{y \mid y > -4\}$, **(d)** $\{t \mid t < 4\}$, **(e)** The set of all positive numbers.

Exercises C   **1. (a)** 14.8, **(c)** $\{y \mid y > 0\}$, **(e)** 2.   **3. (a)** The set of all real numbers, **(c)** By hypothesis, $q = G(p) = 2e^{-p}$. Since $q = 2 \exp(-p) = H(-p)$, $(-p, q) \in H$. **(g)** The curves are reflections of each other in the y-coordinate axis.   **4.** $(h, k) = (e^{-3}, 1)$. **5. (b)** Decreasing, **(c)** Increasing, **(d)** Neither.   **7. (b)** The domain of the function is the set of all real numbers; the range is the set $\{t \mid t \geq 1\}$; the function attains its minimum value at the point $(0, 1)$; the function is an even function; the function is a (rapidly) increasing function on the set $\{x \mid x \geq 0\}$; large numbers in the domain have very large images. [REMARKS: The function is known as the hyperbolic cosine function; its graph is a catenary; as an elementary application, a perfectly flexible cord, hanging freely between its two supported ends, illustrates an arc of a catenary.]

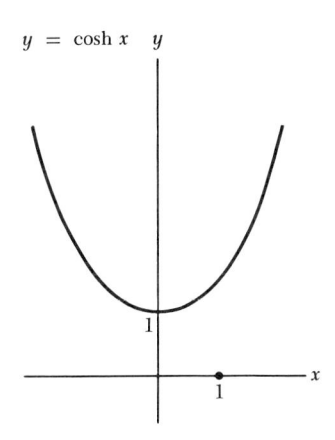

$y = \cosh x$

(d) The domain of the function consists of all real numbers; the range consists of all positive numbers; the function is a decreasing function; the image of each large negative number is a very large positive number; the image of 0 is 1; the image of a large positive number is very near 0. (The graph is shown in Fig. 10 in Section 4 of Chapter 12.) **(f)** Every real number belongs to the domain of the function; the range is the set

$\{y\,|\,0 < y \leq 1\}$; the function contains the ordered pair $(0, 1)$; the function is an even function; the function is decreasing on the set $\{x\,|\,x \geq 0\}$; the images of numbers that are large in absolute value are extremely near zero. (The sketch is not shown here.)
8. (b) 0.865, **(d)** 0.233, **(f)** $\exp(-bc) - \exp(-bd)$, **(i)** $\{y\,|\,0 \leq y < 1\}$.
9. (c) 2,520,000.   **10. (a)** $30.45, **(b)** $80.80.

Exercises D   **1. (a)** 3, **(d)** $5e^2$, **(f)** 2.   **2. (c)** $\{x\,|\,x > 3\}$, **(d)** $\{x\,|\,x < 0\}$.   **5. (a)** $\{e\}$, **(d)** $\{x\,|\,x > \ln \frac{3}{2}\}$, **(e)** $\{x\,|\,e^3 < x < e^7\}$, **(h)** $\{x\,|\,x \leq \ln \frac{1}{2}\}$, **(j)** $\{\ln \ln 2\}$.
6. (d) $\{x\,|\,-1 < x < 2\}$, **(g)** $\ln \frac{9}{4}$.   **7. (a)** $2 \ln x$, **(d)** $-e^x$.   **8. (d)** Let a, c be any two members of Dom F such that $a < c$. Then $-a > -c$. Since ln is an increasing function, $\ln(-a) > \ln(-c)$. That is, $F(a) > F(c)$. Thus F is a decreasing function.

Exercises E   **1. (a)** 2.64, **(d)** -0.73, **(f)** 0.51, **(h)** 2.76, **(k)** 2.30.   **2. (b)** -1, **(d)** 7, **(f)** $\frac{2}{3}$.   **4.** $-\frac{1}{2} \ln c$.   **6.** $-\ln x$.   **8. (b)** $\frac{1}{3}(5 + \ln x)$, **(d)** $\ln 2 - \ln x$, **(f)** $8 + \exp(2x)$.
9. t^{-1}.   **10. (b)** Every negative number belongs to the domain of the function $\ln x^2$, but no negative number has an image under the function $2 \ln x$.   **12. (a)** ac^2/b, **(c)** $(a + b)e^{2c}$.   **13. (c)** $2 + \frac{1}{2} \ln 3$, **(d)** $1 - (\exp 6)/2$.

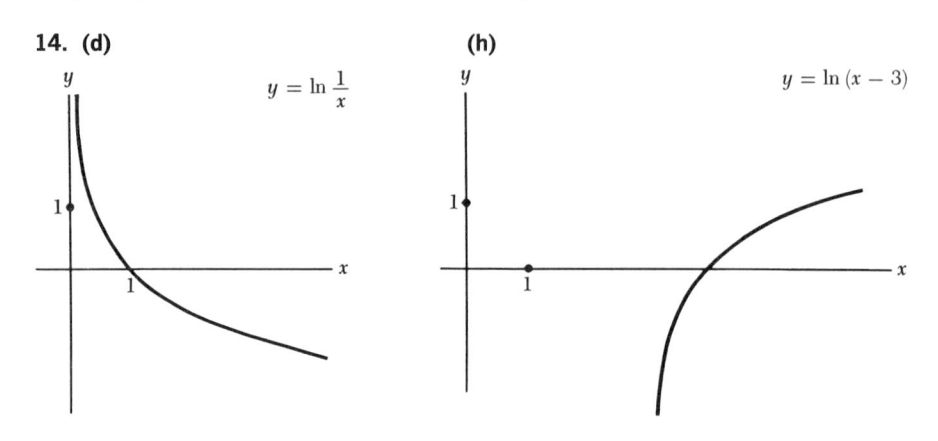

14. (d)   **(h)**

$y = \ln \frac{1}{x}$

$y = \ln(x - 3)$

15. (b) $\{x\,|\,x \neq \frac{3}{2}\}$, **(d)** $\{x\,|\,(x < 0) \vee (x > 1)\}$.   **16. (b)** $x < -\frac{2}{5}$, **(e)** $x \leq 1 - e^{-1}$, **(g)** $x \geq e^2$, or $x \leq -e^2$, **(j)** $0 < x < \frac{1}{2} \exp(-\frac{1}{4})$.   **18. (a)** Every nonzero number greater than -1 is a root.   **20.** By hypothesis, there are a nonzero constant m and a constant b such that $w = mx + b$. Since $\ln y = mx + b$, we find that $y = \exp(mx + b) = e^b \cdot e^{mx}$. Since e^b and m are nonzero numbers, $e^b \cdot e^{mx}$ is the formula for a function of exponential type.   **23. (e)** $y = 45 \exp(-0.012x)$.

Exercises F   **1. (b)** 0.054.   **4.** 0.376 [using $\phi(0)$ and $\phi(\frac{1}{2})$].   **6. (b)** 0.276, **(d)** 0.159, **(f)** 0.728.   **7. (a)** $\frac{3}{2}$, **(d)** $\frac{3}{4}$, **(e)** $-\frac{1}{2}$, **(h)** 0.674.   **10.** 0.818.

Exercises G   **1. (a)** 0.691, **(c)** 0.378, **(f)** 0.841, **(i)** 0.290, **(k)** 0.046.   **2. (a)** 1.282, **(c)** -1.645.

Exercises H   **1. (e)** $F(0.16)$, **(g)** $F(0.29) - F(0.21)$.   **2. (a)** $1 - F(1.68)$, **(c)** $2 - 2F(0.68)$, **(d)** $F(1.32) - F(0.02)$.   **3. (a)** $\sum_{k=978}^{1035} \binom{10000}{k} \cdot 9^{10000-k} \cdot 10^{-10000}$, **(b)** 0.654.   **5. (a)** 0.682, **(d)** 0.998, **(e)** 0.477, **(f)** 1.645, **(h)** 0.674.   **7. (b)** 0.087 (see the Appendix).

Exercises I   **1.** 0.6.   **3.** $-2^{3/2}/3$.   **4. (b)** 1, **(d)** $\frac{1}{6}$.   **6. (b)** $\left(-\frac{1}{2}, -\frac{\sqrt{3}}{2}\right)$, **(c)** $\sqrt{3}$,

(f) $\frac{2\pi}{3}$ (There are many other acceptable answers.) **7. (b)** $\left(\frac{1}{2}, -\frac{3^{1/2}}{2}\right)$, **(c)** $\left(-\frac{3^{1/2}}{2}, -\frac{1}{2}\right)$, **(f)** $\left(-\frac{1}{2}, -\frac{3^{1/2}}{2}\right)$. **8. (a)** The images are the same. **(d)** The images are the same. **(e)** The images are opposite endpoints of a diameter of the unit circle. **(k)** The images are reflections of each other in the x-coordinate axis. **9. (c)** The points are reflections of one another in the horizontal coordinate axis.

Exercises J **1. (a)** 0, **(b)** $2^{-1/2}$, **(e)** 2^{-1}, **(f)** 1, **(i)** $-\frac{1}{2}$, **(k)** $-2^{-1/2}$. **2. (a)** -1, **(c)** $1/\sqrt{2}$, **(d)** 1, **(e)** 0, **(g)** $-\frac{1}{2}$, **(i)** $-3^{1/2}/2$. **3. (b)** $5^{1/2}/3$, $-5^{1/2}/3$. **7. (a)** $(2^{-1/2}, 2^{-1/2})$, $(-2^{-1/2}, 2^{-1/2})$. **8.** (Each part of this exercise has many acceptable answers.) **(a)** $\pi/6$, $5\pi/6$, $-7\pi/6$, $-11\pi/6$, $13\pi/6$, **(b)** 0, π, -3π, 8π, -49π, **(d)** $3\pi/4$, $5\pi/4$, $-3\pi/4$, $-5\pi/4$, $21\pi/4$, **(f)** $4\pi/3$, $-\pi/3$, $11\pi/3$, $-8\pi/3$, $35\pi/3$, **(j)** $7\pi/18$, $-\pi/18$, $-5\pi/18$, $11\pi/18$, $31\pi/18$. **10.** All three numbers are equal to each other.

Exercises K **1. (a)** 0.54, **(c)** 0.54, **(f)** 0.71, **(h)** -0.84. **2. (b)** -0.91, **(f)** -0.42, **(g)** -0.42. **3. (a)** The numbers $\cos(r + \pi)$ and $\cos r$ are negatives of each other. **4. (c)** -0.96, **(e)** -0.28, **(f)** 0.28. **5. (a)** Let a, c be numbers such that $-\pi/2 < a < c < \pi/2$. Then the images of a and c under the winding function are points on the semicircular arc of the unit circle in the right coordinate halfplane, with the image of a nearer the lower endpoint of the arc than the image of c. Thus the y-coordinate of the image of a is less than the y-coordinate of the image of c. That is, $\sin a < \sin c$. In other words, $\sin$ is an increasing function on $\{r \mid -\pi/2 < r < \pi/2\}$.

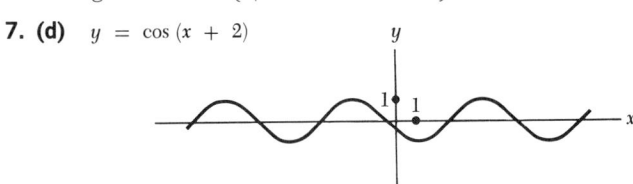

7. (d) $y = \cos(x + 2)$

8. (a) $2^{1/2}$, **(c)** $(8 + 2\sqrt{6} - 2\sqrt{2})^{1/2}/2$, **(e)** $(2 - 2\cos r)^{1/2}$, **(f)** $2|\sin r|$.

Exercises L **1. (c)** $\cos(2\pi/3) = -\frac{1}{2}$, $\sin(-\pi/6) = -\frac{1}{2}$. **2.** 0.54 [Recall that $\pi/2$ is approximately 1.57 and, therefore, 0.57 is approximately $(\pi/2) - 1$.] **3.** HINT: -0.43 is approximately $(\pi/2) - 2$. **4. (a)** $\cos\left(\frac{\pi}{6} + \frac{2\pi}{3}\right) = \cos\frac{5\pi}{6} = -\frac{\sqrt{3}}{2}$ and $\left(\cos\frac{\pi}{6}\right)$ $\cdot \left(\cos\frac{2\pi}{3}\right) - \left(\sin\frac{\pi}{6}\right)\left(\sin\frac{2\pi}{3}\right) = \left(\frac{\sqrt{3}}{2}\right)\left(-\frac{1}{2}\right) - \left(\frac{1}{2}\right)\left(\frac{\sqrt{3}}{2}\right) = -\frac{\sqrt{3}}{2}$ are the same number. **6. (a)** $(6^{1/2} + 2^{1/2})/4$, **(b)** $(6^{1/2} - 2^{1/2})/4$. **7. (a)** -0.76, **(c)** 0.14, **(d)** 0.29. **8. (a)** 0.44, **(c)** HINT: 1.12 is approximately $(\pi/2) - 0.45$. **9. (a)** $\sin t$, **(b)** $-\cos t$. **10. (b)** $\{y \mid -5 \leq y \leq 5\}$, **(c)** $\{y \mid 2 \leq y \leq 4\}$, **(e)** $\{t \mid 3 \leq t \leq 7\}$. **11. (b)** 0.9968. **12. (a)** 0.1247, **(c)** -0.0998.

Exercises M **1. (a)** $\sin 7$, **(c)** $\cos 0.6$. **2. (a)** $\sin 6$, **(b)** $\cos\frac{2}{3}$, **(d)** $\cos b$. **3. (a)** $\sin^2 6$, **(d)** $1 + \cos\frac{6}{5}$, **(e)** $(1 - \cos 2a)/2$. **4. (b)** $-(2 - 2^{1/2})^{1/2}/2$, **(c)** 0.48. **5. (a)** $-\frac{12}{13}$, **(b)** $-\frac{120}{169}$, **(d)** $-3/13^{1/2}$. **6. (a)** $-2^{3/2}/3$, **(c)** $\frac{7}{9}$, **(f)** $-[6(3 + 2^{3/2})]^{1/2}/6$.

Exercises N **1.** $13 \cos(4 - b)$. **2. (a)** $2^{1/2} \sin\left(x + \frac{\pi}{4}\right)$, **(b)** $2^{1/2} \sin\left(x - \frac{\pi}{4}\right)$. **5.** For every number r, $\sin[2\pi(r + 1)] = \sin[2\pi r + 2\pi] = \sin 2\pi r$. Furthermore, the two smallest positive roots of the equation $\sin 2\pi x = 1$ are determined by $2\pi x = \pi/2$ and $2\pi x$

$= 5\pi/2$ and therefore are $\frac{1}{4}$ and $\frac{5}{4} = \frac{1}{4} + 1$. Thus the period of the function is 1. **6. (b)** $2\pi/k$. **7. (c)** The set of all real numbers, **(d)** $\{y \mid m - |h| \leq y \leq m + |h|\}$. **8. (b)** The domain contains all real numbers; the range consists of all numbers between -2 and 2, inclusive; the period is 2π; the function is an even function; the function is a decreasing function on $\{x \mid 0 \leq x \leq 2\pi\}$; the image of 0 is 2. **(d)** All real numbers belong to the domain of the function; the range is $\{t \mid |t| \leq 2\}$; the function is periodic with period 2π; the function is increasing on $\{x \mid -\pi/6 < x < 5\pi/6\}$ and decreasing on $\{x \mid 5\pi/6 < x < 11\pi/6\}$; the image of 0 is $-3^{1/2}$. **(e)** The domain consists of all real numbers; the range consists of all numbers not exceeding one in absolute value; the period is 1; the function is an odd function; the function is increasing on $\{x \mid 0 \leq x \leq \frac{1}{4}\}$ and decreasing on $\{x \mid \frac{1}{4} \leq x \leq \frac{1}{2}\}$; the function contains the ordered pair $(0, 0)$. **(i)** Every real number is in the domain of the function; since the function has the formula $2 \sin(x + \frac{1}{3}\pi)$, the range is $\{y \mid -2 \leq y \leq 2\}$; the period is 2π; the function is increasing on $\{x \mid -5\pi/6 < x < \pi/6\}$ and decreasing on $\{x \mid \pi/6 < x < 7\pi/6\}$; the function assigns the image $3^{1/2}$ to the number 0. **(l)** The domain is the set of all real numbers; the range is the set $\{y \mid 0 \leq y \leq 1\}$; from the equation $\sin^2 x = \frac{1}{2} - \frac{1}{2} \cos 2x$, we recognize that the period of the function is π; the function contains $(0, 0)$, is even, and is increasing on the set of numbers between 0 and $\pi/2$.

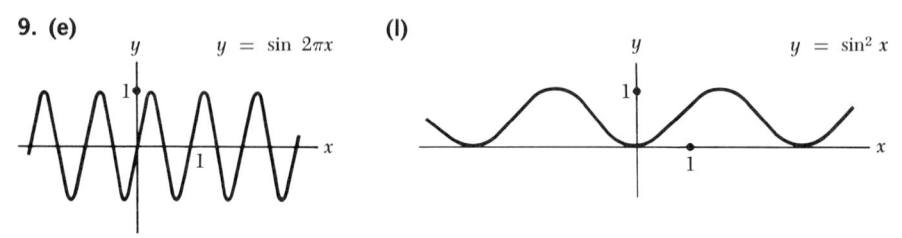

9. (e) $y = \sin 2\pi x$ **(l)** $y = \sin^2 x$

Exercises O   **1. (b)** 0, **(c)** $-\sqrt{3}$, **(d)** $-3^{-1/2}$.   **2.** $-2 \cdot 5^{-1/2}$.   **4.** $y = -1 + (x - 2) \cdot \tan 3$.   **6. (a)** $-\frac{3}{2}$, **(c)** $2(13)^{-1/2}$, $-2(13)^{-1/2}$.   **7. (a)** $\pi/6$, $5\pi/6$, **(c)** $3\pi/4$, **(e)** $-\pi/3$, $-\pi/6$, $2\pi/3$, $5\pi/6$.   **9. (a)** $\pi/6 < x < \pi/2$, **(c)** $-\pi/4 < x < \pi/4$.
10. (c) $\tan\left(\frac{\pi}{4} + \frac{\pi}{2}\right) = -1$ and $\tan\frac{\pi}{4} = 1$ are not equal to each other.   **12.** π.
13. (b) The domain does not contain an odd integral multiple of π but does contain

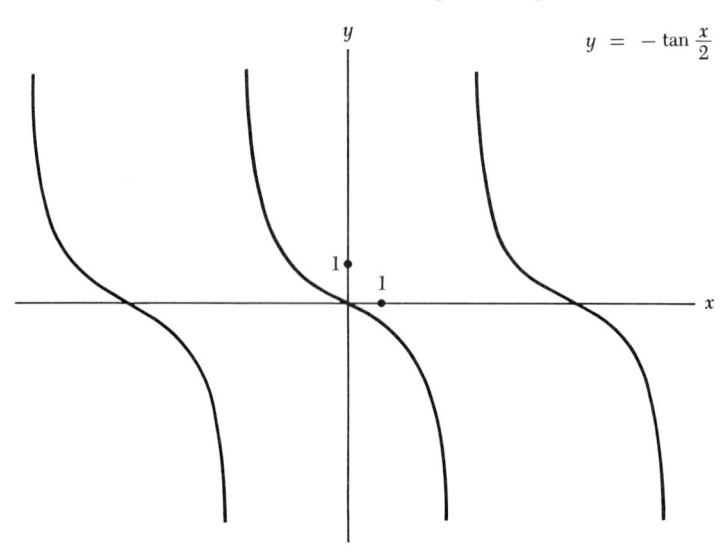

$y = -\tan\frac{x}{2}$

538   Appendix

every other real number; the range is the set of all real numbers; the function is an odd function, is periodic with period 2π, and is decreasing on each set consisting of the real numbers between two consecutive odd integral multiples of π; the images of 0 and $\pi/2$ are 0 and -1, respectively.

Exercises P   **1.** **(b)** $150°$, **(d)** 50π radians, **(g)** $\pi/360$ radians, **(i)** $45°$, **(l)** $(9/5\pi)°$.
2. **(b)** 0, **(e)** $-3^{1/2}$, **(j)** $\frac{1}{2}$, **(m)** $(2 + 2^{1/2})^{1/2}/2$, **(o)** $(2 + 3^{1/2})^{1/2}/2$.
3. $\begin{bmatrix} 0 & -1 \\ 1 & 0 \end{bmatrix}$.   **5.** **(b)** $\begin{bmatrix} -a \\ -b \end{bmatrix}$,   **(c)** Under the linear transformation each vector is rotated halfway around the origin.   **7.** **(a)** By Exercise 6a, $M(-p) \cdot M(p) = M(0) = I$. Hence $M(-p)$ is the inverse of $M(p)$.

Exercises Q   **1.** **(a)** -6, **(e)** $-(37)^{-1/2}$, **(f)** 52.   **2.** **(b)** $5\sqrt{2}$, **(c)** $2^{1/2}/10$.
3. **(c)** $(585 - 168 \cdot 5^{1/2})^{1/2}/3$ meters.   **5.** **(d)** The lengths of the medians from F, from G, and from H are $5\sqrt{31}$ feet, $5\sqrt{10}$ feet, and $5\sqrt{46}$ feet, respectively.   **(f)** $75\sqrt{15}$ square feet.   **6.** **(e)** 8 meters, **(f)** 240 square meters.   **7.** $\left(\dfrac{9 - 7\sqrt{3}}{2}, \dfrac{-25 + 5\sqrt{3}}{2} \right)$.

Index

A posteriori probability, 292
A priori probability, 292
Absolute value, 348
Activities, sequence of, 212
Activity, 175, 260
Addition
   in Boolean algebra, 182
   of matrices, 54
   of vectors, 48
And, 164
Angle, 496
   cosine of, 497
   sine of, 497
   tangent of, 497
Annuity, 255
Arithmetic sequence, 245, 451
Associative, 105, 182
At random, 270
Augmented matrix, 65
Average, sample, 322

Base, 238
Belongs to, 6, 150
Bernoulli experiment, 295, 464
Biconditional statement, 173
Binary operation, 182
Binomial coefficient, 220, 235
Binomial coefficients, properties of, 225
Binomial distribution, 303, 320, 464
Binomial theorem, 232
Boolean algebra, 182
   duality in, 196
   models of, 186
Boundary, 127
Brace notation, 151, 152

Cell, 199
Center
   of circle, 388
   of sphere, 395

Circle, 388
   interior of, 399
   unit, 468
Closed halfline, 121
Closed halfplane, 127
Closed halfspace, 134
Column matrix, 54
Commutative, 182
Complement
   of element in Boolean algebra, 183
   of subset, 158
Complementation in Boolean algebra, 183
Complete the square, 378
Component, 47
Componentwise, 50, 54
Composite of linear transformations, 103
Compound interest, 236, 255, 441
Conditional probability, 285
Conditional statement, 172
Conjunction, 164, 275
Convex polyhedral set, 137
   extreme point of, 138
Coordinate axis, 4, 43
Coordinate plane, 44
Coordinate system
   three-dimensional, 42
   two-dimensional, 4
Cosine function, 472
Cosine of angle, 497
Cosines, law of, 505
Counting principles, 211

De Morgan's laws, 194
Deck of playing cards, 179
Decreasing function, 352
Degree measure, 496
Deviation, 309
   root-mean-square, 310
   standard, 310
Directrix of parabola, 368

Disjoint subsets, 157
Disjunction, 166, 271
Dispersion, 316
Displacement, 47
Distance
   between line and point, 358
   between parallel lines, 358
   between point and origin, 26
   in one-dimensional space, 357
   in three-dimensional space, 367
   in two-dimensional space, 362
Distribution
   binomial, 303, 320, 464
   probability, 303
   uniform, 303
Distributive, 182
Domain, 342
Duality in Boolean algebra, 196

Element
   of matrix, 53
   of set, 150
Elementary row transformation, 66
Empty set, 157
Equation, linear, 16, 30, 69
Equiprobable measure, 270
Even function, 424
Expected value, 303
   summary of, 319
Experiment, 260
Exponent, 238
   irrational, 433
   negative, 241
   rational, 241
Exponential function, 436
Exponents, properties of, 239
Extreme point, 138

Factorial, 216
Failure, 293
False, 163
Favorable outcome, 270
Focus of parabola, 368
Force, 47
Function, 338
   cosine, 472
   decreasing, 352
   even, 424
   exponential, 436
   graph of, 346
   inverse of, 405
   linear, 6
   logarithmic, 445
   odd, 428

   of exponential type, 438
   periodic, 472
   quadratic, 376
   sine, 472
   tangent, 492
   winding, 468

Gauss elimination method, 35, 63
   theory of, 88
Geometric sequence, 247, 451
Geometry of system of linear equations, 75
Graph
   of equation, 346
   of function, 346
   of linear equation, 16, 68
   of linear function, 7
Greater than, 120
Greatest value, 141

Halfline, 121
Halfplane, 127
Halfplanes, intersection of, 130
Halfspace, 134
Halfspaces, intersection of, 137
Hyperbola, 336

Idempotent, 182
Identity element, 182
Identity matrix, 89, 113
Identity transformation, 107
If and only if, 173
If . . . , then . . . , 172
Image, 6, 95, 338
Increasing function, 352
Independent pair
   of random variables, 317
   of statements, 287
Independent trials, 293
Inequality, 120
   sense of, 123
Integer, 151
Interest, compound, 236, 255, 441
Interior
   of circle, 399
   of sphere, 399
Intersection
   of halfplanes, 120
   of halfspaces, 137
   of subsets, 155
Inverse
   of function, 405
   of linear transformation, 111
   of matrix, 114
Investment, 236, 255, 441

Law of cosines, 505
Least value, 141
Less than, 122
Linear equation, 16, 30, 69
   graph of, in three dimensions, 68
   graph of, in two dimensions, 16
Linear equations, system of, 35, 75
Linear function, 6
   graph of, 7
Linear inequalities, 129
Linear programming, 144
Linear transformation, 95
   inverse of, 111
   matrix associated with, 95
Linear transformations, composite of, 103
Logarithmic function, 445
Logarithms, properties of, 448
Logically contradictory pair of statements, 178
Logically equivalent pair of statements, 177
Logically false statement, 177
Logically inconsistent pair of statements, 270
Logically true statement, 177

Matrices
   equality of, 54
   inequality of, 132
   product of, 59
   sum of, 54
Matrix, 54
   augmented, 65
   column of, 53
   element of, 53
   inverse of, 114
   product of number and, 55
   rotation, 499
   row of, 53
   zero, 54
Matrix associated with linear transformation, 95
Maximum, 143
Mean, 309
Measure, 260
   equiprobable, 270
Median of sample, 325
Member of set, 150
Midpoint, 382
Minimum, 143
Mode, 325
Models of Boolean algebra, 186
Multiplication
   in Boolean algebra, 183
   of matrices, 59
   of row matrix and column matrix, 57

n-dimensional vector, 47
Negation, 165, 266
Neither . . . nor, 195
Noncommutative, 60
Normal probability density function, 457
Normal probability distribution function, 458
Not, 165, 266
Number of permutations, 209
Number of subsets, 222
Numbers, sequence of, 245

Odd function, 428
Open halfline, 121
Operation
   binary, 182
   unary, 182
Or, 166
Ordered n-tuple, 47
Ordered pair, 6
Ordered partition, 217
Ordered triple, 42
Orthant, nonnegative, 129
Outcome, 175, 260
   favorable, 270
   logically possible, 176, 260

Pair of statements
   independent, 287
   logically contradictory, 178
   logically equivalent, 177
   logically inconsistent, 270
Parabola, 332
   definition of, 368
   vertex of, 369
Parallel lines, 18
Parameter, 83
Partition, 199
   ordered, 217
Period, 472
Periodic function, 472
Permutations, number of, 209
Perpendicular-bisector, 384
Perpendicular lines, 27
Plane, 69
Playing cards, deck of, 179
Point
   sample, 322
   in three-dimensional space, 43
   in two-dimensional space, 4
Polyhedral set, convex, 137

Positive integer, 152
Power, 238
Probability, 260
   a posteriori, 292
   a priori, 292
   conditional, 285
   summary of, 299
Probability distribution, 303
Probability function, 342
Probability of statement, 260
Product
   of elements in Boolean algebra, 183
   of matrices, 59
   of number and vector, 50
Programming, linear, 144
Proof, 187
Proportional, 23
Proportionality, constant of, 23
Pythagoras, theorem of, 26, 360

Quadratic function, 376
   graph of, 378

Radian measure, 496
Radius
   of circle, 388
   of sphere, 395
Random, at, 270
Random variable, 302
   standard, 319
   standard normal, 462
Random variables, independent pair of, 317
Range, 343
Rational number, 238
Ray from origin through point, 487
Reflection, 414
Root, 15, 89
Root-mean-square deviation, 310
Rotation matrix, 499
Row matrix, 54

Σ, 251
Sample, 322
Sample average, 322
Sample point, 322
Sample standard deviation, 323
Sample variance, 323
Sense of inequality, 123
Sequence
   of activities, 212
   arithmetic, 245, 451
   geometric, 247, 451
   of numbers, 245
   sum of numbers in, 253
Set, 150
   convex polyhedral, 137
   element of, 150
   empty, 157
   member of, 150
   truth, 179, 260
Sine function, 472
Sine of angle, 497
Slope, 9
Space, vector, 94
Sphere, 395
   interior of, 399
Standard deviation, 310
   sample, 323
Standard normal variable, 462
Standard random variable, 319
Statement, 163
   biconditional, 173
   conditional, 172
   logically false, 177
   logically true, 177
   probability of, 260
   truth set for, 179, 260
Statements
   independent pair of, 287
   logically contradictory pair of, 178
   logically equivalent pair of, 177
   logically inconsistent pair of, 270
Steepness, 10
Subset, 152
   complement of, 158
Subsets
   intersection of, 155
   number of, 222
   union of, 156
Success, 293
Sum
   of elements in Boolean algebra, 182
   of matrices, 54
   of numbers in sequence, 253
   of vectors, 49
Summation notation, 251
Symmetric
   with respect to origin, 428
   with respect to y-axis, 424

Tangent function, 492
Tangent of angle, 497
Theorem of Pythagoras, 26, 360
Three-dimensional coordinate system, 42
Translation, 375

Tree diagram, 207
Trial, 293
Trials, independent, 293
Triangle, measurement of, 502
Trigonometric identities, summary of, 496
Trigonometry, 467
True, 163
Truth set, 179, 260
Truth value, 164

Unary operation, 182
Uniform distribution, 303
Union of subsets, 156
Unit circle, 468

Variance, 309
  sample, 323
  summary of, 319
Vector, 47
  product of number and, 50
  zero, 50
Vector space, 94
Vectors
  equality of, 47
  sum of, 49
Venn diagram, 153
Vertex, 138
  of parabola, 369

Winding function, 468
Witch of Agnesi, 424

Zero matrix, 54
Zero vector, 50